Joint Annual Meeting of the International Society of Exposure Science and the International Society of Indoor Air Quality and Climate (ISES-ISIAQ 2019)

The built, natural, and social environments: impacts on exposures, health and well-being

Kaunas, Lithuania
18-22 August 2019

ISBN: 978-1-5108-9756-4

Printed from e-media with permission by:

Curran Associates, Inc.
57 Morehouse Lane
Red Hook, NY 12571

Some format issues inherent in the e-media version may also appear in this print version.

International Society of Exposure Science (ISES)
c/o Infinity Conference Group
1035 Sterling Road, Suite 202
Herndon VA 20170
USA

Phone: (703) 925-9620
Fax: (703) 925-9453

info@intlexposurescience.org

Additional copies of this publication are available from:

Curran Associates, Inc.
57 Morehouse Lane
Red Hook, NY 12571 USA
Phone: 845-758-0400
Fax: 845-758-2633
Email: curran@proceedings.com
Web: www.proceedings.com

Revised August 16, 2019

ISESISIAQ 2019

KAUNAS · LITHUANIA

Joint meeting of the
INTERNATIONAL SOCIETY OF EXPOSURE SCIENCE
and the
INTERNATIONAL SOCIETY OF INDOOR AIR QUALITY AND CLIMATE

August 18 - 22, 2019

The built, natural, and social environments: impacts on exposures, health and well-being

PROGRAM

Table of Contents

PLENARY ABSTRACTS

Sunday: Communities creating health

Monika Petraite

RG Innovation and Entrepreneurship Kaunas University of Technology Lithuania

Communities creating health: innovating for better environmental and public health

Within the sustainable development goals (SDG) of the United Nations Organization the revolution of healthcare is discussed and support for new health management models is expressed. In their 2030 agenda a general shift towards a more holistic vision of health and wellbeing is observed [1], since it requires that all three aspects of sustainable development - economic, social and environmental - work in an integrated way. The implementation of the SDGs implies a positive change that would radically increase prospects for peaceful, just and inclusive societies through collaboration and cooperation between communities with creativity and innovations. SDG also indicates the need for the research that would closely examine the connection of knowledge dissemination between communities and the ability to achieve the Sustainable Development Goals and achieve universal healthcare [2]. The role of local communities here cannot be understated, as the community organizations, civil society and collaborative neighborhoods stand at the initiation and implementation of actual change around their own, and also global problems. The innovation ideology has proven that society reacts best to the well-perceived and visible problems that have an immediate impact on living, health, and environmental conditions. Community-driven collaborative innovation and actions thus can be extremely efficient in achieving sustainability of environmental and human health. In this talk, I will synthesize the existing knowledge and best practices on the community driven innovation in health, including environmental health, supported by smart solutions and collective actions that address immediate socially and culturally embedded challenges associated with health. Reliability of knowledge and knowledge sources, lack of empathy and translational knowledge from the formal knowledge sources, as well as lack of proximities in knowledge bases all lead to the diversified trust and resulting community choices that sometimes are extremely dangerous. We will also provide examples for successful incentives of community mobilization and participation in health initiatives, in collaboration with institutional systems, that help to save lives, co-create and innovate around prenatal and female reproductive health.

References [1]United Nations Organisations. Transforming our world: the 2030 agenda for sustainable development. 2015; Available at: https://sustainabledevelopment.un.org/post2015/transformingourworld

[2.] Center for Global Initiatives, Johns Hopkins University School of Nursing. Utilization of communities of practice for ongoing learning and knowledge dissemination: Making the case for the Global Alliance for Nursing and Midwifery (GANM) https://doi.org/10.1016/j.nepr.2017.07.001. Nurse Education in Practice 2017;26:64-67.by

Monday: Balancing between health and energy consumption for healthy and sustainable buildings

Elisa Van Kenhove

Ghent University, Belgium

How not to waste energy on Legionella

Hot water is produced at temperatures above 60°C, while for most of the domestic hot water applications (like taking a shower, washing our hands) temperatures of only 30 to 40°C are required. However, at temperatures below 60°C there is a risk of contaminating the system with Legionella pneumophila, a bacterium that, upon exposure, can cause severe pneumonia. A large potential exists to improve the design and operation of domestic hot water systems in buildings with respect to energy use and health. The continuously high temperature of 60°C impedes the optimization of domestic hot water system design, both on a principal and a case by case basis, resulting both in poor energy efficiency due to strict defaults of high design temperatures and unaccounted risk taking by blindly choosing the lowest 'acceptable' cutoff temperatures in practice. Additionally, in contaminated cases, decontamination is carried out more or less on a trial and error basis, making it very time consuming, costly and above all unsafe. The overall social purpose of this research is to lower the energy demand for domestic hot water whilst obtaining a low Legionella pneumophila contamination risk. This can be done by improving design and control processes of domestic hot water systems through the use of simulation models. A coupled thermohydraulic and biological simulation model is developed that allows to predict Legionella pneumophila growth in systems and to test the effectiveness of decontamination techniques on a contaminated system. Based on this model a control optimization algorithm is invented that allows a lower set point temperature in combination with a dynamic heat shock regime without compromising on health.

Yuguo Li

University of Hong Kong, China

Respiratory infection in buildings: Transmission and roles of environment intervention

Lower respiratory infections still remain the world leading cause of death. Respiratory diseases are spread between people in indoor environment via close contact, air, and contaminated surfaces, and indoor environment also affects pathogen spread by modifying the interaction of inhalation/exhalation between people, and air and surface hygienic conditions. In this talk, based on two recent completed literature reviews (Wei and Li, 2017 and Zhang et al., in preparation), we shall synthesize existing knowledge on the release of respiratory droplets, their transport and dispersion in the indoor environment via close contact, airflows and surface touch, and the ultimate exposure of a susceptible host by both short-range (airborne, large droplet and immediate surfaces) and distant routes (airborne and fomites). Infectious agents are embedded in respiratory mucus, or expired mucus droplets, which evaporate, disperse, and deposit, and their size as they exit the mouth or nose in the indoor environment is an important parameter. We shall clarify the possible misleading information in the literature, including the threshold size of large droplets. Our discussion on the mechanisms of close contact transmission reveals the mechanisms of intervention, i.e. avoidance or control of close contact, avoidance or control of expired jets, removal of inhaled/expired droplets, and environment (air, skin and inanimate surface) cleaning. Building ventilation is shown to be an effective intervention using the existing data. We shall discuss about the WHO recommendation on the use of natural ventilation. Economic constraints often lead to lack of adequate hospital facilities for the emerging infection challenges not only in resource-limited countries, also in developed economies. Questions remain on the relative effectiveness, roles and costs of environment intervention, such as HVAC.

References • Wei J, Li Y. Airborne spread of infectious agents in the indoor environment. American Journal of Infection Control, 2016, 44(9): S102-S108. • Zhang, N, Tang J and Li Y. Close contact transmission of respiratory infection: a review. In preparation. • Gao X, Wei J, Xu, P, Cowling BJ, Li Y. Building ventilation as an effective disease intervention strategy on a large and dense social contact network. PLOS One. 2016 Sep 9;11(9):e0162481

Tuesday: Challenges in monitoring exposure in the 21st century: Harmonization, collaboration and legal requirements

Corinne Mandin

Scientific and Technical Building Centre (CSTB), France

Challenges in monitoring exposure to air pollutants: what, how and what for?

We have made a lot of progress in monitoring airborne pollutants over the past century and in reducing chemical emissions into the air. Nevertheless, air pollution, both outdoor and indoor, remains a public health issue worldwide. While indoor concentrations of so-called classical volatile compounds have decreased over time, chemicals emitted by new products or associated with new living habits may be increasing and are of concern. How is it possible to identify these chemicals? Can they be measured with traditional methods? Moreover, low-cost sensors coupled with massive measurement campaigns from citizen science initiatives may help to better monitor air pollution. Where are we now with these opportunities? What kind of information is being provided to researchers, decision makers and citizens? Monitoring campaigns and studies performed in France in outdoor and indoor air will be used to illustrate these different points and highlight our challenges.

Marike Kolossa-Gehring

Umweltbundesamt (German EPA), Berlin, Germany

Coordinating and advancing human biomonitoring in Europe with HBM4EU - Pulling together cultures, legislations and perspectives

People in Europe are substantially exposed to various chemicals from different sources and via different pathways. Some of these chemicals have already been restricted or cannot be reasonably further regulated, others raised concern just recently or are not even known yet as pollutants relevant for human health. Targeted and non-targeted human biomonitoring (HBM) provide information on the aggregate exposure from all sources and via all pathways. HBM is therefore an excellent basis to identify the real exposure of European citizens, to improve risk assessment procedures for chemicals, and thus to protect people's health. More than 110 partners from 28 countries cooperate in the European Joint Programme HBM4EU to generate the knowledge that decision makers need to improve Europe's environment and health actions and programs. As an innovative approach, HBM4EU was designed after consultation with European and national policy makers to identify the policy needs and priority chemicals. Inclusion of stakeholders aims at facilitating the use of scientific results by different target groups, including the general public. This talk will focus on how HBM4EU intends to improve the health and well-being of European citizens. Since HBM data are currently fragmented in Europe, HBM4EU develops a network of analytical laboratories qualified to deliver high-quality and comparable exposure biomarker results for 18 priority substances. A tailored training programme shares knowledge on how comparability between studies can be increased and which components need harmonization. Accordingly, study centers in 21 countries collect and analyze new samples from three age groups to deliver comparable HBM data for the general population supplemented by occupational studies. The benefit of HBM is maximized by toxicological and epidemiological interpretation of results. Different approaches, inter alia, derivation of European HBM Health Based Guidance Values (HBGV) and identification of appropriate effect markers will be presented as examples for HBM4EU work on effects.

Wednesday: Keeping ecosystem services safe: Spotlight on climate change and water supply

Pekka Pietilä

Tampere University, Finland

Political and scientific challenges around water supply

The nature and severity of challenges in water supply differ from country to country due to a number of reasons: availability and quality of water resources, level of water pollution, population density, agricultural and industrial development and practices, etc. On the world scale, agriculture is the largest user of water and is responsible for 70 per cent of the total withdrawal of fresh water used mainly for irrigation, while industry accounts for 20 and domestic use slightly less than 10 per cent. Water services are commonly regarded as services of necessity which should not be left for private actors to exploit. One should keep in mind that a clear fundamental difference exists between the basic philosophies of public and private enterprises. Public undertakings are established to serve the needs of the population, typically in a certain geographical area, and they do not aim to substantially increase their share of the market as private companies do. Of course, private companies also aim to serve the needs of their clients and to produce attractive services because winning customers is undoubtedly in their interest. Yet, the prime motive of private companies is to make a profit for their owners. How water and wastewater services are organised is a highly political decision: The extremes are full privatisation, as in England and Wales in 1989, and full public control without any element of commercialisation as was the case in the former Soviet Union. Far more common options are public-private partnerships of various forms and with varying degrees of public vs. private involvement. This talk will explore the reasoning and political intentions behind the different forms and the possibilities involved in transferring more responsibility from central to local political entities.

Thursday: Excellence in exposure and indoor air quality sciences

ISES AWARDS PRESENTATION
and a Lecture by the ISES Excellence in Exposure Science Award winner

Antonia Calafat
U.S. Centers for Disease Control and Prevention (CDC), Atlanta, Georgia, USA

Biomonitoring for public health action: challenges and opportunities

Carl-Gustaf
Bornehag Karlstad University, Sweden; Icahn School of Medicine at Mount Sinai, New York, USA

Interdisciplinarity for better risk assessment of indoor chemicals. A lecture in honor of Prof. Jan Sundell for his lifelong work on Indoor Air Quality and Health

The nature and severity of challenges in water supply differ from country to country due to a number of reasons: availability and quality of water resources, level of water pollution, population density, agricultural and industrial development and practices, etc. On the world scale, agriculture is the largest user of water and is responsible for 70 per cent of the total withdrawal of fresh water used mainly for irrigation, while industry accounts for 20 and domestic use slightly less than 10 per cent. Water services are commonly regarded as services of necessity which should not be left for private actors to exploit. One should keep in mind that a clear fundamental difference exists between the basic philosophies of public and private enterprises. Public undertakings are established to serve the needs of the population, typically in a certain geographical area, and they do not aim to substantially increase their share of the market as private companies do. Of course, private companies also aim to serve the needs of their clients and to produce attractive services because winning customers is undoubtedly in their interest. Yet, the prime motive of private companies is to make a profit for their owners. How water and wastewater services are organised is a highly political decision: The extremes are full privatisation, as in England and Wales in 1989, and full public control without any element of commercialisation as was the case in the former Soviet Union. Far more common options are public-private partnerships of various forms and with varying degrees of public vs. private involvement. This talk will explore the reasoning and political intentions behind the different forms and the possibilities involved in transferring more responsibility from central to local political entities.

SESSION KEY:

First two letters indicate day:

SU=Sunday, MO=Monday, TU=Tuesday, WE=Wednesday, TH=Thursday

Next two letters indicate the type of presentation:

PO=Poster; SY=Symposium; PL=Platform

Next letter indicates room:

A=Basic Hall
B=Nemunas
C=Ampitheatre
D=Baras 11
E=VIP Hall

First numeric digit indicates order of session in the day:

1=early morning, 2=late morning, 3=early afternoon, 4=late afternoon

Topics:

SEM: Sources and emissions
EHR: Concentrations, exposure, hazard, and risk assessment
EAM: Exposure assessment methods, models, simulation
HBM: Health, individuals, biomonitoring, and biological response
CPC: Comfort, productivity, community health and wellbeing
SSS: Advancing research/knowledge in an imperfect world: strategies, sampling, statistics
ECS: Exposure control strategies, practices
PCP: Perception, communication, policy

As an example:

MO-PL-A1-EHR: indicates a general platform session on Monday early morning that will take place in the basic hall and cover the topic "concentrations, exposure, hazard and risk assessment"

TU-SY-C2: indicates a symposium on Tuesday late morning that will take place in the Amphitheatre (no specific topics for symposia)

STUDENT POSTER COMPETITION

Sunday, August 18th

SU-PO-01

Comparison of AOD from CALIPSO, MODIS, and AERONET

Falah S.; Technion, Haifa, Israel

Numerous studies used satellite-based Aerosol Optical Depth (AOD) as an explanatory variable in different statistical models that predict ground level particulate matter (PM) concentrations. This work examines relationships between satellite-sensed and ground-observed AOD, and will use them in the future for predicting PM at the surface. At first, we evaluated the satellite AOD retrievals against ground-based AOD measurements by the AERONET sun photometers. Specifically, AOD retrievals from MODIS onboard Aqua and CALIOP onboard CALIPSO were evaluated against AERONET data. The AOD from the latest MODIS (MAIAC) AOD at 470nm and 550nm, retrieved for 3x3, 5x5, 9x9, 20x20, and 30x30 pixels (each of 1x1 km in size) and Calipso (CALIOP) AOD at 532nm retrieved for 5x5, 10x10, 20x20, and 30x30 pixels (each of 0.1x5 km in size) were centered over each of the AERONET sites in the study areas (Northern Africa, California, Germany and IGP, India) for the years 2007-2017. The AOD in the MAIAC-AERONET and CALIOP-AERONET collocated observations were compared, with the intercept of the linear regression indicating the uncertainty in the surface reflectance estimation and the slope associated with errors in the aerosol model assumptions. We analyzed the root mean square error, mean absolute error, relative mean bias, and the fraction of MAIAC and CALIOP AOD that falls within the reported expected error (EE). Good correlations were observed between MODIS AOD and collocated hourly AOD measurements from two AERONET stations in Tunisia. The different between the two AERONET stations can be attributed to variations in surface reflectance, differences in meteorological conditions (e.g. planetary boundary layer height and relative humidity profile along the atmospheric column), presence of aerosols aloft (e.g. long-range transport of mineral dust), and geographical differences in PM composition. All these will be tested in the following stages of the research.

Keywords: air

SU-PO-02

HCHO, NO2 and SO2 vertical column density across the Caspian Sea: The contribution of five neighbor countries

Yerkenov T. [1], Kuspangaliyeva B. [1], Altazhanova K. [1], Amouei Torkmahalleh M. [1], Darynova, Z. [1]; [1]Nazarbayev University, Kazakhstan

Vertical column density (VCD) of HCHO, NO2, and SO2 over the atmosphere of Caspian Sea during 2005-2017 years period was investigated in order to identify the possible sources of these pollutants and to assess their potential transport to five surrounding countries across the sea. The most polluted coastal area of each country was chosen in this study. The VCD of gases were retrieved from the satellite data derived by Ozone Monitoring Instrument (OMI) which has been built in Aura Satellite board. This study addresses the impact of industrial, social and geopolitical development in the five Caspian littoral states on HCHO, NO2 and SO2 concentrations. Overall NO2 and HCHO VCDs in all coastal countries over the Caspian Sea increased gradually by time due to the elevated oil and gas productions and their consumption, also a boost of population and rise of a number of vehicles. For instance, from 2005 to 2013 NO2 increased up to 4.5×1015 molecules/cm2. However, SO2 VCD showed opposite behavior. The value decreased from 2.5 x 10-6 kg/m2 to 2x10-6 kg/m2 over the study period. HCHO VCD was found to be higher in Sari city located North of Iran compared to other countries due to the present of biogenic VOCs emitted from a forest land located in North of Iran. Among multiple factors affecting the trends, the temperature has an essential effect on the formaldehyde level: the countries closer to the equator have higher temperature thus larger HCHO VCD trend. The results of this study could be a reference for the ongoing negotiations among the five Caspian Sea countries that assigns the Sea share.

Keywords: VOCs

SU-PO-03

Impact of the deployment of electric vehicles on population exposure to air pollution

Minet L.[1], Gai Y.[1], Wang A.[1], Posen D.[1], Hatzopoulou M.[1]; [1]University of Toronto, Toronto, Canada

The transportation sector contributes 35% of Ontario's total greenhouse gas (GHG) emissions. Similarly, it is also a large contributor to criteria air contaminants. For example, on-road vehicles contribute 34% of total nitrogen oxide (NOx) emissions. Since electric vehicles do not generate any operating emissions, vehicle electrification has been promoted to reduce traffic-related air pollution and GHG emissions. However, in the context of a scenario analysis of the impact of electric vehicles on air quality, it is necessary to consider emissions from surrounding power plants. We propose an integrated methodology to investigate the exposure of the Greater Toronto and Hamilton Area (GTHA) population to air pollution under different scenarios expressing various penetration rates for electric vehicles, which would impact the production of ten power plants present in the GTHA. Expected emissions from these power plants are modeled using current data from the National Pollutant Release Inventory (NPRI) database. Marginal emission factors are generated to take temporal variations into consideration, as only certain types of power generators respond to a marginal increase in demand (e.g. electric vehicle charging). Traffic operating emissions are modeled by combining regional travel demand and traffic assignment results with speed-based emission factors. Finally, a plume-in-grid model, which combines a chemical transport model and a Gaussian puff dispersion model, is used to analyse air quality under various meteorological conditions. Different scenarios are examined in terms of equity with respect to air pollution exposure.

Keywords: aggregate exposure, air, environmental justice, exposure models

SU-PO-04

Land Use Regression models for Ultrafine Particles: development and transferability within a mega-city

Yang ZC.[1], Freni-Sterrantino A.[1], Fuller G.[1], Gulliver J.[2]; [1]MRC-PHE Centre for Environment and Health, Imperial College London, London, United Kingdom; [2]Centre for Environmental Health and Sustainability, University of Leicester, Leicester, United Kingdom

Background: A lack of routine monitoring for ultrafine particle (UFP) means that bespoke monitoring (e.g. mobile measurements on vehicles or repeated short-term monitoring over a spatially distributed network of sites) has to be undertaken to develop land use regression (LUR) models. This is especially challenging in mega-cities due to the large spatial area. Methods: As an alternative to monitoring across a mega-city, we developed single- and two-district models in London, and then tested the transferability (i.e. generalisability) of models between districts. In each district, repeated 30-min monitoring was undertaken at 40 sites in three different seasons (2016-2018) to estimate annual mean UFP (particles cm-3). We constructed ten single- and two-district LUR models by allocating monitoring sites to one of ten groups and then used nine groups of sites to develop each model, iteratively, with the remaining group held-out for model evaluation (e.g. groups 1-9 to predict for group 10; groups 2-10 to predict for group 1, etc.). We assessed model performance by comparing R2 and the variables selected, and pooled the 10% of sites held-out each time to produce an overall R2 to assess model robustness. The transferability of models was tested by applying single- and two-district models to the other districts. Results: For the ten models for single districts, R2 ranged from 60% to 73% and model variables were similar and all included nearby road traffic. Model performance and structure was different between areas. R2 for two districts ranged from 50% to 68%. Pooled holdout validation had R2 values of 37%- 59%. Performance in transferring models to other districts within London was overall lower with R2 ranging from 10% to 39%. Conclusion: UFP LUR models may perform well within the confines of a monitoring network but transferring models within a city may have substantially lower performance.

Keywords: exposure models

SU-PO-05
NO2 Air Pollution Exposure Assessment in Urban Mysore, India

Nori-Sarma A. [1], Thimulappa R.K., Venkataraman G., Fauzie A., Venkareddy L. [2], Berman J. [3], Lane K. [4], Fong K. [1], Bell M. [1]; [1]Yale University, New Haven, CT, USA, [2]University of Illinois at Chicago, Chicago, IL, USA, [3]University of Minnesota, Minneapolis, MN, USA, [4]Boston University School of Public Health, Boston, MA, USA

Introduction: In developing countries, rapid urbanization has led to decreased air quality. Few studies have applied spatially heterogeneous sampling to assess ambient air pollution levels in cities in India, to understand the important local predictors of air pollution exposure, which may be unique to Indian urban environments. Methods: Seasonal sampling was conducted in four campaigns (2016-2017) at 150 sites throughout Mysore, a small and rapidly growing city in south India. Nitrogen dioxide (NO2) levels were assessed using passive Palmes tubes and Ogawa badges. Results: Annual average NO2 ranged from 0.8 to 31.2ppb (1.5 to 58.7 ug / m3) for the study area, with higher concentrations in the center of the city. In the final Land Use Regression (LUR) model, proximity to major roads, point sources of pollution, and religious complexes, as well as high human activity land uses and population density were positively associated with air pollution, while proximity to minor roads and low human activity land uses were inversely associated with air pollution (R-squared = 0.535). Cross-validation of the results confirmed the reliability of the model. Kriging improved estimates by accounting for spatial correlation. Conclusions: The combination of passive NO2 sampling and LUR / Kriging modeling techniques allowed for characterization of NO2 patterns in Mysore. While traffic pollution has been indicated in the past as a major contributor to ambient air pollution levels in urbanizing centers of Asia, results indicate that other pollution factors (e.g., point sources), as well as highly localized characteristics of the urban environment (such as proximity to religious buildings), influence air pollution exposure patterns in urban India. Despite a reputation as one of the cleanest cities in India, areas of Mysore consistently experience pollution in excess of WHO health-protective guidelines for NO2.

Keywords: air, exposure models, geospatial analysis/GIS

SU-PO-06
Using AOD retrievals and enhanced modeling techniques to estimate ground-level PM2.5 in Southern California

Stowell J. [1], Alhamadan M. [2], Meng X. [1], Bi J. [1], Lee S. [3], Garcia C. [4], Kinney P. [5], Freedman F. [6], Liu Y. [1]; [1]Rollins School of Public Health, Emory University, Atlanta, GA, USA, [2]National Space Science and Technology Center, Huntsville, AL, USA; [3]Air Quality Management District, [4]Diamond Bar, CA, USA, California Air Resources Board, Sacramento, CA, USA, [5]Boston University School of Public Health, Boston, MA, USA, [6]San Jose State University, San Jose, CA, USA

Background: Increasing economic growth and climate change add to concerns regarding the effects of anthropogenic pollution. Studies of PM2.5 are limited by accuracy of exposure levels. Hence, the use of satellite data in exposure science has increased, with studies highlighting the benefits of using AOD to estimate PM2.5. However, few studies have addressed this issue in southern California. Methods: We estimated the PM2.5/AOD relationship using a two-stage model. The first stage estimated temporal PM2.5/AOD relationships using a linear mixed effects model. The second stage accounted for spatial variation using geographically weighted regression. Goodness of fit for the final model was evaluated by comparing the fitted and observed results using R2 and RMSE statistics. Results: Mean predicted PM2.5 concentration for the study domain was 12.04μg/m3. The first stage performed well with R2=0.74 and RMSE=3.29μg/m3. The second stage yielded a R2 increase to 0.77 and decreased RMSE to 3.11μg/m3. Overall, the second stage improved performance (R2 increase=0.03, RMSE decrease=0.18μg/m3). During model fitting, the presence of larger sizes of PM proved an important variable in the AOD/PM2.5 relationship (β=14.88, p<=0.00). Including a ratio of these parameters improved performance, resulting in 0.06 increase in R2 and a 0.34 decrease in RMSE. Discussion: Utilizing the high-resolution MAIAC AOD, fine–resolution exposure data can be estimated where measurements are sparse. The inclusion of a second stage increases accuracy, accounting for spatial variance in the dataset. This study adds to the current research using remote sensing data to achieve better exposure data in the understudied region of southern California. Overall, we demonstrate the usefulness of MAIAC AOD and the importance of considering larger particulate matter in airborne dust prone areas.

Keywords: air, environmental health, exposure models, geospatial analysis/GIS, particles

SU-PO-07
The Feasibility of Identifying and Quantifying Worker Exposures to Volatile Organic Chemicals in Beauty Salons and Auto Shops in the Southwestern USA

Lothrop N.[1], Sandoval F., Cortez I., Wagoner R., Lopez-Galvez N., Parra K., Bell M., Griffin S., Beamer P.[1], Wolf A., Lee A., Carvajal S., Ingram M.; [1]Mel and Enid Zuckerman College of Public Health, University of Arizona, Tucson, AZ, USA

Workers in beauty salons and auto repair shops are regularly exposed to volatile organic chemicals (VOCs), which have well-documented, adverse health effects. This occupational risk is increased in small businesses due to their limited access to occupational health resources, resulting from a complex interaction of physical, social, and economic factors. Through a community-university partnership, this pilot study examined the feasibility of identifying and quantifying VOCs exposures to predominantly Spanish-speaking workers in beauty salons and auto repair shops in Tucson, AZ. Businesses were recruited door-to-door by trusted community health workers (CHWs). Using a bilingual (Spanish-English), web-based form to minimize shop inconvenience, CHWs documented workplace practices, chemical usage, and controls. CHWs photographed products, which were later referenced by university personnel to build chemical inventories. To quantify personal total VOC exposures, workers wore a VOC monitor in a sling-pack or on a belt custom-fitted for comfort during their shift (total of 4 worker-shifts/business). On a paper form, university personnel recorded worker activities and relevant exposure factors. CHWs visited 16 beauty salons and 23 auto shops, enrolling 10 of each. Participating businesses had 2-3 persons working at a time, often necessitating 2 visits per business. All beauty salons and 9/10 auto shops completed the study. Workers from both shop types changed from sling-pack to belt (or vice versa) more than once per shift, with 3 beauty salon workers hanging the monitor next to them. This pilot study offered valuable lessons for future work. Tracking participation activity through observation was problematic, necessitating asking the participant directly, when appropriate. Entering paper activity logs into an electronic database produced substantial data entry issues, and auto shop recruitment proved difficult. Nonetheless, identifying and quantifying worker VOC exposures is feasible through a community-university partnership, and this pilot study will inform a randomized trial aimed at reducing VOC exposure.

Keywords: VOCs, air, built/indoor environment, community, environmental justice

SU-PO-08 WITHDRAWN

SU-PO-09
Seasonal spatial and temporal variability of Black Carbon in an elementary school catchment area of Milan, Italy

Boniardi L. [1], Dons E. [2], Campo L. [3], Van Poppel M. [4], Int Panis L. [2], Fustinoni, S. [5]; [1]Università degli Studi di Milano, Milano, Italy, [2]Hasselt University; Flemish Institute for Technological Research (VITO), Hasselt, Belgium, [3]Fondazione IRCCS Ca' Granda Ospedale Maggiore Policlinico, U.O.S Tossicologia, Milano, Italy, [4]Flemish Institute for Technological Research (VITO), Moi, Belgium, [5]Università degli Studi di Milano; Fondazione IRCCS Ca' Granda Ospedale Maggiore Policlinico, U.O.S Tossicologia, Milano, Italy

Introduction. The city of Milan is a high-traffic area with an estimated population of about 3.2 million inhabitants. The health impact of traffic-related air pollution is reported to be larger for the more susceptible population groups, such as children. Morning rush hour (MRH) is the most congested period and it is also the time during which schoolchildren commute to school and are directly exposed to traffic pollution. The aim of this study was to analyse the temporal behaviour and the determinants of Black Carbon (BC) during warm and cold seasons with a focus on MRH in an elementary school catchment area. Methods. Eight micro-aethalometers measured BC concentrations on 35 sites classified as street (S), urban traffic (UT) or urban background (UB) according to the influence of traffic sources. The monitoring campaigns lasted from May to June 2017, and from January to February 2018. Every site was monitored for one week during each season, while one reference site, located in the school courtyard, sampled for the entire period of both monitoring campaigns. Four seasonal Land Use Regression (LUR) models were developed. Results. In both seasons, MRH BC average was higher than the daily average: 2.6±0.7 μg/m3 (mean±SD) versus 1.5±0.4 μg/m3, and 4.3±0.7 μg/m3 versus 3.5±0.7 μg/m3, during the warm and cold season respectively. The highest spatial contrast was observed during the warm MRH with 3.7±0.5 μg/m3 for S-sites and 1.9±0.3 μg/m3 for UB-sites. LUR seasonal models explained from 56% to 65% of the BC variability and MRH-focused models performed better than models for daily average. Conclusion: MRH is the most critical time window for air pollution in Milan. The four developed LUR models show that traffic sources play a major role in explaining BC concentrations. Selecting low-pollution routes could lead to an important reduction in the exposure of school-aged children.

Keywords: air, children, geospatial analysis/GIS, particles, statistical methods

SU-PO-10
Breath Volatile Organic Compounds Profile After Short Exposure to High Concentrations of Indoor Fragrance Chemicals

Vicent-Claramunt A. Center for Physical Sciences and Technology, Vilnius, Lithuania

Nowadays people spend most of their time indoors; sleeping at home, working in the offices, eating in canteens or restaurants, or spending free time in cinemas, shopping centers and other stores. It is in those places where we are exposed to a high content of volatile organic compounds (VOCs). In this work, we focused on places where the concentration is especially high like cosmetic stores, drug stores and bathrooms. Fragranced products emit dozens of different volatile compounds. As some VOCs might be harmful for human health, the aim of the work was to monitor how people respond to a short exposure to environments with very high concentration of fragrances. Air samples were taken from the selected places, to benchmark the background compounds. Samples were collected using thermal desorption (TD) tubes and purged into a gas chromatography with mass spectrometer (GC-MS). Then, breath samples from participants were collected before and after being exposed to places with high fragrance content. Optimization of system parameters was performed to get good compound separation and identification.The results shows the high concentration of several VOCs in a study place. If people are exposed, some of them are incorporated into respiratory system and their variation in breath could tell us how those compounds behave inside our body, some will be exhaled, others might be retained and different metabolites can be formed. In the next stage we will look into the possible pathways of these VOCs through the body, as some of them might be very dangerous for our health. Knowing the effect of those compounds inside our body can give us an understanding of the toxicity levels we are exposed to in our every day life.

Keywords: VOCs, air, biomarkers, consumer and personal care products, environmental health, respiratory health, workplace

SU-PO-11 WITHDRAWN

SU-PO-12
Effects of Biophilic Interventions in Office on Stress Reaction and Cognitive Function: A Randomized Crossover Study in Virtual Reality

Yin J. [1], Arfaei N. [1], MacNaughton P. [1], Catalano P. [2,3], Allen, J. [1], Spengler J. [1]; [1]Department of Environmental Health, Harvard T.H. Chan School of Public Health, Boston, MA, USA, [2]Department of Data Sciences, Dana-Farber Cancer Institute, Boston, MA, USA, [3]Department of Biostatistics, Harvard T.H. Chan School of Public Health, Boston, MA, USA

Biophilia hypothesis suggests humans have an innate connection to nature. Yet we currently live in a world that is rapidly urbanizing with people spending more time indoors than ever before. We simulated one open and one enclosed office space in virtual reality (VR) and quantified the acute effects on stress reaction and cognitive function in 30 participants who each visited twice to experience three biophilic environments (natural elements, natural analogs, and combo) compared to a non-biophilic environment in both spaces, with the order randomized. While in VR, we measured blood pressure, heart rate variability and skin conductance level, and administered cognitive tests to measure their reaction time and creativity. These three biophilic interventions affect lower stress levels, higher creativity scores (0.25, 0.15, and 0.20 standard deviation increase in creativity task scores), yet longer reaction times (5.5%, 1.0% and 2.8% increases respectively) on focused tasks, comparing to the non-biophilic environment. Natural elements intervention had more prominent effects on stress reduction and cognitive function than the natural analogues intervention. In addition, effects of virtual exposure to biophilic elements on reduction of stress level and improvement of cognitive function were not linear, with optimal effects seen when the percentage of time participants looked at biophilic elements was around 50%. Biophilic interventions could help reduce stress and improve creativity; however, the effects may be different based on the workspace type (open vs. enclosed). The effects are related to both the type and amount of biophilic elements, with natural elements and 50% space saturation being optimal for stress reduction and improved creativity. This research is essential to help developers, architects and designers know what are the appropriate biophilic design interventions to use, and how these vary from one workspace type to another.

Keywords: biomonitoring, built/indoor environment, environmental health, productivity, workplace

SU-PO-13
Hypertonic Saline Nasal Lavage Intervention in Dairy Workers

Erlandson G.[1], Magzamen S.[1], Carter E.[1], Sharp J.[1], Reynolds, S.[1], Schaeffer J; [1]Colorado State University, Fort Collins, CO, USA

Livestock workers experience an increased burden of bioaerosol-induced respiratory disease. We have previously demonstrated that dairy operations generate bioaerosols that span the inhalable size fraction. These aerosols contained complex bacterial communities and inflammagens. Because those particles with an aerodynamic diameter between 10-100 µm are known to deposit in the nasopharyngeal region, we believe that exposure health outcomes in the nose are equally important and need consideration. We conducted a pilot study to evaluate the effectiveness of a hypertonic saline nasal lavage in reducing inflammatory responses in dairy workers. We recruited a total of ten workers from a high-volume operation in the High Plains region. We collected inhalable personal breathing zone (PBZ) samples (SKC button) and pre-/post-shift nasal lavage samples from each participant over five consecutive days. The treatment group (n=5) received hypertonic saline (HTS) while the control group (n=5) received normotonic saline. The PBZ samples were analyzed for endotoxin using the Pyrogene assay (Lonza) and gravimetrically for particulate. Pro-inflammatory cytokines (i.e., for IL-8, IL-6, and TNF-⍰) were measured from nasal lavage samples using a multiplex assay (Meso Scale Discovery). Inhalable dust concentrations ranged from 0.15mg/m3 to 1.9mg/m3. Cytokine concentrations were analyzed with a hierarchical linear effects model. Significant attenuation of TNF-a, IL-6, and IL-8 was observed among workers receiving HTS compared to those in the control group ($\alpha = 0.05$). For TNF-a, there was a significant day by group interaction and the expression of TNF-a in the treatment group was significantly lower than expression in the control group at the end of the study ($p < 0.001$). There was also an overall significant treatment effect for IL6 and IL8 cytokines. Funding for this research is provided by the National Institute for Occupational Safety and Health U01 grant #: 1U01OH010840.

Keywords: biomarkers, microbial agents, occupational, respiratory health

SU-PO-14
Serum concentrations of polybrominated biphenyls (PBBs) and the other persistent organic pollutants (POPs) among the Michigan PBB cohort

Chang C.J. [1], Terrell M. [1]L, Marcus M. [1], Marder M.E. [1], Panuwet P. [1], Ryan, P.B. [1], Barr, D.B. [1]; [1]Department of Environmental Health, Rollins School of Public Health, Emory University, Atlanta, GA, USA

Polybrominated biphenyls (PBBs) were highly produced and widely used compounds added as flame retardants in various consumer products. Reported health effects associated with exposure to PBB include alterations to timing of pubertal development, adverse reproductive outcomes, thyroid dysfunction and certain cancers. From 1973 to 1974 widespread PBB contamination occurred in Michigan when the commercial PBB mixture was accidentally added to livestock feed instead of a nutritional supplement. Consequently, people in Michigan were exposed to PBBs via consuming contaminated meat, egg, and dairy products. The Michigan PBB Registry was established by the Michigan Department of Health to monitor exposure and health effects among Michigan residents. PBB stores in adipose tissues and is only slowly eliminated. Additionally, PBBs can be transferred from a mother to her fetus through the placenta and to her developing child through breastfeeding. Due to the long biological half-lives of PBBs, as well as the concerns from the exposed community, we carried out biomonitoring measurements from 2012 – 2015. Because of their similar structures, this study measured serum PBB, polychlorinated biphenyl (PCBs), and polybrominated diphenyl ether (PBDEs) levels 40 years after the industrial incident. We found that the serum PBB-153 among the original highly-exposed groups (chemical workers, the family of chemical workers, and individuals who lived on or received food from the contaminated farms) remains significantly higher than other Michigan residents. Overall, the serum PBB-153 concentrations [geometric mean 25.69 ng/g lipid (95% CI 22.01-29.98 ng/g lipid)] are significantly higher than 2003-2004 NHANES data [geometric mean 2.29 ng/g lipid (95% CI 1.82-2.87 ng/g lipid)], and those who were born after the contamination incident (after 1973) also have higher serum PBB levels when compared with similar age groups in the US population, which suggests transgenerational exposure and/or continued environmental exposures following the contamination period.

Keywords: biomonitoring

SU-PO-15
Systematic Identification of Allostatic Load Components Associated with Mortality

Nguyen V. [1,2], Chung J. [1], Pho N. [2], Patel C. [1], Colacino J. [3], Jolliet O. [3]; [1]Harvard University, Boston, MA, USA, [2]University of Michigan, Ann Arbor, MI, USA, [3]University of Washington, Seattle, WA, USA

Physiological stress response can be characterized by measurements of health biomarkers, which represent various biological pathways including neuroendocrine, cardiovascular and metabolic system. These components can be summed as binary variables into a uniformly weighted sum known as the allostatic load (AL) to represent the aggregated burden of chronic stress. AL score components vary widely between studies. Further, many calculations of AL do not quantify biomarker measurement deviance from clinical thresholds as a measure of risk. To generate a more robust AL score, here we identify components significantly associated with mortality to develop a weighting method based on an individual's biomarker distance from a clinical threshold. We used the National Health and Nutrition Examination Survey of biomarker measurements for 41 components in 62,160 participants 1999-2010. We used the mortality data linked with the National Death Index to ascertain mortality status and time to death. Ultimately, we plan to validate our weighted AL scores by evaluating the prediction performance on mortality, so AL components were excluded here based low overlap with mortality data. We conducted survey weighted cox proportional-hazards regression models with mortality status and time to death as the outcome variables and the AL component as the main predictor while adjusting for age, sex, race/ethnicity, smoking, and socioeconomic status. We identified 22 AL components that are significantly associated with mortality including creatinine, glycohemoglobin, and C-reactive proteins that increase the risk of death in the study period by 33%, 17%, and 14%, respectively, for a one-unit increase, while albumin decreases the risk by 68%. Here, we present a framework to systematically identified AL components that are associated with mortality for calculating the AL score. Using the remaining AL components, we are developing methods to incorporate weights based on distance from clinical thresholds to calculate a weighted AL score.

Keywords: biomarkers

SU-PO-16
Contributions of exposure-related behaviors and dietary factors to organophosphate pesticide exposure in Japanese toddlers

Oya N.[1], Ito Y.[1], Ebara T.[1], Kato S.[1,2], Hioki K.[1], Aoi A.[3], Ueyama J.[3], Oguri T.[1,4], Shoji N.[1,5], Sugiura-Ogasawara M.[6], Saitoh S.[7], Kamijima M.[1]; [1]Department of Occupational and Environmental Health, Nagoya City University Graduate School of Medical Sciences, Nagoya, Japan, [2]Department of Pediatrics and Neonatology, Nagoya City University Graduate School of Medical Sciences, Nagoya, Japan, [3]Department of Pathophysiological Laboratory Sciences, Field of Radiological and Medical Laboratory Sciences, Nagoya University Graduate School of Medicine, Nagoya, Japan [4]National Institute of Advanced Industrial Science and Technology (AIST), [5] Asahi University, School of Health Sciences, Mizuho, Gifu Prefecture, Japan, [6]Department of Obstetrics and Gynecology, Nagoya City University Graduate School of Medical Sciences, Nagoya, Japan, [7]Department of Pediatrics and Neonatology, Nagoya City University Graduate School of Medical Sciences, Nagoya, Japan

Several epidemiological studies suggested that exposure to organophosphate (OP) pesticides could elicit adverse health effects on children. The objectives in this study were to reveal the OP exposure levels in Japanese toddlers and to explore the possible contributions of their exposure-related behaviors and their mothers' considerations of food selection and preparation to the exposure levels, considering one-day intake of foodstuffs. We recruited children participating in Japan Environment and Children's Study at Aichi regional center, and collected used disposable diapers from 1037 children (16–23 months old, 531 boys and 506 girls) from June 2015 to August 2016. Six urinary dialkylphosphates (DAPs) were measured using urine extracted from diapers by ultra-performance liquid chromatography with tandem mass spectrometry. The information about exposure-related behaviors, considerations of food selection and preparation, and foodstuffs were obtained from questionnaires. The geometric mean of ΣDAP (total DAPs) concentrations and amounts excreted onto diaper during a night were 120 nmol/L and 13.4 nmol. In binary logistic regression analyses, use of an insecticide, an herbicide, an insect repellent spray, and a fragrance or a deodorant increased odds ratios (ORs) of the exposure levels (top 5% of urinary DAP concentrations or amounts) (ORs: 2.1–2.6). Receiver-operating characteristics curve analysis revealed that discriminatory power to determine whether or not ΣDAP concentrations and the amounts exceeded their 95th percentile values was lower for the respective propensity scores for the exposure-related behaviors (area under the curve 0.72 and 0.74) and the considerations of food selection and preparation (0.60 and 0.67) than that for the foodstuffs taken on the survey day (0.89 and 0.91). We revealed the exposure levels of OP pesticides in Japanese toddlers. The contributions of the exposure-related behaviors and the considerations of food selection and preparation to the exposure levels were likely much smaller than that of foodstuffs.

Keywords: biomonitoring, children, epidemiology, exposure factors, pesticides

SU-PO-17
Imidacloprid Exposure in Grape Workers

Lopez-Galvez N.[1], Wagoner R.[1], Canales R.[1], de Zapien J.[1], Rosales C.[1], Beamer P.[1]; [1]University of Arizona, Tuscon, AZ, USA

Imidacloprid is a neonicotinoid insecticide commonly injected through agricultural drip irrigation systems to reduce the population of vine mealybugs in grape farms. In the present study, we evaluated the exposure to imidacloprid in 20 grape workers two weeks after imidacloprid injection in the irrigation system. We administered a questionnaire on work activities, exposure characteristics, and socio-demographics. We collected personal air and hand wipe samples to determine possible inhalation and dermal exposure. We collected urine samples to quantify metabolite concentrations and we measured hand surface temperatures. Most participants migrated from the state of Chiapas, Mexico, all participants had less than high school education and about half identified an indigenous language as their first language. Imidacloprid was not detected in air samples, but it was detected in 85% of the hand wipe samples (GM: 0.41 μg/wipe). The majority of participants (75%) had detectable concentrations of imidacloprid as a parent compound in their urine (GM: 0.15 μg/g), and almost all of the participants (95%) had detectable concentrations of 5-hydroxy-Imidacloprid, a urine metabolite, (GM: 1.79 μg/g). There was a significant correlation ($p < 0.001$) of the imidacloprid concentration found in hand wipes with the concentrations found in urine (rho: 0.67, and 0.79 for imidacloprid and 5-hydroxy-Imidacloprid). Hands' temperature was significantly correlated ($p < 0.05$) with the pesticide concentration found on hands (rho: 0.70), and the urine biomarkers (rho: 0.68, and 0.60 for imidacloprid and 5-hydroxy-Imidacloprid). No levels of exposure via the inhalation route and higher levels via dermal route were found. Our results suggested that hand wipe can be use as less intrusive measurement to assess imidacloprid exposure in farm workers. Imidacloprid urine and hand wipes concentration were significantly higher during the summer than during the spring season. Further research needs to be conducted on microclimatic conditions to evaluate and reduce dermal pesticide exposure.

Keywords: biomarkers, biomonitoring, cumulative exposure, environmental health, exposure factors, occupational

SU-PO-18

Improved methods of assessing the impact of air pollution on public health.

Morhulova V.[1], Petrosian A.[1], Maremukha T.[1]; [1]State Institution – "O.M. Marzeiev Institute for Public Health of the National Academy of Medical Science of Ukraine", Kyiv, Ukraine

Air pollution is a major threat to health, resulting in seven million premature deaths around the world according to WHO in 2018. The quantitative determination of the air pollutants dose requires the use of regression analysis based on remote sensing data, digital relief models, meteorological data, and infrastructure of study area data. Aim. Provide a comparative analysis of the tools for assessing the air quality to scientific justify using of tools for different scenarios of distribution and impact of pollution, development of the foundations for the introduction of new tools for assessing the impact of air pollution on public health. Methods. Mathematical modeling methods (TM5-FASST, ISC-AERMOD View average concentrations calculation of air pollutants and Transboundary transfers); statistical methods; mapping (ArcGis 10.2, ArcMap Esri); dispersion and regression analysis; assessing the impact of air pollution on public health (AirQ+ and Air Quality Index). Results. Creation algorithms of work with software complexes of air pollution modeling and impact on the public health. Using the prognostic models of the influence the pollutants concentrations on the studied areas with consideration their peculiarities. Conclusions. Based on the analysis of field data and modeling will be deepened theoretical understanding of the air pollution influence mechanisms on public health, will be developed the scientific basis for the use of advanced technologies in air quality management, will be researched the influence of territorial features of air pollution on public health.

Keywords: air, environmental health, exposure models, geospatial analysis/GIS

SU-PO-19

Insecticide use on pets is a strong predictor of pyrethroid exposure in an urban population of young adult men in Poland

Rodzaj W. [1], Wileńska M. [1], Dziewirska E. [2], Klimowska A. [1], Jurewicz J. [2], Radwan M. [3], Radwan P. [4], Hanke W. [2], Wielgomas B. [1]; [1]Department of Toxicology, Faculty of Pharmacy, Medical University of Gdańsk, Gdańsk, Poland, [2]Department of Environmental Epidemiology, Nofer Institute of Occupational Medicine, Łódź, Poland, [3]Faculty of Health Sciences, State University of Applied Sciences, Plock, Poland, [4]Department of Gynecology and Reproduction, "Gameta" Hospital, Rzgów, Poland

Pyrethroids are synthetic insecticides widely used in agriculture, households and veterinary medicine to kill and control pests. It is well documented in the literature that exposure to these pesticides may pose a threat to human health. A deleterious effect on male fertility is one of concerning issues. The aim of the study was to determine levels of exposure to pyrethroids through measurements of five pyrethroid metabolites in urine: 3-phenoxybenzoic acid (3PBA), cis-2,2-dimethyl-3-(2-chloro-3,3,3-trifluoro-1-propenyl)-cyclopropanecarboxylic acid (BIF), cis/trans-3-(2,2-dichlorovinyl)-2,2-dimethylcyclopropane-1-carboxylic acids (cis/trans-DCCA), and cis-(2,2-dibromovinyl)-2,2-dimethylcyclopropane-1-carboxylic acid (DBCA) in the population of young men living in urban area of Poland (n=242) and to investigate the relationship between dietary, lifestyle and sociodemographic factors and urinary levels of these biomarkers in enrolled group. The study cohort was recruited in Lodz, central Poland, and in nearby areas. Along with spot urine sample, the participants provided a completed questionnaire regarding selected exposure factors. Concentrations of pyrethroid metabolites in urine were measured using a validated GC-MS method and adjusted for specific gravity (SG). The limit of detection (LOD) for all analytes was 0.1 ng/mL. BIF, cis-DCCA and DBCA were detected in less than 50% of the samples, whereas 3PBA and trans-DCCA in 67% and 78%, respectively. The geometric mean (GM) concentrations (SG-adjusted) for 3PBA and trans-DCCA were 0.229 and 0.274, respectively. These two metabolites were also positively correlated ($r = 0.4236$, $p < 0.001$), but the strongest correlation was noticed between cis- and trans-DCCA ($r = 0.7828$, $p < 0.001$). Although analysis of dietary habits, sampling time and sociodemographic data failed to identify significant predictors of pyrethroid exposure, we found that insecticide use on pets was the strongest predictor ($p < 0.01$) of 3PBA concentration. In conclusion, a magnitude of pyrethroid exposure in studied population was assessed and factors contributing to this phenomenon were identified but require further investigation.

Keywords: biomarkers, biomonitoring, pesticides

SU-PO-20

Enhancement of thermal comfort in summer conditions by the use of personalized ceiling fans

Rissetto R. [1], Wagner A. [1], Schweiker, M. [1]; [1]Karlsruhe Institute of Technology, Karlsruhe, Germany

Due to increasing summer outdoor temperatures, the cooling demand for active cooling of buildings is expected to increase. Alternatively, the implementation of low-energy cooling strategies has shown great potential, especially by the use of ceiling fans. Additionally, to increase thermal comfort while maintaining energy consumption as low as possible, the attention has been drawn to the use of personalised comfort systems. This study focusses on the impact of different air direction, occupants' activity level and control opportunities of a personalized ceiling fan on occupants' thermal comfort and satisfaction in a working environment. We conducted a 3-week laboratory experiment where 52 participants took part. Two age groups were examined at different ceiling fan's configurations (six different air directions), two indoor temperature settings (28°C and 31°C), two subject's activity levels (sitting and standing) and two air speed control possibilities (fixed and adjustable by the subject). Participants' thermal comfort responses were investigated. Results show that 70% of the subjects perceived the environment at 28°C as comfortable when the fan was in use – in comparison to 22% without fan use. According to both the fan configuration and the control possibility over the air speed, no significant difference in terms of improved thermal comfort was found, while significant differences were found between the sitting and the standing position, under different temperature conditions and between morning and afternoon conditions. The results suggest that the cooling effect of personalized ceiling fans can fulfill participants' thermal comfort at moderately high indoor temperatures. Moreover, this technology could adapt to different building configurations, especially in existing buildings, due to its flexibility and guarantee of thermal comfort, as the air blowing from different directions showed no negative impact on users' thermal satisfaction. On the other hand, occupants' activity level, indoor temperature and subjects' thermal history influence significantly occupants' thermal comfort.

Keywords: air, behavior, built/indoor environment, ventilation, workplace

SU-PO-21
Agricultural Pesticide Exposure and Congenital Abnormalities (CA) in Mexico: A Systematic Review

Trejo B.; Boston University, Boston, MA, USA

Background: Pesticide exposure has been associated with congenital abnormalities (CA), which are a public health concern in Mexico. Objective: This systematic review identified literature that addressed the relationship between parental exposure to agricultural pesticides via agricultural work and CA among infants of Mexican parents. The systematic literature review sought to communicate the findings of the association, identify strengths and limitations, and make recommendations for further research. Methods: Articles were identified using PubMed with secondary searches of Scopus, PubMed Central, ProQuest Environmental Science, Academic Search Complete, Popline and hand search of references of the identified articles. Results: Seven studies were identified that investigated the association between agricultural pesticide exposure and CA in Mexico. Six studies observed a positive association; one study found no association. Five studies investigated exposure to specific pesticide(s) and two investigated general pesticide exposure. Five studies relied on self-reporting for ascertainment of exposure and/or outcome. Two studies used more objective methods, such as biomarkers and clinical diagnosis for both the exposure and outcome of interest. Five studies investigated specific CA; two studies investigated CA generally. Conclusion: The reviewed literature supported the positive association between pesticides and CA while demonstrating a need for improved exposure and outcome ascertainment.

Keywords: pesticides

SU-PO-22
Land Use Regression Models of Traffic-Related SVOC Pollutants in an Urban Area with Elevated Prevalence of Pediatric Asthma

Esenther S.[1], Lin E.[1], Pollitt K.; [1]Yale University, New Haven, CT, USA

While land use regression models have been commonly used to model inorganic gaseous traffic emission marker pollutants such as nitrogen dioxide (NO2), the difficulty of deploying equipment to measure SVOCs across a study area has limited the number of LUR models of SVOCs associated with traffic emissions. Deployment of the PDMS sorbent material mounted in PTFE housing case has enabled the measurement of PAHs at 40 sites across the greater Springfield, MA area (~185 km2). The objective of this study was to assess concentrations and determinants of PAHs traffic emission pollutants in Springfield. The PDMS bars were deployed for five consecutive weekdays (Monday to Friday) at a sites across the urban and suburban area under summer conditions. PDMS sorbent material was thermally desorbed and analysed by gas chromatography time-of-flight mass spectrometry to determine time-integrated exposure measures. Detectable concentrations were measured for six PAHs (isophorone, benzene,1-bromo-4-phenoxy, benzene, hexachloro, pyrene, Benzo[b]fluoranthene, Benzo[k]fluoranthene) at 20 or more of the sample locations. LUR models will be developed using built environment and land use characteristics and compared with NO2 LUR model of the area using ambient exposure data collected through deployment of 40 Ogawa passive sampler badges. Future work will extrapolate these models to estimate exposures for children with asthma.

Keywords: SVOCs, air, children, community, geospatial analysis/GIS

SU-PO-23

Impact of Parental Migration on Environmental Exposure Related Time-Activity Patterns of Left-behind Children in China: Results from the First National Population-based Survey

Xiangyu X. [1], Qian G. [1], Beibei, W. [2], Suzhen C. [1], Ning, Q. [1], Xiuge, Z. [2], Ying, L. [2], Chunrong, J. [3], Xiaoli, D. [1]; [1]School of Energy and Environmental Engineering, University of Science and Technology of Beijing, Beijing, China, [2]Chinese Research Academy of Environmental Sciences, Beijing, [3]University of Memphis, Memphis, TN, USA

Internal migration in China has separated large numbers of rural children from their parents, with the advancement of China's urbanization process. Understanding the patterns of time activity of left-behind children in China helps to make reliable exposure assessments that have not been noticed or confused in the past. Our study based on the first national wide survey, the Children's Environmental Exposure Related Activity Patterns (0-5yeasrs), investigates the distribution of left-behind children in China, outdoor and indoor activity time, traffic use and transportation of left-behind children We found that nearly one third rural children(29.7%) under 6 years old were left-behind children separated with one or both parents. After controlling the confounding factors we found that separation from parents significantly affects children's time activity patterns in rural areas. For different age groups, the increase effect and 95% confidence interval of separation from parents for the outdoor activity time of children in rural are: <1 year: 16.1min/day (95%CI:10.6, 21.6) , 1 year: 16.1min/day (95%CI:10.6, 21.6) , 2 year: 16.6min/day (95%CI:8.9, 24.4) and 3~5year: 16.6min/day (95%CI:8.9, 24.4), respectively. The reduce effect on the indoor activity time are: <1 year: -16.1 min/day(95%CI:-21.6, -10.6), 1 year: -10.3min/day (95%CI:-17.7, -2.9) , 2 year: -16.0min/day (95%CI:-23.7, -8.3) and 3~5year -19.8min/day (95%CI:-24.2, -15.5), respectively. Separation from parents increases the incidence of traffic behavior in rural children which means left-behind children emerge 1.15 (1.02, 1.30)(OR) more times than non-left-behind children on traffic and 1.17 (1.09, 1.33)(OR) more times in vehicle on transportation. The proportion of stunted children left behind is statistically significantly higher than that of non-left-behind children. We quantified the impact of the separation of parents and children on their time activity patterns, which can help us to improve the assessment of children's air pollution exposure without neglecting left-behind children.

Keywords: activity patterns

SU-PO-24

Fractured effects of solid fuel use in early child development: analyses of urban and rural territories in Ghana

Nazif-Munoz J.I.[1], Oulhote Y.[2], Spengler, J.[1]; [1]Harvard University, Boston, MA, USA, [2]University of Massachusetts, Amherest, MA, USA

Background: About 41% of households globally, mainly in developing countries, rely on solid fuels for cooking with consequences for health outcomes in populations. In Ghana, more than 77% of the population depends on biomass fuels for cooking and heating. Previous studies posit that solid fuel use (SFU) in Ghana has had adverse effects on child health outcomes, stillbirth, and pregnancy. Yet these studies have not considered the effect on early child development (ECD), nor how SFU' potential effects vary across rural and urban areas. Objective: To investigate the potential effect of SFU on ECD, and whether it varies across rural and urban territories in Ghana. Methods: Using the 2011-2012 Ghana's Multiple Indicator Cluster Surveys – UNICEF (N = 2972 3-4 years children). A standardized global ECD index (ECDI) was calculated. A two-multilevel logistic regression model conditioning on household characteristics (second level), and individual-level covariates, corresponding to child and mother's characteristics (first level) were used to estimate the association between SFU and ECDI. Results: After adjustment for covariates, SFU was associated higher odds of not being developmentally on track (Odds ratio [OR]= 2.6; 95% confidence interval, CI:1.2, 5.6) compared to non SFU in all children of Ghana. This association was significantly different between rural and urban children, with an OR of 3.8 (95% CI:1.2, 12.3) in rural areas, compared with an OR of 1.5 (95% CI: 0.5, 4.1) in urban areas.

Discussion: The use of solid fuels is an independent predictor of a children's early development in rural Ghana, however this association in urban territories was not observed. This may point to specific characteristics of rural kids that play a synergistic role with solid fuel use. Policy efforts which tackle solid fuel use should be mindful of these differences. More research is needed to assess what rural characteristics are drive this association.

Keywords: built/indoor environment

SU-PO-25

The effects of global efforts of promoting cleaning cooking practices on solid fuel use and early child development

Nazif-Munoz J.I.[1], Oulhote Y.[2], Spengler, J.[1]; [1]Harvard University, Boston, MA, USA, [2]University of Massachusetts, Amherest, MA, USA

Background: Household air pollution from solid fuel burning is a contributor to delay early child development (ECD). A global effort to decrease solid fuel use (SFU) by promoting cleaner burning stoves, cleaner fuels, and/or providing chimneys or other ventilation changes, has been introduced since 2010. This program, called the 'Clean Cooking Alliance' (CCA), has targeted 100 countries, eight of which have received a more in-depth treatment. Thus, countries are not randomly exposed to global efforts to tackle SFU, and thus SFU variation may be dependent on their relative position to this program, and in turn affecting differently ECD. Objective: To investigate the effect of SFU on ECD and explore the possible influence of CCA's global efforts on SFU and ECD. Methods: A cross-sectional analysis of 30 countries with data on their relative position to CCA efforts--used as a proxy for CCA's differentiated efforts--linked to socio-demographics and early development of 76576 3-4 years children from the Multiple Indicator Cluster Surveys. A standardized global ECD index (ECDI) was calculated. Two-stage least squares regression models were conducted, using countries' relative position to CCA's efforts as an instrumental variable, to study the effect of SFU on ECDI. Results: After adjustment for covariates, absence of SFU was associated with an increase of 0.16 in ECDI (95% confidence interval, CI: 0.10-0.20). Countries relative position to CCA's efforts was an independent predictor of SFU (beta: 0.54, CI:0.25-0.84), F value= 14.52. When the relative position to CCA's efforts was used as an instrumental variable, the association between SFU and ECDI weakened but remained significant (beta: 0.15, CI:0.07-0.24). Conclusion: The use of solid fuels in the household is an independent predictor of a child's early development. Global efforts of the Clean Cooking Alliance appear to be improving early childhood development by facilitating better cleaning cooking practices.

Keywords: environmental policy

SU-PO-26
Estimation of life-time exposure to nitrogen dioxide and volatile organic compounds according to similar time-activity groups.

Heo J.[1], Ryu H.[1], Park J.[1], Kim E.[1], Choe Y.[1], Cho M.[1], Kim Y.[1], Sung K.[1], Yang W.[1]; [1]Daegu Catholic University, Gjongsanas, South Korea

Personal or population exposure to hazardous air pollutants has been assessed by personal sampling. however, it has many problems such as cost and time. thus, exposure is assessed by the concentration of pollutants in microenvironments and the time-activity patterns. in addition, similar time activity groups(STG) were categorized through the time activity patterns. In this study, we examined the time activity patterns of the citizen in metropolitan in Korea and classify with STG. In addition, exposure assessment of the life cycle of nitrogen dioxide (NO2) and volatile organic compounds(VOCs) is performed, and the exposure characteristics of each substance are identified. The research subjects were designated aged 0 to 80 and the survey areas were limited to Seoul, South Korea. The research period was conducted on 1,000 people residing in Seoul for two months. The time activity patterns were conducted in the form of interview investigation by an individual visit through the time-use survey. The questionnaire was developed by dividing it into adults, teenagers and infants. In this study, the sampling of NO2 and VOCs was carried out using a passive sampler. Concentration analysis of NO2 was conducted using an ultraviolet spectrometer, and for VOCs , it was desorbed with Carbon disulfide(CS2) and analyzed using GC/MS. Of the 1,000 people surveyed for the time-activity pattern, 493 men and 507 women. On the basis of time-activity patterns, they were divided into infants, preschool child, elementary school students, middle and high school students, college students, office workers, production workers, sales and service workers, housewives, and senior citizens. Infants, housewives, and the elderly were spent more than 80% of their time at home. As a result of the analysis of NO2, housewives and sales worker showed a high concentration of exposure level each of 34.7 ppb, 34.9 ppb.

Keywords: VOCs, activity patterns, cumulative exposure, exposure factors, lifecycle analysis

SU-PO-27
Exposure assessment of VOC emissions from polymer-based consumer products

Even M.[1], Girard M.[1], Rich A.[1], Roloff A.[1], Hutzler C.[1], Luch A.[1]; [1]German Federal Institute for Risk Assessment, Berlin, Germany

The emissions of volatile organic compounds (VOCs) from building materials have been studied in detail for some time. This resulted in the development of standardized analytical methods and health-related indoor air concentration guidelines. Polymer-based consumer products such as toys may also emit harmful substances that could be inhaled and potentially influence the occupant's health. However, emissions from such articles have not been systematically studied so far. After the identification of polymer-derived products with a significant release of VOCs and the characterization of their emission profiles, we provide a realistic evaluation of the inhalation exposure of consumers. First, the polymer type was identified by pyrolysis coupled online to a gas chromatograph with mass spectrometric detection (py-GC/MS). In parallel, a qualitative assessment of emitted VOCs from the same samples was performed using GC/MS in dynamic headspace mode. Subsequently, emissions of relevant samples were quantified in emission chambers with varying dimensions. To this end, the emitted VOCs were sampled on TENAX® TA tubes and analyzed after thermal desorption by GC/MS. Indoor air concentrations were derived and compared to the guideline values, considering different exposure scenarios. It was observed that soft plastics like polyvinyl chloride emit a broader range of VOCs with higher concentrations compared to harder polymers like polypropylene. VOC concentrations in the emission chambers decreased rapidly after loading. Therefore, the first hours after unpacking are of special relevance for the exposure assessment of such compounds. Corresponding indoor air concentrations (30 m3 room) of single target analytes were found to be at least 35-fold lower than thresholds provided by indoor air guidelines. However, when multiple samples or smaller breathing volumes are considered, the derived concentrations can exceed guidelines values. We emphasize that consumer products should be considered as an additional source of VOC emissions indoors beside well-known sources such as building materials.

Keywords: VOCs, analytical methods, children, consumer and personal care products, risk assessment

SU-PO-28
Exposure to Volatile Organic Compounds (VOCs) in Feminine Care Products

Lin N.[1], Ding N.[1], Park, S.K.[1], Batterman, S.[1]; [1]School of Public Health, University of Michigan, Ann Arbor, MI, USA

Feminine care products are used on highly permeable and sensitive vaginal and vulvar tissues. Few regulations require disclosure of ingredients in these products. This study identifies VOCs and quantifies exposures and health risks resulting from using these products. We tested 20 different tampons and 12 feminine washes using purge and trap sampling, thermal desorption, and gas chromatography/mass spectroscopy analysis. Exposure dose and risks were modeled using typical or worst assumptions. We detected 51 and 38 VOCs in tampons and washes, respectively, and the mean total VOC concentrations were 0.3 µg/g and 39 µg/mL, respectively. Chloroform was found in 14 tampons (<6×10-3 µg/g) and 3 washes (<0.14 µg/mL); 1,4-dichlorobenzene was found in 11 tampons (<2×10-3 µg/g); 1,4-dioxane was found in 9 washes (<25 µg/mL); and p-isopropyltoluene was found in 19 tampons (<0.02 µg/g) and 9 washes (<3.1 µg/mL). Other VOCs detected included aromatics, alkanes, aldehydes, halohydrocarbons, ketones, esters, and terpenes. VOC composition tended to depend on the manufacturers. Scented tampons contained more p-isopropyltoluene and limonene. Washes for "sensitive skin" had few differences, except for higher levels of 1,4-dioxane. "Organic" washes had higher VOC concentrations, especially of aldehydes and terpenes, than other washes (97 vs. 27 µg/mL, p=0.03). For washes, the estimated total VOC dose was mainly due to inhalation (2.9 □g/m3), the hazard ratio was 2.3×10-2, and the life time cancer risk was 1.9×10-6 (highest among products tested). For tampons, the primary exposure pathway was dermal exposure via vaginal mucosa. The VOC dose was 7×10-5 mg□kg-1□day-1, the hazard ratio was 6.8×10-5, and the lifetime cancer risk was 2.8×10-9. All of these products contained some hazardous VOCs, and risks from the use of some products was not negligible. At present, consumers should not assume that labels regarding chemical composition in these products are meaningful. The ingredients in these products should be disclosed.

Keywords: VOCs, consumer and personal care products, environmental health, risk assessment

SU-PO-29
Development of new analytical and measurement methods for characterizing the emission of semi-volatile organic compounds (SVOCs) from building and consumer materials

Braish T.[1], Nicolas M.[2], Maupetit F.[2], Desauziers V.[3]; [1]CSTB and IMT Mines Ales, Saint-Martin-d'Heres, France; [2]CSTB, Saint-Martin-d'Heres, France; [3]IMT Mines Ales, Pau, France

Phthalates and organophosphate flame retardants (OPFRs), two families of SVOCs, are ubiquitous indoor pollutants that are added to various indoor building and consumer materials to enhance their properties. However, these compounds are of great research interest due to the severe health effects they cause. The gas-phase concentration of SVOCs in equilibrium with the material surface (y0) is a key parameter in determining indoor emission of materials and estimating the risk of human exposure to these compounds. However, their high molecular weight and low vapor pressure make them hard to study due to their adsorption on different indoor surfaces after emission from the hosting material (sink effect). Few studies are available on the presence of SVOCs in indoor air compared to those that study their presence in settled dust. Moreover, most of the developed y0-measurement methods are of certain limitations. Therefore, the objectives of this study are to develop analytical and measurement methods for characterizing the emission of phthalates and OPFRs emitted from vinyl floorings and polyurethane foams (PUF), respectively into indoor air at room temperature. A robust ATD-GC-MS (Automated Thermal Desorption - Gas Chromatography - Mass Spectroscopy) method has been developed and validated for characterizing eight chosen phthalates and seven OPFRs with LOD ranging from 0.004 to 0.019 µg/m3 for an air sampling volume of 72 L. For the measurement method, the micro-chamber recommended by ISO 16000-25 for studying SVOCs is used as the emission device. y0 is calculated at different elevated temperatures then is determined at room temperature by extrapolation according to van't Hoff's equation [ln(y0) versus 1/T]. Therefore, developing fast and robust method for measuring y0 of emitted SVOCs into indoor air at ambient temperature is a necessity for estimating the risk of human indoor exposure to these compounds, and the micro-chamber extrapolation method is a promising one.

Keywords: SVOCs, analytical methods, building materials, sampling methods, thermal environment

SU-PO-30
Effects of Particle Volatility to Particle Size Distribution of Exhaled Aerosol when Using New Nicotine Containing Devices

Meisutovic-Akhtarieva M.[1], Ciuzas D.[1], Tichonovas M.[1], Misevicius S.[1], Martuzevicius D.[1], Kauneliene V.[1]; [1]Kaunas University of Technology, Kaunas, Lithuania

The rapid spread of new nicotine containing devices as alternative to traditional cigarettes raises many discussions on the potential reduction of harm to both mainstream smoke users as well as bystanders. The exhaled particles following the use of such devices have been indicated as highly volatile by several chamber studies, thus indicating potentially lower impact to particulate matter concentrations indoors compared to exhaled aerosol of traditional cigarettes. This research analysed particle size distributions (PSDs) in exhaled breath using an electronic cigarette and electrically-heated tobacco product. The inhaled main stream aerosol has been exhaled to either non-heating conductive tube (length of 2 m) or heated tube (heater temperature 150 oC, exiting air flow temperature 105 oC). The PSDs were determined at 10 Hz temporal resolution using electrical low pressure impactor (ELPI+, Dekati, Inc., Finland). The exhaled aerosol was diluted 2 times with dry filtered air just before entering the impactor. The particle size distributions based on number concentration indicated bimodal distributions, having majority of particles in 10-20 nm range, and the second mode at 80-200 nm. As expected, exhaled particles proved to be more volatile than mainstream particles, especially in case of heated tobacco product. At the same time, the accumulation mode of exhaled particles was heavily affected by the heating, i.e., almost completely removed in the aerosol denuding system, indicating high volatility. At the same time, the highest mode at 10-20 nm did not seem to be impacted by the heating, indicating either the presence of non-volatile primary particles, or the sampling related artifacts.

Keywords: built/indoor environment, nanoparticles, particles, secondhand smoke

SU-PO-31
Impact of healthier building materials on dust levels of PFASs, flame retardants, and hormone-disrupting potency

Young, A.,Hauser, R.,James-Todd, T.,Coull, B.,Zoeller, T.,Sunderland, E.Pickard, H.Kannan, K.Behnisch, P.Brouwer, A.; [1]Harvard T.H. Chan School of Public Health, Boston, MA, USA, [2]University of Massachusetts Amherst, Amherst, MA, USA, [3]Harvard John A. Paulson School of Engineering and Applied Sciences, Cambridge, MA, USA, [4]Wadsworth Center, New York State Department of Health, Albany, NY, USA, [5]BioDetection Systems, Amsterdam, The Netherlands

Per- and polyfluoroalkyl substances (PFASs) and flame retardants (FRs) are two classes of chemicals ubiquitously used in furniture, carpet, and other building materials. Many of these chemicals have been associated with endocrine disruption, poorer pregnancy outcomes, immune suppression, cancer, and high cholesterol. To our knowledge, no studies have evaluated the exposure or health benefits of interventions to renovate buildings with healthier materials specified as FR- and PFAS-free. In this study, we collected dust samples from 24 building rooms (e.g. common spaces and offices) renovated with furniture and carpet specified as healthier and 24 comparable controls. The dust samples will be analyzed for levels of PFASs and FRs to compare spaces with healthier-specified materials to traditional spaces. We will also perform novel cell assays to evaluate differences in hormone- and immune-disrupting potency of the dust in the spaces. Cell assays are a relatively inexpensive method to assess the health of chemical mixtures in a targeted way instead of trying to analyze many individual chemicals separately. Laboratory results are pending but will be available for the conference.

Keywords: SVOCs, building materials, built/indoor environment, flame retardants, workplace

SU-PO-32
Soil and dust ingestion by children comparing urban and suburban areas in Korea

Kim E.[1]; S, Ryu H, Park J, Yoon H, Kim T, Seo J, Heo J, Choe Y, Cho M, Yang W.; [1]Department of Occupational Health, Daegu Catholic University, Gyeongsan, Republic of Korea

Soil and dust ingestion may be a potentially important exposure pathway to environmental contaminants because they contain hazardous heavy metals. Children, in particular, may ingest significant quantities of soil and dust due to their tendency to play on the floor indoors and ground outdoors and their tendency to mouth objects or their fingers. The aim of this study was to estimate soil and dust ingestion rates of Korean children by geographical and cultural characteristics. The ingestion rates per day were estimated using the EPA/600/R-17/384F method and analyzed concentrations using aluminum(Al) as an indicator. In general, Al is a substance found in soil and dust at a high concentration but not decomposed or affected by living things. This study involved 113 children aged 0-6 years. Fecal samples were obtained from 56 children in urban areas in 2017 and 57 children in suburban areas in 2018, and soil samples near their residence. The soil and dust ingestion rates were 57.5 mg/day in urban areas and 84.2 mg/day in suburban areas. The mean ingestion rate was 70.6 mg/day. These were lower than the recommended values, 118 mg/day by The Korean Ministry of Environment(2007) and 80 mg/day by the U.S.EPA(2017). These results can be explained the children corresponded with the decreased outdoor activities and increased indoor activities compared with the previous study. Also, the playground was built with urethane, not ordinary sand and it was difficult to find artificial green spaces in the cities during sampling soils. It may be another reason that children play phone and internet games at home. Accordingly, these environments could be decrease soil and dust ingestion rates for children. This study is important to access the health risk resulting from exposure to environmental contaminants via soil and dust ingestion pathway by children in Korea as well as other countries.

Keywords: children, environmental health, exposure factors, metals, soil

SU-PO-33
Spatiotemporal distribution of particulate air pollution in a pub

Drahler, D.[1], Fishbain, B.[1]; [1]Dept. of Environmental, Water and Agricultural Engineering, Faculty of Civil & Environmental Engineering, Technion – Israeli Institute of Technology, Haifa, Israel

Environmental tobacco smoke (ETS) or passive smoke, is the side product of smoking cigarettes, cigars, pipes and other smoldering tobacco products (STP). The negative health outcomes of exposure to ETS are well-known and include: lung cancer, cardiovascular disease, stroke and asthma. In an attempt to control the non-smokers' exposure to ETS, governments around the world have restricted or prohibited smoking in public places, particularly in hospitalities, such as restaurants and pubs. A number of laws were updated through the years to contain some area outside of public places in the smoking ban, such that smoking is forbidden within a few meters of a pedestrian access to a building; The Israeli law has not been updated in this way. This study aims to investigate the spatiotemporal distribution of ETS in an Israeli pub, caused by smoking activity that is concentrated adjacently to the front door. The study was hosted in a local pub, at downtown Haifa, Israel. The estimation of ETS concentrations was done by measuring particle number concentration (PNC) as an indicator, using low cost sensors (Plantower PMS5003) and small single-board computers (Raspberry-Pi 3). At the indoor area of the pub, 8 sensors were placed throughout the pub, next to the potential sources and sinks of particles. Another sensor was placed outside of the pub, next to the front door, where most of the smoking activity was expected to take place. Preliminary results showed association between smoking activities that occurred outside the pub, and increased particulate concentrations that were measured in the indoor environment. The preliminary findings add to the importance of expanding the legislation dealing with smoking in and nearby public places. Furthermore, it seems that exposure assessments in a complex indoor environment with high spatiotemporal resolution, can be performed using low-cost PNC micro-sensing units.

Keywords: built/indoor environment, environmental policy, environmental regulation, particles, secondhand smoke

SU-PO-34

The impact of regional weather on residential energy consumption and indoor air quality changes due to energy retrofits

Connolly C.[1], Milando C.[1], Dols W.[2], Levy J.[1], Fabian M.[1]; [1]Boston University, Boston, MA, USA, [2]NIST, Gaithersburg, MD, USA

Climate action plans in cities in the United States include updated requirements for building retrofits that decrease energy demand but without considering the subsequent changes on indoor air quality (IAQ) for residents. While previous studies have shown that energy retrofits paired with proper ventilation and filtration can improve health outcomes for residents, little is known of how the energy/IAQ tradeoffs change in different climate regions. In this study, we used a coupled energy-IAQ model to quantify these tradeoffs in multiple regions of the US. We modeled the effects of regional climate differences on residential energy consumption and indoor air quality (e.g. ambient and indoor sourced PM2.5) using a coupled model integrating an energy simulation program (EnergyPlus) and a multi-zone indoor air quality and ventilation program (CONTAM). The co-simulated framework incorporates multiple factors that affect indoor air quality and energy use such as indoor human behavior, weather, ambient air pollution, and building characteristics. Energy and IAQ were estimated for a residential building located in northeast, southern, northwest, and southwest regions. We found that regions with warmer summers (i.e. Florida, Arizona) had better annual indoor air quality from increased use of the HVAC cooling system in the summer, but at the expense of requiring significantly more energy to cool the residence. Regions with colder winters (i.e. Boston) had worse average yearly IAQ and higher energy consumption due to heating demands in the winter. Indoor sources of PM2.5 had a greater impact on indoor air pollution compared to outdoor sourced ambient pollution, even with increased exhaust ventilation. This co-simulation provides a novel way to account for the multi-faceted reasons for human exposure to pollutants in their residences. With increasing energy demands due to climate change, we can use this structure to model these energy/IAQ tradeoffs.

Keywords: air, built/indoor environment, climate change, ventilation

SU-PO-35
Characteristics of co-use patterns to cosmetic products in high-use groups for Korean population

Lim M. [1], Applicant 2: Park J.Y., Lee K.; [1]Seoul National University, Seoul, Korea, South

Cosmetic product is one of the major chemical exposure sources in real life. The use rates of most cosmetics were different by population group. People used many cosmetics simultaneously. To accurate aggregate exposure assessment, it is necessary to determine co-use patterns of cosmetic products by population group. The purpose of this study was to determine co-use patterns of cosmetics and characterize high use group in Korea. The national representative exposure factor database was established for Korean adult population (n = 1,001). The individual's simultaneous use patterns for 31 cosmetic products included actual co-use and non-use information. The co-use patterns of cosmetic products by population groups were analyzed by number of product used, Ward's hierarchical clustering method and Cohen's kappa correlation coefficient. Korean population used 8.5 ± 5.1 cosmetic products. When the population was divided into 10 groups by gender and age, 20's female was the highest cosmetic use group with 14.5 ± 3.9 products. 60's male was the lowest use group with 3.8 ± 1.6 products. The use pattern of 31 cosmetics in 20's female group clustered into five groups of cosmetics. Ten cosmetic products with high user rate were used together and they were skin toner, face cleanser, facial sunscreen, face lotion, foundation, face cream, body lotion, hand cream, hair treatment and lipstick. Four eye makeup products (eyeliner, eyeshadow, eyebrow and mascara) were used together with moderate correlation (kappa > 0.40). Nail polish and nail remover were used together (kappa = 0.95). Aggregate exposure assessment for cosmetic products might be necessary because Korean used multiple cosmetics. The cosmetic co-use patterns were different by population group. The young female group could be high-exposure group for chemicals in cosmetics. The co-use patterns could be important input data for aggregate exposure assessment.

Keywords: aggregate exposure, consumer and personal care products, environmental health, exposure factors, multimedia

SU-PO-36
Estimation of Potential Dermal Exposure to Organophosphorus Flame Retardants via Direct Contact with Products Using Silicone

Shindo M. [1], Tokumura M. [1], Wang Q. [1], Miyake Y. [1], Amagai T. [1], Makino M. [1]; [1]University of Shizuoka, Shizuoka, Japan

Organophosphorus compounds are presented in indoor products at high concentration, and some of them have toxic consequences like carcinogens. Although the dominant exposure route to organophosphorus compounds is considered as the dust ingestion, recent researches suggest that the dermal exposure rate via direct contact with products could be expected to be higher than those via other exposure routes. As conventional devices to estimate the potential dermal exposure rate, human skin and EpiSkin epi dermal model have been used. However, they have disadvantages (e.g., ethical problems, expensive, and hard-to-use). Therefore, they seem to be not suitable for testing numerous products. In this study, to develop the simple and cost effective devise which can estimate a potential dermal exposure to organophosphorus flame retardants via direct contact with products, we carried out the migration test of phosphorus compounds from a product to the devise made of silicone. The amount of the phosphorus compounds in the silicone increased with the contact time. A higher phosphorus compound concentration in the product was likely to result in a faster migration rate. Phosphorus compounds with a lower n-octanol/water partition coefficient (log Kow) were likely to show a faster migration rate. The average migration rates of phosphorus compounds for 48 hours from the product to the silicone were in the range of 0.07–2.2 ng/cm2/h. We compared the migration rates of phosphorus compounds from the products to the devises (silicone and EpiSkin).

Keywords: SVOCs, flame retardants, risk assessment

SU-PO-37

Transdermal uptake of two phthalates and a paraben from simulated lotion using a chemical activity approach

Eftekhari A. [1], Morrison G. [2]; University of North Carolina, Chapel Hill, NC

Personal care products (PCPs) can be an important source of dermal exposure to chemicals of concern. Transdermal permeation models compliment in vitro and in vivo experiments in assessing risk of environmental exposure such as application of cosmetics and personal care products. Recent approaches characterize dermal transport using Fick's first law of diffusion. In these models, steady state permeation rate is proportional to the "effective concentration", "fugacity" or "activity" difference across the skin. In this research, we characterize this chemical driving force and use it to model uptake from lotion. A basic lotion (containing no target analytes) was mixed with 6% wt/wt each of diethylphthalate (DEP), di-n-butylphthalate (DnBP), and butylparaben (BP). The solution was sequentially diluted to make lotions with concentrations ranging from 0.025% to 6%. The fugacity, relative to the pure compound, was measured in headspace of lotions using solid-phase micro extraction (SPME). The fugacity of chemicals in 2% wt/wt formulation was used in a diffusion-based model to predict transdermal uptake of each chemical and compared with excretion data of human subjects who participated in a whole-body topical application of 2% formulation. We found that relationship between the concentration in lotion and fugacity is highly non-linear and cannot be predicted from Henry's or Raolt's laws. Dynamic simulations of excretion are qualitatively similar to the results of the human subject experiment. Quantitative predictions are similar but very sensitive to choice of vapor pressure.

Keywords: SVOCs, consumer and personal care products, environmental health, exposure models, phthalates

SU-PO-38

Using a passive sampler to measure the dermal migration of di(2-ethylhexyl)phthalate from polyvinylchloride flooring at different skin condition

Kim D.Y. [1], Kwon, J.H.[1]; [1]Division of Environmental Science and Ecological Engineering, Korea University, Seoul, Korea

Di(2-ethylhexyl)phthalate (DEHP) is still widely used as plasticizer in polyvinylchloride (PVC) products, although it is known to be toxic to human. One of the exposure pathways of DEHP from PVC products is the dermal migration. Generally, the dermal migration is measured from the transferred mass of a chemical per contact area and time. The dermal migration can occur through direct contact between human skin and PVC products or deposition of gaseous DEHP to human skin. In the case of the direct contact between skin and PVC products, the presence of sebum should be considered because sebum has high affinity to DEHP and might affect to the dermal migration rate. In this study, the dermal migration of DEHP from a PVC coated heating pad in direct contact condition with and without artificial sebum between a polydimethylsiloxane (PDMS) passive sampler and the heating pad were measured. The pre-cleaned PDMS sampler was attached on the heating pad or over filled artificial sebum in the gap between the sampler and the heating pad using custom made glassware. Those samples were stored at constant temperature 60 °C to simulated heating condition, taken after pre-designated sampling time, and DEHP was extracted using n-hexane and was analyzed with GC-MS. The dermal migration rates of DEHP from PVC with and without artificial sebum were 0.0378 and 0.00122 ug cm^-2 min^-1 respectively, showing that artificial sebum accelerates the process by a factor of 30. Artificial sebum might be absorbed into PVC and PDMS, mimicking the real skin condition better than dry skin condition, without artificial sebum. Considering the covering matrices such as sebum is needed when assessing the exposure to DEHP through direct contact.

Keywords: SVOCs, phthalates, sampling methods

SU-PO-39

In-vehicle airborne fine and ultra-fine particulate matter exposure: the impact of emission reduction technologies

Campagnolo D. [1], *Cattaneo A.* [1], *Borghi F.* [1], *Keller M.* [1], *Rovelli S.* [1], *Spinazzè A.*[1], *Cavallo D.* [1]; [1]*Department of Science and High Technology, University of Insubria, Como, Italy,*

Traffic-related air pollutant (TRAP) concentrations inside vehicle cabins are often high compared to background levels. The present study was focused on airborne particulate matter (PM) exposure and was motivated by the implementation of new emission standards that has led to the reduction of vehicle particle emissions. The objective of this work is to investigate the relationship between leading vehicle (LV) emissions and in-cabin PM exposure levels in the immediately following vehicle (henceforth called the study vehicle - SV), with particular emphasis on the role of the LV's emission reduction technologies (e.g., diesel particulate filter-DPF) as an effective risk management measure. The research was performed using an instrumented study vehicle (always to be considered as the following vehicle) on a 26-km fixed route where 10 repeated tests were conducted during 60-minute trips. On-line monitoring of the fine (i.e., PM0.3-1 and PM1-2.5) and ultra-fine particle (UFP) concentrations was carried out inside the SV's car cabin with fixed ventilation settings (i.e., windows closed, air conditioning off, recirculation fan off). Simultaneously, the license plate numbers of the LVs along the route were recorded to collect information pertaining to their fuel type and Euro emission standard category. The results clearly showed that the in-cabin PM exposure levels were significantly affected by the LV's Euro emission standard. In particular, a strong and significant decrease in the in-cabin median exposure levels was observed (up to -62%, -44%, and -48% for UFPs, PM0.3-1, and PM1-2.5, respectively) when following diesel-fuelled LVs equipped with recent-emission standards (i.e., Euro 5-6) than when following older-emission standard LVs (i.e., Euro 0-4). An in-depth analysis revealed that the in-cabin median exposure concentrations of PM were highly and significantly reduced by DPF-equipped LVs. The most important reduction (about 80%) was observed for PM0.3-1 compared to both petrol-fuelled and diesel-fuelled (non-DPF) LVs.

Keywords: air, environmental policy, particles, sampling methods, statistical methods

SU-PO-40

Organophosphate flame retardants: a review of indoor contamination and human exposure in Europe, and human health effects.

Chupeau Z. [1], *Bonvallot N.* [1], *Mercier F.* [1], *Le Bot B.* [1], *Chevrier C.* [1], *Glorennec P.* [1]; [1]*Univ Rennes, EHESP, Inserm, Irset, Rennes, France*

Organophosphorus Flame Retardants (OPFRs) are increasingly used as in replacement of Brominated flame retardants. The literature published from 2000 to 2018 investigating their presence in indoor environments and human exposures in Europe, as well as worldwide epidemiological evidence of adverse effects was reviewed. Considering studies with at least 10 samples, 30 OPFRs were detected in air and dust of homes, day care centers, offices and cars from 11 European countries. The most concentrated compounds in both air and dust were tris(2-butoxyethyl)phosphate (TBOEP) , tris(2-chloroisopropyl)phosphate (TCIPP), tris(2,3-dichloropropyl)phosphate (TDCIPP) and tris(chloroethyl)phosphate (TCEP) .Concerning biological measurements of exposure , OPFRs or their metabolites were measured mainly in urine, but also in hair and breast milk. The most consistently detected in humans were TBOEP, TCIPP, TDCIPP, TCEP, tris(butyl)phosphate (TBP) and tris(phenyl)phosphate (TPHP), which appears consistent with air and dust concentrations except for TBP and TPHP. TBOEP and TPHP found in paints, lacquers and construction material were the most prevalent in general population. Concentrations of TDCIPP and TPHP were found to be higher in children than in adults. Regarding epidemiological evidence, the most studied OPFRs were TDCIPP, TPHP, mono-isopropylphenyl phenyl phosphate (IP-PPP) and TBP. The most consistent associations were found between TDCIPP and decrease of thyroid hormone levels, TDCIPP and decrease of sperm quality, IP-PPP and decrease of cognitive abilities, and TNBP and increased risk of allergy. Finally OPFRs are very prevalent in indoor environment and human populations. Given the data analyzed in this work, future epidemiological studies are required to focus on TBOEP, TCIPP, TPHP and IP-PPP, a metabolite of isopropylated triarylphosphate isomers (ITPs) used in Firemaster® 550.

Keywords: air, biomonitoring, built/indoor environment, epidemiology, flame retardants

SU-PO-41
Proof of principle testing for using small black carbon monitors for real-time PAH measurement

Jaramillo J. [1]*, Yan B.* [2]*, Ross J.* [2]*, Pitiranggon M.* [2]*, Yang Q.* [2]*, Blair J.* [3]*, Perzanowski M.* [4]*;* [1]*Hunter College, Manhattan, NY, USA,* [2]*Lamont-Doherty Earth Observatory at Columbia University, Palisades, NY, USA,* [3]*NYAethLabs, San Francisco, CA, USA,* [4]*Columbia University Mailman School of Public Health, New York, NY, USA*

Exposures to PAH compounds have been associated with various adverse health outcomes. Real-time PAH measurements would allow a better understanding of temporal and spatial variability in urban settings. Black carbon is a reliable tracer of incomplete combustion emissions, and optical methods exist for making real-time measurements. The purpose of this work is to explore whether BC data from residential microAeth® monitors (AethLabs, MA350) can be used to estimate real-time levels of high molecular weight PAHs. We collected PM2.5 onto Teflon filters at 4 LPM co-located with a MA350 which collected particles at two spots simultaneously at two flow rates (0.1 and 0.05 LPM) onto a Teflon tape roll. BC measurements made on the co-located Teflon filters agreed well with the mean BC levels from the MA350 (R2 = 0.90 and slope = 0.99). Mean BC concentrations from the MA350 were associated with levels of individual PAH compounds extracted from the Teflon filters such as B[k]F (R2 = 0.94), providing a basic premise for using the MA350 to estimate levels of PAH compounds. PAHs were measured with a highly sensitive fluorescent detector linked to LC/MS. Direct PAH measurements from the used MA350 tape were impacted by high blank levels, whereas initial experiments suggested low blank levels from unused MA350 tapes. Future work is needed to determine whether the high blank levels are due to tape-to-tape variability or contamination during sampling. If successful, site- and time-specific calibrations for MA350s could be developed for high molecular weight PAHs without the need for co-located filters, thus expanding the possibility for high-quality, real-time data for these pollutants.

Keywords: air, analytical methods, particles, sampling methods

SU-PO-42
Study of a Saharan Dust Intrusion into the Colombian Atmosphere

Pinto Herrera, L.C. [1]*, Mendez Espinosa, J.F.* [1]*, Belalcazar Ceron, L.C.* [1]*;* [1]*University of Los Andes, Bogota, Columbia*

Sahara desert is one of the main sources of particulate matter (PM) in the world. This Particulate Matter has the potential to affect the air quality, thousands of kilometers away from the source. The dust intrusions in the Caribbean are well documented; however, there are no reports of this pollutant in Colombian large cities. In this article we analyze and establish the possible relation between the Sahara's dust emissions and an episode of elevated PM levels registered at the end of June 2014 in Bogota and Medellin. For this purposes we used data collected by the automatic air quality monitoring stations installed in these cities, the column mass density of the MERRA-2 project, and results from the NMMB/BSC-Dust model. We also implemented the HYSPLIT model to determine the origins and trajectories of air masses during the episode.The impact of this type of episodes affects vulnerables comunnities, this is assessed through incidence of respiratory illnesses and ecological studies.

Keywords: air, ecological exposure, emergency response, geospatial analysis/GIS, respiratory health, risk assessment, sensor technology, source control, statistical methods

SU-PO-43

The effect of wind velocity on airborne release of ochratoxin A from cultivated Aspergillus niger

Lee S.A. [1], Chen J.C. [1]; [1]Department of Environmental Engineering and Science, Feng Chia University, Taichung, Taiwan, [2]Department of Environmental Engineering and Science, Feng Chia University, Taichung, Taiwan

Mycotoxins are secondary metabolites produced by fungi and might cause serious health problems in human. Among mycotoxins, ochratoxin A is carcinogenic, teratogenic, genotoxic, and immunosuppressive, and has been classified as "possibly carcinogenic to humans" (group 2B) by the International Agency for Research on Cancer (IARC). This study investigated the effect of wind velocity on airborne release of ochratoxin A from cultivated Aspergillus niger as well as particles concentrations and size distribution, viability and culturability of airborne fungi. Aspergillus niger was dispersed using our newly developed automated fungal particle aerosolizer (AFPA) and the air samples were collected using NIOSH two-stage cyclone bioaerosol samplers for subsequent microbial analysis. The particle concentrations and size distribution of airborne fungi were measured using an electrical low pressure impactor (ELPI). The results show that no ochratoxin A was found at wind velocity < 1.7 m/s and only particles of 11.6 m/s, we began to see airborne release of fungal spores and ochratoxin A, which mostly fell in the particle size >1.8μm. The concentrations of airborne fungi and ochratoxin A, viability, and culturability significantly increased with increasing wind velocity. This study shows that airborne fungi and ochratoxin A might not be a respiratory issue indoors in terms of wind velocity.

Keywords: air, built/indoor environment, dampness/mold, exposure factors, respiratory health

SU-PO-44

The Hopi Environmental Health Project (HEHP)- Measuring Concentrations of PM2.5 in Households on Hopi Lands

Hadeed S.[1], O'Rourke M.K.[1], Alshammari M.[1], Blohm J.[1], Canales R.[1], Paukgana P.[2], Olivas G.[3], Carl A.[3], Lugo-Lerma V.[1], Gonzalez-Figueroa E.[1]; [1]The University of Arizona College of Public Health, Tuscon, AZ, USA, [2]The Hopi Tribe, Kykotsmovi, AZ, USA, [3]The University of Arizona, Tuscon, AZ, USA

Despite an estimated 6.5 million people in the United States burning solid fuels for heating, respiratory health effects from indoor exposure to smoke and particulate matter (PM) have focused on populations in developing nations, and have been largely ignored in developed nations. The Hopi Tribe of Northern Arizona heavily relies on burning of coal and wood for heating and cooking during winter months. PM2.5 was measured in Hopi households for 24-hours during heating and non-heating seasons using real-time area monitors. Primary heating fuel type was recorded from questionnaires completed by primary respondents. Indoor mean concentrations of PM2.5 were assessed across seasons and household fuel types. Fifty-one households were sampled, 36 of which completed follow-up sampling during both heating and non-heating seasons. Across all households and seasons, mean (SD) concentrations of PM2.5 were 36.8 μg/m3 (175.3); 51.13 μg/m3 (218.7) during the heating season, and 18.6 μg/m3 (91.4) during the non-heating season. When considering fuel type, households burning wood and coal (n=17) had highest indoor PM2.5 concentrations (53.6 μg/m3 (136.0)) during the heating season. Homes using a combination of electricity, coal and wood (n=11) during the heating season had an indoor concentration of 46.6 μg/m3 (155.7). During the heating season, concentrations were 4.16 times greater for coal and wood homes, 2.52 times greater in electric, coal and wood homes, and 2.03 times greater in homes burning coal, wood and natural gas, compared to the non-heating season. PM2.5 concentrations were elevated in coal and wood burning households. Although slight reductions were observed in homes that used a combination of electricity or natural gas with coal and wood, concentrations exceeded the USEPA ambient standard of 35 μg/m3 and the WHO guideline of 25 μg/m3. Occupants of these households may be overexposed to PM2.5 and at greater risk for developing respiratory and cardiovascular diseases.

Keywords: air, built/indoor environment, community, environmental health, particles

SU-PO-45

A study on the Evaluation and Validation of Plasma Sterilizing Dust Collector Unit

Jeon H.[1], Sohn, J.[2], Seo J.[1], Choi J[1]; [1]Naver Corporation, Seongnam, South Korea, [2]Korea University, Seoul, Korea

With the continuous progress of urbanization and the improvement of economic standards, modern people spend most of their day indoors as well as changes in the lifestyle of modern people. According to a U.S. EPA report, modern people have high indoor activity times, with an estimated 90 percent of the time spent indoors. Interest in indoor air quality has been extended to the problem of indoor air pollution in office spaces. Accordingly, air purifiers and dust collectors that can clean up office space's indoor air have been developed and distributed continuously, and a dust sterilizer that introduces plasma has been also developed. Since the impact of outdoor air penetration in Pohang Steel Mill in Korea is expected to be significant due to the characteristics of the workshop, it is deemed necessary to verify the performance through field application evaluation to assess the performance change according to the location characteristics. For further evaluating and analyzing the efficiency reduction caused by the number and activity of the office workers, an on-site application assessment is also necessary to scientifically demonstrate the performance of Plasma Sterilizing Dust Collector Unit through verification of performance in actual buildings. Accurate assessment will be carried out, considering performance of actors in the office such as commuting/retiring, lunchtime, and other events, analysis of the reduction efficiency of fine dust by time. The purpose of this research is to evaluate the performance of Plasma Sterilizing Dust Collector Unit installed directly in buildings and verify their efficiency in the field, as the office space at the center of Pohang Steel Mill has many limitations, including the wide range of places to apply general air `purifiers, and the fact that indoor air quality needs certain improvement due to the characteristics of the place.

Keywords: activity patterns, air, air cleaning, built/indoor environment

SU-PO-46

Home Environmental Intervention to Reduce Exposure Level to PM2.5 for Children with Asthma using Clustered data of ESCORT Child Asthma Panel Study, South Korea

Kim S[1], Seeun O.[1], Lim D.[1]; [1]Inha University, Incon, South Korea

BACKGROUND: Children with asthma in inner-city communities may be particularly vulnerable to adverse effects of indoor air quality because they spend most of their time in indoor, yet little is known about the effect of environmental interventions in the Korean home setting. OBJECTIVE: We piloted the feasibility and effectiveness of a home-based air cleaner intervention to reduce particulate matter ≤ 2.5 μm (PM2.5) in homes of children with asthma. METHODS: In this pilot randomized controlled trial, with a case-crossover design, we assessed the effect of air cleaners on indoor PM2.5 concentrations at 30 homes (15 control, 15 intervention) in the metropolitan city of Incheon, South Korea. We obtained 867 episodes from children with asthma (15 control, 15 intervention) aged 7 to 12 years. Daily home air pollutant measurements and peak expiratory flow rate were completed over 6 weeks (3 weeks before and another 3 weeks twice after randomization). Indoor air quality were compared between control and interferon group and using mixed models the association of fluctuations of daily indoor air quality and PEFR. RESULTS: Indoor PM2.5 concentration levels (median (Q1~Q3) for overall intervening period was 8.8(3.2~12.8) ug/m3 for intervention group and 15.8(8.5~20.5) for case group. Unit increase in concentration of PM2.5 significantly reduced the PEFR by 0.8 l/min (95% confidence interval [CI]: 0.1–1.5) after adjusting for indoor CO2, temperature, relative humidity, sex, age and body mass index (BMI). In this study, we found that fluctuation of indoor PM2.5 were also associated with PEFR and a home-based air cleaner intervention was effective to reduce indoor PM2.5 levels.

Keywords: environmental health

SU-PO-47

Inter and intra-variability of personal exposure to air pollution in different microenvironments

Hossain M.[1], Che W.[1], Lau A.[1]; [1]The Hong Kong University of Science and Technology, Hong Kong, China

Ambient concentrations of air pollution measured at fixed-site monitor are typically used as a surrogate for exposure in estimating health effects in epidemiological studies. However, people spend more than 85 percent of their time in indoor microenvironments which may have different exposure concentrations than ambient levels. To better estimate of health risk, it is important to know the variability of personal exposure to air pollution in different microenvironments. This study was carried out to investigate the inter and intra-variability of personal exposure to air pollution in different microenvironments including home indoor, office indoor, others indoor, outdoor and transportations (e.g., MTR, bus, minibus and car/taxi). Personal measurements with time-location information were performed to determine the concentrations of particles (PM2.5, PM10, PM1) and gaseous pollutants (NO2, O3, CO2, CO) using portable air sampler (PAS). The results revealed the significant differences of the concentrations and exposure among microenvironments visited by a participant due to the variation in local emission sources. The study also demonstrated the substantial variations in daily personal exposure patterns among the participants owing to the difference of the concentrations and time spending by the participants in different microenvironments. Understanding the variability of the personal exposure to air pollution among the microenvironments will help in reducing exposure errors in health estimates and can help to mitigate the air pollution exposure for individuals and at population level. Further study is needed to evaluate the health risk of air pollution considering personal exposure variability among microenvironments.

Keywords: activity patterns, aggregate exposure, air, built/indoor environment

SU-PO-48

Risk assessment by exposure to nitrogen dioxide and ozone during lifetime

Park J.[1], Ryu H.[1], Heo J.[1], Kim E.[1], Choe Y.[1], Cho M.[1], Kim Y.[2], Yang W[1]; [1]Department of Occupational Health, Daegu Catholic University, Gyeongsan, Republic of Korea, [2]EH-Service Center, Daegu Catholic University, Gyeongsan, Republic of Korea

Although the risk assessments for air pollutants such as nitrogen dioxide (NO2) and ozone (O3) has been extensively studied, most of them were limited mainly to the indoor environment. Therefore, integrated risk assessment considering exposure in all microenvironments including outdoor is needed. The purpose of this study was to assess risk by population group according to time-activity patterns, and the lifetime risk of Korean. In this study, we calculated time-weighted average concentration for preschool child, student, housewife, worker and the old using residential time and concentrations of indoor (house, workplace or school, others), outdoor and transport by meta-analysis and data offered from the Korean National Statistical Office. The analyzed data were compared with measured data for validation. The risk for NO2 and O3 were assessed by hazard quotient using reference concentration, 30 ppb and 60 ppb respectively. The risk assessments were conducted by 1,000,000 times of Monte-Carlo simulation for probabilistic analysis. Preschool child, student, housewife, worker and the old spent 91.9%, 86.0%, 79.8%, 82.2% and 77.3% of their day at house and workplace or school respectively. The R2 value of meta-analyzed and measured NO2 concentration by population group was 0.7793. The risk assessment for the lifetime showed 28.4% and 33.8% of hazard quotient of NO2 for worker and housewife exceed 1, and more than 99% of hazard quotients of O3 were less than 1. The risk of NO2 and O3 by sub-population groups and for the lifetime of housewife and worker were assessed. The risk of NO2 was higher than O3 and showed different risk by sub-population groups. Since both NO2 and O3 showed a higher risk in the housewife than the worker. This study can be used as a basis of exposure and risk assessment for the lifetime to NO2 and O3.

Keywords: activity patterns, aggregate exposure, environmental health, exposure models, risk assessment

SU-PO-49

Seasonal difference of indoor-outdoor relationship of PM2.5 in various microenvironments

Guak S. [1], *Lee K.* [1]; [1]*Seoul National University*

Background/Aim: Exposure to fine particle (PM2.5) has been associated with increased morbidity and mortality in numerous epidemiologic studies. Since people spend most of their time indoors, outdoor concentrations are not sufficient to assess personal exposure to PM2.5. Indoor-outdoor relationship could provide valuable information of emission and exchange of air pollution. The purpose of this study was to determine I/O ratio of PM2.5 in various microenvironments in two seasons. Methods: Field technicians measured daily personal exposure to PM2.5 for 50 person-day both in summer and winter of 2017. PM2.5 concentrations were continuously measured using MicroPEM (RTI incorporated, USA). Microenvironments were classified into six categories including home, workplace/school, restaurant, other locations, private transportation, and public transportation. Outdoor concentrations were obtained from ambient monitoring stations in Seoul (AirKorea). I/O ratio of PM2.5 was calculated by microenvironmental concentration and outdoor concentration. Results: Average daily personal PM2.5 exposures were 20.8±30.7μg/m3 in summer and 22.2±35.3μg/m3 in winter, while average outdoor concentrations were 22.2±12.3μg/m3 in summer and 30.1±20.7μg/m3 in winter. The highest PM2.5 I/O ratio in summer and winter were 2.5±3.5 and 3.8±8.3 in restaurant, respectively. The lowest PM2.5 I/O ratio in summer and winter were 0.8±0.7 in workplace/school and 0.6±0.5 in home, respectively. Other indoor locations, restaurant, and public transportation had PM2.5 I/O ratio higher than 1. Home, workplace/school, and private transportation had PM2.5 I/O ratio lower than 1. Conclusions: I/O ratio of PM2.5 was different by microenvironments. I/O ratio of PM2.5 in microenvironments was not changed by seasons. Indoor-outdoor relationship could indicate characteristics of the microenvironments.

Keywords: particles

SU-PO-50

Low-cost Sensor Array Devices as a Method for Reliable Assessment of Exposure to Traffic-related Air Pollution

Mykhaylova N.; WeavAir, Toronto, Canada

The exposure to air pollutant mixtures is a well-known risk factor for inducing and increasing the severity of diseases. Combining low-cost sensors in arrays holds great potential for real-time monitoring of pollutant exposure because of versatility and aptitude for tracking multi-pollutant exposure. While many low-cost air pollution monitoring devices have been proposed, several underexplored opportunities remain, including sensor-derived pollution indices, source analysis and exposure assessment. A thorough investigation of different low-cost commercial gas and particulate matter sensors from 5 manufacturers has been conducted and best-performing sensors were identified. A device for monitoring air quality has been developed and tested. Each device consists of an array of commercially available metal oxide semiconductor for monitoring NOx and O3, CO, CO2 and optical sensors for monitoring PM2.5. Level of pollutant exposure has been characterized at different locations in Toronto over 3 different campaigns between 2013 and 2016. These deployments allowed long-term sensor performance to be evaluated under different meteorological conditions as well as different ranges of pollutant concentrations. Analysis of a large range of gas sensors revealed several key challenges, including high intra-sensor variability, interference from temperature and nonlinearity. Air quality health index estimation from sensor readings was successfully demonstrated. Three aspects of device reproducibility were evaluated: drift over time, impact of interferences and impact of site-specific mixtures. Three categories of approaches for improving sensor accuracy and reproducibility were tested: nonlinear calibration models, variable transformations and training data selection. Model reproducibility, ability to adjust for multiple combinations of interferences and ability to resolve sites was improved when devices were calibrated at multiple sites. Analysis showed that both short-term and long-term temporal patterns could be resolved and compared at different sites. Background subtraction helped further emphasize the differences and rank sites in terms traffic-related pollution.

Keywords: VOCs, air, analytical methods, behavior, built/indoor environment, environmental health, exposure factors, sensor technology, ventilation

SU-PO-51
A review on light in the built environment and its relationship to mental health outcomes

Kohl N. [1], Hoisington A.[1]; [1]Air Force Institute of Technology, Dayton, OH, USA, [2]2Military and Veteran Microbiome: Consortium for Research and Education (MVM-CoRE), Dayton, OH, USA

With an increased awareness of public mental health issues, more investigation is warranted in the mechanisms of how the built environment may influence an individual's mental health. Specifically, light research has shown progress in reducing negative mental health outcomes and lessons learned in that research may be informative in the built environment. For example, bright light therapy, bright light in the morning, low light in the evening, and exposure to natural light over artificial light positively influences mental health outcomes. Bright light in the morning and low levels of light in the evening reduces depression symptoms within a week. In contrast, bright light in the evening has an adverse effect on depression through modulation of individual's melatonin levels that alters the natural circadian rhythm affecting mood, sleep, and mental agility. The excretion of melatonin in individuals can be suppressed by light when its intensity exceeds approximately 2500 lux. In addition to the intensity, individuals are affected psychologically by the different spectrums of light. Artificial lighting does not usually contain the full spectrum lighting required for positive mental health outcomes, although full spectrum fluorescent lights contain the necessary color spectrum to cause positive mental health outcomes. Research suggest that natural light and full spectrum fluorescent lights have roughly the same color spectrum and both produce positive mental health outcomes. Due to the rise in urbanization and time in the urban built environment, architects and engineers may need to consider optimizing the built environment to include proper light exposure influencing positive mental health outcomes. Prior to design and modifications, it is recommended that can controlled studies to target specific light parameters and their relation to mental health outcomes be conducted.

Keywords: behavior, built/indoor environment, health, other/general, workplace

SU-PO-52
Analysis of Thermal Comfort in Residential Structures for the Elderly in Atlantic Climate

Torres P. [1], Aguiar L. [2], Pereira C. [1,2], Forcada N.[3], Tejedor B. [3], Teixeira J.P. [1,2], Mendes A. [1,2]; [1]Instituto de Saúde Pública, Universidade do Porto, Porto, Portugal, [2]National Health Institute - Environmental Health Dept., Porto, Portugal, [3]Group of Construction Research and Innovation, Department of Project and Construction Engineering, Universitat Politècnica de Catalunya, Terrassa, Barcelona, Spain.

Europe has the highest percentage of people aged 60 or over (25%), expecting that the number of people 80 and older will triple by 2050, explaining the growing demand for long-term care services, such as elderly care centers (ECC). Determining influential parameters in thermal comfort (TC) is necessary both to design pleasant spaces and to ensure the well-being and health of buildings occupants. Efficient design and management of the buildings not only offer a comfortable thermal sensation to the occupants but also the amount of energy consumed by the building's cooling and heating systems. This ongoing project focus on the analysis of TC in ECC in the Atlantic climate zone and on the prediction of acceptable and preferred thermal conditions for older people. For this, environmental parameters such as air temperature, average radiant temperature and air velocity, external conditions such as relative humidity and atmospheric temperature, metabolic rates, thermal insulation of the clothing and thermal sensation of the users, will be analyzed in 8 ECC, making a longitudinal analysis across all seasons over a year. The ECC have been analyzed for TC 3 entire days by each season during normal activities, starting from February to November 2019. At the same time, residents will be surveyed weekly to determine the thermal sensation within the ASHRAE range. In order to determine TC, the predicted mean vote (PMV) and the adaptive model will be taken into consideration. These data will later be analyzed and used in the application of environmental comfort models for the elderly. From the resulting analytical model, it will be possible to create more efficient TC control practices, both at the level of building structures and at the level of implementation of new practices, improving people's quality of life and the energy efficiency of buildings.

Keywords: aging/elderly, analytical methods, built/indoor environment, environmental health, thermal environment

SU-PO-53 WITHDRAWN

SU-PO-54
Effect of ambient air pollution on gestational diabetes mellitus and potential biological linkage: A review

Zhang H.; Department of Health Policy and Management, School of Public Health, Sun Yat-sen University, Guangzhou, China

Gestational diabetes mellitus (GDM) is a serious complication of pregnancy, which could have adverse health effects on both mothers and fetuses. Air pollution is a growing global issue. Plausive links has been established between air pollution and GDM. To better understand the association between maternal exposure to air pollution and GDM, we summarized the recent data on how ambient air pollution impact GDM and its potential mechanisms. Pollutants include particulate matter ≤2.5 μm (PM2.5), particulate matter ≤10 μm (PM10), nitrogen dioxide (NO2), nitrous oxides (NOX), black carbon, sulfur dioxide (SO2), carbon monoxide (CO) and ozone (O3). Studies varied refer to assessment of exposure to pollution, definition of GDM, and the variables selected in the model. The results showed that air pollution, especially for PM2.5, has adverse effect on the incidence of GDM in first trimester, second trimester and the entire pregnancy. It indicates β-Cell dysfunction, chronic insulin resistance, neurohormonal dysfunction, inflammation, oxidative stress and gut microbiome may all play important roles in air pollution induced GDM.

Keywords: air, climate change, environmental health, epidemiology, prenatal

SU-PO-55
Exposure to household air pollution associated with adverse immune responses in rural women in northwest China

Yao, Y.[1], Wang D.[1], Ma H.[1], Li C.[1], Wang, J.[1], Chang X.[1], Low P.[2], Liu S.[3]; [1]School of Public Health, Lanzhou University, Lanzhou, Gansu Province, China, [2]School of Public Health, University of California, Berkeley, Berkeley, CA, USA, [3]School of Health Sciences, Purdue University, West Lafayette, IN, USA

Background: The associations between impairments of immune functions and exposure to polycyclic aromatic hydrocarbons (PAHs) in air pollution remain inconclusive due to the limitations of previous studies. In this study, we aimed to examine the direction and magnitude of immune changes related to PAH exposure from household air pollution among rural women living in Gansu, China. Methods: Healthy village women (n = 34) were recruited and enrolled in the study. Blood and urine samples were collected and analyzed during non-heating and heating seasons. Urinary 1-hydroxypyrene (1-OHP) was quantified as the biomarker of PAH exposure. Immunoglobulin E (IgE), percent of T-regulatory (Treg) cells, and gene expression of forkhead box transcription factor 3 (Foxp3), transforming growth factor-β (TGF-β), interleukin 10 (IL-10), and interleukin 35 (IL-35) were analyzed to evaluate immune responses. Urinary 8-hydroxy-2-deoxyguanosine (8-OHdG) was measured as a marker of oxidative DNA damage. Results: the concentration of 1-OHP increased almost 20 times from 0.90 to 17.4 μmol mol-Cr -1 from summer to winter ($p < 0.001$). Meanwhile, average percent of Treg cells decreased from 5.01% to 1.15% ($p < 0.001$); IgE and mRNA expressions of Foxp3, TGF-β, IL-10, IL-12α and EBi3 all significantly decreased ($p < 0.001$); Urinary 8-OHdG increased from 12.7 to 30.3 ng mg-Cr -1 ($p < 0.001$). The changes in 8-OHdG, Foxp3 and TGF-β were significantly associated with the increase of 1-OHP. Conclusions: We observed a substantial increase of PAH exposure in winter, which was significantly associated with the repression on Treg cell function and oxidative DNA damage. Exposure to PAHs in household air pollution possibly induced immune impairments among rural women in northwest China.

Keywords: epidemiology

SU-PO-56
Research on Human Thermal Adaptive Behavior in the Vernacular Housings: A case study in Sandaoguai Traditional Settlements, Chongqing, China

Zhao Y.[1], Liu K.[1], Lin B.[1]; [1]Tsinghua University, Beijing, China

Vernacular architecture, unlike many modern buildings, mainly relies on various bioclimatic design strategies and passive methods to create and regulate the indoor physical environment rather than mechanical equipment. Therefore, the types and patterns of human thermal adaptive behavior are different and varied, which is worthy of further study. However, most studies that related to thermal adaptive behavior are focused on modern buildings, such as offices, schools and apartments, only few aimed at vernacular architecture. Based on the theoretical framework of thermal adaptation comfort, this paper, taking Sandaoguai traditional settlements（Chongqing, China） as a case study, conducting a long-term field investigation on the behavioral adjustments in three aspects: personal adjustments, technical/environmental adjustments and cultural adjustments. Firstly, in the aspect of personal adjustments, the average thermal resistance of clothes in this area is counted. Combined with the calculations in other studies in the same area, this paper reveals the characteristics and causes of clothing adjustments. Secondly, in terms of technical/environmental adjustments, more than 10 behavior patterns that related to cooling and heating, as well as the regulation rules of typical modes such as door opening, window opening and air conditioning use are analyzed. In addition, this paper finds there are different patterns of reconstruction behavior, which are unique in vernacular architecture. It shows that the technical/environmental adjustments in Sandaoguai vernacular architecture are not only acted on the use of the building, but also on the partial renovation of the building. Thirdly, in cultural adjustments, the patterns and regularities of work and rest time, spatial use and social activities are discussed. Finally, based on the characteristics and patterns of human thermal adaptive behavior, this study also puts forward suggestions for the sustainable design and renovation of vernacular housings.

Keywords: behavior, built/indoor environment, statistical methods, sustainability, thermal environment

SU-PO-57
The connection between indoor air quality and mental health outcomes

Taylor W.[1], Hoisington A.[1]; [1]Air Force Institute of Technology, Wright Patterson AFB, Dayron, OH, USA

In the United States, nearly one in five adults live with a mental illness, and 44,000 individuals committed suicide in 2016. Neuropsychiatric disorders trail only cancer and cardiovascular diseases as the highest cause of disability adjusted life years in Europe. Such profound health outcomes drive a need for greater understanding of mechanisms that might influence mental health, including possibly the built environment. To better understand the connection between mental health and the built environment, we administered a self-reported survey to United States Veterans. The survey included aspects of the built environment such as: sources of indoor particulate matter, residence proximity to a busy roadway, and recent occurrence of water damage. Particulate matter is generated by common household activities like smoking, cooking, and burning candles or incense. Studies of outdoor particulates have shown that individuals have an increased risk of suicide when exposed to increased levels of particulate matter. This opens the possibility that indoor generated particulate matter could have an affect on mental health outcomes as well. Additionally, the location of a residence can strongly influence particulate matter concentrations within the home, with homes within 100 meters of a busy roadway having increased concentrations. Finally, water damage in a residence can also have harmful mental health repercussions. Housing quality has been shown in literature to be inversely related to psychological distress, suggesting that individuals unable to afford the repair of water damage within their homes will suffer from increased psychological distress. The literature surrounding the physical health effects of indoor air quality is thorough and well studied, yet the connection to mental health is lacking. Future studies of indoor air quality should seek to identify factors that have a direct impact on mental health.

Keywords: built/indoor environment, health, chronic diseases, health, other/general, ventilation

SU-PO-58

The Impact of Cooking Aerosol on Human Brain Wave Pattern

Gabdrashova R.[1], Naseri M.[2], Jouzizadeh M.[3], Tabesh M.[3], Malekipirbazari M.[4], Nurzhan S.[1], Farrokhi H.[3], Khanbabaie R., Mehri-Dehnavi, H.[3], Bekezhankyzy Z.[1], Gimnkhan A.[1], Dareini M.[2], Kurmangaliyeva A.[1], Islam N.[1], Crape, B.[5], Buonanno G.[6], Cassee F.[7], Torkmahalleh M.[2, 8]; [1]Nazarbayev University, Nur-Sultan, Kazakhstan, [2]Chemical and Aerosol Research Team, School of Engineering, Nazarbayev University, Nur-Sultan, Kazakhstan, [3]Department of Physics, Babol Noshirvani University of Technology, Babol, Iran, [4]Department of Industrial Engineering, Bilkent University, Ankara, Turkey, [5]School of Medicine, Nazarbayev University, Nur-Sultan, Kazakhstan, [6]Department of Civil and Mechanical Engineering, University of Cassino and Southern Lazio, Cassino, Italy, [7]National Institute for Public Health and the Environment, Bilthoven, The Netherlands, [8]The Environment and Resource Efficiency Cluster, Nur-Sultan, Kazakhstan

The cooking emission in residential or commercial kitchens is one of the main indoor PM sources that results in human exposure to high concentrations of ultrafine particles (UFPs) (Amouei Torkmahalleh et al., 2017). He et al. (2004) found cooking as the largest source of indoor air pollution with the highest level of PM2.5. So far there exist only two exposure studies in literature that investigated the impact of combustion fumes from cigarette smoking (Wilbanks et al., 2016) and diesel exhaust (Cruts et al., 2008) on the human brain. The main objective of this study was to investigate the brain's wave pattern changes during and after acute exposure to cooking aerosol. Twelve non-atopic, non-smoking, and healthy adults (18-46 years old) were monitored for brain wave pattern before, end of, and 30 minutes after cooking, using electroencephalography (EEG). Frying ground beef meat in sunflower oil using electric stove was conducted. UFPs, particulate matter, CO2, indoor temperature, indoor RH, oil and meat temperatures were monitored continuously throughout the experiments. The UFP peak concentration was recorded to be approximately 2.0×105 part.cm−3. Brain activity had statistically significant changes during the post-exposure compared to the before exposure suggesting the translocation of UFPs to the brain. This change mainly occurred in the frontal and temporal lobes of the brain. The brain slow-wave band (Delta) decreased while the fast-wave band (Beta2 and Beta3) increased similar to the pattern found in the literature for the exposure to smoking fumes and diesel exhaust. The brain pattern observed in this study showed no similarity to neurodegenerative diseases such as Alzheimer, Parkinson, dementia and schizophrenia which were found to decrease the fast–wave band Beta. The second phase of this study was conducted with 30 volunteers and measurements up to 120 minutes after cooking. The results will be presented during ISES/ISIAQ 2019.

Keywords: air, biomonitoring, built/indoor environment, health, other/general, particles

SU-PO-59
The plants affected by air pollution have impact on human health

Budreviciute A. [1], *Plakys G.* [2], *Katileviciute A.* [2], *Kodzius R.* [2]; [1]*Lithuanian University of Health Sciences, Kaunas, Lithuania,* [2]*Kaunas University of Technology, Kaunas, Lithuania*

The outdoors air pollution affects both humans and other life forms such as plants. The grass, especially it's kind, the food crops are of enormous importance to our human lives, as we daily consume the crops (such as wheat) as food. The burning fossil fuel generates sulfur dioxide and sulfuric acid, which in its turn affects leaves and plant tissues. The same effect is observed for the ozone influence. The car combustion engines produce nitrogen oxides, slowing growth of the plants. However, generated nitric acid is a source of nitrogen, which is essential for plant growth. The carbon monoxide from combustion engines halts the growth of the plants as well. The secondary pollution effect on plants is the diminished immune system, causing easier infestation by various viral, bacterial or fungi infections, including the insect infestation. The small particles generated by anthropogenic or non-anthropogenic sources are especially damaging to the plants. The plant growth is challenged, and the small particles are absorbed by leaves. In this way, the pollutants are removed from the air. Some of the contaminants, such as sulfur dioxide, can be metabolized by the plants. Other pollutants, however, remain in the plants and can enter the food chain (i.e., wheat). Processing the food crops, theoretically, it is possible to mitigate or completely remove the pollutants by modifying the industrial processes. Human health is of enormous importance, and the pollutions ways are affecting us through the plants should be investigated and followed.

Keywords: biomonitoring, ecological exposure, food, health, other/general, particles, sustainability

SU-PO-60
Measurement of ambient ozone (O3) levels, correlation with the NOx levels and the development of the monitoring network for O3

Turos O. [1], *Petrosian A.* [1], *Kobzarenko I*[1]*., Kharchenko K.* [1], *Maremukha T.* [1], *Ananyeva O.* [1]; [1]*State Institution – "O.M. Marzeiev Institute for Public Health of the National Academy of Medical Science of Ukraine", Kyiv, Ukraine*

Background/Aims. The link between increase O3 ambient air pollution and health issues has been well documented, prompting the introduction of strict rationing standards of O3. This study aims to conduct a quantitative assessment of the level of the ambient air pollution with O3 in order to prove that there is a necessity of organizing of the modern governmental system of monitoring of ambient O3 and substances from which it derives, as well as systems for notification of the population about high level of ambient air pollution in Ukraine. Methods. Concentrations of the O3 and NOx in the air were measured by the gas analyzers (Horiba APOA-370 and APNA-370). The obtained data was processed using the software package Statistica, results were compared with the limit values recommended by WHO. Results. According to the data of the completed research concentrations of tropospheric O3 in Kyiv significantly exceed the established norms. Daily concentration of the O3 for the summer period varies from 127.7818±3.76 µg/m3 to 202.4±20.7 µg/m3, and during spring period – from 200.87±3.61 µg/m3 to 222.83±6.72 µg/m3. Data from semi-annual monitoring indicates that the chemistry of ambient O3 in Kyiv confirm to NOx-sensibility mode, what is verified by the occurrence of photochemical smog in Kyiv with high O3 concentrations. What is more, there was found a stable anticorrelation between O3 and NOx, what is also an evidence of the above-mentioned. Conclusions. The continuation of the monitoring for ambient air quality and prompt analysis of the monitoring data will allow not only to fix the dangerous concentrations of the tropospheric O3, but also justify the organization of the activities for the decreasing of its level, and forecast the ecological situation in the city.

Keywords: industrial hygiene

SU-PO-61

Risk assessment of chronic exposure to antibiotics in rural population

Ali A. [1], Malik A. [2], Uddin I. [1], Rashid A. [1]; [1]Department of Environmental Sciences, Arid Agriculture University, Rawalpindi, Pakistan, [2]Civil Aviation Authority, Islamabad International Airport, Islamabad, Pakistan

The rural environment receives antibiotics and their residues from agriculture, livestock and human that may persist for long-time. Hence public exposure to antibiotic residues is multifaceted and need health risk assessment. This study aimed to quantify antibiotic usage, their residues in environment and assess chronic exposure of rural population consequent health risks analysis. Field soil, fecal and irrigation water samples were analyzed for fluoroquinolones and quinolone antibiotics. Target rural population (n=234) were surveyed which includes livestock handlers, agricultural farmers, field workers, female households and children, to assess extent of exposure to antibiotic persist in their immediate environment and epidemiological symptoms. Significant use of ciprofloxacin as self-medication (unethical) was strongly correlated with flu and gastrointestinal infections. Among females gastro (OR=1.51; 95% CI=1.16–1.95) whereas in livestock workers both flu (OR=1.39; 95% CI=1.07–1.80) and gastro (OR=2.11; 95% CI=1.62–2.74) incidences were recorded as consequent impact of chronic exposure to antibiotics. In feces wide range of ciprofloxacin was detected (97.5–188.2 mg kg–1) whereas amount of levofloxacin and ofloxacin were within the WHO regulatory limits. Wastewater channels from rural dwellings and livestock farms were identified as potential hotspots for chronic exposure. Indiscriminate use of antibiotics to enhance dairy produce has forced rural people to indulge in unethical drug administration which has increased their susceptibility to further exposure to antibiotics and their residues in rural environment. We conclude that chronic exposure of rural population to antibiotics provoke health risks and this exposure is on rise due to continuous add-up of sub-lethal levels linked with self-administrated usage which must be strictly checked to minimize their release in environment.

Keywords: community

SU-PO-62

Urinary pesticide metabolite concentrations in pregnant women from Suriname

Alcala C.; Tulane University, New Orleans, LA, USA

Pesticides are widely used in Suriname, a middle-income Caribbean country. The primary agricultural area is district Nickerie. Exposure to pesticides leads to adverse health effects and pregnant women and children are vulnerable. The Caribbean Consortium for Research in Environmental and Occupational Health (CCREOH) assesses the impact of environmental exposures on birth outcomes and neurodevelopment in 1000 mother/child dyads. This study represents a subset of the CCREOH cohort and is aimed at comparing pesticide metabolite concentrations in pregnant women from Nickerie and the capital Paramaribo. Urine samples from 220 participants were collected. A total of 50 samples (42 from Paramaribo and 8 from Nickerie) were recently analyzed at the US Centers for Disease Control and Prevention's environmental health laboratory for three pesticide classes, including the commonly used 2,4-D, organophosphate insecticides, and pyrethroid insecticides. Pregnant women in Nickerie had the highest geometric mean urinary metabolite concentrations of 2,4-D and pyrethroids: 3-phenoxybenzoic acid compared to those living in Paramaribo, (0.56 vs. 0.16 ug/L), (0.52 vs. 0.24 ug/L), respectively (p<0.05). Pregnant women living in Paramaribo had significantly higher geometric mean urinary metabolite concentrations of organophosphates: para-Nitrophenol, diethylthiophosphate and diethylphosphate compared to those living in Nickerie, (0.44 vs. 0.22 ug/L), (0.73 vs. 0.22 ug/L), (1.54 vs. 0.54 ug/L), respectively. The above-mentioned levels are higher in Suriname compared to those in the United States based on data from the National Health and Nutrition Examination Survey. Urinary metabolite concentrations of pesticides varied in Paramaribo and Nickerie. This could be due to residential use, intensive spraying during mosquito control programs in Paramaribo and excessive use in Nickerie, as it is a rice production area, respectively. Analysis of an additional 170 urine samples is underway.

Keywords: environmental health, epidemiology, occupational, pesticides, prenatal

SU-PO-63
Influence of the renovation on ultrafine particle concentrations in occupied Swedish residences

Omelekhina Y.[1], Pedersen E.[1], Nordquist B.[1], Wallenten P.[1], Nilsson P.T[1]., Wierzbicka A.[1]; [1]Lund University, Lund, Sweden

A number of health effects have been linked to exposure to airborne particles, however the consequences of exposure to ultrafine particles is largely unknown. Ultrafine particles can be emitted from different indoor activities or be of the outdoor origin. This is further complicated by the fact that we are spending majority of the time indoors. The aim of this work was to assess the influence of renovations on indoor-to-outdoor (I/O) ratios of ultrafine particles (UFP) in 6 occupied Swedish residences. Simultaneous indoor and outdoor measurements were performed in 6 residences for seven consecutive days in urban area in southern Sweden. The apartments were mechanically ventilated and had volumes between 115-218 m3. The renovation activities included adjusting air exchange rates (AER) to meet Swedish standards (0.5 AER h-1) and installation of the new kitchen extraction fans. NanoTracers (Oxility Aerasense) were used to measure particle number concentrations and mean particle diameters (10 - 300 nm). Based on the averages of particle number concentrations (PNC) during occupancy period, indoor-to-outdoor ratios (I/O) were calculated for each home. AER have been increased in the apartments during renovations to 0.4-0.7 h-1 in comparison to values before the renovation 0.2-0.4 h-1. After renovation I/O ratios of PNC during occupancy periods varied between 0.2-3.8 and were lower in comparison to the values before renovations 1.0-11.9. This could indicate increased efficiency of particles removal after renovation but also can be influenced by lower amount of performed activities by occupants, which were 2-3 times lower after in comparison to before. Results have shown the positive changes of the renovation processes on AER and particle removal. Nevertheless, the frequency and type of indoor activities performed by residents at home influence the indoor concentrations. The Swedish Research Council FORMAS and the Swedish Energy Agency financed the project.

Keywords: activity patterns, air cleaning, built/indoor environment, nanoparticles, particles, ventilation

SU-PO-64
Addressing Water Quality Alongside Community Health Workers in a Farm Worker Community

Wagoner R.[1], Verhougstraete M.[1]; [1]University of Arizona, Tuscon, AZ, USA

The community of Winchester Heights, Arizona is made up of approximately 400 migrant farm workers and has been identified as a community near the US-Mexico border lacking accessible potable water, adequate sewer services, and other infrastructural issues. Previous work has illustrated community concern over environmental health issues, including water quality. The objectives of this project were to increase community capacity to address environmental health concerns through the establishment of a community health worker (CHW) environmental health (EH) program, measure fecal indicators and metal concentrations in community water samples, and identify community water usage patterns and barriers to improvement. Five community members were recruited as CHWs and trained in environmental health principles, human subject research, water collection, sample processing, and result-reporting methods. The trained CHRs collected tap water samples from 43 households and administered questionnaires to 43 adults. Morning and evening tap water samples were analyzed for E.coli using compartmental bag test as well as membrane filtration followed by incubation on modified mTEC media. Water samples were analyzed for lead and arsenic via inductively-coupled plasma mass spectrometry. Water samples from all households were negative for E.coli. The concentration of lead and arsenic in all water samples were below the EPA Maximum Contaminant Level (MCL) for both contaminants. The majority of respondents used bottled water as their primary drinking water source and listed lack of regular testing and acceptable consumer confidence reports as barriers to improvement. This project demonstrates the utility of CHW programs in addressing community environmental health concerns along the US-Mexico border region. The CHWs recruited participants, conducted water sampling, and reported of results to their own community, which are transferrable skills that can be used to address other community issues. Furthermore, the project can be used as a model in other underserved border communities.

Keywords: community, environmental health, environmental justice, metals, microbial agents, susceptible/vulnerable, water

SU-PO-65

Influence of house types on the prevalence of chronic respiratory diseases in Ho Chi Minh City

Tran N.T.[1], Tran B.N.[1], Tran T.M.H.[1], Tang K.H.[1], Ngo M.X.[1], Godin I.[2], Michel O.[2], Bouland C.[2]; [1]Ho Chi Minh City, Viet Nam, [2]Brussels, Belgium

Exposure to pollutants is related to house typology. Besides smoking, indoor air pollution is a significant risk factor of chronic respiratory diseases (CRD). The prevalence of CRD in different house types was studied in Ho Chi Minh City. 1561 people living in four house types, were enrolled. Those house types are apartment, rental house (a very small house with ground floor, toilet and main room, with or without windows, often built into row of rental houses), rural house (a one floor and separate house which is built in rural and suburban district by wood or semi-solid materials) and tube house (several floors, built close together, very small in width and deep in length). Information on respiratory health, lung function, residential characteristics, and indoor environmental sources was obtained by questionnaire and spirometry. The two main respiratory health outcomes studied are clinical CRD (CCRD) (symptom of CRD) and chronic obstructive respiratory diseases (CORD) (FEV1/FVC < 0.7). The prevalence of CCRD (24.3%) and CORD (5.3%) did not equally distribute among studied house types ($p<0.0001$, Chi-square). It is equal in tube house and apartment. Comparing to people living in apartment, those who live in rental house had a 46% higher risk of developing CCRD. The odds of CORD for people living in rental house and in rural house were 4.64 (1.97-10.5) folds and 2.99 (1.21-7.37) folds higher. Controlled factors were age, sex, in home time, smoking level, occupational risk, previous tuberculosis and presence of pet, rat, cockroach, flooded wall, biomass usage and air-conditioner. House type was associated with a likelihood of CCRD and CORD morbidity.

Keywords: respiratory health

SU-PO-66

Evaluating vertical concentration variation of non-polar organic compounds for PM2.5 in Taipei metropolis

Tsai W.C.[1], Liao, H.T.[2], Wu, C.F.[1]; [1]Institute of Environmental Health(IEH), National Taiwan University, Taipei, Taiwan, [2]Research Center for Environmental Changes, Academia Sinica, Taipei, Taiwan

Numerous people live in multistory buildings in Taipei metropolis. However, vertical variation in exposure to air pollutants is not well characterized. This study was conducted in Taipei, Taiwan from September to October 2018. Three sampling sites were set up on balconies at three different floor levels (low-, mid-, and high-level sites) of one multistory building. Twenty-four-hour integrated PM2.5 samples were collected on quartz-fiber filters simultaneously at all sites twice a week, making a total of 36 samples. Non-polar organic compounds (NPOCs), including 24 polycyclic aromatic hydrocarbons (PAHs), 16 hopanes, 12 steranes, 23 n-alkanes, and 10 Iso-/anteiso-alkanes, were identified and quantified using Thermal Desorption Gas Chromatography/Mass Spectrometry (TD–GC/MS) method. At the sampling building, n-alkanes showed highest concentrations [10.12±0.96 ng/m3 (Mean±SD)] that accounted for 62% in the analyzed NPOCs, followed by PAHs (3.88±0.33 ng/m3, 24%), and Iso-/anteiso-alkanes (1.86±0.25 ng/m3, 11%). The Heneicosane (n-C21) to Tritriactotane (n-C33) were the most abundant components of n-alkanes, whereas the four- and five-ring elements were dominant in total PAHs. The average concentration of all analyzed species decreased as the height increased. However, there were no statistically significant differences ($p>0.05$) in the average concentration of the three floors. The preliminary results in this study reveal the vertical distribution of air pollutants in the Taipei metropolitan area. Nevertheless, long-term monitoring campaigns are needed to further evaluate the seasonal effects.

Keywords: SVOCs, air, environmental health, particles, sampling methods

SU-PO-67

Examine the Impacts of Human Activities on Indoor Exposure Profile Using a Low-cost Passive Air Pollutant Monitor

Lin E. [1], Pollitt K.; [1]Yale University, New Haven, CT, USA

In most developed countries, people spend 90% of their time indoors. Personal activity patterns will determine an individual's exposure profile. Cooking, cleaning and personal care products are major emission sources in the indoor environment. The HomeChem study was conducted in June 2018 in Austin, Texas to explore these indoor exposure sources. As part of this study, non-selective passive air samplers were deployed in the model house to profile organic contaminants associated with common indoor activities, including chrysene, Benzo[a]pyrene, fluoranthene. Sampling devices were placed in the kitchen over a 24-hour period and analyzed using thermal desorption gas chromatography time-of-flight mass spectrometry to quantify time-weighted average exposures to a panel of 36 volatile and semi-volatile organic compounds during typical daily activities. Detailed activity logs were recorded within 1 hour on individual experimental day. Experimental days were designed to imitate typical human activity patterns. Those experimental days can be categorized into four types of days, they are Friend Visit, Cleaning, Cooking and Thanksgiving Day. On each of the experimental days, three sampling replicates were conducted. Principal Component Analysis was applied to exposure results to examine relationships between experimental days and individual analytes. Comparisons were made between the length of activity and the exposures to demonstrate the does-response relationship. At the end of the study, key factors that determine the exposure profile were identified; does-response relationships between exposure and activity were established among EPA identified carcinogenic organic compounds.

Keywords: SVOCs, built/indoor environment, cumulative exposure, exposure factors, sampling methods

SU-PO-68

PFAS and total fluorine levels in fire station dust

Young A. [1], Sparer-Fine E. [1], Peaslee G. [2], Sunderland E. [3], Pickard H. [3], Allen J.[1]; [1]Harvard T.H. Chan School of Public Health, Boston, MA, USA, [2]University of Notre Dame, Notre Dame, IN, USA, [3] Harvard John A. Paulson School of Engineering and Applied Sciences, Cambridge, MA,

Per- and polyfluoroalkyl substances (PFAS) are a class of over 4,000 highly persistent, fluorinated chemicals used in consumer products, building materials, outdoor gear, and firefighting foams. Firefighters are particularly vulnerable to PFAS exposure due to inhalation of smoke during combustion of consumer products, use of firefighting foams, and contact with turnout gear (which may potentially be treated with PFAS for water resistance). Studies on firefighter exposures to PFAS have focused on exposures encountered while fighting fires, yet most career firefighters spend the majority of their shifts at the fire station itself. In addition, the limited studies available on firefighter PFAS exposure have only analyzed fewer than 15 specific PFAS of the thousands of possible regrettable substitutes to the two legacy PFAS phased out of firefighting foam. Total fluorine can provide a proxy for cumulative levels of fluorinated chemicals. In this study, we measured total fluorine levels with PIGE and analyzed for 25 PFAS compounds in dust samples from six room types (truck bays, turnout gear locker spaces, sleeping quarters, kitchens, living rooms, and gyms) in 15 fire stations in Massachusetts. Fluorine was detected in every dust sample collected, ranging from 7.33 ppm (kitchen) to 953 ppm (truck bay) with a median of 158 ppm. Our multilevel model showed that dust in the truck bays and gear locker spaces had significantly higher total fluorine levels compared to the reference living rooms ($p<0.01$). Our PFAS analysis will allow us to determine if the 25 targeted PFAS account for a significant fraction of the high total fluorine detected, to compare PFAS profiles by room type and corresponding potential sources, and to estimate health risks for firefighters. These results will help establish whether firefighting activities and turnout gear use are contributing sources of firefighter exposure to PFAS.

Keywords: SVOCs, building materials, built/indoor environment, occupational, workplace

SU-PO-69

SO2 and CHCO vertical column densities in major cities of Kazakhstan from 2005 to 2016

Amouei Torkmahalleh M.[1], Darynova Z.[1], Sabyrzhan A.S.[1], Sagynov S.[1], Abdrakhmanov T.[1]; [1]Nazarbayev University, Astana, Kazakhstan

Tropospheric SO2 and CH2O vertical column densities were analyzed for period 2005-2016 over the atmosphere of Kazakhstan using the Ozone Monitoring Instrument onboard the NASA Aura satellite. We found main hot spots, seasonal trends and overall trend over 11 years. Major hot spots in terms of SO2 relate to regions with high population and industrial objects. For example, the city Ekibastuz has the biggest thermal power plants in the country. Determinants of the hot spots are fuel sources and electricity usage. Power plants in Ekubastuz burned brown coal that contains 30% toxic minerals with long lifetimes. The annual average concentration of SO2 for this city was 2.5x10-5 kg/m2, whereas for the rest of the country it was 6 times lower. Other hotspots includes megacities such as Almaty and Astana experienced high concentrations of SO2 in 2008. One of the main reasons for that is the financial crisis in 2008, as the cities used cheap low-quality coal. Regarding CH2O VCD, one of the hot spots was the city Atyrau, the major oil producer of the country with two massive oil fields. The highest VCD of formaldehyde (9 x 1015 molecules/cm2) is appeared in the summertime due to secondary production in the presence of hydroxyl radicals as a result of photo-oxidation of VOCs emitted from industrial objects and vehicles. Moreover, stimulation of CH2O emissions by wildfires was observed in the capital city Astana in 2012, when the abnormal number (37) of forest area fires happened. Nevertheless, two significant steps were done for improvement of the air quality in 2013: the introduction of the emission trading system (ETS) and Euro-4 standard. It led to the negative trend in CH2O concentration between 2012 and 2015 in the two highly populated cities.

Keywords: SVOCs

SU-PO-70

Fugitive Emissions from a Breath Actuated Jet Nebuliser and a Vibrating Mesh Nebuliser for a Paediatric Patient

O'Toole C.[1], McGrath J.[1], Bennett G.[2], Joyce M.[2], MacLoughlin R.[2], Byrne M.[1]; [1]NUI Galway, Galway, Ireland, [2]Aerogen, Galway, Ireland

Introduction Recent studies have identified the potential for fugitive emissions to be released during nebulisation treatments. Here we quantify fugitive emissions from a commercially available breath actuated compressor driven jet nebuliser (JN) designed to reduce emissions, and a vibrating mesh nebuliser (VMN) and valved spacer. Methodology A breath actuated jet nebuliser with a mouthpiece (Aeroeclipse2, Monaghan Medical, Canada) was operated at 8 L/min. Two modes on the nebuliser were tested during simulated paediatric breathing (Vt 250 mL, breath rate 25 BPM, I:E ratio 1:2); breath-actuated nebulisation (BAN) and continuous nebulisation (CN) (n = 5). A vibrating mesh nebuliser (VMN) (Aerogen Solo with Aerogen Ultra, Aerogen, Galway, Ireland) was used with a mouthpiece in CN mode only. A standard prescription dose of 3 mL of 2 mg/mL salbutamol was nebulised. An Aerodynamic Particle Sizer (TSI Inc., US) placed at 0.8 m (arm's length) from the nebuliser was used to characterise the fugitive aerosol concentration (mg drug/m3) over a 45 minute period. Results Fugitive emissions were released during nebulisation treatment for both nebuliser types and both modes. CN mode on the jet nebuliser had a higher fugitive aerosol concentration of 0.061 ± 0.048 mg/m3 compared to the BAN and the VMN (0.044 ± 0.030 mg/m3 and 0.011 ± 0.006 mg/m3). MMAD averaged 1.21 ± 0.05 for JN CN, 1.22 ± 0.04 for JN BAN and 1.20 ± 0.12 µm for VMN. Conclusion The results confirm that fugitive emissions are released into the surrounding environment during respiratory therapy using both CN and BAN mode on the jet nebuliser as well as CN mode for the vibrating mesh nebuliser. However, across all tests, the vibrating mesh nebuliser in combination with a valved spacer recorded the lowest secondary emissions.

Keywords: air, built/indoor environment, children, cumulative exposure, environmental health, exposure factors, health, chronic diseases, health, other/general, particles, respiratory health

SU-PO-71 - WITHDRAWN

SU-PO-72
Novel air filtration materials via combined melt-solution electrospinning

Buivydiene D[1]., Krugly E. [1], Todea A. [1], Asbach C. [1], Kliucininkas L. [1], Martuzevicius D.; [1]Kaunas University of Technology, Kaunas, Lithuania

Hierarchically structured nanofibers are gaining attention in the fields of energy and the environment, particularly for air filter and fluid membrane applications. Non-woven fibrous media have been used extensively in air filtration applications and nanofibrous air filters have previously shown grate results for high efficiency air filtration. However, the durability is still questionable because nanofibers form high packing density filtersfeaturing high pressure drop even with the presence of slip flow phenomena. Mixing micro and nano sized structures has been proven to increase the performance of the air filter by allowing a higher surface area to trap more particles and introducing the presence of slip flow at the same time. Synthesis of dual-sized fibrous material using the method of combined melt-solution electrospinning is reported here for the first time. The formation of well-mixed nano- and submicron-size Polyamide (PA 12 and PA 6/6) composites was accomplished simultaniously utilizing melt and sollution electrospinning via custom made electrospinning apparatus. By manipulating the process conditions and the ratios between melt and solution electrospinning, different structures of polimeric mats were achieved. Air filtration parameters were tested and with the particle size of 300 nm filtration efficiency reached up to 89.9 % and pressure drop was 55 Pa. In comparison, uniform nano-fibrous mat formed using only solution electrospinning had higher filtration efficiency of 99.5 % but presure drop was almost 3 times higher at 161 Pa.

Keywords: air, air cleaning, nanoparticles, nanotechnology, particles

SU-PO-73
Particulate science dynamics in indoor micro-environments of Agra, India

Rohra H. [1], Taneja A. [1]; Dr. B.R. Ambedkar University, Agra, India

Improved understanding of particle size distribution of residential air particles is valuable owing to a larger fraction of time spent in homes. Size segregated aerosol measurements was conducted using a multi staged cascade impactor in residential microenvironments of Northern India. Particles with diameter<2.5μm contributed about 64-80% of PM10 in which over 29% of mass was shared by ultrafine PM0.25. Increased significant inter-correlation among adjacent sized PM fractions was further noted. The accumulation mode dominated the lognormal particle size distribution during the complete monitoring campaign whereas those of elemental constituents reflected the distinct emission sources. These findings in future will be used to classify the impactor aerosol samples for toxicological studies and would provide source information to state agencies responsible for development of effective pollution control strategies.

Keywords: metals

SU-PO-74
Source characterization and apportionment of river pollution in central Taiwan

Lin[1], Wang[1], Huang[2], Wang[3], Wang[3], Hsu[3], Wu[1]; [1]National Taiwan University, Taipei, Taiwan, [2]Academia Sinica, Taipei, Taiwan, [3]Environmental Analysis Laboratory EPA, Taoyuan, Taiwan

The systematic forensic technology and pollution source tracking model have been widely used in analyzing sources of air pollution. However, these methods were not commonly used to trace sources of water pollution. In this study, receptor modeling was applied to investigate the spatial and temporal distributions of water pollution sources in Wu River in Taiwan by the chemical mass balance (CMB) model. Six types of industrial manufacturing plants are located along the river, including brewing industry (BI), agricultural and environmental pesticide manufacturing (AEP), leather manufacturing (LM), chemical manufacturing (CM), wastewater treatment (WT) and electroplating industry (EI). Two monitoring sites at a distance of 7.3 km were selected as the as the receptor sites. Two water samples at each site were taken in July and September, respectively. In addition, wastewater samples including the effluent and process wastewater at 24 industrial plants near the Xinan Bridge and Neixin Bridge were collected to represent the potential source profiles. The sample collected were analyzed for fifty-five heavy metals were analyzed by ICP-MS. The main characteristic species were identified for different types of industry, e.g. Mn and Sn in CM, Fe in EI-process, and Al in LM. The CMB modeling results showed that in July the main sources at Xinan Bridge and Neixin Bridge were CM and EI process wastewater with contribution rates of 49% and 47%, respectively. The sources at these two sites of the samples collected in September were both LM (66% and 85%). As the results showed, the sources of pollution in rivers exhibited temporal and spatial variation.

Keywords: environmental health, metals, statistical methods, water

SU-PO-75
CFD simulation of bacteria distribution in an operating room with mixing ventilation at St. Olavs hospital

Zhang Y., Cao G., Pedersen C., Sternstad L., Drangsholt F., Feng G., Skogås G.

Background: Transmission of airborne bacteria is one of these main factors causing surgical site infection (SSI), which is harmful to patient's health and even life, and would cause great economic burden. Many earlier studies have provided evidence of relationships between ventilation of operating room and incidence of SSI. Method: This paper focuses on the airborne transmission of bacteria under mixing ventilation in hospital operating room by using CFD simulation, which is carried out to calculate the bacteria trajectories using the large eddy simulation (LES) coupling with Lagrangian method. Under operating conditions, one patient, two surgeons and three nurses were simulated by relatively realistic three-dimensional thermal manikins. In the simulation, the bacteria are produced from the skin and nose breathing. Field measurements were conducted at St. Olavs hospital to validate the simulation. The influence of other parameters on the bacteria distribution in the wound zone are analyzed, including ventilation rate, supply air velocity and supply vent area. Results: The numerical simulation results of wound zone are analyzed by using velocity vector diagram, temperature contour diagram and bacteria distribution. The results indicate that the interaction of thermal plumes from patient and surgeon, the use of surgical light and the downward airflow may dominate the operating microenvironment and influence the bacteria distribution. Conclusions: The bacteria distribution in the wound zone may be influenced by both the surgical facility and the presence of medical staff. Simply increasing supply air velocity and supply airflow area may reduce bacteria deposition on wound surface, but its effect may be less evident than increasing clean air supply directly to the operating microenvironment.

Keywords: air, particles, ventilation

SU-PO-76
Development of predictive models for PM2.5 concentration in office

Wang W. [1], Jung C. [2], Su H. [1]; [1]Department of Environmental and Occupational Health, College of Medicine, National Cheng Kung University, Tainan, Taiwan, [2]Research Center of Environmental Trace Toxic Substances, National Cheng Kung University, Tainan, Taiwan

Background: Studies have shown that fine particulate matter (PM2.5) exposure increases both the mortality and morbidity of cardiovascular or respiratory diseases. People, meanwhile, often spend more than 85% of their time indoors, and for most adults, office setting becomes the predominant type of indoor environment where they reside. This study aimed to develop a predictive model for PM2.5 concentration in offices to better facilitate the needs for exposure assessment and epidemiological studies. Method: We collected indoor PM2.5 concentration from 39 offices from 2001 to 2010 in Taiwan and used a Geographic Information System Kriging model to estimate the outdoor concentration around these offices. We also acquired the building characteristics, indoor pollution sources (such as printer usage) by questionnaire, and outdoor pollution sources from land uses data of different buffer zone analysis through GIS. Finally, a stepwise regression model was used to establish the predictive model for PM2.5 concentrations in office. Results: Results showed that kriging maximum outdoor PM2.5 concentration and restaurant within 500 meters of radius can explain the most of 8-hour average PM2.5 concentration ($R2 = 0.66$). We also establish a predictive model for 24-hour average PM2.5 concentration and found that medium outdoor PM2.5 concentration, numbers of express delivery services industry within 3000 meters of radius, and the curtains with metal were major effect factors ($R2 = 0.88$). Conclusions: We suggest that it is necessary to collect outdoor PM2.5 concentrations, sources of outdoor pollution and indoor decoration, in order to establish accurate 24-hour average PM2.5 concentration in office. However, more information is warranted for increasing the predictive power for 8-hour average PM2.5 concentration in office, such as the frequency of occupants' activities indoors.

Keywords: air, behavior, built/indoor environment

SU-PO-77
Biophilic adaptation at the workplace: Multidimensional impact on human's behavior

Sun S.[1], Ordieres Meré J.[1], Yin J.[2], Zheng X.[1]; [1]Universidad Politécnica de Madrid, Madrid, Spain, [2]Harvard T.H. Chan School of Public Health, Boston, MA, USA

Background: Productivity is vital in the workplace. Numerous studies endorsed by scientific community affirm that biophilic design in workplace improves productivity. In previous conducted research studies, biophilic design was only evaluated with well control Indoor Environmental Quality (IEQ). The World Health Organization (WHO) reported that nine out of 10 people breathe polluted air. However, how biophilic design behaves in an inadequate IEQ workplace is still not well addressed. Objective: To gain additional knowledge about the impact of biophilic design on productivity (creativity) and cognitive performance in a working environment with limited quality. Methods: The dissatisfactory IEQ site was selected through on site data analytics, of which data were collected by a low-cost IoT based indoor air quality monitoring system. Recruited participants will be engaged in this study, with a standard body, non-smokers and non-sensitive groups. Selected plants are applied as biophilic design. The environmental parameters and human behaviours are measured by the developed indoor air quality monitoring system and various wearable devices. The productivity ability is measured by exercises designed by Cambridge University Brain Science lab. The simulation task could be typing (information processing work). The creativity is measured by Taxonomy of Creative Thinking Method. The simulation task could be mind map (knowledge creation work). Cognitive performance is assessed by visual reaction time task, stroop task and visual backward digit span task etc. Results: Multiple data sources collected from various wearable devices are analyzed to measure human reaction to biophilic design. The adopted methodology to assess biophilic design on site will be also defined. An IoT based indoor air monitoring system was developed for environmental assessment. Conclusions: Effects of the biophilic design for the people's effectiveness at work is discussed, by considering multidimensional data collection involving IoT technologies. Limitations for the ongoing research are presented and future research is foreseen.

Keywords: productivity

SU-PO-78
Ambient concentration of airborne microbes and endotoxin in rural households of southern India

Muthalagu A. [1], Eerappa R. [1], Qureshi A. [1]; [1]Hyderabad, Telangana, India

Investigation of microbial communities in the inbuilt environment is critically important as 90% of human life is indoor. Airborne microbial flora and their pyrogenic substances in the living spaces can impact human health, and only little is known of their identity and the mechanism of their assembly. Since 70% of Indian population lives in villages, this study provides baseline information about the primary biological aerosol particles (PBAPs) concentration, diversity and abundance, indoor-outdoor relationship and endotoxin concentration from the particulate matter in the indoor and outdoor region of thirty-six rural household settings. While most of the indoor air quality studies focused on the concentration of the bioaerosol in different household settings, this study provides the first-hand information about the bioaerosol concentration and diversity, endotoxin concentration along with the microbial risk assessment for pathogenic bacteria B. anthracis for the study region. Bacterial and fungal aerosols concentration in both the sites of all the thirty-six homes are ranged from IB-822 CFU m-3, IF– 650 CFU m-3, OB-937 CFU m-3, OF– 616 CFU m-3 with a size specific fungal and bacterial composition comprising of Agaricomycetes, Dothideomycetes, Euascomycetes, Eurotiomycetes, Leotiomycetes, Sordariomycetes, Zygomycetes for fungi and Actinobacteria, Alphaproteobacteria, Bacilli, Betaproteobacteria, Flavobacteria and Proteobacteria for bacteria. The I/O ratio for total bacteria and fungi was 1.38 and 0.790. The results suggest in addition to outdoor sources, indoor sources such as home ecology and metrological factors play a role in elevated bioaerosol concentration. Endotoxin concentration ranged from 2-21 EU m-3 and the population living in higher endotoxin sites have asthma and allergy related issues. Quantitative microbial risk assessment (QMRA) model was developed for the pathogen B. anthracis which was found in 8 homes among 36. The probability of infection by B. anthracis was found 11% in one sampling site where already people have been hospitalized due to fever.

Keywords: built/indoor environment

SU-PO-79
Experimental study on evaporation mechanism of microbial droplets

Liu X. [1], Liu L. [2], Wang Y. [1], Zhang X. [3]; [1]State Key Laboratory of Green Building in Western China, Xi'an, China, [2]School of Architecture, Tsinghua University, Beijing, China, [3]School of Environmental and Municipal Engineering, Xi'an University of Architecture and Technology, Xi'an, China

Infectious diseases such as influenza and SARS can be spread by droplets exhaled by humans. Evaporation of exhaled droplets is affected by its composition and environment, droplet temperature, pH, and evaporation also affect its activity of carrying microorganisms. The pathogens carried in the droplets affect its evaporation or not is not clear. In this study, different genus species bacteria were used to replace motile pathogens, and high-precision balances were used to measure mass changes during droplet evaporation. The effect of bacterial species on the evaporation rate of droplets was studied. The experimental results illustrate that after adding bacteria of different genus in 0.9% (w/v) NaCl droplets, bacterium can slow down the evaporation rate and prolong the evaporation time. Bacillus accelerates the evaporation rate and shortens the evaporation time. The evaporation law of Bacillus aeruginosa is similar to that of NaCl solution, and its evaporation rate is between the globular bacteria and Bacillus, but the evaporation rate of the bacterial liquid droplets is faster.

Keywords: infectious disease

SU-PO-80
Occupational Noise Burden of Disease

Chen Y.[1], Styliano K.[1], Jolliet O.[1], Neitzel R.L.[1]; [1]Environmental Health Sciences, School of Public Health, University of Michigan, Ann Arbor, MI, USA

According to NIOSH, approximately 22 million U.S. workers are exposed to hazardous occupational noise leading to irreversible hearing loss. Therefore, there is a need to quantify occupational noise impacts and noise-induced hearing loss (NIHL) on a national scale. This study aims to estimate health impacts induced by occupational noise in the US economy by occupation and sector. We developed a novel burden of disease framework for occupational noise, based on specific job occupations. NIHL burden rates (BRs) were estimated in 443 occupations accounting for 6 disease severities and 13 age groups, expressed in disability adjusted life years (μDALY/pers- yr). To provide comparable estimates and identify those with the highest impacts, we multiplied occupation and sector specific BRs by the number of persons employed to yield the annual occupational health impact due to noise, in DAYLs/yr. From the 126 million US workers, 51% were exposed to 80-85 dBA (minimum), 14% to 85-90 dBA (moderately high) and 1.1% to >90 dBA (high). From an occupational perspective, NIHL BRs range from 400 to 7200 μDALY/pers-yr. Accounting for employment, health impacts by occupation ranged from 3.6× 〖10〗 ^(-1) to 1.0 × 〖10〗 ^4 DALYs /yr. The total national burden in the US was estimated 1.7 × 〖10〗 ^4 DALYs/yr. The highest cumulative annual impact was experienced by "Building Cleaning Workers," a group exposed to a moderately high level, which had the 6th largest worker population. From a sector perspective, NIHL BRs range from 160 to 3500 μDALY/pers-yr. The highest impact was experienced by the "Agriculture, Forestry, Fishing and Hunting" sector, representing 13% of total cumulative impacts for only 0.3% of total employment. This study provides a reference for minimizing the NIHL burden by pinpointing areas to reduce the amount of noise in priority.

Keywords: acoustics/noise, health, other/general, lifecycle analysis, occupational, risk assessment

SU-PO-81
Online (continuous) measurement of indoor/outdoor reactive oxygen species (ROS) in a US home

Eftekhari A.[1], Morrison G.[1]; [1]University of North Carolina at Chapel Hill, Chapel Hill, NC, USA

Excessive reactive oxygen species (ROS) can disturb pro-oxidant/antioxidant balance and lead to oxidative stress causing different chronic obstructive pulmonary, cardiovascular, and metabolic diseases. ROS is thought to be associated with particulate matter toxicity in ambi-ent (outdoor) air. There are few measurements of ROS in indoor air, despite infiltration of outdoor ROS and generation indoors by oxidative chemistry. In this research, we designed, tested and field-deployed an online (continuous) instrument based on particle collection after condensational growth using low temperature water vapor. Continuous measurement of in-door and outdoor paricle phase (denuded) and gas+particle phase ROS (undenuded or "to-tal") was conducted in an unoccupied residential house in St. Louis, MO during August 2018. The system was operated under conditions of open and closed windows. The fluorescent in-tensity of ROS-2',7'- dichlorodihydrofluorescin diacetate (DCFH) in the presence of Horse-radish peroxidase (HRP) was measured using a fluorescence detector and calibrated by hy-drogen peroxide solutions. Results of laboratory tests showed a particle collection efficiency of 90±5% using ammonium sulfate seeds. During field sampling, we measured total phase indoor ROS and total phase outdoor ROS in the range of 0.9-4.1 and 1.5-5.1 nmol H2O2 equivalents m-3 air respectively. On separate measurement days, particle phase ROS of in-door and outdoor ranged 2.6-3.6 and 2.6-3.3 nmol H2O2 equivalents m-3 air; respectively. Both indoor and outdoor total peaked late in the afternoon. Indoor particulate phase ROS was 90-120% of outdoor ROS, and fairly insensitive to window opening or air exchange rate. This suggests most ROS resides in particles that are not readily removed by deposition in the building shell or on indoor surfaces.

Keywords: air, built/indoor environment, particles

SU-PO-82
PM2.5 exposures in different routes for Bike and Scooter Commuters in Taipei, Taiwan

Wu T.G.[1], Chang J.C.[2], Wu C.F.[2], Chan C.C.[3]; [1]Institute of Environmental Health, College of Public Health, National Taiwan University, Taipei, Taiwan, [2]Institute of Occupational Medicine and Industrial Hygiene, College of Public Health, National Taiwan University, Taipei, Taiwan, [3]Innovation and Policy Center for Population Health and Sustainable Environment, College of Public Health, National Taiwan University, Taipei, Taiwan

In urban cities, fine particle (PM2.5) is a major pollutant emitted from the vehicle engine combustion that causes adverse effects on human beings. To reduce vehicular emissions, promoting bike commuting is considered a healthy way to reach the goal. Furthermore, the survey from Taiwan EPA showed that there are about 36% of people who own the scooters in Taipei in 2017. In this study, we aimed to evaluate the PM2.5 exposure of the bike and scooter commuters in Taipei. Two types of mobile platform, bicycle and electric scooter based, were used to monitor the PM2.5 along three paired routes in Taipei, Taiwan. A total of 82 runs with approximate 68 riding hours were conducted in morning rush-hours and afternoon non-rush hours in different paired routes. The mean PM2.5 measurements for bike and scooter platforms ranged between 16.01 and 25.24 μg/m3 and 16.52 and 29.26 μg/m3, respectively. By applying the generalized linear models, it was found that riding in routes in small lane had lower PM2.5 exposures, as compared to riding along major roads. After controlling other effects in the models, the level of exposure reduced ranged between 1.36 to 1.48 μg/m3. Moreover, riding scooter could have higher PM2.5 concentrations than riding bike in the routes that the level of exposure increased ranged between 1.15 to 1.16 μg/m3. The study results suggest that commuters could choose the small lanes to avoid riding in the routes frequently with traffic congestions in the rush hours.

Keywords: air, exposure factors, particles, sampling methods

SU-PO-83
Indoor measurements and chemistry of chlorine bleach-related compounds

Mattila J.[1], Farmer D.[1], Lakey P.[2], Shiraiwa M.[2], Wang C.[3], Abbatt J.[3], Ampollini L.[4], DeCarlo P.[4], Zhou S.[5], Kahan T.[5]; [1]Colorado State University, Fort Collins, CO, USA, [2]UC Irvine, Irvine, CA, USA, [3]University of Toronto, Toronto, Canada, [4]Drexel University, Philadelphia, PA, USA, [5]Syracuse University, Syracuse, NY, USA

We measured indoor and outdoor gas-phase compounds with time-of-flight chemical ionization mass spectrometry (TOF-CIMS) during the House Observations of Microbial and Environmental Chemistry (HOMEChem) field campaign at the University of Texas Austin in the USA in June 2018. One set of experiments included sequential bleach mopping, while another set involved bleach mopping after large cooking events. Indoor gas-phase mixing ratios of hypochlorous acid (HOCl), chlorine gas (Cl2) and nitryl chloride (ClNO2) increase to part-per-billion by volume (ppbv) levels during these periods, substantially higher than outdoor concentrations. These compounds and their associated chemistry influences organic chemistry and radical oxidation in indoor air. The photolysis of these compounds can produce hydroxyl (OH) and chlorine atom (Cl) radicals indoors, thereby impacting the oxidative capacity of air in the built environment. We observe stark decreases in indoor isoprene, α-pinene, and limonene mixing ratios during periods of bleach mopping, consistent with this oxidation chemistry. Indoor multi-phase reactions of ammonia (NH3) and nitrous acid (HONO) with bleach-related compounds produce chloramines (NH2Cl, NHCl2, NCl3) and ClNO2, respectively. We observe elevated indoor concentrations of potentially toxic compounds from bleach mopping, including isocyanates (HNCO, CH3NCO, C2H5NCO; toxic and irritating to a variety of organs), nitrogen trichloride (NCl3; respiratory irritant), and chloroform (CHCl3; carcinogenic). Future directions include chemical modeling to better understand the oxidant budget and multi-phase chemistry during bleach cleaning.

Keywords: air, analytical methods, built/indoor environment, health, other/general

SU-PO-84

A community-based probabilistic risk assessment for the Diné people impacted by the Gold King Mine Spill

Ornelas Van Horne Y. [1], Chief K. [1], Canales R. [1], Begay M. [2], Beamer, P. [1]; [1]The University of Arizona, Tuscon, AZ; [2]Nation Department of Health, Window Rock, AZ, USA

On August 5th, 2015, the Gold King Mine released 3 million gallons of acid mine drainage-containing arsenic and lead into a tributary of the San Juan River, which flows through the Navajo Nation. The Diné (Navajo) rely on this River and use it for a variety of purposes lending to unique exposure pathways. We administered questionnaires in three Navajo communities in collaboration with the Navajo Nation Community Health Representatives to obtain frequency and duration data on 42 activities (e.g., praying with River water, fishing). These activities were utilized to conduct a community-based probabilistic risk assessment from exposure to As and Pb: 1) before the Spill, 2) during the peak of the Spill, and 3) one year after the Spill. Utilizing the Lifeline Community-Based Assessment Software we incorporated distributions for different exposure factors (e.g., activity duration, activity frequency, hand-to-mouth contacts) along with Diné specific activities (e.g., using the sediment as sunscreen) to estimate a distribution of dose. There is a significant difference in exposures (mg/kg/day), with the highest occurring during the peak Spill period (mean= 1.1 x 10-8, sd = 3.9 x 10-9) followed by post (mean= 9.4 x 10-9, sd = 3.9 x 10-9) and pre (mean= 6.9 x 10-9, sd = 2.5 x 10-9) Spill periods. While exposure from non-dietary activities is below the arsenic reference dose (RfD = 3 x 10-4 mg/kg/day), the exposures for the Diné community are higher when compared to those who would only recreationally interact with the San Juan River (mean= 1.6 x 10-13, sd = 9.6 x 10-4). This research provides the first documentation of unique exposure pathways of the Diné people and raises the need to understand and document unique exposure pathways of indigenous communities with subsistence and land-based livelihoods.

Keywords: risk assessment

SU-PO-85 WITHDRAWN

SU-PO-86

Public Health Risk attribution due particulate matter from Biomass Burning Emissions

Pinto Herrera L.C. [1], Hernandez Florez L.J. [1], Morales Betancourt R. [1]; [1]University of Los Andes, Bogota, Columbia

Background: Biomass burning emissions have a substantial impact on regional air quality and climate, we evaluated the short-term effect of particulate matter on acute respiratory disease morbidity and mortality, determinated by the risk of childs under 5 years old and adults over 60 years old in three n cities at 2016. Methods: Daily Emergency Department (ED) visit records for respiratory illnesses and selected diagnosis, besides daily concentrations of PM2.5 obtained in Bogota as a high altitude city and other cities as Bucaramanga and Medellin, therefore we selected the back-trajectories of air masses reaching the receptor, that were linked with fire spots. Colombia: Bogota, Medellin, Bucaramanga during 2016. Using conditional Poisson time series analysis we assessed the effect of air pollutants on health outcomes using single pollutant, controlling for meteorology and time trends. The percentages of change in the rate of ED visits and their 95% confidence interval were estimated for the joint effect of particulate matter. Results and Conclusions: On one hand in dry season, the amount of firespots increases as single-pollutant model increases particulate matter were associated with 15% increases in ED visits for respiratory diseases. On the other hand the during the wet season the incidence of ED visits decreases. The health charge of concentrations above 50 Micrograms per cubic meter of air represent the 17,6 % of Chronic Respiratory illneses. The cost linked with air pollution is US $1431, this amount increases when the poor communities are included.

Keywords: soil

SU-PO-87

Risk Assessment of Polycyclic Aromatic Hydrocarbons and their Chlorinated Derivatives Unintentionally Produced During Cooking via Exhaust Gas

Masuda M.[1], Wang Q.[1], Tokumura M.[1], Miyake Y.[1], Amagai T.[1]; [1]University of Shizuoka, Shizuoka, Japan

Cooking exhaust gas includes polycyclic aromatic hydrocarbons (PAHs) unintentionally generated during cooking foods, which is exposure to a cook and residents via inhalation. PAHs are carcinogens and/or mutagens. As a typical example, benzo[a]pyrene (BaP) is classified as Group1 by International Agency for Research on Cancer (IARC). On the other hand, chlorinated PAHs (ClPAHs), which are compounds whose one to several hydrogen atoms of PAHs are substituted to chlorine atom, some studies reported that some ClPAHs have more toxicities than their parent PAHs. Given that the generation mechanism of ClPAHs is similar with that of PAHs and cooking processes fulfill the conditions to generate ClPAHs (i.e., sufficient organic substances, heat, and chlorine source), ClPAHs can be generated during cooking processes such as grilled foods. Pacific sauries, which are a typical fish consumed in Japan, were targeted in this study and gas-grilled in a typical fish oven in an indoor kitchen. In this study, the PAH/ClPAH concentrations indoors from cooking exhaust were estimated using National Institute of Advanced Industrial Science and Technology – Indoor Consumer Exposure Assessment Tool (AIST-ICET). The risk assessment for the exposure to them was carried out. The exposure amounts of PAHs/ClPAHs from oral exposure and respiratory exposure were comparable. The results revealed that the inhalation of indoor air including cooking exhaust was the major pathway of PAHs/ClPAHs in addition to the consumption of cooked food.

Keywords: food

SU-PO-88

Secondhand smoke exposure in rural households in northwest China: a pilot investigation

Wang D.[1], Yao Y.[1], Ma H.[1], Li C.[1], Wang J.[1], Liu S.[2]; [1]School of Public Health, Lanzhou University, Lanzhou, Gansu Province, China, [2]School of Health Sciences, Purdue University, West Lafayette, IN

Objectives: The prevalence of exposure to secondhand smoke (SHS) in rural China is high. However, few studies have focused on measuring SHS exposure in rural households in China. This abstract reports SHS levels in rural village households, measured as a potential confounding factor in a study conducted to assess health effects of exposure to household air pollution from using solid fuel in rural northwest China. Methods: Thirty-three village women were enrolled in the study. The field sampling was conducted in September, 2017 and January, 2018. Questionnaires were administered to collect information on demographics, household smoking status and the awareness of SHS exposure. Passive samplers were used to measure nicotine air concentrations over a period of one month in these women's home. Urine samples were collected and analyzed for 8-hydroxy-2-deoxyguanosine (8-OHdG, as a marker of oxidative DNA damage) and 1-hydroxypyrene (1-OHP, as an indicator of exposure to household air pollution from solid fuel use). Results: Seventy-six percent (25/33) of women live with active smokers, with 52% living with one, 18% with two and 6% with three in their households. The median (average; range) households' nicotine concentration (as one-month average) was 0.126 μg/m3 (0.326; 0.019 - 1.88) in summer and 0.731 (1.25; 0.010 – 5.79) in winter. The standard deviation was 0.44 μg/m3 in summer and 1.61 in winter. Paired t test indicated that nicotine levels were significantly higher in winter than in summer ($p = 0.0016$). However, nicotine concentration was not associated with 8-OHdG measured in the village women in summer nor in winter, after controlling for 1-OHP. Conclusion: Household SHS exposure was high in the rural village in northwest China. Variations are also large among households and from season to season. SHS exposure was not the main contributor to the oxidative stress among the village women.

Keywords: secondhand smoke

SU-PO-89

The Safe Urban Harvests Study: Assessing exposure to six toxic metals among urban agriculture participants in Baltimore, MD

Lupolt S.[1], Santo R.[2], Kim B.[2], Green C.[3], Codling E.[3], Chen R.[1], Rule A.[1], Nachman K.[4]; [1]Department of Environmental Health & Engineering, Johns Hopkins Bloomberg School of Public Health, Baltimore, MD, USA, [2]Center for a Livable Future, Johns Hopkins Bloomberg School of Public Health, Baltimore, MD, USA, [3]Agricultural Research Service, US Department of Agriculture, Beltsville, MD, USA [4]Department of Environmental Health & Engineering, Center for a Livable Future, Risk Sciences and Public Policy Institute, Johns Hopkins Bloomberg School of Public Health, Baltimore, MD, USA

Background: Urban agriculture (UA) is increasing in popularity due to its numerous sociocultural, health, environmental, and economic benefits, but few groups have evaluated exposure to soil contaminants among urban agriculture participants (UAPs). The community-driven Safe Urban Harvests Study aims to characterize potential exposures to six toxic metals (arsenic, barium, cadmium, chromium, lead, and nickel) in soil, irrigation water, and urban grown produce among UAPs in Baltimore, MD. Methods: We recruited, surveyed, and collected samples of water and soil from 104 (70% of eligible) farms and community gardens growing food during the 2017 season. At 72 sites, we collected 1-7 of the thirteen most commonly grown produce items (e.g., kale, carrots, sweet potatoes, tomatoes, peppers, and cucumbers) and purchased those same items from grocery stores and farmers markets in Baltimore. We measured total metal concentrations in water using ICP-MS and in soil using ICP-Optima. Produce analysis is in progress at the time of abstract submission. Results: Five (<5%) water samples (n=114) exceeded US EPA's drinking water standard for lead (15ppb). We compared concentrations of all six metals in soil to several health-based guidelines. Thirty eight (6.2%) of soil samples (n= 616) exceeded the New York Soil Cleanup Objective, (NY SCO) for lead (400ppm) and 238 (45%) of samples exceeded the NY SCO for chromium (III) (36ppm). Measured produce concentrations will be compared to concentrations reported in US FDA's Total Diet Study. Discussion: These results are the first to evaluate metal contamination among UA sites in Baltimore to address safety perceptions and will inform specific recommendations for reducing exposure. Our long-term goal is to describe and quantify UA activities that may increase soil exposure, in order to evaluate the adequacy of existing guidelines for soil contaminants to protect UAPs who have frequent, direct contact with urban soil and consume urban-grown produce.

Keywords: community, exposure factors, food, metals, risk assessment, soil

SU-PO-90 WITHDRAWN

SU-PO-91

Transgenic pollen corn dispersion at Cordoba, Colombia

Mazo Castaño C.[1], Pinto Herrera L.C.[1]; [1]University of Los Andes, Bogota, Columbia

Transgenic crops were created in order to supply the high food demand as a result of population growth. Through biotechnology, different ways have been developed to satisfy this need, among them, the creation of transgenic seeds, capable of being resistant to pests and offering greater yield and productivity. Since its first sale in the 90s, several countries in the world have planted these seeds with the aim of increasing the potential of their crops; however, different authors and organizations have created campaigns against their commercialization, since, according to them, they have impacts on the environment, health and the social sector. Based on this concern, several researches have been made with the intention to figure out the effects of their use, such as possible genetic contamination by altering the native ecosystems, the effects on the use of pesticides, the tearing of cultural traditions and other types of affectations. This research work seeks to study one of its characteristics and its possible problem: the dispersion of pollen from a batch of transgenic corn crops in the municipality of Tierralta, Córdoba. Several models of pollen dispersion were carried out with the Hysplit program, changing the sedimentation rate and the density of the particle, in order to conclude the deposition speed and the viability of the transgenic corn pollen. The results showed that the greatest dispersion occurs in the first months of the year, causing the pollen particles to move away up to 23 km from the source. This study has great relevance, it serves as a basis for future research on genetic contamination of traditional crops.

Keywords: air, analytical methods, community, exposure factors, food, geospatial analysis/GIS, soil

SU-PO-92
Can we really promote air quality by essential oil diffusion?

Verriele M. [1]*, Angulo Milhem S.* [1]*, Nicolas M.* [2]*, Thevenet F.* [1]*;* [1]*IMT Lille Douai, Douai, France,* [2]*CSTB, Grenoble, France*

Essential oil blend diffusers create the buzz on "green lifestyle" or "diy" blogs! Inconsistent posts promote air cleansing and toxin-neutralizing by the diffusion of "purifying essential oil blends". Although their antibacterial and sedative properties can be undoubtedly acknowledged, the terpenoid compounds they contain are characterized by significant allergenic and irritant effects. These species are highly reactive with air oxidants and may even lead to fine particle formation which is worrying regarding indoor air quality and occupant exposure. Our study addresses the question of terpene emissions from the use of essential oil diffusers indoors. Experiments implementing several types of diffusers have been designed to assess pollutant concentration levels resulting from the vaporization of tee-tree oil in a 1-m3 experimental chamber. The emission dynamics and the duration of the impact on air quality have been recorded using a custom compact GC-FID analyzer. Contrasted behaviors have been evidenced. (i) Diffusing oil by an ultrasonic mist diffuser or by a soft oil warmer induces similar terpene concentration trends: compounds are efficiently and rapidly emitted leading to a concentration peak, before being purged by the chamber air renewal. The maximum concentration is reached between 2 and 5 hours depending on the compound volatility; however the concentration peak is dependent on the compound volatility and on its mass fraction in the pure oil. (ii) In contrast, a capillary diffuser leads to long-term emissions: after several hours, steady state concentrations are achieved and maintained for several days, depending on the volume of the flask. Moreover, analyzing the relative abundances between emitted terpenes, it was noticed that the footprint of the pure diffused oil is here more consistent using capillary diffusers. The final purpose of this kind of study is to propose effective recommendations on the use of such substances indoors.

Keywords: VOCs, activity patterns, consumer and personal care products

SU-PO-93 WITHDRAWN

SU-PO-94 - WITHDRAWN

SU-PO-95

Evaluation of an Improved Receptor Model for Analyzing Data with Spatial Variation: A Simulation Study

Lee C-L. [1], *Liao C-F.* [2], *Chen H-T.* [1], *Wu, Y-D.* [1]; [1]*Institute of Occupational Medicine and Industrial Hygiene, College of Public Health, National Taiwan University, Taipei, Taiwan,* [2]*Research Center for Environmental Changes, Academia Sinica, Taipei, Tawain*

The concentrations of air pollutants show horizontal and vertical variations. In previous receptor model studies, data collected from multiple sites would be analyzed individually or be pooled before analysis. However, it is still controversial which method is better because it would cause smaller sample size if data from different locations was analyzed separately. In contrast, analyzing pooled data may cause unrealistic results because profile variation in horizontal domain could be large. In addition, correlations of source contributions in vertical aspect were not considered. This study was conducted to evaluate the performance of a modified source apportionment model that is suitable for data with spatial variation. The evaluation was achieved by creating synthetic data sets that simulated PM2.5 and VOCs from five sources: petroleum refinery, vehicle exhaust, industrial coating, coal combustion, and natural gas. Daily PM2.5 and VOCs in 3 floors were simulated for a 45-day period. Positive matrix factorization (PMF) is commonly used to quantify the contribution of the sources at a receptor site. The Multilinear Engine (ME) program, a more flexible tool for solving PMF problems, was applied to deal with the vertical issue. It was set that time series of resolved factor contributions in different floors of a building should have similar trends. The average absolute error (AAE) is calculated to examine how profile variation affected the model performance. The correlation coefficients between the input and retrieved source contribution and profiles were above 0.98 when data from different floors of a building was analyzed. Sensitivity analyses results indicated that analyzing data pooled from 3 buildings was more fitted than analyzing them separately with one source having profile variation below 20% or with all source profiles varied below 10%. Therefore, it is recommended that the profile variation should be considered before analyzing data with spatial variations.

Keywords: VOCs, air, particles, statistical methods

SU-PO-96

Exposure assessment of urban greenness using satellite-derived normalized difference vegetation index (NDVI): the impact of spatial resolution on exposure misclassification

Jimenez R. [1]*, Lane K.* [1]*, Fabian, P.* [1]*;* [1]*Boston University School of Public Health, Boston, MA, USA*

Epidemiological studies analyzing the association between urban greenspace and health are increasingly using satellite-derived normalized difference vegetation index (NDVI) to assess greenness exposure. In this study we quantified the effect of using NDVI data at different spatial resolutions and aggregation areas to assess its potential for greenness exposure misclassification. We estimated NDVI from satellite data at different spatial resolutions across different levels of geographic aggregation (census block, block group, and tract) for the 2016 summer season in the metro Boston area. NDVI was calculated using remote sensing data from MODIS (250m x 250m), Landsat 8 OLI/TIRS (30m x 30m) and Sentinel 2 (10m x 10m). We calculated mean NDVI at each census geography class, and analyzed their distributions as continuous and categorical variables (quantile). We explored patterns and differences in continuous variables using descriptive statistics, t-tests and Pearson correlation analysis, and compared NDVI categories to identify discrepancies in exposure classification. The largest differences in NDVI range and distribution were observed at the block level, followed by block group and census tract. Mean estimates of NDVI from MODIS were consistently higher than estimates from Landsat 8 and Sentinel 2 across census geographies in the study area. NDVI quantiles varied across spatial resolution and census geography, yielding substantial differences in categorical exposure classification. The level of agreement between NDVI exposure categories for the three satellite products and census geographies ranged from 48% (quintiles, census blocks) to 89% (tertiles, census tracts). Greenness exposure estimates from satellite-derived NDVI are not invariant to spatial resolution of remote sensing data. Researchers studying the relationship between greenness and health should consider the appropriate NDVI spatial resolution in relation to the size of aggregation area in order to reduce the risk of exposure misclassification in epidemiologic studies.

Keywords: environmental health, geospatial analysis/GIS

SU-PO-97

Relationship between exposure to humidifier disinfectants and lung disease by analyzing the errors of exposure classification

Ryu H. [1]*, Park J., Heo J., Kim E., Choe Y., Choi Y.* [1]*, Yang W.* [1]*;* [1]*Catholic University of Daegu, Gyengsan, South Korea*

Lung disease caused by chemical substances (PHMG, PGH, CMIT/MIT) used in humidifier disinfectants (HDs) occurred in 2011, Korea. The Ministry of Environment of Korea has started to accept damage request for victims, and has investigated the causal relationship between HDs and lung injury using environment exposure assessment and medical records. For the HDs victims, environmental exposure assessment and clinical diagnosis were conducted to determine the levels of damage by HDs. However the results of environmental exposure and medical assessment are different from each other. It is needed to supplement for this method. The affecting factors of exposure levels were analyzed by characterizing exposure and demographics. Exposure concentration and cumulative time, exposure level were classified and compared by percentage of clinical diagnosis rating. The high exposure and low clinical diagnosis rating group, and low exposure and high clinical diagnosis rating group were analyzed to consider the limitation of past exposure assessment such as recall bias. The damages by HDs exposure were concentrated on these susceptible populations such as under 10 years, pregnant women with many deaths and showed low exposure and high clinical diagnosis rating. On the other hand, the groups classified by high exposure and low clinical diagnosis rating showed different characteristics such as never smokers, and a large number of elderly people. In cases of victim was child at the time of the injury or death, the environmental exposure assessment and causal relationship investigation between exposure and lung injury were getting harder over time by recall-bias. However, the dose-response relationship between classified exposure levels and clinical diagnosis rating showed positive correlation if the exposure assessment errors were analyzed and controlled. The relationship between dose(exposure level) and response (diagnosis rating) was significant (adjusted R2= 0.58)

Keywords: epidemiology, exposure factors, exposure models, respiratory health, risk assessment

SU-PO-98

Using continuous exposure in-vitro model to estimate percutaneous absorption of different characteristics chemical exposure

Zhu G. [1]*, Nien Y-T.* [1]*, Wang Y-H.* [2]*, Hsu J-F.* [3]*, Chou T-C.* [1]*;* [1]*Department of Public Health, China Medical University, Taichung, Taiwan,* [2]*Department of Nursing, Yuanpei University of Medical Technology, Hsinchu, Taiwan,* [3]*National Institute of Environmental Health Sciences, National Health Research Institute, Miaoli, Taiwan*

The skin is the main barrier to protect the body and continuous exposure of chemicals to the skin is extremely common in daily life or occupational settings. Continuous exposure refers to the exposure of chemicals in a sequential manner at different times. In the previous in-vitro studies, co-exposure of two or more chemicals were common exist, but few studies have been conducted on the assessment of percutaneous absorption kinetic parameters after continuous exposure in-vitro. The purpose of this study was to evaluate and to compare the percutaneous absorption of target exposed substances with different characteristics after the pre-exposed substance in the continuous exposure in-vitro model. In this study, the pre-exposed substance was linalool (hydrophobic), and the target exposed substances were salicylic acid (hydrophobic) and caffeine (hydrophilic). We used the pig's side abdomen skin and the dynamic diffusion cell was the testing materials. The pre-exposure time was 18 hours, and the sample collection time of the target exposed substance was 24 hours. All samples were analyzed by HPLC-UV. Cumulative mass and kinetic parameters are presented to compare the different characteristics exposed substances. Under 18 hours pre-exposure conditions, we found that the flux of salicylic acid was 25.0±13.6 μg/cm2/h, the permeability coefficient was 0.02 cm/h. Moreover, the flux of caffeine was 180.7 ± 63.8 μg/cm2/h, the permeability coefficient was 0.05 cm/h. It shows that the flux and permeability coefficient of caffeine greater than that of salicylic acid. After exposure to linalool, the absorption of caffeine is higher than salicylic acid. The results show that the pre-exposure to hydrophobic substances can cause the absorption of hydrophilic target substances to be higher than that of hydrophobic target substances on skin. Based on the above research results, the mechanism of percutaneous absorption status of this phenomenon should be further studied.

Keywords: cumulative exposure, exposure factors, exposure models, occupational

SU-PO-99

Contribution of Low-Cost Sensor Measurements to the Prediction of PM2.5 Levels: A Case Study in Imperial County, California, USA

Bi J.[1], Stowell J.[1], Seto E.[1], English P.[1], Freedman F.[1], Al-Hamdan M.[1], Liu Y.[1]; [1]Emory University, Atlanta, GA, USA, [2]University of Washington, Seattle, WA, USA, [3]California Department of Public Health, Sacramento, CA, USA, [4]San Jose State University, San Jose, CA, USA, [5]NASA Marshall Space Flight Center, Arsenal, AL, USA

Current regulatory monitoring is usually not sufficient to reflect PM2.5 details at the community level which is important to small-scale PM2.5 exposure assessment. The emergence of low-cost PM2.5 monitors provides a possible way to fill in the gaps as they can be densely deployed due to their lower cost, ease of use, and portability. As of now, very few studies focused on using the low-cost PM2.5 measurements to improve the statistical model-based PM2.5 prediction, which is a promising way to refine the PM2.5 exposure assessment. This case study evaluated the contribution of Identifying Violations Affecting Neighborhoods (IVAN) low-cost measurements to the PM2.5 estimation in Imperial County in California with sparse Environmental Protection Agency (EPA) Air Quality System (AQS) monitors. The PM2.5 prediction was conducted by the random forest algorithm at 1-km resolution on a daily scale based on three sets of PM2.5 measurements: 1) AQS measurements (AQS-only model), 2) IVAN measurements (IVAN-only model), and 3) AQS and IVAN combined measurements (AQS/IVAN model). The AQS-only model had a 10-fold cross-validation (CV) R2 of 0.53 and the addition of the IVAN measurements improved the CV R2 to 0.73. The IVAN measurements also contributed to more robust and reasonable PM2.5 pollution patterns reflecting elevated PM2.5 levels from the major sources of the region, showing that the dense deployment of low-cost monitors may be a cost-effective way to significantly improve the quality of PM2.5 predictions. On the other hand, the validation of the IVAN-only model by the AQS measurements showed a decreased CV R2 of 0.43 compared to the AQS-only model, indicating the importance and necessity of keeping high quality regulatory measurements. The random errors in calibrated IVAN low-cost measurements still caused apparent random prediction outliers, which highlighted the need of more effective calibrations to reduce the variances in the low-cost measurements.

Keywords: air, community, environmental health, exposure models

SU-PO-100

Estimation of ventilation rate in Korean residential house using IoT sensor, time-activity patterns and CO2 generation rate from breathing

Choe Y.[1], Ryu H.[1], Heo J.[1], Kim E.[1], Park J.[1], Cho M.[1], Kim Y.[1], Yang W.[1]; [1]Department of Occupational Health, Daegu Catholic University, Gyeongsan, Republic of Korea

The purpose of this study was to estimate the ventilation rate of residential house in Korea using indoor and outdoor concentrations of carbon dioxide (CO2) and CO2 generation rates using breathing rate and time-activity patterns of Korean adults. Indoor and outdoor CO2 concentrations, indoor CO2 generation rate by breathing, volume of the house were applied to mass balance model for residential indoor environments. Indoor CO2 generation rate was calculated using time-activity patterns, the average number of people per household from the Korean National Statistical Office and the amount of CO2 generation by breathing. The amount of CO2 generation rates by breathing were measured by activity loads for 194 Korean adults. The average volume of residential house was provided by the Korean National Statistical Office, indoor CO_2 concentrations were measured by IoT sensors, and outdoor CO_2 concentrations provided by the Korea Meteorological Administration. The average number of people were 2.53 per household and Koreans spend 61.0% of their day at home. The average volume of Korean residential house was 172.6 m^3. The CO_2 generation rate from breathing was 13.9±5.3 L/h during sleep and 15.1±5.7 L/h in a sedentary state. Indoor and outdoor CO_2 concentrations were 849 ppm and 407 ppm, respectively. The ventilation rate in Korean residential houses calculated by the mass balance model were 42.1 m^3/h and 0.71 air change per hour. The estimated ventilation rate tended to increase with an increase in the number of occupants. Since the IoT sensor devices were used to collect data, sustainable data could be collected to estimate the ventilation rate of Korean residential houses, which enables further studies such as on changes in the ventilation rate by season resulting from the activities of occupants. The results of this study could be used as a basis for exposure and risk assessment modeling.

Keywords: activity patterns, air, environmental health, exposure factors, ventilation

SU-PO-101
Evaluation of Porous Tube-type Passive Samplers for Volatile Organic Compounds according to the ISO 16107

Oishi J.[1], Nishio H., Wang Q., Tokumura M., Miyake Y., Amagai T., Fukushima Y., Suzuki Y., Enomoto T.; [1]University of Shizuoka, Japan

Volatile Organic Compounds (VOCs) are widely used as adhesives, solvents, and paints. Some VOCs cause adverse effects to human health depending on the exposure amount of VOCs. It is necessary to measure the concentrations of VOCs in the work environment according to Industrial Safety and Health Act in Japan. The active sampling method is commonly used to measure VOC concentrations in the work environment. However, the active sampling method requires a sampling pump, electrical power, and so on. Although the passive sampling method is useful for measuring VOCs, the accuracy of this method was often insufficient since sampling rates (SRs) of VOCs with passive samplers were affected by various environmental factors. In this study, we have developed a chamber to evaluate the performance of porous tube-type passive samplers under various environmental conditions according to the ISO16107 method. The chamber developed in this study was capable of controlling the temperature, humidity, wind speed, and VOC concentrations in the chamber by real-time monitoring. Our results showed that the sampling rates of eighteen VOCs targeted in this study were almost constant under the different conditions. In addition, we investigated the relationship between the environmental factors and physico-chemical properties of target chemicals. It was suggested that the validated passive sampling methods enabled us to measure the accurate concentrations of VOCs under the various conditions operated in this study.

Keywords: VOCs

SU-PO-102
Low-cost Sensor Array Devices as a Method for Reliable Assessment of Exposure to Traffic-related Air Pollution

Mykhaylova N.; University of Toronto, Toronto, Canada

The exposure to air pollutant mixtures is a well-known risk factor for inducing and increasing the severity of diseases. For real-time detection and monitoring of pollutant exposure, sensor arrays are an optimum choice because of versatility and aptitude for tracking composite multi-pollutant exposure. A thorough investigation of 30 different low-cost commercial gas and particulate matter sensors from 5 manufacturers has been conducted and 16 best-performing sensors were identified. A novel device for monitoring air quality has been developed and tested. Each device consists of an array of commercially available metal oxide semiconductor for monitoring NOx and O3, CO, CO2 and optical sensors for monitoring PM2.5. Level of pollutant exposure has been characterized over 3 different campaigns between 2013 and 2016. Analysis of a large range of gas sensors revealed several key challenges, including high intra-sensor variability, interference from temperature and nonlinearity. Air quality health index estimation from sensor readings was successfully demonstrated. Three aspects of device reproducibility were evaluated: drift over time, impact of interferences and impact of site-specific mixtures. Three categories of approaches for improving sensor accuracy and reproducibility were tested: nonlinear calibration models, variable transformations and training data selection. Model reproducibility, ability to adjust for multiple combinations of interferences and ability to resolve sites was improved when devices were calibrated at multiple sites. Analysis showed that both short-term and long-term temporal patterns could be resolved and compared at different sites. Background subtraction helped further emphasize the differences and rank sites in terms traffic-related pollution.

Keywords: VOCs, aggregate exposure, air, analytical methods, built/indoor environment, sensor technology, statistical methods, ventilation

SU-PO-103
Low-cost, high-resolution sensor networks for air quality monitoring in logisitically difficult environments: Ghana case study

Hodoli C.G.; Cranfield University, Cranfield, England

Pollution is a major cause of death in both low and middle income countries. In 2012 alone, exposure to polluted soil, water and air contributed to an estimated 8.4 million deaths in these countries.1 Low-cost feasible new approaches is required for air quality monitoring in these regions. Ghana, among others, is an emerging economic powerhouse in Sub-Saharan African (SSA) region featuring rapid industrialization and urbanization associated with a high demand for energy, food and water that has repercussions on local air quality. Appropriate air quality management plan is required to inform and guide environmental policy and pollution mitigations measures. This study seeks to demonstrate that low-cost miniaturised environmental sensing technologies can be used to develop cheaper and regionally specific methodologies for air quality monitoring in the SSA region by providing fast, detailed, accurate and near real-time data. In this work, a deployment in Ghana will be used as a regional case study. Currently, the use of multiple low-cost sensors within a network is being validated for the routine analysis of selected key atmospheric species, including NO, NO2, CO, O3, and particulates matter (PM1, PM2.5 and PM10). Other key environmental parameters, such as relative humidity and temperature, are also measured by the sensor nodes and will be used, along with auxiliary data, to establish the influence of meteorology on the observed air quality. The analysis of results and validation of these deployments will allow the implementation of appropriate air quality networks initially in Ghana, and later across the wider SSA region.

Keywords: air, children, climate change, ecological exposure, environmental health, environmental justice, environmental policy, environmental regulation, epidemiology, respiratory health, sensor technology

SU-PO-104
Evaluation on the Suitability of Different Ventilation Strategies for Operating Rooms

Fan M.[1]*, Cao G.*[2]*, Pedersen C.*[2]*, Stenstad L.*[2]*, Lv S.*[1]*, Skogås J.G.*[2]*;* [1]*Tianjin University, Tianjin, China,* [2]*Norwegian University of Science and Technology, Trondheim, Norway*

The performance of ventilation system in operating room is closely related to the quality of operations, since it not only controls the concentration of bacteria and particles around the wound area, but also dedicates to create an appropriate indoor environment for patients and doctors. However, high energy consumption and costs caused by the large ventilation rate are also problems that trouble the users and energy-saving department. This study aims to develop a comprehensive evaluation framework on the suitability of different ventilation strategies for existing operating rooms. Evaluation indices are selected from the aspects of ventilation effectiveness, energy consumption, lifecycle costs and users' satisfaction. Analytic hierarchy process (AHP) and fuzzy comprehensive evaluation (FCE) methods are adopted in this framework. To verify the applicability of this evaluation framework, operating rooms with laminar air flow (LAF) and mixing ventilation (MV) system in St. Olavs Hospital, Norway are used as case studies. Measurements including bacteria, particle concentration, air velocity, temperature as well as humidity etc. are conducted through mock surgeries. Also, questionnaires about thermal and acoustic comfort are distributed to surgical team staffs and patients. Meanwhile, it is expected that energy consumption benchmark of the ventilation system in the operating room in St.Olavs Hospital will be established through the analysis of actual working statistics. The significance of this study is that it may help to provide guidance for the adjustment of ventilation system in operating rooms and the selection of reasonable ventilation system for future designers.

Keywords: ventilation

SU-PO-105 WITHDRAWN

SU-PO-106
Field investigation of thermal comfort level of patients and surgical staff in operating rooms at St Olavs hospital

Kuivõgi H.[1]*, Cao G.*[2]*, Bogdan A.*[3]*, Thalfeldt M.*[1]*, Pedersen C.*[2]*, Zhang Y.*[4]*, Fan M.*[6]*, Stenstad L-I.*[5]*, Skogås, J.G.*[5]*;* [1]*Department of Civil Engineering and Architecture, Tallinn University of Technology, Tallinn, Estonia,* [2]*Department of Energy and Process Engineering, Norwegian University of Science and Technology, Trondheim, Norway,* [3]*Department of Air-Conditioning and Heating, Warsaw University of Technology, Warsaw, Poland,* [4]*Urban Construction and Environmental Engineering, Chongqing University, Chongqing, China,* [5]*St. Olavs Hospital Operating Room of the Future, Trondheim, Norway,* [6]*School of Environmental Science and Engineering, Tianjin University, Tianjin, China*

In healthcare facilities and hospital environments, it is essential to be able to predict thermal comfort as a parameter of indoor air quality. Unstable thermal conditions are critical in the operating room (OR) environment and may reduce the work efficiency and increase the possibility of surgical errors. At the same time, the main purpose of ventilation systems in operating rooms is to enable clean and comfort indoor environment. Likewise, the metabolic rates of occupants have a substantial impact to thermal comfort and it is completely different for each occupant in operating room. Same is for surgical clothing parameters (as vapor and thermal resistance) – clothing, made by qualified material, should protect wound from infection. Although, there is an increasing number of studies covering the area of thermal comfort, there have been few literature reports written on the comfort of healthcare staff and patients in hospitals and in the OR environment. Therefore, the purpose of this study is to estimate the thermal comfort level for every occupant and analyze the thermal environment in the operating room. The target is to base on PMV-PPD level and deduce the thermal sensation from environmental conditions (temperature, humidity and airflow), and from personal conditions (their metabolic rate and clothing insulation). Research methods includes the thermal environment measurements in hospital operating room after mock surgeries, a survey among surgical staff and patients, to get information about real sensation, and as well, the estimation of the activity level and clothing vapor and thermal resistance of every occupants in operating room. The expected result to be obtained, is the PMV-PPD theoretical and real level of four main occupant group in operating room. As a conclusion, the purpose is to provide suggestions and guidance of indoor environment control in hospitals improving thermal comfort level of patients and surgical staff.

Keywords: built/indoor environment, thermal environment, ventilation, workplace

SU-PO-107
Measurement of Local Clothing Properties for Human Thermal Comfort Evaluation

Nomoto A.[1], Takahashi Y.[1], Yoda S.[1], Ogata M.[1], Tanabe S.[1], Ito S.[2], Aono Y.[2], Yamamoto Y.[2], Mizutani K.[2]; [1]Waseda University, Tokyo, Japan, [2]Tokyo Polytechnic University, Kanagawa, Japan

To accurately simulate skin, core temperatures, and thermal comfort, certain human physiology and comfort models categorize the human body into multiple body parts. Most of these parts are normally clothed, which must be quantified in the simulation. Unfortunately, existing databases of clothing evaporative resistance only characterize the evaporative resistance for the whole body and not for individual body parts. In other words, each body part has the same level of clothing evaporative resistance. However, using a constant value as input to each body part of a multi-segmented human physiology and comfort model may be one of the causes of simulation error. Therefore, while evaluating the human thermal physiology and comfort in detail, it is desirable to evaluate it by measuring the local clothing evaporative resistance. In this research, we measured local clothing insulation and clothing evaporative resistance for 8 typical summer clothing ensembles in a climate chamber using a 20-segment sweating thermal manikin and presented the clothing insulation and evaporative resistance value for each body part as well as those for the whole body. Further, the moisture permeability index for each body part was calculated using the obtained two values. As a result, it was observed that the local clothing insulation, clothing evaporative resistance, and clothing moisture permeability index had comparative part differences for any clothing ensemble measured in this study because of the material of clothing and the state of overlaying. Hence, it was quantitatively shown that when we evaluate the human thermal physiology and comfort in detail for each part, it is necessary to consider the clothing thermal characteristics of each body part. The data of the local thermal characteristics of clothing obtained via this study is useful for the multi-segmented model of human physiology and comfort model.

Keywords: thermal environment

SU-PO-108
Overheating Risk Assessment: a Real-life Case of a Public Library in Auckland

Lesley M.; Victoria University Wellington, Wellington, New Zealand

Very few data and scientific studies on measurements of public library environments, especially on thermal conditions in relation to the risk of overheating are available. Also, the indoor environments in public libraries are rarely studied in comparison to other building types such as offices. This raises issues on how to further improve indoor thermal condition and productivity. This paper examines the thermal condition of occupants and the risk of overheating in an eco-friendly public library in Auckland region of New Zealand. The study utilises long term environmental monitoring of the building performance during the summertime and occupants' survey of staff who have occupied the buildings for more than two years. The results obtained with the overheating assessment demonstrate that the library has low risk of overheating. These results were aligned with the survey responses since all the responses coincide that the environment is slightly warm which is acceptable. Furthermore, results show that a considerable amount of hours would be below the lower temperature and humidity ratio limit although none of the occupants claimed the environment as slightly cold. The study stressed the need to assess the thermal comfort responses of people who are not adapted to the building thermal condition. In this regard, commonly used methods might not be the most appropriate tool since they are based on the adaptive assumption that occupants will adapt their behavior according to previous experiences. Therefore, in-depth interviews might shed light of how people who are not adapted to the building finds the thermal sensation of the building.

Keywords: thermal environment

SU-PO-109 WITHDRAWN

SU-PO-110
Investigating the influence of human activities on exposure to semi-volatile organic compounds through multimedia modeling

Kvasnicka J. [1], Diamond M. [1]; University of Toronto, Toronto, Ontario, Canada

Multimedia models have proven to be useful predictive tools for assessing the fate of neutral organic compounds released in indoor environments and associated human exposures. The characterization of human occupants in these models, however, has been limited to treating them as passive individuals with static and continuous behaviors. Using an indoor multimedia model, we investigated how typical human activities, such as household cleaning and use of furniture, influence exposure to semi-volatile organic compounds under static conditions. We then conducted a dynamic simulation by incorporating a time-activity schedule for the human occupant. The results of these simulations present a tentative picture of the differences in exposure estimation that may arise from different modeling assumptions pertaining to the influence of human occupants on the indoor system. Further research is underway, aimed at better accounting for temporal variation in exposure-relevant human behaviors through dynamic, multimedia modeling.

Keywords: SVOCs, activity patterns, behavior, built/indoor environment, exposure models

MO-PL-A1-EHR: Human Exposure to SVOCs

MO-PL-A1-01
Levels and sources of SVOCs in homes near a former PCB-producing facility in Slovakia

Demirtepe H. [1], *Melymuk L.* [1], *Diamond M.* [2], *Bajard L.* [1], *Vojta S.* [1], *Prokes R.* [1], *Sanka O.* [1], *Klanova J.* [1], *Murinova L.* [3], *Richterova D.* [3], *Rasplova V.* [3], *Trnovec T.* [3]; [1]*RECETOX, Masaryk University, Brno, Czech Republic,* [2]*Department of Earth Sciences, and Dalla Lana School of Public Health, University of Toronto, Toronto, Canada,* [3]*Department of Environmental Medicine, Slovak Medical University, Bratislava, Slovakia*

Semivolatile organic compounds (SVOCs) cover a wide range of compound classes of importance in the indoor environment, such as flame retardants (FRs), pesticides, and plasticizers, among many others. They are emitted from consumer products, building materials, and indoor and outdoor activities, and persist indoors due to low removal and degradation rates relative to emission rates. Considering the length of time we spend indoors, the indoor environment is a significant contributor to overall human exposure to many SVOCs. Environmental contamination with polychlorinated biphenyls (PCBs) was previously reported in a region near a former PCB production facility in Slovakia, with associated high human exposure. This study investigated the indoor levels and possible sources of seven SVOC groups in this region. We found that the indoor PCB levels were not correlated with the distance from the former PCB production facility. Rather, PCB levels were correlated with historic PCB use in the country, and the presence of primary sources of PCBs in the homes. Organochlorine pesticide levels were correlated to agricultural land area around the homes and building age. Indoor FR levels and distributions were consistent with typical levels reported for Europe. Polycyclic aromatic hydrocarbon levels were related to wood burning for home heating but not to smoking. Comparison with previously published indoor levels showed that Slovakian SVOC levels were similar to those measured in other countries, except for higher levels of hexachlorobenzene in air and dust and lower levels of a few FRs in air. A classification framework to prioritize SVOCs for evaluating risks from indoor exposures was proposed by considering human intake via air inhalation, dust ingestion and dermal contact and toxicity expressed as highest recommended intake per day. According to this classification, PCB-11, HCB, γ-HCH, PCB-28, PCB-52 and benzo(a)pyrene should be prioritized for indoor monitoring and evaluation of associated risks.

Keywords: SVOCs, air, chemical prioritization, flame retardants, pesticides

MO-PL-A1-02
Oral bioaccessibility of SVOCs in indoor settled dust: a simplified in vitro method for polybrominated diphenyl ethers (PBDEs) and pyrethroids

Raffy G. [1], *Mercier F.* [1], *Mandin C.* [1], *Le Bot B.* [1], [1]*EHESP/Irset-Inserm, Rennes,* [2]*CSTB, Marne-la-Vallée, Paris, France*

Dust ingestion is a pathway of human exposure to many indoor chemical pollutants including several semi-volatile organic compounds (SVOCs), such as polybrominated diphenyl ethers (PBDEs) and pyrethroid insecticides. Children are particularly concerned because their specific behavior (crawling on the floor, hand to mouth contact, object to mouth contact) may contribute to a higher ingestion of dust. To assess human exposure to SVOCs, their oral bioaccessibility, i.e. the fraction of the pollutant released from the dust matrix into the digestive fluid following dust ingestion, should be considered. Here, a simple method for the extraction of the bioaccessible fraction of SVOCs is proposed, consisting of 200 mg of dust incubated continuously for 20 hours at 37°C in 40 mL of neutral pH water containing bile and pancreatin. An insert made of 500 mg of sorptive material (XAD®-2) is added, acting as an adsorption sink trapping the SVOCs released into the gut solution. Method performance was assessed using SRM 2585, producing comparable results to those acquired with physiologically base extraction tests (PBET). The new method was then used to measure the bioaccessibility of PBDE flame retardants (BDE 47, 85, 99, and 100) and pyrethroids insecticides (cyfluthrin, cypermethrin, permethrin, and tetramethrin) in 30 settled dust samples collected in French schools. Permethrin and BDE congeners 47 and 99 were the most frequently detected compounds in dust samples and their median bioaccessibilities ranged from 37 to 48%. Our data show that doses considered for human exposure could be potentially lowered by a minimum of 50%, thus emphasizing the importance of considering oral bioaccessibility when assessing human exposure to organic pollutants via dust ingestion.

Keywords: SVOCs, analytical methods, built/indoor environment, flame retardants, pesticides

MO-PL-A1-03

Temporal trends of phthalate exposures in California's pregnant women: Comparison with NHANES data

Dhar U. [1], Bennett D. [2], Calafat A. [3], Shin H. [1]; University of Texas-Arlington, Arlington, TX, USA, [2]University of California-Davis, Davis, CA, USA, [3]Centers for Disease Control and Prevention, Atlanta, GA

Background: Phthalates are ubiquitous chemicals in personal care products and indoor residential materials. Because of potential health concern over exposure to some phthalates, product reformulation and legislation may result in temporal changes in phthalate exposures for the general population, but little is known about trends for pregnant women. Objective: We examined temporal trends of pregnant women's phthalate exposure using urine samples available in the MARBLES (Markers of Autism Risk in Babies – Learning Early Signs) study. Methods: We quantified 14 metabolites of eight phthalates in 770 urine samples collected from 2nd and 3rd trimesters of 188 pregnancies of MARBLES during 2007–2013. We used multivariate regression to estimate percent changes and least square geometric means (LSGMs) with adjustment for age, race/ethnicity, education, homeownership, and pre-pregnancy body mass index (BMI). We also compared geometric means (GMs) of our study with those of the general female population in the National Health and Nutrition Examination Survey (NHANES) over the same study period. Results: LSGMs of monoethyl phthalate (MEP), monobenzyl phthalate, mono(2-ethylhexyl) phthalate (MEHP), mono(2-ethyl-5-hydroxyhexyl) phthalate, mono(2-ethyl-5-oxohexyl) phthalate (MEOHP), and mono(2-ethyl-5-carboxypentyl) phthalate decreased over the study period ($p < 0.05$) [percent change (95% confidence interval (CI)): -8% (-12, -4); -6% (-9,-2); -23% (-26, -20); -21% (-48, 19); -20% (-23, -17); -21% (-23, -17), respectively]. In contrast, LSGMs of monohydroxy-isobutyl phthalate and mono-carboxyisooctyl phthalate increased ($p < 0.05$) [percent change (95% CI): 4% (1, 7); 13% (8, 19)] over the study period. During 2011–2012, compared to NHANES, GM concentrations of mono-isobutyl phthalate, MEHP and MEOHP were approximately 30% higher in our study while GM concentration of MEP was approximately 51% lower in our study. Conclusions: Pregnant women's exposure to phthalates has changed over the study period. Further studies assessing exposure to replacement phthalates are needed.

Keywords: SVOCs, biomarkers, biomonitoring, phthalates, prenatal

MO-PL-A1-04

Temporal trends and predictors of exposure to per- and polyfluoroalkyl substances among Northern California mothers with a young child, 2009-2016

Kim K. [1], Bennett D. [2], Calafat A. [3], Shin H. [1], [1]University of Texas-Arlington, Arlington, TX, USA, [2]University of California-Davis, Davis, CA, USA, [3]Centers for Disease Control and Prevention, Atlanta, GA, USA

Background: Human exposure to per- and polyfluoroalkyl substances (PFAS) has changed since the early 2000s due, in part, to the phase-out and replacement of long-chain PFAS. Despite widespread recognition that gestation and early postnatal life are critical time windows of exposure to PFAS, studies of PFAS exposure and its temporal changes have to date been limited mostly to the general population or pregnant women. Objective: We examined temporal trends and predictors of PFAS exposure among mothers with a young child using maternal serum samples available in CHARGE (CHildhood Autism Risk from Genetics and Environment), a population-based case-control study. Methods: We quantified nine PFAS in 448 maternal serum samples collected from 2009 to 2016 in Northern California when their child was 2 to 5 years old. We used multivariate regression to estimate least square geometric means (LSGMs) with adjustment for age, race/ethnicity, education, homeownership, and breastfeeding duration. We also used the estimated regression coefficient (?) to calculate percent changes over the sample collection period. Results: LSGM concentrations of perfluorooctanoate (PFOA), perfluorooctane sulfonate (PFOS), and perfluorohexane sulfonate (PFHxS) decreased over the study period [percent change (95% confidence interval (CI)): - 10.8% (-12.8, -8.8); - 10.2% (-12.4, -8.0); - 7.3% (-9.8, -4.7), respectively]. On the other hand, perfluorononanoate (PFNA) and perfluorodecanoate (PFDA) did not show a clear trend over the study period. Sampling year and breastfeeding duration were the two strongest predictors of PFAS serum concentrations (i.e., p-value for ? was less than 0.05). Conclusions: Our study further demonstrated that exposure to common long-chain PFAS among California mothers with a young child decreased over the study period and that breastfeeding appears to contribute to the elimination of PFAS in lactating mothers. Further studies comparing PFAS concentrations from mothers and children during multiple critical time windows of exposures are warranted to confirm our findings.

Keywords: SVOCs, biomarkers, biomonitoring, exposure factors, other (specify)

MO-PL-A1-05
Indoor and personal exposures to semi-volatile organic compounds in Central Appalachia

Salamova A. [1], Wang S. [1], Romanak K. [1], Hendryx M. [2], Venier M. [1]; [1]O'Neill School of Public and Environmental Affairs, Indiana University, Bloomington, IN, USA, [2]School of Public Health, Indiana University, Bloomington, IN, USA

Health disparities persist in rural Appalachian coal mining communities. Environmental pollution from activities of the mining industry can contribute to population health inequities. We used passive air samplers using polyurethane to investigate the levels of several groups of semi-volatile organic compounds (SVOCs), including brominated flame retardants (BFRs) and organophosphate esters (OPEs) in indoor air and silicone wristbands to examine personal exposure to these chemicals in 8 communities in Central Appalachia. Passive air samplers were deployed indoors for 30 days in 101 residential homes during March 2017 – July 2018. One resident in each of these homes wore a silicone wristband for 7 days. Samples were extracted and analyzed for BFRs, including 10 novel BFRs (nBFRs), 37 polybrominated diphenyl ethers (PBDEs), and 19 OPEs using gas chromatographic mass spectrometry (GC-MS). The median total concentrations for PBDEs, nBFRs, and OPEs were 211, 58.5, and 24,107 pg/m3 in indoor air and 50.5, 115, and 2,818 ng/g wristband in wristbands, respectively. The most abundant chemicals in both air and wristbands were BDE-47 and BDE-99 among PBDEs, 2-ethylhexyl 2,3,4,5-tetrabromobenzoate (EHTBB) and bis(2-ethylhexyl) tetrabromophthalate (BEHTBP) among nBFRs, and tris[(2R)-1-chloro-2-propyl] phosphate (TCIPP) and tris(2-butoxyethyl) phosphate (TBOEP) among OPEs.

Keywords: SVOCs, air, built/indoor environment, flame retardants

MO-PL-A1-06
Reactivity of semivolatile organic compounds (SVOCs) with hydroxyl radicals, nitrate radicals, and ozone in indoor air

Wei W. [1], Mandin C. [1], Ramalho O. [1]; [1]Scientific and Technical Center for Building (CSTB), Paris, France

Some reactive compounds in indoor air, such as hydroxyl radicals (OH), nitrate radicals (NO3), and ozone can either come from outdoor air due to indoor/outdoor air exchange or be formed indoors. These oxidants react with indoor volatile organic compounds (VOCs) and produce secondary organic aerosols and compounds. Recently, we have calculated the reaction rate constants of 72 SVOCs in indoor air and suggested that the reactivity of some SVOCs is competitive with the air ex-change rate and has a non-negligible influence on the removal of indoor SVOCs. The objectives of this paper are to study quantitatively the reactivity for more SVOCs and to identify the most reactive SVOCs for future measurements and models. The second-order reaction rate constants of SVOCs in indoor air under room temperature and ambient air pressure were searched in the literature. Both measured and predicted values were retrieved. Studies addressing reactions with a catalyst, artificial UV light, or in aqueous solution were then discarded. The pseudo-first-order reaction rate constant for the reaction of the retrieved SVOCs with the three oxidants were calculated, respectively, by multiplying the second-order rate constant with the concentration of the oxidants. Ninety-nine research articles were retrieved, which included studies on the second-order reaction rate constant for 171 SVOCs (135 SVOCs in the gas phase and 47 SVOCs in the airborne particles). The k1 values were > 1 h-1 for 19 polycyclic aromatic hydrocarbons (PAHs) and pesticides (4 SVOCs in the gas phase and 16 SVOCs in the airborne particles). The k1 values were < 0.09 h-1 for bis(2-ethylhexyl) phthalate, < 0.05 h-1 for polybrominated diphenyl ethers, < 0.60 for organophosphates, and < 0.07 h-1 for polychlorinated biphenyls. The study suggests that some SVOCs, particularly PAHs and pesticides, may be highly reactive in indoor air.

Keywords: SVOCs, air, built/indoor environment, flame retardants, phthalates

MO-SY-B1: Improving Indoor Air Quality by Controlling Sources of Pollutants: Progress, Opportunities and Challenges of the EU-LCI Concept

MO-SY-B1-07
EU-LCI Harmonization Framework for the Health-related Evaluation of Product Emissions

Scutaru A.; German Environment Agency, Berlin, Germany

Low emitting materials are a basic prerequisite in reduction strategies to mitigate the concentration of chemicals in the indoor environment. In Europe, the Construction Products Regulation sets harmonized conditions for the marketing of construction products and this includes the objective of protecting the building users' health. Currently only three European countries have implemented mandatory requirements to evaluate VOC emissions from construction products using different approaches and lists of substances. In addition, a number of voluntary, health related quality labels with long tradition offer help for the consumer to choose low emitting products. In ECA Report No 27, the major European labels agreed core criteria for a harmonized evaluation of chemical emissions from indoor products. The most important element of the core criteria was a harmonized list of assessed substances with so-called EU-LCI values for an appropriate number of substances encountered in emission tests. The procedure used by the EU-LCI Working Group to derive these European reference values is based on toxicological and risk assessment principles and is published in the ECA Report No 29. The European Commission is currently planning the issue of a delegated act on the communication of VOC emissions from construction products in the form of "VOC Classes". A key prerequisite for defining the VOC classes is the completion of the EU-LCI list. This presentation highlights important milestones of the harmonization framework for the health-based evaluation of chemical emissions to indoor air from construction products in Europe, describes the principles and protocol for the establishment of EU-LCI values and practical implications and implementation of the EU-LCI concept.

Keywords: VOCs, building materials, health, other/general, risk assessment, source control

MO-SY-B1-08
Emission Testing for Indoor Products

Wilke O.; Federal Institute for Materials Research and Testing (BAM), Berlin, Germany

The chemical emissions from products are tested by means of emission test chambers under defined conditions (climate, loading, air change rate). The standard method for the determination of volatile organic compounds (VOC) is the sampling onto Tenax-tubes followed by thermal desorption (TDS) and gas chromatography-mass spectrometry (GC-MS) analysis. The EU-LCI list includes some very volatile organic compounds (VVOC) and some VOC for which there are limitations when using the standard method. For VVOC additional sampling is required using stronger absorbers like Carbotrap or multi-bed adsorption tubes. The analysis of VVOC also requires a different GC oven program and a different column for the separation. For the determination of formaldehyde and other low boiling aldehydes (e.g. acetaldehyde, acetone, propanal, propenal) DNPH-cartridges are used which are extracted with acetonitrile followed by liquid chromatography (HPLC-UV) analysis. The derivatisation of propenal and other unsaturated aldehydes (e.g. 2-butenal) with DNPH might lead to lower findings due to incomplete derivatization and forming of by-products. For a better quantification of acetic acid the use of ion chromatography (IC) is recommended because the analysis of acetic acid with the standard method (TDS) leads to lower findings due to break through during sampling. The use of ion chromatography for the analysis of organic acids requires a third sampling technique. The acids are adsorbed onto silica-gel and extracted with water.

Keywords: building materials

MO-SY-B1-09
The development of exposure scenarios for testing consumer products within the EPHECT project; the EU-LCI concept for exposure scenarios

Stranger M.[1], Goelen E.[1], Bartzis J.[2], Wolkoff P., De Oliveira Fernandes E.; [1]Flemish Institue for Technological Research, Flanders, Belgium, [1]University of Western Macedonia, Kozani, Greece

The EPHECT project (Emissions, exposure patterns and health effects of consumer products in the EU; www.ephect.eu) was focussed on the assessment of exposures resulting from the household use of personal care products, air fresheners, and cleaning agents; products causing an exposure related to the use scenario, being potential sources of health relevant air pollutants in households. To establish emission test scenarios characterised by a maximum agreement with the existing standards for building material emission testing, as well as a maximum similarity between the protocols for the different product classes, the EPHECT umbrella for consumer product testing was developed. In this umbrella, the product test scenario was determined by (1) the product's aggregation state (liquid, gel, solid), and (2) its package, which is on its turn related to the use scenario (to be sprayed, to be volatilized, etc.).The EPHECT umbrella for consumer product testing was found to be a suitable tool to assess emission profiles and emission factors. Emission characteristics varied between products of the same product class and reached TVOC ideal room concentrations up to 960 $\mu g/m^3$ for continuous sources. Complex use scenarios were found to increase uncertainly of emission factors and chamber concentrations, yet more work needed to develop a standardized protocol for consumer product emission testing. EU-LCI (Lowers Concentration of Interest) values have been proven to be of significant value for building product emission evaluations, based on chamber measurements of single products under defined conditions. The use scenarios of consumer products however, in a complex indoor environment, characterised by a multitude of sources, physical parameters and occupancy patterns, complicate the exercise of mirroring the EU-LCI principle to consumer product emission evaluations. However, the application of the EU-LCI concept on the emissions assessed within the EPHECT project, illustrates its potential value for consumer product evaluation.

Keywords: VOCs, built/indoor environment, consumer and personal care products, environmental health, environmental regulation

MO-SY-C1: Holistic View on the Interaction Between Residents and Indoor Air Quality in Dwellings

MO-SY-C1-10
Indoor environments as complex systems and the need for metatheories

Stålne K. [1], *Pedersen E.* [2], *Borell J.* [2], *Li Y.* [1]; [1]*Malmo University, Malmo, Sweden,* [2]*Lund University, Lund, Sweden*

The following contribution aims at introducing a metatheoretical approach to bridging physical, technical and human factors of the indoor environment system. The study is grounded in research experiences within the transdisciplinary project PEIRE (People – Environment – Indoor – Renovation – Energy; www.peire.lth.se) in which we study tenants and their dwellings in multi-family housing built in 1960 – 1970. The study is performed before and after renovation in terms of measuring air quality, noise, temperature, moisture, lighting, investigating the technical system in terms of heating and ventilation system and the building envelope, and conducting interviews of the tenants on their experience, understanding and behaviour in relation to the indoor environment. Certain behaviours relevant for indoor environment are also technically measured. Typically, the different factors are studied from different disciplines that can be considered as incommensurable being based on different research paradigms and practices with different underlying assumptions, for instance in terms of ontology (e.g. differing views on causality and agency of the tenants), methods (e.g. qualitative versus quantitative) and the purpose of theories (in terms of describing reality, coherence with the basic sciences or applicability for the researchers or the tenants). This study is based on a metatheoretical approach, where a metatheory can be expressed as a generalization of different theories from different scientific disciplines based on the respective concrete measurement situations. The metatheoretical approach aims at firstly making such assumptions explicit, secondly identifying missing perspectives, e.g. sociological or economical, thirdly identifying common patterns between disciplines and theories, fourthly offer a systematic procedure of triangulating between different methods, e.g. by combining qualitative and quantitative data, and fifthly to offer conceptual frameworks for organizing the different scientific perspectives and the tenants' differing views of the indoor environment. Such metatheoretical frameworks will also be subject to systematic validation and empirical testing.

Keywords: VOCs, acoustics/noise, activity patterns, air, behavior, built/indoor environment, climate change, dampness/mold, health, other/general, light/lighting, occupational, sensor technology, sustainability, thermal environment, ventilation

MO-SY-C1-11
Measuring the occupant behaviour aspect of indoor air in apartments

Borell J. [1], *Alce G.* [1]; [1]*Lund University, Department of Design Sciences, Lund, Sweden*

People's behaviors affect the indoor air quality in their homes. For example, airing and cooking behaviors have a direct impact on air quality and air exchange. However, excessive airing beyond what is needed to achieve adequate air quality might waste heat energy. It is therefore valuable to identify the relations between crucial behavior patterns and resulting effects on indoor air quality, and therewith provide a knowledge basis for optimized design of buildings and their installations, of information and education to residents, and of behavior-guiding feedback to residents. In the PEIRE project, we have studied ten apartments before and after a renovation that aimed at reducing the energy use of the buildings and improve comfort. We have developed and used a Behavioral Measurement System (BMS), consisting of sensors that can detect relevant behaviors and wirelessly communicate with a central unit, which locally logs time-stamped data. The BMS captures behaviors affecting indoor air exchange rate and quality (airing, presence, and cooking). Beside behaviors, aerosols, carbon dioxide, temperature and relative humidity were technically measured. Through algorithms combining several types of sensor data, including presence, temperature and relative humidity, technical behavior logs were produced, showing when it was likely that no one was present in the apartment, and when cooking (of relevance for indoor air quality) was taking place. The study gave valuable insights in the large influence behavior has on indoor air quality, and identified crucial relationships. The study also showed that BMS is useful in research, especially when analyzed and discussed together with physical and technical characteristics of the indoor environment, such as the quality of ventilation and air movement patterns, and as a basis for behavior recommendations. Within PEIRE we will demonstrate how BMS data can be used as input to tailored behavior feedback, aimed at optimizing human control of airing.

Keywords: behavior, exposure models, sensor technology, ventilation

MO-SY-C1-12
Influence of the renovation on ultrafine particle concentrations in occupied Swedish residences

Omelekhina Y.[1], Pedersen E.[1], Nordquist B.[1], Wallenten P.[1], Nilsson P.T[1]., Wierzbicka A.[1]; [1]Lund University, Lund, Sweden

A number of health effects have been linked to exposure to airborne particles, however the consequences of exposure to ultrafine particles is largely unknown. Ultrafine particles can be emitted from different indoor activities or be of the outdoor origin. This is further complicated by the fact that we are spending majority of the time indoors. The aim of this work was to assess the influence of renovations on indoor-to-outdoor (I/O) ratios of ultrafine particles (UFP) in 6 occupied Swedish residences. Simultaneous indoor and outdoor measurements were performed in 6 residences for seven consecutive days in urban area in southern Sweden. The apartments were mechanically ventilated and had volumes between 115-218 m3. The renovation activities included adjusting air exchange rates (AER) to meet Swedish standards (0.5 AER h-1) and installation of the new kitchen extraction fans. NanoTracers (Oxility Aerasense) were used to measure particle number concentrations and mean particle diameters (10 - 300 nm). Based on the averages of particle number concentrations (PNC) during occupancy period, indoor-to-outdoor ratios (I/O) were calculated for each home. AER have been increased in the apartments during renovations to 0.4-0.7 h-1 in comparison to values before the renovation 0.2-0.4 h-1. After renovation I/O ratios of PNC during occupancy periods varied between 0.2-3.8 and were lower in comparison to the values before renovations 1.0-11.9. This could indicate increased efficiency of particles removal after renovation but also can be influenced by lower amount of performed activities by occupants, which were 2-3 times lower after in comparison to before. Results have shown the positive changes of the renovation processes on AER and particle removal. Nevertheless, the frequency and type of indoor activities performed by residents at home influence the indoor concentrations. The Swedish Research Council FORMAS and the Swedish Energy Agency financed the project.

Keywords: activity patterns, air cleaning, built/indoor environment, nanoparticles, particles, ventilation

MO-SY-C1-13
Measurements of how users affect air flows between rooms in renovated apartments with exhaust ventilation.

Wallentén P.[1], Nordquist B.[1], Harderup L-H.[1]; Lund University, Lund, Sweden

When investigating the ventilation efficiency after a renovation measure, it is not enough just to focus on the total exhaust air flow from an apartment. The efficiency of a ventilation system is mainly dependent on the amount of outdoor air supply and how this air is distributed. Fourteen residential buildings were renovated 2018 in the south of sweden and detailed measurements of indoor air quality, perceived thermal comfort, ventilation etc were performed before and after the renovation. The exhaust ventilation system was affected by the renovation by adjustment and by modifications of the inlet vents to reduce excessive window opening. The ventilation in the presented study was measured through exhaust vents measurements and tracer gas measurements. The tracer gas measurements were performed by releasing NO2 in a room in an apartment and tracking how the concentration changed in the other rooms. By using a reverse engineering non-linear regression technique, the air flows between the rooms could be calculated. The main results from the study were that: the chosen method was difficult since the tracer gas was not completely mixed in the rooms but the air flows could still be estimated, the air flows were much dependent on how the window vents were positioned and that the effect of mixing between rooms was much bigger than the actual flow of outdoor air.

Keywords: air, analytical methods, built/indoor environment, health, symptoms, sampling methods, ventilation

MO-SY-C1-14
Does a building renovation improve the indoor thermal comfort? - A thermal environment evaluation before and after renovation

Lundgren-Kownacki K. [1], Halder A. [1], Petersson J. [1], Kuklane K. [1], Wierzbicka A. [1], Pedersen E. [2], Gao C. [1]; [1]Department of Design Sciences, Lund University, Lund, Sweden, [2]Department of Architecture and Built Environment, Lund University, Lund, Sweden

A sustainable renovation results in both a good indoor environment and high-energy efficiency. However, contemporary renovations often focus on energy and environmental performance, leaving out other aspects, such as the thermal comfort. The aim of the ongoing study is to compare the results of an extensive thermal environment evaluation before and after major renovation of ten typical 1970's rental apartments in multi-family buildings located in Southern Sweden. The data collected is comprehensive and includes measurements of air temperature, relative humidity (RH), air velocity, plane radiant and globe temperature, draught rate, turbulence intensity, operative temperature, PMV/PPD indices and thermal sensation (thermal comfort evaluation) using a LumaSense INNOVA 1221 Thermal Comfort data logger. MSR Temp/RH data logger sensors were also placed at four different heights. The outside weather data and individual factors such as clothing, activity, gender, age were also collected. Measurements were taken in the living room of each apartment for 2 hours during three winter seasons: one measurement session before and two after renovation resulting in 30 measurements in total. The preliminary results from the first two winter seasons for draught rate, PMV/PPD, RH and radiant temperature all showed slight improvements after renovation. Further, the study results show that the individual perceived thermal comfort does not always agree with the measured and calculated thermal comfort. The data is currently under analysis and final results will be presented.

Keywords: built/indoor environment, green buildings, sustainability, thermal environment

MO-SY-C1-15
Measured and perceived thermal comfort in apartments, before and after renovation

Petersson P. [1], Wallentén P. [2], Nordquist B. [1], Harderup L-E. [2]; [1]Building Services, Lund University, Lund, Sweden, [2]Building Physics, Lund University, Lund, Sweden

When energy saving measures are implemented during a renovation it is important that the indoor environment is not impaired. A renovation of 50 year old apartment buildings have been performed in the southern part of Sweden. Several energy saving measures have been implemented e.g. reducing the heat transfer trough the facades, changing thermostat valves on the hydronic heating system and changing the out door air supply devices in the façade. Some of the measures can be expected to also improve the thermal climate inside the apartments. Temperature measurements of the operative temperature during one week and also temperature measurements during a longer time period were performed before and after the renovation. The residents' perception were studied by questionnaires before and after. The results show that the thermal conditions during the heating season were improved after the renovation. The temperatures were distributed in a more narrow interval after the renovation compared to before. The temperature levels increased somewhat after renovation. The physical measurements supported the perceived thermal comfort reported by the residents in many cases. The study also illuminated the fact that physical measurements can not always be used as a prediction of the preferences of a resident on the individual level. Generally the thermal conditions were improved after the renovation.

Keywords: built/indoor environment, thermal environment

MO-PL-D1-EAM: Sensors and Tools

MO-PL-D1-16
Accuracy of sensor-based measurement and modeling of PM2.5 and black carbon in assessing exposure to indoor aerosols

Cox J. [1]*, Cho S.* [2]*, Ryan P.*[3]*, Isiugo K.* [1]*, Ross J.* [4]*, Chillrud S.* [4]*, Zhu Z*[1]*., Jandarov R.* [1]*, Grinshpun S.* [1]*, Reponen T.* [1]*;* [1]*University of Cincinnati, Cincinnati, OH, USA,* [2]*RTI International, Research Triangle Park, NC, USA,* [3]*Cincinnati Children's Hospital Medical Center, Cincinnati, OH, USA,* [4]*Columbia University, Palisades, NY, USA*

Accurate, cost-effective methods are needed for rapid assessment of traffic-related air pollution (TRAP). Typically, real-time data of particulate matter (PM) from portable sensors have been adjusted using data from reference methods such as gravimetric measurement to improve accuracy. The objective of this study was to develop linear regression models for the real-time measurements of the RTI's Micro Personal Exposure Monitor (MicroPEM™) and microAeth® black carbon (AE51) sensor to generate accurate real-time data for PM2.5 (PM2.5RT) and black carbon (BCRT). The linear regression models were developed by using an all subset regression test and Akaike information criterion for model selection. The two sensors and an SKC PM2.5 Personal Modular impactor were collocated in 44 indoor sampling events for 2 days. The reference filter-based analyses included particle mass (SKC PM2.5 and MicroPEM™ PM2.5) and black carbon (SKC BC); these analyses are more accurate than real-time sensors but are also cumbersome and costly. The mean values for MicroPEM™ PM2.5, SKC PM2.5, PM2.5RT, SKC BC and BCRT were as follows, respectively: 12.67, 12.93, 16.28, 0.49 and 0.68 µg/m3. The data from both MicroPEM™ PM2.5 ($r=0.62$, $p<0.001$) and SKC PM2.5 ($r=0.77$, $p<0.001$) correlated significantly with PM2.5RT. The final model MicroPEM™ PM2.5 and SKC PM2.5 improved the correlations ($r=0.77$ and 0.83, $p<0.001$, respectively). The correlation between SKC BC and BCRT was significant (0.77, $p<0.001$) and output improved with the final model for BC ($r=0.84$, $p<0.001$). The NIOSH criterion that the 95% confidence interval of the sensor readings or model outputs be within ±25% of the filter-based data was nearly met by the MicroPEM™ PM2.5 model with an error range of -26% to 9%. No other models were close to meeting this criterion. The results from this study will improve the accuracy of real-time exposure monitors and can provide a rapid cost-effective alternative for estimated PM2.5.

Keywords: built/indoor environment, exposure models, sampling methods, sensor technology, statistical methods

MO-PL-D1-17
ForHealth platform: a researcher's tool to make smartphone-based data collection simple, safe, and cost effective

Cedeno Laurent J.G. [1]*, Allen J.* [1]*, MacNaughton P.* [1]*, Jones E.* [1]*;* [1]*Havard T. H. Chan School of Public Health, Boston, MA, USA*

Smartphones have an 85% penetration in the U.S. market. This represents a unique opportunity to utilize their data acquisition capacity for health research. While the two most adopted device types (i.e., Android and Apple) offer app-development research platforms, development and maintenance costs have precluded the mainstream adoption of customized apps in health research. We present the ForHealth platform, a solution to reduce the developing time and costs of research-oriented apps with advanced exposure assessment capabilities. Its modularized architecture allows the immediate deployment of a growing library of validated survey tools, and the integration of peripheral data-collection devices. An instance template includes the basic mobile app with hardcoded features and a webApp managing data transmission, data storage, and connection to available API modules (e.g., electronic survey software, environmental sensors, wearable devices) or integration of newly developed modules. Available modules include libraries with cognitive tests and validated surveys from a wide spectrum of health studies. All data transmission and storage is HIPAA compliant. Apps based on our platform are currently being used in studies of indoor environmental quality in offices, benefits of exposure to nature, and environmental assessments of clinical trial sites. These studies have used existing modules such as multi-language survey administration, ecological momentary assessment based on real-time environmental data, and participant compensation, and have developed custom modules now available to new platform users. Study administrators report reductions in the time and effort to manage the study panels, thanks to the provided participant management and data exporting dashboards. The ForHealth platform offers a cost-effective, secure, and easy-to-use solution to use smartphone-apps for research. Platform users will benefit from the expanding number of modules and the efficiencies of centralized maintenance to a single app ecosystem.

Keywords: behavior, built/indoor environment, health, other/general, sensor technology, thermal environment, ventilation

MO-PL-D1-18

Ultra-compact personal exposure kit for high performance exposure assessment of PM and gas pollutants (ozone and nitrogen dioxide)

Zhi N. [1]*, Dane W.* [1]*, Qing Z.* [1]*, Kai Z.* [1]*;* [1]*The Hong Kong University of Science and Technology, Hong Kong, China*

Exposure to air pollution is a critical health risk, especially for the susceptible population in polluted regions. There have been extensive investigations on the particulate matter (PM) exposure in urban areas including fine particles, and black carbon etc due to the readily available sensors and monitors for PM measurements. However, epidemiologic studies have shown criteria gas pollutants, such as ozone, nitrogen dioxide etc, have significant association with the human adverse health effects. Unfortunately, only very few studies are reported in literature on exposure assessment of these important gas pollutants due to their relatively low concentration in ppb levels posing challenges of the detection limit for sensor-based monitors. The fast-changing microenvironment during daily activities also affect the sensor performance with unstable temperature and humidity. This study presents the investigation of a novel personal exposure kit that can measure PM size distributions and ppb level NO_2/O_3 gases for air pollution exposure, together with CO_2 for indoor ventilation indicator, and GPS, temperature, relative humidity, light, noise and motion for environment and activities logging. The multi-function sensor is integrated in compact package with size of an iPhone and weighed less than 600g. The NO_2 and O_3 gas sensors were designed with virtual baseline tracking that is self adapted to the change of temperature and humidity conditions to allow stable measurements even at ppb level. The performance was evaluated in different microenvironments with reference monitors for quality control. The results showed very high potential for these sensors to be used as a personal exposure assessment tool to tackle the important knowledge gap in exposure sciences on the criteria gas pollutants impact o the human health effect.

Keywords: built/indoor environment, community, exposure factors, sensor technology

MO-PL-D1-19

Novel approaches to analyzing personal exposure to air pollution: Employing wearing compliance to estimate the effect of community-level air pollution and the effect of cooking interventions on exposure.

Gould C.F., Owusu-Agyei S., Chillrud S., Ae-Ngibise K.A., Burkart K., Boama- Kaali E., Kinney P.L., Mujtaba M., Quinn A., Manu G., Jack D.W., Asante K.P.

The Ghana Randomized Air Pollution and Health Study randomized 1,414 households with pregnant mothers into a control arm (N = 515), a fan-assisted biomass-burning cookstove arm (N = 519), or an arm with liquefied petroleum gas (LPG) stoves and a supply of free LPG (N = 355). Exposure in baseline and six post-intervention 72-hour sessions was primarily assessed using personal CO monitors; two maternal sessions also included personal PM2.5 monitors with accelerometer measurements for wearing compliance. Data were truncated at 48 hours for comparability in the number of cooking events captured across samples. Daytime wearing compliance (percent of hours worn between 6 am and 10 pm) vs mean 48-hour personal exposure plots inspired an assessment of the relative change in exposure in the intervention groups as compared to the control group in the post-intervention period. The predicted y-intercept at 0% compliance provides estimates of the average contribution of ambient/community air pollution to each study arm. We estimated the difference between study arm exposure estimates from 0% to maximum compliance, interpreted as the contribution of capturing cooking events and other point sources to personal exposure. The three study arms had similar average community-levels of air pollution at 0% compliance for both CO (0.65 ppm) and PM2.5 (45 μg/m3). We estimated the contribution of cooking to exposure as 52 ± 12 μg/m3 in the control arm and 8 ± 33 μg/m3 in the LPG arm (85% less). The improved biomass stove had negligible impact. These slope-intercept inspired results agree well with mixed effect model approaches that assess the effect of the intervention on overall personal exposure. Although low wearing compliance creates error and bias, a continuous measure of wearing compliance provided the ability to derive the average of community derived exposures and the effect of the intervention on exposure.

Keywords: built/indoor environment, exposure models, particles, sampling methods, sensor technology

MO-PL-D1-20

Contribution of Low-Cost Sensor Measurements to the Prediction of PM2.5 Levels: A Case Study in Imperial County, California, USA

Bi J. [1], Stowell J. [1], Seto E. [1], English P. [1], Freedman F. [1], Al-Hamdan M. [1], Liu Y. [1]; [1]Emory University, Atlanta, GA, USA, [2]University of Washington, Seattle, WA, USA, [3]California Department of Public Health, Sacramento, CA, USA, [4]San Jose State University, San Jose, CA, USA, [5]NASA Marshall Space Flight Center, Arsenal, AL, USA

Current regulatory monitoring is usually not sufficient to reflect PM2.5 details at the community level which is important to small-scale PM2.5 exposure assessment. The emergence of low-cost PM2.5 monitors provides a possible way to fill in the gaps as they can be densely deployed due to their lower cost, ease of use, and portability. As of now, very few studies focused on using the low-cost PM2.5 measurements to improve the statistical model-based PM2.5 prediction, which is a promising way to refine the PM2.5 exposure assessment. This case study evaluated the contribution of Identifying Violations Affecting Neighborhoods (IVAN) low-cost measurements to the PM2.5 estimation in Imperial County in California with sparse Environmental Protection Agency (EPA) Air Quality System (AQS) monitors. The PM2.5 prediction was conducted by the random forest algorithm at 1-km resolution on a daily scale based on three sets of PM2.5 measurements: 1) AQS measurements (AQS-only model), 2) IVAN measurements (IVAN-only model), and 3) AQS and IVAN combined measurements (AQS/IVAN model). The AQS-only model had a 10-fold cross-validation (CV) R2 of 0.53 and the addition of the IVAN measurements improved the CV R2 to 0.73. The IVAN measurements also contributed to more robust and reasonable PM2.5 pollution patterns reflecting elevated PM2.5 levels from the major sources of the region, showing that the dense deployment of low-cost monitors may be a cost-effective way to significantly improve the quality of PM2.5 predictions. On the other hand, the validation of the IVAN-only model by the AQS measurements showed a decreased CV R2 of 0.43 compared to the AQS-only model, indicating the importance and necessity of keeping high quality regulatory measurements. The random errors in calibrated IVAN low-cost measurements still caused apparent random prediction outliers, which highlighted the need of more effective calibrations to reduce the variances in the low-cost measurements.

Keywords: air, community, environmental health, exposure models

MO-PL-D1-21

Low-cost Sensor Array Devices as a Method for Reliable Assessment of Exposure to Traffic-related Air Pollution

Mykhaylova N.; University of Toronto, Toronto, Canada

The exposure to air pollutant mixtures is a well-known risk factor for inducing and increasing the severity of diseases. For real-time detection and monitoring of pollutant exposure, sensor arrays are an optimum choice because of versatility and aptitude for tracking composite multi-pollutant exposure. A thorough investigation of 30 different low-cost commercial gas and particulate matter sensors from 5 manufacturers has been conducted and 16 best-performing sensors were identified. A novel device for monitoring air quality has been developed and tested. Each device consists of an array of commercially available metal oxide semiconductor for monitoring NOx and O3, CO, CO2 and optical sensors for monitoring PM2.5. Level of pollutant exposure has been characterized over 3 different campaigns between 2013 and 2016. Analysis of a large range of gas sensors revealed several key challenges, including high intra-sensor variability, interference from temperature and nonlinearity. Air quality health index estimation from sensor readings was successfully demonstrated. Three aspects of device reproducibility were evaluated: drift over time, impact of interferences and impact of site-specific mixtures. Three categories of approaches for improving sensor accuracy and reproducibility were tested: nonlinear calibration models, variable transformations and training data selection. Model reproducibility, ability to adjust for multiple combinations of interferences and ability to resolve sites was improved when devices were calibrated at multiple sites. Analysis showed that both short-term and long-term temporal patterns could be resolved and compared at different sites. Background subtraction helped further emphasize the differences and rank sites in terms traffic-related pollution.

Keywords: VOCs, aggregate exposure, air, analytical methods, built/indoor environment, sensor technology, statistical methods, ventilation

MO-PL-E1-SEM: Monitoring of VOCs Indoors

MO-PL-E1-22

Developing passive flux sampler based on a commercially-available cartridge for on-site measurement of formaldehyde emissions

Doyun W.[1], Nong G.[1], So S.[1]; [1]National Research Council of Canada, Ottawa, Canada

Passive flux samplers (PFS) are intended to measure formaldehyde emission rates from existing surfaces on-site and passively. A majority of PFS reported in the literature is composed of a Petri-dish lined with quartz fiber filter and coated with 2,4-dinitrophenylhydrazin (DNPH). Its relatively large surface area for sampling allows for a short-term sampling, typically, for 6 to 8 hours. However, there is no information in the literature whether it can be applicable to a longer-term sampling period (e.g., 4 – 7 days), which is commonly used in characterizing levels of formaldehyde under occupancy in residential buildings. Additionally, the need for coating filters with DNPH in house is too labor-intensive for the field application involving a large number of homes. A PFS that is suitable for 4- to 7-day deployment in the field and that uses a commercially available DNPH cartridge was developed in this study. The performance of the PFS was characterized in the laboratory setting in comparison with the conventional 50-L chamber method (ASTM D5116). The formaldehyde emission rates from PFS were in good agreement with those from the conventional chamber method, as shown with a linear relationship between two data sets (Slope= 0.9634 and R2 = 0.96) obtained from the concurrent chamber test with 8 representative composite wood materials. Low method detection limit, good precision and good repeatability supported the suitability of PFS for measuring emission rates on-site and investigating major sources of formaldehyde in the field. The performance of PFS was further validated in a newly-built research house, where PFS was deployed onto 6 interior surfaces. The predicted concentration based on the measured emission factors with PFS showed good agreement with the measured formaldehyde concentration with a maximum difference of 24%.

Keywords: VOCs, source control

MO-PL-E1-23
Indoor air pollutants determinants in modern office buildings - the OFFICAIR project

Campagnolo D.[1], Spinazzè A.[1], Cattaneo A.[1], Urso P.[2,3], Sakellaris I.[4], Saraga D.[1], Mandin C.[5], Canha N.[6], Mabilia R.[7], Perreca E.[7], Mihucz V.[8], Szigeti T.[9], Ventura G.[10], de Oliveira Fernandes E.[10], de Kluizenaar Y.[11], Cornelissen E.[11], Hänninen O.[12], Carrer P.[2], Wolkoff P., Cavallo, D.[1], Bartzis J.[4]; [1]Department of Science and High Technology, University of Insubria, Como, Italy, [2]Department of Biomedical and Clinical Sciences-Hospital "L. Sacco", University of Milan, Milano, Italy, [3]Radiotherapy Department, Clinica Luganese Moncucco, Lugano, Switzerland, The Netherlands, [4]Department of Mechanical Engineering, University of Western Macedonia, Kozani, Greece, [5]Scientific and Technical Centre for Building, University Paris Est, Marne-la-Vallée, France, [6]Centro de Ciências e Tecnologias Nucleares, Instituto Superior Técnico, Universidade de Lisboa, Bobadela, Portugal, [7]Department of Biology, Agriculture and Food Science of National Research Council, Roma, Italy, [8]Cooperative Research Centre for Environmental Sciences, Eötvös Loránd University, Budapest, Hungary, [9]National Public Health Center, Public Health Directorate, Budapest, Hungary, [10]Institute of Science and Innovation in Mechanical Engineering and Industrial Management, Porto, Portugal, [11]The Netherlands Organization for Applied Scientific Research (TNO), The Hague, The Netherlands, [12]Department of Health Protection, National Institute for Health and Welfare, Kuopio, Finland, National Research Centre for the Working Environment, [13]Copenhagen, Denmark

Objective. The aim of this study was to define the determinants of aldehyde and volatile organic compound (VOC) indoor concentrations in a sample of more than 140 modern office rooms, in the framework of the European Union OFFICAIR research project. Methods. A large field campaign was performed, which included (i) the air sampling of aldehydes and VOCs in 37 modern office buildings across 8 European countries in summer and winter, and (ii) the collection of information on offices' characteristics via checklists. Linear mixed models for repeated measurements were applied to identify the main factors affecting the measured concentrations of selected indoor air pollutants. Results. Several correlations between aldehyde and VOC concentrations and structural or occupants' activity-related factors were defined. The aldehydes and VOCs determinants in modern office buildings include building and furnishing materials (floor covering, ceiling, furnishings), indoor climate characteristics (room temperature and relative humidity), the use of consumer products (e.g., cleaning and personal care products, office equipment), as well as the presence of outdoor sources in the proximity of the buildings (i.e. vehicular traffic). Results also showed that determinants of indoor air concentrations varied considerably among different type of pollutants. Conclusions. This study provided an up-to-date picture of the main environmental and behavioural factors affecting indoor concentrations of aldehydes and VOCs in European modern offices, and a global insight on the most common determinants of IAQ in modern office environments. This insight could be useful for the development of guidelines and recommendations for improving IAQ in office buildings - and thus reducing exposures of occupants - acting on the building design, construction and management, operation and cleaning services, and by modification of the individual activity patterns and to develop public health policies.

Keywords: VOCs, building materials, built/indoor environment, consumer and personal care products, occupational

MO-PL-E1-24
Concentration of Volatile Organic Compounds in Small Auto Repair Shops in a Latino Community

Beamer P. [1], Wagoner R. [1], Lopez-Galvez N. [1], Lothrop N. [1], Sandoval F. [2], Parra K. [1], Wertheim B. [1], Lee A. , Griffin S. [1], Cortez I. [2], Bell M. [1], Wolf A.M. [2]; [1]Mel and Enid Zuckerman College of Public Health, University of Arizona, Tuscon, AZ, USA, [2]Sonora Environmental Research Institute, Inc., Tuscon, AZ

Much research has focused on reducing volatile organic compounds (VOCs) exposures to auto painters. Yet, only a few exposure assessments have been conducted in auto repair shops, even though mechanics have increased rates of cancer and heart disease. Often in low-income minority areas, auto body work is conducted in auto repair shops, and current occupational health interventions and guidelines are limited due to a complex interaction of physical, social, and economic factors. Our objective was to characterize workers' exposure to VOCs in auto repair shops in the a predominantly Latino area of southern Arizona. Businesses were recruited via community health workers. A photo-ionization detector, calibrated with isobutylene, measured real-time VOC concentration in 20 second intervals over 40 work shifts in 10 auto repair shops. Summa canisters were used to measure 70 specific VOCs in workplace air. In half of the businesses studied, Summa canister measurements were taken on two separate days. Workers' activities, ventilation conditions, and other business characteristics were recorded. Chemical inventories were compiled at each business. Mixed models were used to estimate VOC exposure and variance components between and within shops and workers. Thirty-one different VOCs were detected in workplace air across all 10 shops with a median of 11 chemicals detected per sampling event (range: 2-22). Distinct mixtures of chemicals were identified in the same shop on different days and between shops. Mixed models showed that the highest VOC exposures occurred during painting/body/collision repair, followed by cleaning parts. During a gas tank repair, exposures reached over 7500 ppm. Variability was greatest within worker, followed by between-worker, -day, and –shop, respectively. Auto repair workers have highly-variable VOC exposures related to their activities. Interventions are needed to reduce these workers' exposures.

Keywords: VOCs

MO-PL-E1-25
Monitoring of VOCs emissions from selected products in 3 different field settings

Paralovo S. [1], Stranger M. [2], Spruyt M. [2], Lauwers J. [2], Lazarov B. [2], Laverge J. [1]; [1]Department of Architecture and Urban Planning, Ghent University; Environmental Risk and Health Unit, VITO, Ghent, Belgium, [2]Environmental Risk and Health Unit, VITO, Mol, Belgium

Three experiments were designed to test products emissions in field conditions, by monitoring the volatile organic compounds (VOCs) room concentration both before and after the placement of emitting products (sources). These experiments also aimed at observing the influence of temperature and relative humidity over the products VOCs emissions under realistic field conditions. The first experiment (Dormitory experiment) took place at a unoccupied dormitory suite for 20 days, the second one at an occupied house for 9 days (House experiment) and the third one at a unoccupied office space for 8 days (Office experiment). The products used as VOCs sources in the Dormitory experiment were OSB plates and floor coating, whereas in the House and Office experiments OSB plates and an air freshener were used. In the House experiment, two sets of sources were used: one with dry and another with wet OSB plates. In all three experiments, VOCs concentrations were monitored simultaneously by a Sift MS monitor and passive sampling. Temperature, relative humidity and CO2 concentration were also monitored in multiple points of the assessed spaces. Results provided by the Sift MS were in general low in the House and Office experiments, but considerably higher in the Dormitory experiment, which may be linked to some equipment issues faced during that experiment. Nevertheless, the VOCs found by the Sift MS consistently in highest concentrations in the three experiments were acetone and hexanoic acid (along with α-pinene in the House experiment). Results provided by passive sampling showed high concentrations of dodecane in the three experiments, of decane in the Dormitory and House experiments and of acetone in the Office experiment. Hexanaldehyde and undecane were also found in higher concentrations in the Dormitory experiment. No clear connection between room conditions and VOCs concentrations was observed.

Keywords: VOCs, air, built/indoor environment, consumer and personal care products

MO-PL-E1-26
Indoor air VOC concentrations in a new log building – a follow-up study

Vaittinen O.[1], Lapinlampi T.[1], [1]Inspector Sec Ltd, Vantaa, Finland

Log construction has become in fashion in Finland during the last years. Currently, one fifth of the new homes in Finland are built of log. One reason for the growing popularity of log construction is the alleged improvement in indoor air quality. High-rise buildings and public buildings are also built increasingly from logs. Pudasjärvi Log Campus is a school building where the main building material has been pine lamella log. Known emissions of coniferous trees include e.g. terpenes and aldehydes. Terpenes have been claimed to have health-promoting effects. On the other hand, aldehyde-related symptoms have been reported even in normal indoor air concentrations. We measured indoor air concentrations of VOCs, formaldehyde and ammonia in Pudasjärvi Log Campus in a two-year follow-up study. The sampling was carried out in four stages: 1) upon completion of the building before furnishing; 2) after furnishing before the regular use of the building; 3) after the building became occupied by users; 4) long-term follow-up two years after completion. Samples were collected from across the building in five different locations. Measurements were performed with standard ventilation settings at a constant indoor air temperature. The obtained results showed that the concentrations of terpenes, especially pinene, were clearly elevated. Pinene indoor air concentration varied between 32 and 739 µg/m3 depending on the room and time point of the sampling. Concentrations of carene, limonene and hexanal were also elevated. The concentrations of VOCs can largely be explained by the proportion of log structures. The highest VOC concentrations were measured in summer 2016 just after construction and they clearly decreased later on. The concentrations were yet again elevated during the follow-up period in summer 2018. The formaldehyde and ammonia concentrations were normal in all measurements.

Keywords: VOCs

MO-PL-E1-27
Source apportionment of ambient volatile organic compounds (VOCs) using hourly-resolved measurements

Zhiyuan L., The Chinese University of Hong Kong, Hong Kong

Detailed information on source contributions to ambient volatile organic compounds (VOCs), precursors of ground-level ozone formation, is critical to understand tropospheric ozone pollution. We utilized the positive matrix factorization (PMF) model to conduct source apportionment of hourly concentrations of thirty VOC species, CO, NO2, and NOx measured at Hong Kong Environmental Protection Department (HKPED) Mong Kok (MK), Yuen Long (YL), and Tung Chung (TC) air quality monitoring stations from January 2010 to December 2014. The preliminary PMF analysis of the hourly data set (PMF1h) shows that six sources, including combustion (with contribution at 25%), paint solvents (4%), vehicle exhaust (14%), liquefied petroleum gas (LPG) (38%), biogenic emission (7%), and gasoline evaporation (12%), were identified. To characterize the benefit gained from time resolution, the PMF analyses of 4-hr and 6-hr averaged data of the hourly-resolved data set (PMF4h and PMF6h) will be presented. The results of this study can provide insights in spatiotemporal variations in contributions of various emission sources to ambient VOCs, providing critical information of source targets for more effective VOCs controls.

Keywords: VOCs, air, statistical methods

MO-PL-A2-EHR: Consumer Products

MO-PL-A2-28

Particle and organic vapor emissions and exposure potential from children's toys based on 3-D printing technology

Stefaniak A. [1]*, Yi J.* [2]*, Duling M.G.* [1]*, Bowers L.N.* [1]*, Knepp A.K.* [1]*, Lebouf R.F.* [1]*, Nurkiewicz, T.R.* [2]*, Ranpara A.* [1]*, Luxton, T.* [3]*, Martin S.B., Burns, D. A.* [1]*, Peloquin, D.M.* [4]*, Baumann E.* [5]*, Virji A.M.* [1]*,* [1]*National Institute for Occupational Safety and Health, Morgantown, WV, USA* [2]*West Virginia University School of Medicine, Morgantown, WV, USA,* [3]*U.S. Environmental Protection Agency, Cincinnati, OH, USA,* [4]*Oak Ridge Institute for Science and Education, Oak Ridge, TN, USA,* [5]*Pegasus Technical Services, Cincinnati, OH, USA*

Three-dimensional (3-D) printing has expanded beyond the workplace to include toy printers and pens for use by children to build objects. Given the vulnerability of children's lungs and other organs during growth, it is important to understand if exposure potential exists from use of these toys. Emissions from two brands of 3-D pens and one brand of 3-D printer were evaluated in a 12.85 m3 chamber while extruding different colors of polylactic acid (PLA) or acrylonitrile butadiene styrene (ABS) filaments. Emissions characterization included particle number concentration, size, and elemental composition and concentrations of individual and total volatile organic compounds (TVOC). Exposure potential and particle lung deposition were modeled using emissions data. Filament feed rate (slow/fast) and color (natural ABS; blue and yellow PLA) influenced emissions ($p<0.05$) from the 3-D pens. PLA filament color (blue, yellow, orange) had a minor influence on TVOC emissions ($p=0.055$) from the 3-D printer (constant feed rate). For the brand of 3-D pen that could extrude ABS (natural color) yields were 109 to 1010 particles/g printed. For either pen or the printer extruding PLA (blue, yellow, orange) yields were 106 to 1010 particles/g printed. Geometric mean particle sizes were 30 to 300 nm depending on the toy and filament. Most particles contained iron and some contained nickel. Dose modeling indicated that 8-year old kids have higher alveolar particle deposition than 3- or 14-year olds. Detectable vapors included styrene (possible asthmagen) while extruding natural ABS and methyl methacrylate (dermal sensitizer and lung irritant), while extruding yellow PLA. Exposure modeling indicated that TVOC concentration from use of a single toy would be 1 – 31 µg/m3 (school classroom) and 3 – 150 µg/m3 (residential living room). Toy 3-D pens and printers emitted particles and chemicals of health interest. If appropriate, control technologies could be implemented to reduce emissions.

Keywords: VOCs, children, metals, nanoparticles, respiratory health

MO-PL-A2-29

Does natural and eco-certified household products promote indoor air quality? From liquid compositions to test chamber emissions.

Angulo Milhem S.[1], Verriele M.[1], Nicolas M.[2], Thevenet F.[1]; [1]IMT - Lille Douai, Douai, France, [2]CSTB, Grenoble, France

Currently, major efforts have been made to address indoor air pollution due to current awareness of health hazards related to the exposure of human to chemicals. Consequently, the popularity of "green" cleaning products has upsurged. These products are formulated with essential-oils relying on their anti-bacterial properties to supposedly improve indoor air quality. Indeed, essential oils might contain a hundred of odorous molecules, mainly terpenes and terpenoids (TerVOCs) which can act as antibacterial agents. This work aims at evaluating the emissions of ecologically-certified and naturally-fragranced cleaning products at various scales. First, the correlation between the liquid compositions of 7 natural cleaning products with their TerVOC emissions is investigated. Among products in the liquid form, a total of 28 terpenes have been quantified. Then, volatile fractions have been evaluated using micro-chamber test. Interestingly, only 22 species are detected in gas samples with volatilization yields ranging from 9.1 % to 99.8 %. Results do not evidenced any direct correlation between liquid concentrations and volatilized concentrations of terpenes. Indeed, chemical affinities between terpenes and the solvent matrix in product formulation can affect their transfer to the gas phase. Second, 3 products, characterized by different cleaning purposes, have been selected to evaluate their emissions in a 1-m3 experimental chamber. The product application patterns are based on a real scenario regarding the applied quantity, the loading factor and the dispersion mechanism. Emitted terpenes and carbonyl compounds are monitored by off-line and on-line chromatographs. Results evidenced that the maximum concentrations of total terpene in the gas phase ranged from 162 to 308 µg/m3. Noticeably, a specific emission dynamic is evidenced for formaldehyde. The continuous increase of HCHO concentration is suggested to be induced by the presence of formaldehyde releasers. This study contributes to the understanding of housecleaning product emissions and the subsequent exposure hazards of users.

Keywords: consumer and personal care products

MO-PL-A2-30

Exposure to Volatile Organic Compounds (VOCs) in Feminine Care Products

Lin N.[1], Ding N.[1], Park, S.K.[1], Batterman, S.[1]; [1]School of Public Health, University of Michigan, Ann Arbor, MI, USA

Feminine care products are used on highly permeable and sensitive vaginal and vulvar tissues. Few regulations require disclosure of ingredients in these products. This study identifies VOCs and quantifies exposures and health risks resulting from using these products. We tested 20 different tampons and 12 feminine washes using purge and trap sampling, thermal desorption, and gas chromatography/mass spectroscopy analysis. Exposure dose and risks were modeled using typical or worst assumptions. We detected 51 and 38 VOCs in tampons and washes, respectively, and the mean total VOC concentrations were 0.3 µg/g and 39 µg/mL, respectively. Chloroform was found in 14 tampons (<6×10-3 µg/g) and 3 washes (<0.14 µg/mL); 1,4-dichlorobenzene was found in 11 tampons (<2×10-3 µg/g); 1,4-dioxane was found in 9 washes (<25 µg/mL); and p-isopropyltoluene was found in 19 tampons (<0.02 µg/g) and 9 washes (<3.1 µg/mL). Other VOCs detected included aromatics, alkanes, aldehydes, halohydrocarbons, ketones, esters, and terpenes. VOC composition tended to depend on the manufacturers. Scented tampons contained more p-isopropyltoluene and limonene. Washes for "sensitive skin" had few differences, except for higher levels of 1,4-dioxane. "Organic" washes had higher VOC concentrations, especially of aldehydes and terpenes, than other washes (97 vs. 27 µg/mL, p=0.03). For washes, the estimated total VOC dose was mainly due to inhalation (2.9 �g/m3), the hazard ratio was 2.3×10-2, and the life time cancer risk was 1.9×10-6 (highest among products tested). For tampons, the primary exposure pathway was dermal exposure via vaginal mucosa. The VOC dose was 7×10-5 mg⍰kg-1⍰day-1, the hazard ratio was 6.8×10-5, and the lifetime cancer risk was 2.8×10-9. All of these products contained some hazardous VOCs, and risks from the use of some products was not negligible. At present, consumers should not assume that labels regarding chemical composition in these products are meaningful. The ingredients in these products should be disclosed.

Keywords: VOCs, consumer and personal care products, environmental health, risk assessment

MO-PL-A2-31 - WITHDRAWN

MO-SY-B2: Cleaning for Healthy Environmental Surfaces

MO-SY-B2-32
Drool, green mucus, and other human-borne gnarly stuff on high-contact surfaces: Any relationship to health?

Ijaz M.K. [1,2]*, Bruning E.* [1]*, Rubino J.* [1]*;* [1]*Research & Development, RB, Montvale, NJ, USA,* [2]*Department of Biology, Medgar Evers College of The City University of New York (CUNY), Brooklyn, NY, USA*

High-touch environmental surfaces (HITES) become visibly contaminated with body fluids released by humans particularly post-acquisition of infectious agents (vomit / diarrheal stool, respiratory secretions, blood, urine etc.). These human-borne gnarly stuff (body fluids), apart from being an excellent source of organic load providing safe shelter to these Germs/infectious agents for survival, also provide a food and/or water source for these pathogens to flourish and protect them against environmental distress to survive on the contaminated HITES. In fact, Germs have been recovered from naturally contaminated HITES. Based on studies carried out in the past few decades, it's well established that the Germs remaining on such HITES, survive, thrive and further spread via hands-surfaces-hands-surfaces and surface-air-surface-air nexuses. Therefore, simple cleaning (not hygienic cleaning) while it removes these gnarly stuff from contaminated HITES may not remove the Germs from the contaminated HITES rather inadvertently spread them in the process to other HITES. To mitigate the risk, prevention-studies (in vitro/in vivo including intervention field studies), focused on using hygiene promotion (education + use of consumer hygiene products), highlights that prudent use of targeted hygiene promotes effective public health intervention in targeted communities studied. It's evident that cleaning / targeted hygiene-intervention may reduce the risk of acquiring infectious agents from contaminated HITES (where the occupants may otherwise potentially acquire these pathogens), and as a consequence, targeted hygiene, can lead to healthy outcomes under both community and institutional settings. The health implications of these studies including the need for identifying / implementing and monitoring the effectiveness of hygienic-cleaning critical to ensure risk mitigation / sustain positive health-outcome will be discussed.

Keywords: infectious disease

MO-SY-B2-33

What does 'Clean' mean; how do we affordably measure what's invisible, and how does the metric actually reflect a concerted custodial cleaning effort in schools

Shaughnessy R., Hernandez M., Haverinen-Shaughnessy U., Nieto M.

Inadequate cleaning of fomites in school settings leads to accumulation of non-visible ecological niches, which can promote biological contaminants including viruses. Significant correlations have been noted in recent studies as to Adenosine triphosphate (ATP) readings being indicative of surface associated bacterial, fungal and human cell levels on school desktops. A pilot project is in the midst of being carried out in a large western school district in the USA. Fundamental questions being addressed in the pilot are: 1) can a concerted custodial effort be sustained in a district to provide reasonable desktop cleaning on a routine basis, at a minimum of two times per week?, and 2) will such an effort reveal a significant reduction in desktop contamination based on the enhanced cleaning efforts? The pilot phase focuses on 4 schools (two study, and two control) pertaining to cleaning practices in classrooms. The main aim of the pilot study has been to develop and test a reasonable, practical cleaning protocol that will result in effective cleaning assessed based on microbial contamination on high touch surfaces throughout the school year. The protocol includes training of custodians, specific cleaning methods and materials used, as well as the frequency of cleaning for desktops. The pilot study is expected to inform researchers from the University of Tulsa, Colorado University, and Yale University on the feasibility and impact of the recommended cleaning protocol that may be engaged for a full implementation of an in-depth intervention across the district. Preliminary results, midway through the pilot, indicate that a dedicated effort by the custodian can effectively reduce ATP levels on the desktops. Because of the depth and magnitude of the robust data set being compiled, this endeavor can expand upon prior findings, and deepen the understanding of environmental stressors in schools and their impact on student health.

Keywords: microbial agents

MO-SY-B2-34

Viruses on high contact surfaces... Are they present for any considerable time, can we capture them, and should we be concerned?

Shaughnessy R., Hernandez M., Haverinen-Shaughnessy U., Peccia, J.;

Formidable knowledge gaps remain with regard to the abundance and persistence of human viruses indoors. As the United State National Research Council recently noted, systematic research has yet to address how common viruses cycle between airborne and surface-associated states; thus, virus reconnaissance has not been considered in exposure assessments to any practical degree. Unlike their bacterial counterparts, viral activity on surfaces cannot be associated with biomarkers of microbial metabolism (i.e. Adenosine Triphosphate (ATP)). Even with the most modern molecular biology methods, indoor surveys of viral abundance and diversity are practically limited to quantifying sentinel sequences of the major classes of pathogenic human viruses—influenzas, rotavirus, astrovirus and adenoviruses. While individual virus particles are in sub-micron size ranges, the literature suggests that viruses in the built environment predominately associate with a variety of human tissue fragments, mucus and inorganic matter in conglomerate supramicron particles associating with surfaces. These fomites can facilitate the transfer of viruses from surfaces to humans by direct contact, as well as by partitioning into the aerosol phase. Understanding the identity and distribution of viruses in the built environment thus requires the time-resolved bioaerosol sampling along with randomized surveys of fomites associated with high-contact surfaces. Characterizing virus occurrence begins with understanding their diversity and abundance using quantitative genetic surveys; these can be obtained from parallel polymerase chain reaction (qPCR) observations of genetic materials extracted from area normalized swab sampling. While these assays do not yield information regarding the infectious potential of viruses recovered from fomites, they can provide reliable evidence of their raw relative abundance. Such observations are valuable in terms of better understanding basic exposure potential, as well as understanding how surface cleaning may impact virus abundance.

Keywords: infectious disease

MO-SY-C2: Holistic View on the Internaction Between Residents and Indoor Air Quality in Dwellings (continued)

MO-SY-C2-35

Correlation between technical indoor air and tenants' perception of their indoor air, health and habits - Results from a Danish multiple site interdisciplinary cross-sectional study

Gabel C.E. [1], Iholm G. [1], Petersen S. [2], Rasmussen M.K. [3], Sigsgaard T. [1]; [1]Aarhus University, Institute of Public Health, Aarhus, Denmark, [2]Aarhus University, Institute of Engineering, Aarhus, Denmark, [3]Alexandra Institute, Aarhus, Denmark

Background: The existing building mass is extensively renovated on a large scale to achieve energy reduction. Renovations will most likely alter the indoor environmental quality (IEQ), where we spend the majority of our time, and nearly 70% of this time in our own homes. The Danish building industry refers to The Danish building regulation codex where IEQ is based on technical simulations. The codex does not account for the people occupying the home, whose lifestyle and practice has a major impact on IEQ along with their health, perception and comfort. It is crucial to study this association in a real-life setting where energy renovation is planned. Material and method: BE-READY and ReVALUE are longitudinal panel studies to achieve causal inference of the complex interplay between tenant, building and IEQ before and after energy renovation. The study includes three different social housing estates in Denmark where energy renovation is planned (N=427). Objective and subjective measures before energy renovation were obtained during the heating season (February-April) and/or non-heating season (August-September). Measurements included a survey on demographics, lifestyle, building material, own practice, perception of IEQ, health, comfort and safety. Objective IEQ was measured using technical sensors logging temperature (°C), relative humidity (%) and carbon dioxide (ppm) every 5 minutes for 14 days. Air flow (l/s) and absolute humidity (%) were modelled, dust was collected on an electrostatic dust collector (EDC) and lung function was measured by spirometry. Selected participants were invited to an extended qualitative interview with an anthropologist (n=29) to achieve in-depth knowledge about indoor climate and everyday practices. All measurements were obtained in the tenants' own home by trained staff. Results and conclusion: Cross-sectional results are expected to reveal the interplay between objective IEQ and tenants' perception of health, comfort and IEQ and display dominant exposures related to poor health.

Keywords: air, behavior, built/indoor environment, epidemiology, health, symptoms

MO-SY-C2-36
Investigation of Indoor Exposures and Indoor Air Quality in Multi-apartment Buildings using Three Different Data Streams

Mainelis G. [1], Thomas N. [1], Calderon L. [1], Senick J. [1], Sorensen-Allacci M.A. [1], Plotnik D. [1], Guo M. [1], Yu Y. [1], Gong J. [1], Andrews C. J. [1]; [1]Rutgers University, New Brunswick, NJ, USA,

Investigation of indoor air quality (IAQ) and assessment of indoor exposures using traditional air sampling techniques are often limited in scope and do not provide information about the state of the building and residents' behavior affecting IAQ as well as their interaction with the building. In a study of two residential multi-apartment buildings in the Northeastern US, we attempted to overcome this limitation by combining three different data streams: traditional IAQ measurements, questionnaire data, and 3D infrared thermography. As part of IAQ measurements, we investigated the presence of ultrafine particles, PM2.5 particles, culturable bacteria, culturable and total mold as well as indoor temperature and humidity. Questionnaires informed us about the residents' behavior, such as smoking, their health, and perception about the buildings; 3D infrared thermography, which combines laser scanner and infrared analysis, informed us about various building deficiencies. A sophisticated analysis of correlations of all the measured variables was performed. Among the observed correlations, missing insulation detected by the 3D infrared thermography had a positive and significant correlation with the presence of ultrafine particles indoors (p=0.013) as well as their indoor/outdoor ratio (p=0.029). The correlation was driven by apartments where no smoking or use candles or incense was reported. Ultrafine particle concentrations in apartments where such combustion sources were reported were higher regardless of the levels of missing insulation. Also, higher levels of missing insulation were detected in apartments where residents reported an asthma attack in the past 12 months. Overall, this pilot study demonstrated the feasibility of integrating different and sophisticated techniques into the investigation of indoor exposures, building conditions, IAQ, and residents' health. Future studies should apply and investigate this methodology on a larger scale.

Keywords: air, built/indoor environment, environmental health, exposure factors, microbial agents, nanoparticles, particles, sampling methods

MO-SY-C2-37
Calculations of how users affects energy use and indoor environment by interaction with the ventilation system in apartments with exhaust ventilation systems

Magnusson R. [1], Selli N. [2], Nordquist B. [3], Wallentén P. [4], Harderup L-E. [4]; [1]WSP, Lund, Sweden, [2]HOAB, Malmö, Sweden, [3]Building Services, Lund University, Lund, Sweden, [4]Building Physics, Lund University, Lund, Sweden

In the effort to reduce energy use in buildings all contributing factors must be observed and addressed. The technical measures aimed at decreasing the energy supply after a renovation have been given focus for a long time. To what extent the users in a building affect the energy use and the indoor environment has so far been given lesser attention. The objective of this study was to examine how much users can affect the energy use through their interaction with the building. An apartment building, built in the 60 ties, has been modelled in the dynamical programme IDA-ICE; every room modelled as a zone. Simulations of the energy use before and after a renovation were performed including the different renovation measures implemented in the real apartment blocks. The user interactions were also simulated by varying the opening degree of the ventilation openings in the façade. The results show that the user interaction affects the energy use in the same order of magnitude as the different technical renovation measures. A satisfactory indoor environment is crucial. Relevant parameters for the indoor environment were calculated. The out door air supply was calculated and operating modes were identified where insufficient levels compared to the building regulation occurred, e.g. when the ventilation openings were completely shut or just ajar. These cases can occur in reality as the resident is controlling the openings. The findings show that the interaction the user have with the technical systems, such as components in the ventilation system, will have a significant effect on both the energy use and the air quality in terms of the amount of outdoor air supply. More attention must accordingly be given to the interaction in existing buildings both for energy reasons as well as for a public health perspective.

Keywords: behavior, thermal environment, ventilation

MO-SY-C2-90

Indoor air quality and occupant satisfaction in BREEAM and WELL certified buildings: A case study

Licina D.[1], Langer S.[2]; [1]Human-Oriented Built Environment Lab, School of Architecture, Civil and Environmental Engineering, École Polytechnique Fédérale de Lausanne, Lausanne, Switzerland, [2]IVL Swedish Environmental Research Institute, Göteborg, Sweden

Air pollutant level in workplace environment and occupant satisfaction have been associated with workers' health & well-being and their cognitive performance. While the green building industry has a long-standing history of attention to human health, there has been a recent shift in the prioritization of this issue relative to the others, with a new emphasis on features that explicitly promote the human experience of building occupants. In this context, the WELL certification is the new and emerging rating system that promotes health-based building design and operation; however; to date, no studies have investigated its effectiveness. Here, we aim to bridge the knowledge gap by quantitatively comparing indoor air quality and occupant satisfaction in two green buildings using the same cohort of >200 occupants. The first building (pre-relocation phase) was BREEAM-certified, and the second building (after relocation) was WELL-certified. Survey responses were collected from >200 respondents and include information about indoor environmental quality, workplace satisfaction, building-related symptoms and self-reported productivity. Physical measurements of air quality at multiple locations included integrated samples of individual VOCs and total VOCs, formaldehyde, NO2, SO2, O3 and longitudinal records of CO2 and size-resolved particles in the diameter range 0.3-10 µm. The results for BREEAM building indicate a relatively good level of air quality, with open plan offices associated with increased level of air pollutants. Nevertheless, 30% of respondents reported dissatisfaction with the air quality, while >50% of respondents did not express satisfaction. Ongoing campaign in the WELL-certified building (post-relocation phase) will create quantitative insights in the role of different green building schemes in influencing indoor air quality, occupants' satisfaction and productivity.

Keywords:

MO-PL-D2-EAM: Characterizing Transportation-Related Exposures

MO-PL-D2-38

Target-driven Substitution of Hazardous Chemicals in Consumer Products

Fantke P.; Technical University of Denmark, Kongens Lyngby, Denmark

Chemicals are the building blocks for the diversity and functionality of modern consumer products and industry processes, but the tens of thousands of marketed chemicals are also causing rapidly rising pollution and health concerns on a global scale. In response, the substitution of hazardous chemicals in products and processes has become a high-priority area in European chemicals management. Unfortunately, effective methods to identify viable alternatives to hazardous chemicals are not available. State-of-the-art tools used by industry and policy makers are limited to manual qualitative assessments of hazard properties of individual chemicals, and are thus unable to consider entire product life cycles from resources extraction to waste handling, and fail to optimise multiple alternatives with respect to human and ecological health targets for chemical pollution. To address this problem, automated and chemical function-based substitution workflows are needed with novel quantitative high-throughput screening methods that are able to consider for each alternative all relevant exposure settings and chemical pressures along complete product life cycles, and to relate these pressures to science-based human and ecological health targets across geographical scales. With such science-based and operational methods at hand, the use of hazardous chemicals can be successfully substituted by fully sustainable alternatives early in the design process. This will ultimately lead to designing products with ever fewer harmful chemicals, providing benefits across human, ecological, and economic landscapes by delivering methods enabling a sustainable consumption and production. In this talk, I will discuss challenges and opportunities to move toward methods for sustainable design that accounts for full life cycles and science-based health targets.

Keywords: acoustics/noise, aggregate exposure, environmental health, exposure factors, exposure models

MO-PL-D2-39

An Integrated Approach for Improving Traffic-Related Air Pollution Exposure Assessment

Ananyeva O. [1], Turos O. [1], Petrosian A. [1]; [1]State Institution "O.M. Marzieiev Institute for Public Health of the National Academy of Medical Sciences of Ukraine", Kyiv, Ukraine

Introduction: Improving traffic-related exposure estimates to air pollutants is critical for epidemiologic and risk health studies. As limited air quality monitoring data is available for Ukrainian cities, the objective of this study was to develop an integrated approach to traffic-related exposure and health risks assessment procedure. Methods: Air pollution sampling was conducted in two stages (2012-2014) at 55 near-roadway locations in Darnytskyi and Dniprovskyi districts of Kyiv city, Ukraine. Field data was used for development and validation of exposure models. A combination of LUR and air pollution dispersion model (ISC-AERMOD View v. 9.6.5) was used to estimate traffic-related residential exposures. Results and Conclusions: The results suggested that dispersion and LUR models can be used as exposure assessment tools when limited monitoring data is available. Both models provided favorable outputs, when compared to the field data, and were suitable for development of predictive surface of traffic-related pollution within a city. R2 values for LUR models were 0.90 for NO2, 0.77 for CO concentrations and 0.59, 0.79 and 0.81 for PM10, PM2.5, PM1 levels respectively. The validation of the prediction models demonstrated good agreement between observed and predicted data (average of 3-10% difference). ISC-AERMOD View evaluation tests revealed strong association between modeled and measured NO2 concentrations (r=0.93: p<0.005) and weaker associations for other pollutants (with p<0,005): r=0.88 for CO, r=0.90 for SO2, r=0.88 for NMHC, r=0.74 for PM10. Based on validation results, an integrated framework for outdoor traffic-related exposure and health risk assessment was introduced, which provided a step by step algorithm, taking into account data availability and country's specifics. The algorithm allows to obtain more accurate traffic-related exposure data and health risks estimates that are further used in the development of rational city planning designs as well as public health interventions.

Keywords: air, exposure factors, exposure models, respiratory health, risk assessment

MO-PL-D2-40

Source-specific Fine Particulate Using Spatiotemporal Concentration Fields Developed using Chemical Transport Modelling and Data Assimilation: Application to North Carolina for Health Associations with Coronary Heart Disease

Russell A. [1], Huang R. [1], Zhai X. [1], Mulholland J. [1], Ivey C. [2], Guo S. [3], Breitner S. [4], Devlin R. [4]; [1]Georgia Institute of Technology, Atlanta, GA, USA, [2]University of California Riverside, Riverside, CA, USA, [3]Nankai University, Tianjin, China, [4]US Environmental Protection Agency, Chapel Hill, NC, USA

A number of studies have found differing associations of disease outcomes with PM2.5 components (or species), and sources (e.g., biomass burning, diesel vehicle and gasoline vehicle). Here, a unique data fusion method has been utilized to generate spatiotemporal fields of major gaseous pollutants and PM2.5 components (e.g., ozone, NO2, SO2, total PM2.5 mass and speciated PM2.5 including crustal metals) over North Carolina for 2002-2010. In a prior study, the PM2.5 total mass field was used as part of the CATHGEN study of associations between PM2.5 and disease associated with cardiac heart disease patients. Here, we extend the exposure method for further health analyses. The method fuses daily CMAQ model observations with observations to develop accurate spatiotemporal maps of pollutant concentrations. Those results are then used in an advanced chemical mass balance source apportionment model, CMBGC-Iteration, that uses both gas and particulate matter concentrations to quantify source impacts. The method, as applied to North Carolina, quantifies the impacts of nine source categories and estimate source contributions of total PM2.5 mass. The nine source categories include sources of both primary (diesel vehicle, gasoline vehicle, suspended dust, biomass burning, and coal combustion sources) and secondary components (ammonium sulfate, ammonium bisulfate, ammonium nitrate and secondary organic carbon). The results show the dramatic decrease in source impacts, e.g., sulfate, primarily from coal-burning, and from mobile sources. Secondary organic aerosol, e.g., from biogenic emissions, is becoming more dominant over the state. This study highlights an advantage of using a chemical transport model to develop spatiotemporal fields of pollutants, i.e., the ability to assess PM components and their sources.

Keywords: epidemiology, exposure models, health, other/general, metals

MO-PL-D2-41

Spatially-explicit characterization of the exposure and health burden of fine particulate matter in the U.S.

Stylianou K. [1], Tessum C. [2], Marshall J. [2], Fantke P. [3], Jolliet O. [1]; [1]Department Environmental Health Science, School of Public Health, University of Michigan, Ann Arbor, MI, USA, [2]Department of Civil and Environmental Engineering, University of Washington, Seattle, WA, USA, [3]Quantitative Sustainability Assessment, Department of Management Engineering, Technical University of Denmark, Kgs. Lyngby, Denmark

Air pollution impacts are estimated using spatialized exposures and dose-response functions. We developed spatially-explicit intake fractions (iF – fraction of the emission taken in by population) and characterization factors (CF – health burden per precursor emission unit) for sector-specific emissions of primary PM2.5, NH3, SO2, and NOx in the U.S. We calculated iFs for 43,304 locations in the contiguous U.S. using a reduced-form chemical transport model, the Intervention Model for Air Pollution (InMAP). For each source location, we integrated iFs multiplied by the non-linear integrated exposure-response function and region-specific burden estimates at each receptor location, both obtained from the Global Burden of Disease, to derive cumulative location-specific CF estimates. Using spatial information of annual emissions for 50 states and 60 sectors, we estimated emission-weighted state and sector-specific iFs and CFs. The emission-weighted national average iF was 0.8 parts per million (ppm) for PM2.5, 0.4 ppm for NH3, 0.3 ppm for SO2, and 0.1 ppm for NOx. Location-specific estimates varied up to three orders of magnitude, driven by population. Using an average exposure-response slope between background and "no risk" PM2.5 levels, national CFs were estimated at 209, 99, 75, and 38 µDALYs/kg for PM2.5,

NH3, SO2, and NOx, respectively. State CFs varied substantially with a precursor-dependent range. For PM2.5, CFs varied by a factor 60 from highest (Arizona) to lowest state (North Dakota). For NH3 CFs varied by factor of 40 by state and for NOx and SO2 by a factor of 10. A marginal slope resulted in twice-lower estimates. For PM2.5 and NH3, CFs for the mobile sector were ~5 times higher than those for the agriculture sector. We found little variability between sectors for SO2 and NOx emissions. Results highlight the importance of spatial and sector-specific estimates in characterizing more accurately the health burden of PM2.5.

Keywords: air, environmental health, exposure factors, geospatial analysis/GIS, health, chronic diseases, lifecycle analysis, risk assessment

MO-PL-E2-SEM: Consumer Products and Building Materials

MO-PL-E2-42
Inadvertent Polychlorinated Biphenyls (PCBs) in Consumer Products

Liu X.[1], Mullin M.[2]; [1]U.S. Environmental Protection Agency, Durham, NC, USA, [2]U.S. Environmental Protection Agency, Seattle, WA, USA

Although production of commercial polychlorinated biphenyls (PCBs) was banned in 1979 under the Toxics Substance Control Act (TSCA), inadvertent generation of PCBs continues during a variety of chemical production processes and can contaminate products as well as waste streams. A well-studied process leading to inadvertent PCB contamination is diarylide yellow pigment production and the resulting creation of PCB-11 within the pigment molecule. Existing research has focused on evaluating PCB concentrations in consumer products, or in the environment and human blood. Data concerning the fate and transport of inadvertent PCBs from consumer products and the exposure pathways to humans is lacking. Our research investigates the concentrations of inadvertently generated PCBs in consumer products typically used by children to better understand migration pathways from these products into the environment and potential routes of exposure for human health risk assessment. As the first step, fourteen consumer products, such as chalk, crayon, paint, glue sticks, foam sheets, and food packages, were purchased from retail stores in the United States between September and October of 2018. Duplicate products were extracted and analyzed for 209 PBC congeners using modified EPA Method 1668C. Sonication and soxhlet extraction methods were compared for several products as well. Our preliminary results show that 4 out of 14 products contained PCB-11 only, in the range of 43 to 168 ng/g and a fifth product had PCB-95, PCB-121, PCB-85, PCB-118, PCB-149, PCB-153, and PCB-138, in the range of 64 to 137 ng/g. More product testing is needed to identify and quantify inadvertent PCBs in consumer products to provide the basis for further study of their migration pathways and potential routes of human exposure.

Keywords: SVOCs, children, consumer and personal care products, exposure factors

MO-PL-E2-43
Development of new analytical and measurement methods for characterizing the emission of semi-volatile organic compounds (SVOCs) from building and consumer materials

Braish T.[1], Nicolas M.[2], Maupetit F.[2], Desauziers V.[3]; [1]CSTB and IMT Mines Ales, Saint-Martin-d'Heres, France; [2]CSTB, Saint-Martin-d'Heres, France; [3]IMT Mines Ales, Pau, France

Phthalates and organophosphate flame retardants (OPFRs), two families of SVOCs, are ubiquitous indoor pollutants that are added to various indoor building and consumer materials to enhance their properties. However, these compounds are of great research interest due to the severe health effects they cause. The gas-phase concentration of SVOCs in equilibrium with the material surface (y0) is a key parameter in determining indoor emission of materials and estimating the risk of human exposure to these compounds. However, their high molecular weight and low vapor pressure make them hard to study due to their adsorption on different indoor surfaces after emission from the hosting material (sink effect). Few studies are available on the presence of SVOCs in indoor air compared to those that study their presence in settled dust. Moreover, most of the developed y0-measurement methods are of certain limitations. Therefore, the objectives of this study are to develop analytical and measurement methods for characterizing the emission of phthalates and OPFRs emitted from vinyl floorings and polyurethane foams (PUF), respectively into indoor air at room temperature. A robust ATD-GC-MS (Automated Thermal Desorption - Gas Chromatography - Mass Spectroscopy) method has been developed and validated for characterizing eight chosen phthalates and seven OPFRs with LOD ranging from 0.004 to 0.019 µg/m3 for an air sampling volume of 72 L. For the measurement method, the micro-chamber recommended by ISO 16000-25 for studying SVOCs is used as the emission device. y0 is calculated at different elevated temperatures then is determined at room temperature by extrapolation according to van't Hoff's equation [ln(y0) versus 1/T]. Therefore, developing fast and robust method for measuring y0 of emitted SVOCs into indoor air at ambient temperature is a necessity for estimating the risk of human indoor exposure to these compounds, and the micro-chamber extrapolation method is a promising one.

Keywords: SVOCs, analytical methods, building materials, sampling methods, thermal environment

MO-PL-E2-44

Quantitative distribution of SVOC substances in a room during and after cleaning measures.

Uhde E.[1], Varol D.[1], Mull B.[1], Salthammer T.[1]; [1]Fraunhofer WKI, Braunschweig, Germany

Semi-volatile organic compounds (SVOC) in the indoor environment are nowadays receiving increased attention. They can occur in a multitude of products and materials, and their release into air is governed by many parameters. Due to the fact that it is both time-consuming and analytically difficult to measure the usually very low concentrations of SVOC in the field, modelling the fate of these substances has become quite popular. Measured data, especially from realistic environments, is still very scarce. In this study, five different SVOC substances were introduced into two test rooms (one reference room, one intervention room). A wall paint (doped with 3 substances) and pre-cleaned house dust (doped with 2 SVOC substances) were used as sources. The indoor air concentrations of all five substances were monitored over a period of 180 days. After an equilibration phase of 80 days, interventions (air cleaning, vacuum cleaning) were performed in one of the rooms, and the effect on air concentrations was monitored. The interventions included short-term operation of an air purifier and continuous operation over 50 hours. During operation of the air purifier, the air concentration of the target substances in the room was found to be reduced by approximately 50%. Following deactivation of the purifier, however, the concentration swiftly returned to its initial value. This is a strong indication of an evaporation-controlled process. The effect of vacuum cleaning on air concentrations was negligible, and even the successive removal of the doped dust in the intervention room during the experiment did not result in a significant decrease of the SVOC concentration compared to the reference room.

Keywords: SVOCs, air cleaning, built/indoor environment, phthalates

MO-PL-E2-45

Deterioration of PVC flooring due alkaline moisture

Leivo V.[1], Sarlin E.[1], Sonketo J.[1], Pikkuvirta J.[1], Pentti M.[1]; [1]Tampere University, Civil Engineering, Tampere, Finland

Dispute constant development towards low-emission or emission free materials and special guidelines to control moisture during construction processes, there are still a lot of moisture and indoor air problems associated with glued PVC floorings in Finland. Concrete as alkaline (pH ~12.5) and moist material can cause deterioration of PVC floor coverings and adhesives through alkaline hydrolysis. Alkaline hydrolysis has been mainly tried to prevent installing PVC floor coverings above ”dry-enough” concrete and by using self-levelling low-alkali (pH ~11...11.5) screeds between concrete and floor covering. However, it is not known how the low-alkali screeds protect the floor coverings if the concrete humidity increases. The objective of the current research was to investigate deterioration processes and related active factors. Screed, adhesive and PVC floor covering combinations have been varied using different floor covering conditions (relative humidity of concrete) in laboratory test series. VOC-emissions have been analysed from material samples (Bulk-VOC, where samples were collected into Tenax-tube using Micro-Chamber/Thermal Extractor and then analysed by TD-GC-MSD) and from surface emission samples (FLEC-VOC, ISO 16000-10). The deterioration was noticed earlier in the material samples, as the emission products of adhesive or floor covering must first emit through PVC floor covering into surface in order to be perceived in FLEC-sampling. The surface emission values were converted into indoor air concentration of European reference room. According to the results, 5 mm layer of low-alkali screed will effectively protect against alkaline hydrolysis. The difference was significant, especially in material samples covered in relative low moisture conditions (RH85%): TVOC of sample without screed was 1900 ⍰g/m^3 g (Bulk) and 50 ⍰g/m^2 h (FLEC) and with low-alkali screed 270 ⍰g/m^3 g (Bulk) and <20 ⍰g/m^2 h (FLEC), respectively. Protection effect of low-alkali screed weakened in higher covering conditions (RH93%).

Keywords: activity patterns

MO-PL-A3-EHR: Exposure to Fine and Ultrafine Particles I

MO-PL-A3-46
ROS production in aerosolized nanoparticles: an innate or environmentally-induced toxicity?

Vernez D. [1], Hirzel H. [1], Pralon J. [1], Goekce S. [1], Suarez G. [1]; [1]Center for Primary Care and Public Health, University of Lausanne (Unisanté), Lausanne, Switzerland

The impact of particulate matter (PM) exposure on respiratory and cardiovascular human health has been abundantly evidenced. It is also widely agreed that cell oxidative stress plays a key role in the observed adverse health effects. Little is known however about the environmental factors that enhance the capacity of PM components to produce Reactive Oxygen species (ROS). Recent studies suggest that ROS production is enhanced by the interaction of nanoparticles, such as TiO_2, with UV light through a photocatalysis or Photo-Fenton mechanism. Indeed, aerosolized TiO_2 nanoparticles have been shown to exacerbate the ROS production by a factor up to 14 in appropriate environmental conditions. The aim of this study was to evaluate the photo-enhanced ROS capacity of two types of particles commonly found in metal processes and in urban environmental - Fe_2O_3 and ZnO particles. We developed an experimental setting to measure the oxidation capacity of Fe_2O_3 and ZnO with or without exposure to UV light. Particles in the nano range were generated using a collison nebulizer and exposed to UV using a solar light simulator. Sensing strategy based on FOX-2 assay was adapted to our experimental setup to detect both highly reactive HO• and H_2O_2 produced along the aerosol generation/UV-exposure pathway. The aerosol stream was flown into two parallel impingers, separately filled with water and sorbitol. While H_2O_2 remained in the water-filled impinger, the sorbitol in the second impinger, converted highly reactive HO• into more stable H_2O_2, allowed the determination of both ROS. The observed enhancement of ROS production under UV light conditions raises, broadly, the question of the role of environmental conditions in the exacerbation of the oxidative potential of some airborne pollutants. How much is due to the innate physico-chemical property of the pollutants and how much is due to the UV and humidity environmental conditions?

Keywords: air, environmental health, industrial hygiene, light/lighting, nanoparticles

MO-PL-A3-47

In-vehicle airborne fine and ultra-fine particulate matter exposure: the impact of emission reduction technologies

Campagnolo D.[1], Cattaneo A.[1], Borghi F.[1], Keller M.[1], Rovelli S.[1], Spinazzè A.[1], Cavallo D.[1]; [1]Department of Science and High Technology, University of Insubria, Como, Italy

Traffic-related air pollutant (TRAP) concentrations inside vehicle cabins are often high compared to background levels. The present study was focused on airborne particulate matter (PM) exposure and was motivated by the implementation of new emission standards that has led to the reduction of vehicle particle emissions. The objective of this work is to investigate the relationship between leading vehicle (LV) emissions and in-cabin PM exposure levels in the immediately following vehicle (henceforth called the study vehicle - SV), with particular emphasis on the role of the LV's emission reduction technologies (e.g., diesel particulate filter-DPF) as an effective risk management measure. The research was performed using an instrumented study vehicle (always to be considered as the following vehicle) on a 26-km fixed route where 10 repeated tests were conducted during 60-minute trips. On-line monitoring of the fine (i.e., PM0.3-1 and PM1-2.5) and ultra-fine particle (UFP) concentrations was carried out inside the SV's car cabin with fixed ventilation settings (i.e., windows closed, air conditioning off, recirculation fan off). Simultaneously, the license plate numbers of the LVs along the route were recorded to collect information pertaining to their fuel type and Euro emission standard category. The results clearly showed that the in-cabin PM exposure levels were significantly affected by the LV's Euro emission standard. In particular, a strong and significant decrease in the in-cabin median exposure levels was observed (up to -62%, -44%, and -48% for UFPs, PM0.3-1, and PM1-2.5, respectively) when following diesel-fuelled LVs equipped with recent-emission standards (i.e., Euro 5-6) than when following older-emission standard LVs (i.e., Euro 0-4). An in-depth analysis revealed that the in-cabin median exposure concentrations of PM were highly and significantly reduced by DPF-equipped LVs. The most important reduction (about 80%) was observed for PM0.3-1 compared to both petrol-fuelled and diesel-fuelled (non-DPF) LVs.

Keywords: air, environmental policy, particles, sampling methods, statistical methods

MO-PL-A3-48

Air Pollution Implications of Energy Transformation Pathways in Disadvantaged Communities

Gonzales D.[1], Jerrett M.[1]; [1]Fielding School of Public Health, University of California, Los Angeles, California, USA

Residential energy transformation pathways accelerate reductions in energy consumption. Four primary pathways being pursued related to residential buildings within the United States are: retrofits for energy efficiency, electrification of natural gas appliances, electrification of vehicles, and deployment of distributed generation and storage. Each pathway has a distinctive set of expected impacts on energy use, but the impacts on greenhouse gas (GHG) emissions and air quality are not well understood. The goal of this research is to holistically assess multiple potential pathways leading towards both increased energy conservation and increased renewable energy generation sources for building energy needs. To understand associated air quality and GHG reduction benefits, we are assessing the indoor and ambient air quality of 70 homes within the environmental justice communities of eastern Los Angeles using particulate matter (PM) and nitrogen dioxide (NO2) measurements and conducting a detailed indoor questionnaire that will provide information on appliance usage and their energy sources. We are also linking to a unique Energy Atlas of Los Angeles that has hourly measurements of electricity and natural gas usage. Our preliminary findings suggest poorly equipped or ventilated homes experience exceptionally high peaks in air pollution concentrations. This presentation will detail our analyses of the transitions. This assessment will inform State and local energy investment priorities and identify possible pitfalls and potential remedies to maximize benefits to disadvantaged communities.

Keywords:

MO-PL-A3-49

Air pollution exposure at Philadelphia Subway Stations

Shakya K. [1], *Saad A.* [1], *Aharonian A.* [1], [1]*Villanova University, Villanova, PA, USA*

Air pollution is one of the leading causes of global deaths and is associated with cardiovascular and respiratory diseases. Previous studies have identified built environments such as subways to be associated with high air pollution levels. Philadelphia is the sixth largest city in the United States with approximately 1.5 million population. About 290,000 people use Philadelphia's SEPTA train systems each week. However, there are no published studies assessing the air pollution exposure levels at Philadelphia's subways. The main objectives of this study are to measure the exposure levels of particulate pollution at Septa subway stations in Philadelphia and to investigate the particulate matter variations with subway building characteristics. We measured PM2.5 and PM10 using nephelometers (pDR-1500, Thermo Scientific Inc.) and particle number concentrations (0.3 µm – 10 µm) using a laser particle counter (GT-526S, MetOne Instruments, Inc.). Replicate measurements were conducted during two winter seasons (March 2018 and March 2019) from 12 underground subway stations along the South Eastern Pennsylvania Transit Association's Broad Street and Market Street lines during morning and evening rush hour periods. Mean PM2.5 was 58 ± 33 µg/m3 in 2018 and 36 ± 17 µg/m3 in 2019; mean PM10 was 62 ± 38 µg/m3 in 2018 and 44 ± 17 µg/m3 in 2019. The station with poor ventilation had the highest particulate level and the station with more direct access to outside environment had the lowest particulate level. Particulate levels at the underground subways were 2 to 7 times higher than the above ground street levels. Based on morning and afternoon measurements, particulate levels seemed to dissipate well at a ventilated station but seemed to accumulate at the station with less direct ventilation. Further analyses will be conducted on particle number concentrations, temporal variation of particle mass and particle number concentrations and the influence of weather.

Keywords: air, built/indoor environment, particles, ventilation

MO-PL-A3-50

The Hopi Environmental Health Project (HEHP)- Measuring Concentrations of PM2.5 in Households on Hopi Lands

Hadeed S.[1], *O'Rourke M.K.*[1], *Alshammari M.* [1], *Blohm J.*[1], *Canales R.*[1], *Paukgana P.*[2], *Olivas G.*[3], *Carl A.*[3], *Lugo-Lerma V.*[1], *Gonzalez-Figueroa E.* [1]; [1]*The University of Arizona College of Public Health, Tuscon, AZ, USA*, [2]*The Hopi Tribe, Kykotsmovi, AZ, USA*, [3]*The University of Arizona, Tuscon, AZ, USA*

Despite an estimated 6.5 million people in the United States burning solid fuels for heating, respiratory health effects from indoor exposure to smoke and particulate matter (PM) have focused on populations in developing nations, and have been largely ignored in developed nations. The Hopi Tribe of Northern Arizona heavily relies on burning of coal and wood for heating and cooking during winter months. PM2.5 was measured in Hopi households for 24-hours during heating and non-heating seasons using real-time area monitors. Primary heating fuel type was recorded from questionnaires completed by primary respondents. Indoor mean concentrations of PM2.5 were assessed across seasons and household fuel types. Fifty-one households were sampled, 36 of which completed follow-up sampling during both heating and non-heating seasons. Across all households and seasons, mean (SD) concentrations of PM2.5 were 36.8 µg/m3 (175.3); 51.13 µg/m3 (218.7) during the heating season, and 18.6 µg/m3 (91.4) during the non-heating season. When considering fuel type, households burning wood and coal (n=17) had highest indoor PM2.5 concentrations (53.6 µg/m3 (136.0)) during the heating season. Homes using a combination of electricity, coal and wood (n=11) during the heating season had an indoor concentration of 46.6 µg/m3 (155.7). During the heating season, concentrations were 4.16 times greater for coal and wood homes, 2.52 times greater in electric, coal and wood homes, and 2.03 times greater in homes burning coal, wood and natural gas, compared to the non-heating season. PM2.5 concentrations were elevated in coal and wood burning households. Although slight reductions were observed in homes that used a combination of electricity or natural gas with coal and wood, concentrations exceeded the USEPA ambient standard of 35 µg/m3 and the WHO guideline of 25 µg/m3. Occupants of these households may be overexposed to PM2.5 and at greater risk for developing respiratory and cardiovascular diseases.

Keywords: air, built/indoor environment, community, environmental health, particles

MO-PL-A3-51

Exposure of pregnant women to particle matter pollution: levels and risks

Paz M.[1], Slezakova K.[1], Costa C.[2,3], Inês Silva A.[2,3], Lage B.[2,3], Aguiar L.[2], Mendes A.[2,3], do Carmo Pereira M.[2,3], Teixeira J.P.[2,3], Joana Madureira J.[2,3]; [1]LEPABE, Departamento de Engenharia Química, Faculdade de Engenharia, Universidade do Porto, Porto, Portugal, [2]Environmental Health Department, National Institute of Health, Porto, Portugal, [3]EPIUnit - Instituto de Saúde Pública, Universidade do Porto, Porto, Portugal

Because of the potential health outcomes, coarse (PM10), fine (PM2.5) particulate matter and ultrafine particles (UFP) have been receiving scientific attention. Information about exposure to these pollutants is especially relevant for susceptible individuals, such as pregnant women who are at greater risks of serious respiratory consequences. To deepen the existent knowledge on associated health risks, this study aims to: (1) evaluate exposure of pregnant women to PM10, PM2.5 and UFP, (2) estimate particle deposition within respiratory system, and (3) assess the respective inhalation risks. Air quality of 65 homes of pregnant women living in Porto Metropolitan Area (Portugal) was evaluated over the course of 9 months (2018/19). Information regarding the pregnant subjects was collected via a detailed questionnaire, after approval of the Ethical Committee of Centro Hospitalar de São João. Specifically, PM10, PM2.5 and UFP were concurrently measured indoors and outdoors (DustTrak DRX Aerosol Monitor 8533, TSI; P-Trak UFP Counter 8525, TSI). For PM10 the respective indoor means were 14-194 µg m–3 and 11-187 µg m–3 for PM2.5 [indoor to outdoor (I/O) ratios of 1.12 and 1.11, respectively]. Indoor UFP ranged between 2.41×103 and 79.9×103# cm–3, with I/O ratio of 1.46. Further analysis will be performed to estimate exposure risks using the USEPA methodology Quantitative Health Risk Assessment for Particulate Matter, while particle deposition will be assessed by the Multiple-Path Particle Dosimetry (MPPD) model. The data obtained here can be used for determining risks associated to particulate matter exposure supporting national regulatory committees and policy makers to take the necessary steps to protect human health. This work is supported by FCT-FAPESP (FAPESP/19914/2014) and by UID/EQU/00511/2019 - Laboratory for Process Engineering, Environment, Biotechnology and Energy – LEPABE, funded by national funds through FCT/MCTES (PIDDAC); Carla Costa and Joana Madureira are supported by FCT grants (SFRH/BPD/96196/2013 and SFRH/BPD/115112/2016, respectively).

Keywords: air, particles, risk assessment, susceptible/vulnerable

MO-SY-B3: National Indoor Air and Health Programme 2018-2028 in Finland

MO-SY-B3-52

Symptoms, environmental sensitivity, and public beliefs about indoor air in Finland

Pekkanen J.[1], Lampi J.[2], Hyvärinen A.[2]; [1]Department of Public Health, University of Helsinki, and National Institute for Health and Welfare, Helsinki, Finland, [2]Finland National Institute for Health and Welfare, Kuopio, Finland

Reporting of symptoms related to poor indoor air quality is frequent in Finland. According to the national FinHealth 2017 Survey, one third of the women and one-fifth of men have ever experienced symptoms related to poor indoor air quality. This presentation discusses some possible contributing factors. Knowledge and perceptions were surveyed in a random national sample of Finnish adults and compared to opinions of 8 national experts (1). Half of the population, but none of the experts, think that symptoms are the best measure of the quality of the indoor air. All of the experts, but only 30% of the population, think that worry can cause symptoms similar to symptoms related to poor indoor air. In Finland, questionnaires on indoor air quality are frequently used at work place and also in schools and the results of such surveys have crucial influence when deciding on building renovations. Self-reported intolerance to indoor air is common in Finland. 7% of Finnish respondents reported intolerance to certain buildings, which is higher than in Sweden. Intolerance to sounds and electrical devices was less common in Finland than in Sweden. Preconceptions and concerns about the harmfulness of a given exposure have an important effect on the amount of symptoms elicited by the exposure. This was seen also in a survey in 5 problem and 5 control schools in Finland. Awareness and worry about indoor air problems in the school appeared to be the main determinants of the symptoms among the pupils. The current emphasis on symptom and misconceptions and concerns about health effects of indoor air exposures may partly explain the frequent reporting of symptoms related to poor indoor air quality in Finland 1) Finnish Medical Society Duodecim. Current care guideline on 'Patient exposed to moisture damage' 22.12.2016. http://www.kaypahoito.fi 2) Karvala et al. 2018

Keywords: behavior, environmental policy, health, symptoms, risk assessment

MO-SY-B3-53
Indoor air exposure in Finland - building stock, guidelines and public discussion

Hyvärinen A. [1], Haverinen-Shaughnessy U. [1], Leppänen H. [1], Salmela A. [1], Lampi J. [1], Pekkanen J. [1], National Institute for Health and Welfare, Kuopio, Finland

The most important indoor air exposures increasing burden of diseases in Finland are outdoor pollutants, radon, passive smoking and moisture damage. Based on a recent review, measured indoor air quality (IAQ) of Finnish building stock is better than in most of European countries except radon concentrations, which are spatially high in Finland. Inadequate ventilation in homes and schools seems to occur in similar proportions as in the other Nordic countries. Issues related to moisture damage seem to be less prevalent in Finland and Nordic countries than in southern countries, but more studies with comparable methodology identifying moisture damage relevant to health are still needed. Problems with indoor air quality are, however, quite common in Finland. For example some of the public buildings such as schools have urgent need for renovations due to the age of the buildings and poor maintenance. This is true despite the fact that recognition of IAQ issues, as well as related legislation and guidance has been developed greatly since early 1990s. The legislation and guidance not only identifies and detects the health-relevant exposure agents, but also describes the processes to prevent and manage IAQ issues and how to renovate buildings to meet the standards. It seems that legislation and guidance is more extensive than in the other Nordic countries and possibly worldwide. Yet, IAQ issues are also very visible in both the media and social media. Time to time this discussion is even fierce and difference of opinions and distrust for authorities, ministries and research institutes are expressed. New means and course of actions are needed to revive the trust, to increase positive communication and interaction between all parties, to fulfill the current needs for reliable information and to help individuals with symptoms and diseases.

Keywords: built/indoor environment, dampness/mold, environmental health, environmental policy, risk assessment

MO-SY-B3-54

National Indoor Air and Health Programme 2018-2028 in Finland

Lampi J.[1], Hyvärinen A.[1], Pekkanen J.[2]; [1]National Institute for Health and Welfare, Kuopio, Finland, [2]University of Helsinki, Helsinki, Finland

Clean and fresh indoor air promotes health and well-being. The primary principle in Finland is to prevent and reduce indoor air exposures with strict regulations and by intervening on damages in the buildings. The levels of most of the relevant indoor air pollutants are lower in Finland as compared to Europe. However, perceived symptoms related to poor indoor air quality are common. The problems related to indoor air are a very complex entity. Indoor air-related problems are common in Finland and sometimes they are made worse by poor communication and lack of trust. Improvements are needed in the prevention, management and communication related to problem situations. There are confusion and conflicting views on the health effects of indoor air exposures. Help is needed for those who suffer symptoms in indoor environments or have illness. Provision of diagnostic and treatment services, social security and other services that support rehabilitation still have shortcomings and must be improved. Because of the scale of the problem, people's needs for information, and the large costs involved, new evidence-based methods and solutions are needed.In spring 2017, the National Institute for Health and Welfare (THL) initiated the preparation of the National Indoor Air and Health Programme 2018–2028 together with a number of collaborators. The primary objective of the programme is to reduce the risks to health and welfare due to poor indoor air environments in Finnish buildings. To fulfil this objective, the programme will engage the health service sector and the prevention of hazards, improvement of communication as well as more comprehensive treatment and support for those who are sick. The progress and meeting the defined objectives will be continuously monitored throughout the programme using e.g. surveys.

Keywords: environmental health

MO-SY-C3: Characterizing Exposures and Potential Health Effects Related to Use of Conventional and Newly Emerging Tobacco Products such as E-Cigarettes

MO-SY-C3-55

Analysis of Urinary Biomarkers of Exposure to Volatile Organic Compounds from the Population Assessment of Tobacco and Health (PATH) Study (Wave 1): Characterization of Tobacco Product Exposure

De Jesus V., US Centers for Disease Control and Prevention, Atlanta, GA, USA

Introduction: Volatile organic compounds (VOCs) are ubiquitous in the environment. Long-term exposure to certain VOCs may increase the risk of cancer, heart disease, and chronic obstructive pulmonary disease. In the United States, tobacco smoke is the major non-occupational source of exposure to a number of harmful VOCs (e.g., acrolein, acrylonitrile, benzene, 1,3-butadiene). The Population Assessment of Tobacco and Health (PATH) Study is a national longitudinal study of civilian non-institutionalized youth and adults that examines tobacco use patterns and related health outcomes. We measured VOC metabolites (VOCM) in urine samples collected from a subset of adult (ages 18+) tobacco users and non-users in Wave 1 (2013-2014). Methods: We developed a mass spectrometry-based method to quantify VOCM levels. It detects 28 VOCM resulting from exposure to 20 parent VOCs. We analyzed data for 20 VOCM from 11,501 PATH Study Wave 1 samples. We defined three classes of tobacco product use: combustible products (e.g., cigarettes, hookah); e-cigarettes, and; non-combustible products (e.g., loose snus, chewing tobacco). Results: Creatinine-adjusted urinary levels of VOCM from exposure to acrolein, crotonaldehyde, isoprene, acrylonitrile, and 1,3-butadiene were higher in exclusive combustible tobacco product users (N=3,283) than in non-users (N=1,698). Similar analyses found minor differences of VOCM levels among exclusive e-cigarette users (N=149) and exclusive smokeless tobacco product users (N=364) for all assayed VOCM when compared to non-users. Conclusions: Combustible tobacco product users showed higher VOCM concentrations than exclusive e-cigarette users, exclusive smokeless tobacco product users, and non-users of tobacco products. We provide reliable estimates of population exposure to 16 harmful and/or carcinogenic parent VOCs resulting in well-characterized VOCM ranges for four tobacco user groups. These data can be used to establish a baseline of exposures to VOCs to identify exposure trends as a result of changes in tobacco use behavior over time.

Keywords: VOCs

MO-SY-C3-56

Exposure to polycyclic aromatic hydrocarbons (PAHs) in adults differ by tobacco product type and frequency of use: Data from the U.S. Population Assessment of Tobacco and Health (PATH) Study

Botelho, J.C.[1], Wang Y.[1], Wong L.Y.[1], Meng L.[1], Pittman E.N.[1], Trinidad D.A.[1], Hubbard K.L.[1], Calafat A.M.[1]; [1]Centers for Disease Control and Prevention, Atlanta, GA, USA

Polycyclic aromatic hydrocarbons (PAHs) are environmental pollutants formed from incomplete combustion of organic matter. PAHs are known carcinogens and exposure can occur through smoking. Information about PAHs exposure by tobacco product type and frequency of use is limited. To address this gap, we evaluated exposure to PAHs among 11,519 adult participants in Wave 1 (2013¬–2014) of the Population Assessment of Tobacco and Health (PATH) Study. From self-reported information, we categorized participants based on their use of tobacco products as never-tobacco user, exclusive current established combustible products user (e.g. cigarette smoker), and exclusive current established non-combustible products user (e.g., e-cigarette user). We further classified tobacco users as exclusive cigarette user, exclusive smokeless product user, and exclusive e-cigarette user. Last, we categorized frequency of product use (everyday vs some days) and time since use. We evaluated associations between tobacco product user categories and PAHs urinary biomarkers geometric mean concentrations determined by liquid chromatography tandem mass spectrometry. Combustible products users had significantly higher concentrations of all biomarkers than non-combustible products users and never users; non-combustible products users had significantly higher concentrations than never users for most biomarkers. Everyday cigarette and smokeless product users had significantly higher concentrations for most biomarkers than some days' users; cigarette and smokeless product users who used the product in the last hour had significantly higher concentrations of most biomarkers than other occasional cigarette or smokeless product users, respectively. By contrast, everyday e-cigarette users' concentrations of most biomarkers did not differ significantly from those in some days' e-cigarette users; we did not observe clear trends by time of last use among e-cigarette users. These data showed that users of tobacco products had higher PAHs urinary biomarker concentrations compared to never users, and concentrations differed by type and frequency of tobacco product used.

Keywords: analytical methods

MO-SY-C3-57

Ethylbenzene and Styrene Exposure in the United States based on Urinary Mandelic Acid and Phenylglyoxylic Acid data from NHANES

Blount, B.; Centers for Disease Control and Prevention, Tobacco and Volatiles Branch, Atlanta, GA, USA

Both ethylbenzene and styrene are air toxicants with widespread non-occupational exposure sources, including tobacco smoke and diet. Ethylbenzene and styrene (EB/S) exposure was quantified from their common metabolites measured in spot urine samples obtained from participants (≥6 years old) in the 2005 – 2006 and

2011 – 2012 cycles of the National Health and Nutrition Examination Survey (NHANES; N = 4690). EB/S metabolites mandelic acid (MA) and phenylglyoxylic acid (PGA) were measured using ultra-high performance liquid chromatography coupled with electrospray ionization tandem mass spectrometry (UPLC-ESI-MS/MS). MA and PGA were detected in 98.9% and 90.6% of tested urine specimens, respectively. Exclusive smokers had 2-fold and 1.6-fold higher median urinary MA and PGA, respectively, compared with non-users. Sample-weighted regression analysis among exclusive smokers showed that smoking 0.5 pack cigarettes per day significantly increased MA (+97.9 μg/L) and PGA (+69.3 μg/L), controlling for potential confounders. In comparison, exposure from the median daily dietary intake of grain products increased MA by 1.95 μg/L and was not associated with statistically significant changes in urinary PGA levels. Conversely, consuming vegetables and fruit was associated with decreased MA and PGA. These results confirm tobacco smoke as a major source of ethylbenzene and styrene exposure for the general U.S. population.

Keywords: VOCs, biomarkers

MO-SY-C3-58

Tobacco-derived aldehydes: Quantification in aerosols and utility of urinary aldehyde metabolites as biomarkers of exposure

Conklin D.[1], Lorkiewicz P.[1], Bhatnagar A.[1], Srivastava S.[1]; [1]American Heart Association, Tobacco Regulation Center, Louisville, KY, USA

Saturated and unsaturated aldehydes are products of incomplete combustion, and thus, many aldehydes are common constituents of tobacco smoke including electronic cigarette aerosols. E-cigarettes have emerged as a popular electronic nicotine delivery system (ENDS). Despite the absence of some toxins such as carbon monoxide (CO) and tobacco-specific nitrosamines (TSNA), carbonyls including short-chain, toxic aldehydes have been detected in e-cigarette-derived aerosols at levels found in tobacco smoke. Given the health concerns primarily cardiopulmonary regarding exposures to toxic aldehydes, understanding both aldehyde generation in e-cigarette and e-cigarette aerosol exposure is critical. Thus, we measured aldehydes generated in aerosols derived from propylene glycol (PG):vegetable glycerin (VG) mixtures and from commercial e-liquids with flavorants using a state-of-the-art carbonyl trap and mass spectrometry and compared these levels with those present in mainstream cigarette smoke (MCS). To track e-cigarette and MCS aerosol exposure in mice, we measured urinary metabolites of 4 aldehydes using ULPC-MS/MS and GC-MS as needed. Aldehyde levels, independent of abundance [saturated: formaldehyde, acetaldehyde >> unsaturated: acrolein, crotonaldehyde], were dependent on the PG:VG ratio and the presence of flavorants. The metabolites of 3 aldehydes - formate, acetate and 3-hydroxypropyl mercapturic acid (3-HPMA; acrolein metabolite) were increased in urine after both e-cigarette aerosol and MCS exposures, but the crotonaldehyde metabolite (3-hydroxy-1-methylpropylmercapturic acid, HPMMA) was increased only after MCS exposure. Interestingly, exposure to menthol-flavored e-cigarette aerosol increased the levels of urinary 3-HPMA and the sum of nicotine exposure (nicotine, cotinine, trans-3'-hydroxycotinine) relative to exposure to Classic Tobacco-flavored e-cigarette aerosol. By comparing these findings with aerosols of other ENDS, by measuring aldehyde-derived metabolites in human urine following exposure to e-cigarette aerosols and by assessing the relationship between aldehyde exposure and cardiopulmonary outcomes, we will further our understanding of the relationship between tobacco product use, aldehyde exposure and cardiopulmonary health risk.

Keywords: SVOCs, biomonitoring

MO-SY-C3-59
Maternal exposure of mice to electronic cigarette aerosols early in life alters neurodevelopment, adult behavior and weight gain

Zelikoff J.[1], Blum J.[1], Rather J.[1], Schwartzer J.[1], Church J.[1], Gordon T.[1], [1]NYU Langome Health, New York, NY, USA, [2]Mount Holyoke College, South Hadley, MA, USA

Electronic cigarettes (e-cigarettes), battery-powered devices containing nicotine, glycerin, propylene glycol, flavorings, and other substances, are increasing rapidly in popularity. Despite the escalating usage of electronic nicotine delivery systems (ENDS) by young adults and the potential health implications, particularly those associated with use during pregnancy, the health implications of vaping are relatively unknown. Murine studies performed in this laboratory have demonstrated that exposure to e-cigarette aerosols during early life stages induce biological changes in the central nervous system (CNS) transcriptome associated with adverse neurobiological outcomes and long-term disease states. Investigations also revealed that maternal and/or early life exposure to relevant levels of e-cigarette aerosols, with and without nicotine, during pregnancy leads to behavioral changes in the adult offspring associated with hyperactivity and depression in a sex-dependent manner. This presentation will discuss the health implications associated with exposure of a mouse model to e-cigarettes during pregnancy/early life on birth outcomes, neurodevelopment and inflammation, adult offspring behavior, and changes in weight gain. Taken together, these studies demonstrate that maternal use of e-cigarettes during pregnancy can produce some adverse outcomes similar to those associated with smoking conventional cigarettes.

Keywords: SVOCs, other (specify)

MO-PL-D3-EAM: Exposure Modeling of Consumer Products

MO-PL-D3-60
Extension and experimental validation of the multi-pathway exposure model DustEx

Eichler C.[1], Delmaar C.[2], Bi C.[1], Little J.[1]; [1]Virginia Tech, Blacksburg, VA, USA, [2]RIVM, Bilthoven, Netherlands

Human exposure to SVOCs present in materials is of great interest because of the large number of SVOCs used for different purposes in different materials. Recent findings greatly advanced estimating critical emission parameters necessary for exposure assessment of chemicals in materials. The inclusion of this new knowledge in existing exposure assessment tools such as DustEx is crucial for their continued reliability and applicability. DustEx simulates potential exposures based on a range of input parameters. To further improve the tool, we integrated a recently established relationship between the material-phase concentration of an SVOC in a material (C0) and the equilibrium concentration in the gas phase in contact with the material surface (y0) for phthalates and phthalate alternatives in PVC products in the DustEx tool. To predict SVOC emissions, transport, and subsequently human exposure, y0 is a critical but often unavailable parameter. C0 is generally more readily available. Therefore, estimating y0 reliably based on C0 is an important step towards more useful exposure predictions. We also included our knowledge of the surface/air partition coefficient KS in the model to more comprehensively describe SVOC partitioning to and from a sink surface. Additionally, we conducted an evaluation for simple indoor conditions using chamber experiments and solving the mass balance with the updated DustEx tool. All necessary parameters for the chambers were known, especially those describing possible sinks. We introduced different, well-characterized phthalate-containing PVC source materials in the chamber and measured the gas-phase concentration in the chamber as well as the surface concentration on the chamber walls. We used the measurements to validate the predictions of the DustEx tool for the specific chamber conditions. Validating the tool for these SVOC/material combinations increases the confidence in the results for other combinations.

Keywords:

MO-PL-D3-61

A simple human physiologically-based pharmacokinetic model for exposure assessment of four groups of environmental phenols

Choi J., Shin M.[1], Lee S.[2], Lee J.[1], Kim S.[1]; [1]Department of Environmental Health Sciences, Graduate School of Public Health, Seoul National University, Seoul, Republic of Korea, [2]Department of Occupational and Environmental Medicine, Ajou University School of Medicine, Suwon, Republic of Korea

Background: Physiologically-based pharmacokinetic model (PBPK) modeling has been utilized for exposure assessment, for instance, back-calculation of exposure amounts from the levels of biomarkers of exposure. However, the modeling needs chemical-specific structure and parameters for each chemical in general. Objectives: Our goals were to develop a simple human PBPK model with the same structure for four groups of environmental phenols found in consumer products such as paraben [methyl-, ethyl-, propylparaben (MeP, EtP, PrP)], benzophenone-3 (BP3), triclosan (TCS), bisphenol analogues (BPA, BPS) and to show the effectiveness in reverse-dosimetry of the exposure biomarkers collected from the general population. Methods: The simple model has four core-compartments and two exposure-routes (ingestion and dermal) and the model parameters were optimized and validated using pharmacokinetic (PK) profiles of human volunteers after exposure to deuterium-labelled chemical standards, respectively. Then, we applied the model to calculated the estimated daily intake amounts (EDIs) from each urinary biomarker of chemical (urinary MeP, EtP, PrP, BPA), collected representative volunteers of nationwide biomonitoring program, Korean National Health Survey (KoNEHS 2015-2017); the estimates were compared with the ones from a conventional method of reverse-dosimetry, fractional urinary excretion (Fue). Results: Most of the model predictions corresponding to the target chemicals showed good agreement with the human PK profiles for 8 to 196 hours after the administration. The EDIs (median (IQR)) of MeP, EtP, PrP, and BPA were 22.0 (6.1–73.5) μg/kg/day, 21.6 (4.4–85.1) μg/kg/day, 2.6 (0.5–21.6) μg/kg/day and 0.06 (0.02–0.12) μg/kg/day in the Korean adult population. Those values were 1.9–4.0 times higher than the EDIs using Fue. Conclusions: The present PBPK model is useful in estimating the exposure amounts from urine biomarkers and to simulation exposure in adult humans. Suggestively, it would be applicable to the risk assessment of other phenolic compounds with similar chemical properties.

Keywords: biomonitoring, environmental health

MO-PL-D3-62

Transdermal uptake of two phthalates and a paraben from simulated lotion using a chemical activity approach

Eftekhari A.[1], Morrison G.[2]; University of North Carolina, Chapel Hill, NC

Personal care products (PCPs) can be an important source of dermal exposure to chemicals of concern. Transdermal permeation models compliment in vitro and in vivo experiments in assessing risk of environmental exposure such as application of cosmetics and personal care products. Recent approaches characterize dermal transport using Fick's first law of diffusion. In these models, steady state permeation rate is proportional to the "effective concentration", "fugacity" or "activity" difference across the skin. In this research, we characterize this chemical driving force and use it to model uptake from lotion. A basic lotion (containing no target analytes) was mixed with 6% wt/wt each of diethylphthalate (DEP), di-n-butylphthalate (DnBP), and butylparaben (BP). The solution was sequentially diluted to make lotions with concentrations ranging from 0.025% to 6%. The fugacity, relative to the pure compound, was measured in headspace of lotions using solid-phase micro extraction (SPME). The fugacity of chemicals in 2% wt/wt formulation was used in a diffusion-based model to predict transdermal uptake of each chemical and compared with excretion data of human subjects who participated in a whole-body topical application of 2% formulation. We found that relationship between the concentration in lotion and fugacity is highly non-linear and cannot be predicted from Henry's or Raolt's laws. Dynamic simulations of excretion are qualitatively similar to the results of the human subject experiment. Quantitative predictions are similar but very sensitive to choice of vapor pressure.

Keywords: SVOCs, consumer and personal care products, environmental health, exposure models, phthalates

MO-PL-D3-63

Evaluating current modelling capabilities for assessing exposures to polycyclic aromatic hydrocarbons

De Brouwere K.[1], Bierkens J.[1], Ketelslegers H.[1], Trantallidi M.[1]; [1]VITO - Flemish Institute for Technological Research, Moi, Belgium, [2]Concawe, Brussels, Belgium

Polycyclic aromatic hydrocarbons (PAHs) are ubiquitous. There are many sources and routes by which human exposure to PAHs occurs. However, the contribution of petroleum substances (PS) to PAH exposures for general population has not been widely characterised. The aim of the project was to test existing integrated multi-source, multi-route (MSMR) exposure model(s) in terms of their applicability for predicting human exposure (direct and indirect) for two PAHs (pyrene and phenanthrene), as well as to assess relative contribution of PS use scenarios into aggregate population exposure. Two exposure models, IndusChemFate and MerlinExpo, were used to predict internal doses to pyrene and phenanthrene metabolites, respectively, in the following five scenarios: filling motor oil of a car; refueling a diesel car, use of personal care products, practicing sport on artificial grass and man via the environment exposure (scenarios cover both consumer and environmental exposure). For each scenario the predicted exposures were compared to human biomonitoring (HBM) data (scenario-specific, where possible). None of the tested models was able to accurately predict in an fully integrated MSMR way the exposure to PAHs. IndusChemFate could estimate on a substance specific basis internal exposure if external exposure was provided as model input. MerlinExpo did not cover dermal exposure route. The match between modelled exposure and HBM data was generally poor across consumer scenarios because 1) the HBM data did not reflect scenario conditions; 2) external exposure models used were designed to be conservative. For Man via the Environment scenario IndusChemFate predictions were comparable to the levels observed in general population, whereas MerlinExpo tended to underpredict exposure to both PAHs. Furthermore, the presentation will discuss sources of variability and uncertainty of input data, conservatism of model defaults, as well as ranking of the scenarios in view of their relative contribution to PAH exposure for the general public.

Keywords: SVOCs, aggregate exposure, consumer and personal care products, exposure models

MO-PL-D3-64

Characteristics of co-use patterns to cosmetic products in high-use groups for Korean population

Lim M.[1], Park J.Y.[1], Lee K.[1]; [1]Seoul National University, Seoul, Korea, South

Cosmetic product is one of the major chemical exposure sources in real life. The use rates of most cosmetics were different by population group. People used many cosmetics simultaneously. To accurate aggregate exposure assessment, it is necessary to determine co-use patterns of cosmetic products by population group. The purpose of this study was to determine co-use patterns of cosmetics and characterize high use group in Korea. The national representative exposure factor database was established for Korean adult population (n = 1,001). The individual's simultaneous use patterns for 31 cosmetic products included actual co-use and non-use information. The co-use patterns of cosmetic products by population groups were analyzed by number of product used, Ward's hierarchical clustering method and Cohen's kappa correlation coefficient. Korean population used 8.5 ± 5.1 cosmetic products. When the population was divided into 10 groups by gender and age, 20's female was the highest cosmetic use group with 14.5 ± 3.9 products. 60's male was the lowest use group with 3.8 ± 1.6 products. The use pattern of 31 cosmetics in 20's female group clustered into five groups of cosmetics. Ten cosmetic products with high user rate were used together and they were skin toner, face cleanser, facial sunscreen, face lotion, foundation, face cream, body lotion, hand cream, hair treatment and lipstick. Four eye makeup products (eyeliner, eyeshadow, eyebrow and mascara) were used together with moderate correlation (kappa > 0.40). Nail polish and nail remover were used together (kappa = 0.95). Aggregate exposure assessment for cosmetic products might be necessary because Korean used multiple cosmetics. The cosmetic co-use patterns were different by population group. The young female group could be high-exposure group for chemicals in cosmetics. The co-use patterns could be important input data for aggregate exposure assessment.

Keywords: aggregate exposure, consumer and personal care products, environmental health, exposure factors, multimedia

MO-PL-D3-65

Multipollutant Models of Mixed Exposures to Cleaning Chemicals and Asthma Symptoms

Virji M. [1], Su F. [1], Liang X. [1], Kurth L. [1], Stefaniak A. [1], LeBouf R. [1], Stanton M. [1], Henneberger P. [1]; [1]National Institute for Occupational Safety and Health, Morgantown, WV, USA

Cleaning products are complex mixtures of chemicals, and their use is associated with elevated prevalence of asthma and respiratory symptoms among healthcare workers. Epidemiologic models of asthma outcomes have typically fit single exposure models, likely leading to false positive discoveries. To better represent workplace mixed exposures, we used a multipollutant model to address variable reduction and collinearity, and to obtain effect-estimates for specific exposures. Healthcare workers from nine occupations working in New York City hospitals and nursing homes completed a questionnaire with modules on cleaning and disinfecting tasks and product use. Cleaning chemical exposure scores were calculated based on the frequency and duration of use of 14 cleaning products. Quantitative exposure to total and eight specific volatile organic compounds (VOCs) were assigned to participants based on predictive statistical models. To select important exposure variables, an elastic-net regression model was fit for a five-level asthma symptoms severity outcome using all exposure predictors, followed by a generalized linear model (GLM) to estimate the effects of the selected variables. The multipollutant model included multiple exposure and hospital-unit-job variables, adjusted for age, gender, education, smoking and atopy status. The elastic-net model selected the following as important variables for at least one level of asthma symptoms severity: nursing jobs in wards and operating rooms; housekeepers in wards; clinical laboratory technicians; respiratory therapists; exposure scores for alcohol, bleach, quaternary ammonium compounds; and exposure to total VOCs, isopropyl alcohol, chloroform, and acetone. In the GLM model for asthma symptoms severity, chloroform and acetone exposures, bleach and alcohol scores, and nursing jobs in wards and operating rooms remained significantly associated with the two most severe asthma symptoms outcome levels. The elastic-net model identified relevant exposures associated with asthma symptoms, and is a more realistic representation of multipollutant workplace exposures than models with single exposure variables.

Keywords: VOCs, epidemiology, occupational, respiratory health, statistical methods

MO-PL-E3-HBM: Biomarkers and Biomonitoring Studies

MO-PL-E3-66

Thinking on related issues about biorhythm synchronization and indoor environment

Chen B. [1], Wang S. [1], Li X. [1], Deng J. [1]; [1]School of Civil Engineering, Dalian University of Technology, Dalian, Liaoning Sheng

With the acceleration of modern lifestyles increasingly featuring full-time life or work activities and responsibilities brought by scientific and technological progress, humans have gradually neglected their interdependence with nature, spending most of their time living and working in artificial environments. Research has shown that when lifestyle and the biological clock deviate, the risk of various diseases rises. This has increased attention towards biorhythm disorders caused by lack of sunlight or inappropriate artificial lighting at night. In this paper, we present a discussion on the literature concerning the indoor environment characteristics required to promote biorhythm synchronization, with emphasis on biorhythm disorders caused by illumination conditions in indoor environments. On this basis, we try to present a method for the evaluation of the relationship between biorhythm synchronization and indoor environments. Finally, we present a case study applying this evaluation method.

Keywords:

MO-PL-E3-67

Bisphenol (A, F, S) exposure levels in the French population aged 6 to 74 years in 2014-2016: The Esteban study

Tagne-Fotso R., Saoudi A. [1], Zeghnoun A. [1], Oleko A. [1], Gane J. [1], Bidondo M.L. [1], Denys S. [1], Fillol C. [1], Balicco A. [1]; [1]Environmental Health Division, Santé publique, Saint-Maurice, FranceEnvironmental Health Division, Santé publique, Saint-Maurice, France

Background: Bisphenol A (BPA) is a chemical compound used in manufacturing products, and now considered as an endocrine disruptor. Its main substitutes are bisphenols S (BPS) and F (BPF). Objective: As part of the French biomonitoring program, Esteban evaluated the urine levels of BPA, BPS and BPF in the French general population and investigated the main determinants of exposure. Method: The first morning urine was collected on a random sub-samples of people living in metropolitan France (500 children and 900 adults aged 6 to 74) and analysed by GC-MS/MS. Results: Total BPA, BPS and BPF were detected in almost all samples in children (99.9%) and adults (100%). The geometric means (µg/g creatinine) for total BPA, BPS and BPF were respectively: i) in children, 2.25 [95%CI: 2.07-2.45], 0.44 [95%CI: 0.36-0.54], and 0.26 [95%CI: 0.23-0.30]; ii) in adults, 2.69 [95%CI: 2.49-2.92], 0.53 [95%CI: 0.46-0.60], and 0.31 [95%CI: 0.29-0.34]. No determinant of BPA exposure has been identified. An increase of BPS and BPF levels were observed in children with the purchase of pre-packaged fish and with the housing ventilation several times a week during the fall/winter, respectively. The purchase of pre-packaged foods were associated with increased BPS levels in adults. Female sex and active smoking were both associated with increased BPF levels in adults. Conclusion: Esteban is the first study to measure the exposure levels of BPA, BPS and BPF in representative samples of children and adults from the French general population. In both children and adults, BPA exposure levels in France were higher than those described to date in the United States and Canada. Due to the recent European regulations on the industrial use of BPA, and the increasing use of BPA substitutes, it will be essential to continue the monitoring of BPA, BPS and BPF in the French population.

Keywords: biomarkers, biomonitoring, children, exposure factors, food

MO-PL-E3-68

Exposure to Bisphenol A and Bisphenol A alternatives and asthma morbidity among inner-city children

Quiros-Alcala L. [1], Hansel N. [2], McCormack M. [2], Matsui E.C. [3]; [1]Department of Environmental Health and Engineering, Johns Hopkins University, Baltimore, MD, USA, [2]Department of Pulmonary and Critical Care, Johns Hopkins School of Medicine, Baltimore, MD, [3]Dell Medical School, University of Texas, Austin, TX, USA

Background: Bisphenol A (BPA) is a chemical used in the manufacture of polycarbonate plastics and epoxy resins. In animals, BPA has been linked with allergic airway inflammation and in, limited studies, with pediatric asthma development. Whether exposure to BPA or its structural analogues, BPF and BPS, is associated with asthma morbidity remains unknown. Methods: We quantified concentrations of BPA, BPF, and BPS in 507 urine samples from 141 inner-city children (5-17 years) with asthma, participating in a 12-month prospective study. Biospecimens and data on asthma symptoms, healthcare utilization, and pulmonary function and inflammation were collected every 3 months. We used generalized estimating equations to examine associations between individual bisphenols and morbidity outcomes, controlling for age, sex, race, caregiver education, atopy, season, and particulate matter. We also assessed heterogeneity of associations by sex. Results: BPA and BPS were detected in >95% of samples; BPF was detected in 66% of the samples. We observed increased prevalence odds of reporting coughing, wheezing or chest tightness per 10-fold increase in BPA concentrations (adjusted prevalence odds ratio, aPOR:1.51;95% Confidence Interval, 95%CI:1.02, 2.24). Similar positive associations, albeit not statistically significant, were observed with other symptoms (e.g., nocturnal wakening, exercise-related symptoms, maximal symptom days). We also observed increased odds of reporting emergency department (ED) visits per 10-fold increase in BPA concentrations (aPOR:2.12, 95%CI:1.17, 3.82). We observed evidence of sexually dimorphic effects, whereby boys generally had higher prevalence odds of asthma symptoms and ED visits associated with increased BPA exposure. No other statistically significant findings were observed. Conclusions: In a cohort of inner-city children with asthma, we found suggestive evidence that BPA may be associated with asthma morbidity and elicit sexually dimorphic effects. Future studies are warranted to confirm these findings and assess biological mechanisms by which BPA could increase morbidity risk and result in sexually dimorphic effects.

Keywords: children, environmental health, epidemiology, health, chronic diseases, respiratory health

MO-PL-E3-69

Neonatal Urine and Cord Blood Metals Analysis in a Cohort

Benson K. [1], Irvin-Barnwell E. [1], Ward C. [1], Jarrett J. [1], Ragin A. [1]; [1]State University of New York (SUNY), NY, USA

Adverse health effects from environmental exposure to heavy metals is well-documented. Specifically, uranium exposure has been associated with adverse effects to the kidneys and urinary system in humans and developmental toxicity in animals. However, there is no information on neonatal urine uranium body burden levels and limited information on cord blood metal levels in neonates in the US. Matched neonatal urine (N=546) and cord blood (N=315) samples from the Global Alliance to Prevent Prematurity and Stillbirth (GAPPS) biorepository were analyzed for 15 elements in urine, including uranium, and five elements in whole cord blood, including cadmium, manganese, lead, selenium, and mercury (total). Urinary uranium and lead had higher geometric means, 0.087 µg/L (95% CI: 0.080 – 0.094) and 0.808 µg/L (95% CI: 0.756 - 0.864), respectively, in comparison with NHANES 2015-2016 geometric means for 3 to 5 year olds, 0.009 µg/L (95% CI: 0.007 – 0.11) and 0.590 µg/L (95% CI: 0.532 – 0.654), respectively. For the cord blood samples, lead had a lower geometric mean, 0.254 µg/dL (95% CI: 0.236 - 0.273), in comparison with the NHANES 2015-2016 geometric mean for 1 to 5 year olds, 0.758 µg/dL (95% CI: 0.675 – 0.850) and is less than the CDC reference value of 5 µg/dL. Blood mercury had a higher geometric mean, 0.670 µg/L (95% CI: 0.612 - 0.732), in comparison with the NHANES 2011-2012 geometric mean for 1 to 5 year olds, 0.262 µg/L (95% CI: 0.237-0.291). Note that the detection of chemicals in neonatal urine or cord blood indicates exposure but does not by itself indicate adverse health effects. Detection of elements may also be an indication of contamination in collection supplies, which were not pre-screened in this study. However, these results will fill existing data gaps and increase our knowledge about heavy metal exposures levels in neonates.

Keywords: environmental health, health, other/general, metals, neonatal

MO-PL-E3-70

Biomonitoring Blood Mercury, Lead, and Cadmium Levels in Susceptible Great Lakes Populations

Li Z. [1], Wattigney W. [1], Irvin-Barnwell E. [1], Hwang S. [2], Lewis-Michl E. [2], Savadatti S. [2], Liu M. [2], Hsu W. [2], Parsons P. [3], Palmer C. [3], Ragin-Wilson A. [1]; [1]Division of Toxicology and Human Health Sciences, Agency for Toxic Substances and Disease Registry, Atlanta, GA, USA, [2]Bureau of Environmental and Occupational Epidemiology, New York State Department of Health, Albany, NY, USA, [3]Wadsworth Center, New York State Department of Health, Albany, NY, USA

The Great Lakes, the world's largest fresh surface water system, has been contaminated with numerous pollutants, including heavy metals. In 2010, the Biomonitoring of Great Lakes Populations program-consisting of a series of cross-sectional studies-was implemented to assess human exposure to legacy and emerging contaminants in susceptible populations. As part of the program, three studies were conducted collaboratively with health departments at New York and Michigan. These studies targeted five populations, including shoreline anglers, licensed anglers, urban subsistence anglers, and Burmese and Bhutanese refugees, who ate fish caught in contaminated waterbodies. The target populations were sampled and recruited independently using different sampling strategies that were appropriate for the specific populations. Mercury and lead were measured in blood samples collected from all five target populations, and blood cadmium was measured in four cohorts. Compared to the U.S. adult levels (NHANES 2013-14), blood metal levels were elevated in most cohorts. For example, median blood mercury was 4.44 ug/L in Burmese refugees in Syracuse, NY (n=182) and 2.4 ug/L in shoreline anglers in Detroit, MI (n=273), compared to 0.74 ug/L in U.S. adults. In most cohorts, blood mercury was associated with eating locally caught fish after adjusting for covariates. Median blood lead levels were 3.09 ug/dL and 2.0 ug/dL in the Buffalo Burmese and Detroit anglers, respectively, compared to 0.94 ug/dL in U.S. adults. Although blood lead and cadmium were generally not associated with fish consumption, multiple demographic and behavior risk factors (e.g. smoking) were identified for each metal in each cohort. In conclusion, the biomonitoring results generated from this program identified elevated metal exposure in several Great Lakes susceptible populations. The information helped guide public health actions at the states, such as targeted education and outreach, to reduce chemical exposures in vulnerable Great Lakes populations.

Keywords: biomonitoring, community, environmental health, metals, susceptible/vulnerable

MO-PL-E3-71

Combination of multi-residue methods based on hair analysis for pollutants exposure biomonitoring in Grande-Synthe, France - Towards a comprehensive assessment of population exposome

Iglesias-González A.[1,2], Schaeffer C. [1], Dahm G. [3], Vanandruel E. [4], FranceHardy E.M. [1], Appenzeller B.M.R. [1]; [1]Luxembourg Institute of Health, Strassen, Luxembourg, [2]University of Luxembourg, Esch-sur-Alzette, Luxembourg, [3]Laboratoire National de Santé, Dudelange, Luxembourg, [4]Espace Santé du Littoral, Grande-Synthe, France

Because the biological effects of several pollutants acting simultaneously may strongly differ from the effects produced by a single chemical, multi-residue analysis is essential to provide the most comprehensive information possible on human exposure. Most biomonitoring studies are however focused on a limited number of pollutants and only give a partial view of the exposome. Hair generally allows the detection of more biomarkers that other matrices, and integrates both parent pollutants and their metabolites. In the present study, hair samples were collected from 117 adults and 40 children living in the city of Grande-Synthe, France, and analysed with 3 multi-residue methods targeting respectively 152 organic pollutants (pesticides, PCBs, bisphenols, PBDEs), 62 polycyclic aromatic hydrocarbons (PAHs) and metabolites (OH-PAHs) and 36 metals. The number of organic pollutants detected in each sample ranged from 33 up to 70 in children and from 34 up to 74 in adults. Fifteen compounds were detected in all the children (13 pesticides, and 2 bisphenols) and 14 in all the adults (13 pesticides and bisphenol A). Regarding PAH, 7 up to 26 compounds were detected per child and 7 up to 21 per adult. 1-OH and 2-OH-naphthalene were detected in all the samples. Nineteen metals were detected in all the samples analysed. This study clearly highlighted the cumulative exposure of the population under study, bringing relevant data on its exposome and standing out the differences this region has with other areas in France and around the world. The results clearly support the use of multi-residue methods to widen the information on human exposure. They also demonstrated the relevance of hair analysis, which allows the detection of pollutants from different classes with a unique biological sample.

Keywords: biomarkers, biomonitoring, children, cumulative exposure, pesticides

MO-PL-A4-72

Phthalates and organophosphates in settled dust and HVAC filter dust of U.S. low-income homes: association with season, building characteristics, and childhood asthma

Xu Y.; The University of Texas at Austin, Austin, TX, USA

Phthalates and organophosphates are ubiquitous indoor semi-volatile organic contaminants (SVOCs) that have been widely used as plasticizers and flame retardants in consumer products. Although many studies have assessed their levels in house dust, only a few used dust samples captured by filters of building HVAC systems. HVAC filters collect particles from large volumes of air over a long period of time (potentially known) and thus provide a spatially and temporally integrated concentration. This study measured concentrations of phthalates and organophosphates in HVAC filter dust and settled floor dust collected from low-income homes in Texas, United States, in both the summer and winter seasons. The most frequently detected compounds were benzyl butyl phthalate (BBzP), di-(2-ethylhexyl) phthalate (DEHP), di-n-octyl phthalate (DnOP), tris (1-chloro-2-propyl) phosphate (TCIPP), triphenyl phosphate (TPHP), and tris (1,3-dichloroisopropyl) phosphate (TDCIPP). The median level of TCIPP in settled dust was 3- to 180-times higher than levels reported in other studies of residential homes. Significantly higher concentrations were observed in HVAC filter dust as compared to settled dust for most of the frequently detected compounds in both seasons, except for several phthalates in the winter. Settled dust samples from homes with vinyl flooring contained significantly higher levels of BBzP and DEHP as compared to homes with other types of floor material. The concentration of DEHP and TDCIPP in settled dust also significantly associated with the presence of carpet in homes. A statistically significant association between DEHP concentration in HVAC filter dust in summer and the severity of asthma in children was observed. These results suggest that HVAC filter dust represents a useful sampling medium to monitor indoor SVOC concentrations with high sensitivity; in contrast, when using settled dust, it is critical to know the sampling location because the type and level of SVOCs may be related to local materials used there.

Keywords: SVOCs, built/indoor environment, children, particles, respiratory health, sampling methods

MO-PL-A4-73

Impact of healthier building materials on dust levels of PFASs, flame retardants, and hormone-disrupting potency

Young A[1]., Hauser R. [1], James-Todd T. [1], Coull B. [1], Zoeller T. [2], Sunderland E. [3], Pickard H. [3], Kannan K. [4], Behnisch P. [5], Brouwer A. [5]; [1]Harvard T.H. Chan School of Public Health, Boston, MA, USA, [2]University of Massachusetts Amherst, Amherst, MA, USA, [3]Harvard John A. Paulson School of Engineering and Applied Sciences, Cambridge, MA, USA, [4]Wadsworth Center, New York State Department of Health, Albany, NY, USA, [5]BioDetection Systems, Amsterdam, The Netherlands

Per- and polyfluoroalkyl substances (PFASs) and flame retardants (FRs) are two classes of chemicals ubiquitously used in furniture, carpet, and other building materials. Many of these chemicals have been associated with endocrine disruption, poorer pregnancy outcomes, immune suppression, cancer, and high cholesterol. To our knowledge, no studies have evaluated the exposure or health benefits of interventions to renovate buildings with healthier materials specified as FR- and PFAS-free. In this study, we collected dust samples from 24 building rooms (e.g. common spaces and offices) renovated with furniture and carpet specified as healthier and 24 comparable controls. The dust samples will be analyzed for levels of PFASs and FRs to compare spaces with healthier-specified materials to traditional spaces. We will also perform novel cell assays to evaluate differences in hormone- and immune-disrupting potency of the dust in the spaces. Cell assays are a relatively inexpensive method to assess the health of chemical mixtures in a targeted way instead of trying to analyze many individual chemicals separately. Laboratory results are pending but will be available for the conference.

Keywords: SVOCs, building materials, built/indoor environment, flame retardants, workplace

MO-PL-A4-74

Developing a long-term dynamic model for the prediction of SVOC concentrations in indoor environments

Wei W.[1]*, Mandin C.*[1]*, Ramalho O.*[1]*;* [1]*Scientific and Technical Center for Building (CSTB), Paris, France*

The partition of indoor environmental semi-volatile organic compounds (SVOCs) among the gas phase, airborne particles, settled dust, and available surfaces leads to the exposure of occupants' to SVOCs via inhalation, dermal contact, and dust ingestion. Mass transfer models were developed in order to predict SVOC concentrations in indoor environments. These models didn't take into account the combined effect of the variations in indoor environmental factors and the influence of the reactive chemistry. Therefore, the objective of this paper is to develop a long-term dynamic model to address the problem. The SVOC mass transfer processes in a single-zone indoor environment were characterized using mathematical equations. The influence of temperature, relative humidity, ventilation, particle concentration, loading factor, and reactive chemistry on SVOC mass transfer processes was addressed in the model. A preliminary version of software was programed in C# to run the model. The predicted concentrations of benzyl butyl phthalate (BBzP) in the air, settled dust, and indoor surfaces were compared to the measured values obtained in a test house. Three environmental factors, i.e., indoor temperature, relative humidity, and air exchange rate, were changed during the modeled and measured period of time. The predicted values are within the ranges of the measured values. Future measurements are needed to validate the prediction of the model when indoor airborne particle concentrations vary and when typical reactants (hydroxyl radical, nitrate radical, and ozone) are present. As a summary, the long-term dynamic model will help design, evaluate, and renovate indoor environments, as well as support the health risk evaluation in relation to SVOC exposures.

Keywords: SVOCs, air, building materials, particles, pesticides

MO-PL-A4-75

Soil and dust ingestion by children comparing urban and suburban areas in Korea

Kim E.[1]*; S, Ryu H, Park J, Yoon H, Kim T, Seo J, Heo J, Choe Y, Cho M, Yang W.;* [1]*Department of Occupational Health, Daegu Catholic University, Gyeongsan, Republic of Korea*

Soil and dust ingestion may be a potentially important exposure pathway to environmental contaminants because they contain hazardous heavy metals. Children, in particular, may ingest significant quantities of soil and dust due to their tendency to play on the floor indoors and ground outdoors and their tendency to mouth objects or their fingers. The aim of this study was to estimate soil and dust ingestion rates of Korean children by geographical and cultural characteristics. The ingestion rates per day were estimated using the EPA/600/R-17/384F method and analyzed concentrations using aluminum(Al) as an indicator. In general, Al is a substance found in soil and dust at a high concentration but not decomposed or affected by living things. This study involved 113 children aged 0-6 years. Fecal samples were obtained from 56 children in urban areas in 2017 and 57 children in suburban areas in 2018, and soil samples near their residence. The soil and dust ingestion rates were 57.5 mg/day in urban areas and 84.2 mg/day in suburban areas. The mean ingestion rate was 70.6 mg/day. These were lower than the recommended values, 118 mg/day by The Korean Ministry of Environment(2007) and 80 mg/day by the U.S.EPA(2017). These results can be explained the children corresponded with the decreased outdoor activities and increased indoor activities compared with the previous study. Also, the playground was built with urethane, not ordinary sand and it was difficult to find artificial green spaces in the cities during sampling soils. It may be another reason that children play phone and internet games at home. Accordingly, these environments could be decrease soil and dust ingestion rates for children. This study is important to access the health risk resulting from exposure to environmental contaminants via soil and dust ingestion pathway by children in Korea as well as other countries.

Keywords: children, environmental health, exposure factors, metals, soil

MO-SY-B4-76

Exposure to pesticides of residents living close to agricultural land: research questions and study design

Scheepers P.; Radboud University Medical Center, Nijmegen, Netherlands

Dutch residents living close to agricultural land expressed concerns about the possibility of enhanced exposure to pesticides. In 2015 a project was started with the aim to study residential exposure to pesticides in homes close to agricultural land. The research question where (a) What are the concentrations of pesticides in the environment residents living near agricultural land compared to residents living further away? (b) What is the personal exposure to pesticides of these residents? (c) What sources and routes contribute to personal and environmental exposures? Agricultural fields were selected, based on crop type and the distance of residences to these fields. Residents living 500m distance. The farmers of the selected fields provided information on their pesticide use. At each residence weather conditions and outdoor air were monitored. In the homes, in addition to air samples, deposited dust, dust from doormats, dust form precipitator plates and vacuumed dust was collected. All participants collected their first urine void in the morning. Questionnaires and diaries were used to collect dietary information and keep track of activities and behavior patterns. Aforementioned measurements were taken during the spray season and during a period of low/no use. Environmental and personal samples were used to model direct and indirect routes of exposure, taking into account both airborne and dust-borne pesticide residues in households at different distances from fields where pesticides were applied. The urine samples were analyzed for biomarkers of five pesticides: asulam, carbendazim, chlorpropham, prochloraz and tebuconazole. These substances were also studied in human volunteers to determine the excretion kinetics and estimate a conversion factor for back-calculation of exposure from urinary metabolite excretion. Protocols for the field and the human volunteer studies were approved by the regional medical-ethical board.

Keywords: exposure models

MO-SY-B4-77

Determination of the toxicokinetics of asulam, carbendazim, chlorpropham, prochloraz, and tebuconazole following oral and dermal administration in human volunteers

Oerlemans A.[1], Mol, H.[2], van Dael M.[1], Nijssen R.[2], Vermeulen R.[3], Russel F.[1], Scheepers P.[1]; [1]Radboud University Medical Center, Nijmegen, Netherlands, [2]Wageningen Food Safety Research, Wageningen, Netherlands, [3]Utrecht University, Utrecht, Netherlands

The toxicokinetics of asulam, carbendazim, chlorpropham, prochloraz, and tebuconazole in humans are not fully described. Knowledge on their metabolism and elimination kinetics is required for proper interpretation of biological monitoring data. Here, we exposed human volunteers to each of the aforementioned pesticides separately, with the aim to study the urinary excretion-time course of their metabolites following controlled oral and dermal administration. For each administration method a fixed dose was calculated to prevent exposure from exceeding the current acceptable daily intake (ADI) for each of the pesticides. Approval of the study was obtained from the regional medical-ethical board. Three male and three female volunteers received a fixed oral dose and on a separate occasion a controlled dermal exposure. In addition to a pre-exposure urine sample, complete urine voids were collected over 48 h, following the start of administration. The main metabolite of each pesticide was quantified in urine using liquid chromatography tandem mass spectrometry (LC-MS/MS). For each pesticide we calculated the ratio of the average amount of main metabolite excreted in urine, relative to the amount administered, either via ingestion or by dermal absorption. Additionally, the time courses of urinary excretion were described and toxicokinetic parameters of the metabolites were derived, such as half-life, elimination rate constant, time point to the maximum concentration and the area under the curve. Overall, metabolites were rapidly excreted into urine after oral exposure and renal elimination was considerably slower after dermal exposure. Significant differences in time courses of excretion and toxicokinetics between individuals could not be observed.

Keywords: biomarkers, biomonitoring, environmental health, pesticides

MO-SY-B4-78
Detection of urinary pesticide biomarkers using liquid chromatography high resolution mass spectrometry (LC-HRMS)

Nijssen R. [1], Oerlemans A. [2], Scheepers P. [2], Vermeulen R. [3], Mol H. [1]; [1]Wageningen Food Safety Research, Wageningen, Netherlands, [2]Radboud Institute for Health Sciences, Radboudumc, Nijmegen, Netherlands, [3]Institute for Risk Assessment Sciences, Utrecht University, Utrecht, Netherlands

Following exposure and uptake, most pesticides are rapidly metabolised and excreted in urine. A major bottleneck in biomonitoring of pesticides is that the optimum biomarkers of exposure in humans are often not known. The improvements in analytical instrumentation enables biomarker discovery/verification in urine from volunteers exposed to pesticides at low (<ADI) levels. The aim of this study is to find the most suitable urinary biomarkers of exposure for pesticides in humans. A workflow was developed which involved analysis of urine samples collected from volunteers before and after exposure, with and without enzymatic deconjugation. Analysis was done by liquid chromatography with full scan high resolution mass spectrometry (LC-full scan HRMS). Multiple data processing workflows were employed for biomarker detection using dedicated software tools. Results are presented for analysis of urine samples from volunteers orally and dermally exposed to either asulam, carbendazim, chlorpropham, prochloraz or tebuconazole. Multiple metabolites were (tentatively) identified. Most metabolites found in human urine have also been reported in animal studies. As was to be expected, higher responses were observed after oral intake compared to dermal application. For the pesticides studied, no clear differences were found between oral and dermal exposure in terms of relative responses of metabolites. Analytical reference standards were purchased or synthesized for confirmation of the most suitable biomarker candidates. The studies show that current LC-HRMS instrumentation and software tools are suited for human biomarker discovery/verification in urine from subjects exposed to low levels of pesticides.The work flow, with slight modifications, has also been applied for pesticide biomarker detection in humans without controlled administration of pesticides. A number of additional pesticide biomarkers were found, demonstrating the feasibility of suspect screening for exposure assessment in the general population.

Keywords: analytical methods, biomarkers, biomonitoring, pesticides

MO-SY-B4-79
Results from a field study into the potentially enhanced internal exposure of pesticides in residents living close to agricultural land

Vermeulen R.; Utrecht University, Utrecht, Olandija

Approximately 5% of the Dutch population, living in less urbanized areas (500m). We successfully measured 46 different pesticides in indoor dust concentrations of 58 residential and 16 control homes and repeated 24hr air concentrations at 71 residential and 16 control homes. In addition, urinary measurements of 5 selected pesticides (i.e. asulam, carbendazim, chlorpropham, prochloraz and tebuconazole) were performed among the 192 residents from the homes in the study. Results of both the environmental- and bio-monitoring were analyzed for patterns over the application season, between residents and controls, and with distance to the fields. Empirical results were corroborated with a novel model framework allowing the estimation of environmental concentrations and personal exposures out and inside the home. Detailed results of these comparisons, currently embargoed, will be presented.

Keywords: pesticides

MO-SY-C4: The Healing Power of Indoor Air and Material Surfaces

MO-SY-C4-80
The healing power of indoor air and material surfaces

Taylor S., Taylor Healthcare Consulting

Biologists know that organisms compete to survive and reproduce in their environment. The physical properties of this environment have profound effects on the shape, physiology and actual viability of all things, from single cell microbes to complex multi-system mammals such as humans. We spend over 80 percent of our time indoors, yet understanding how the indoor built-environment influences acute and chronic diseases that we struggle with has lagged behind other research. Thankfully, our understanding has grown rapidly from new genetic analysis tools. These techniques have informed us that humans are vastly outnumbered by large indoor communities of viruses, bacteria, and fungal organisms that can promote either health or disease. We also are learning that indoor air management as well as the surface properties of indoor materials can select the bacteria and viruses that cause disease, while simultaneously weakening the human immune system. While this is alarming, it also reveals a whole new, powerful strategy to help curtail viral and bacterial epidemics. Understanding this intersection of buildings, indoor microbes and humans is not simply theoretically interesting; it has become an urgent topic for all of us. Hospitalized patients are at serious risk of new infections, often from antibiotic resistant bacteria, which claim the lives of hundreds of thousands of patients per year in the US alone. In addition, seasonal influenza, or more alarmingly – the specter of a viral pandemic, affects all of us who reside in buildings. Get ready to be totally amazed by the preventive power of properly managed indoor air and surfaces to suppress infectious microbes, support human immunity and enhance productivity and learning. For those still unconvinced, the business case for these interventions will be presented. Participants will be excited to learn actionable steps for indoor-air management that improves health, productivity and learning.

Keywords: built/indoor environment

MO-SY-C4-81
Mankind moved from “Ecosystem Nature” into “Sealed Building Biotopes”

Hugentobler W.; Inst. of Primary Care, University Zurich, Gerra-Gambarogno, Switerland

Mankind moved from “Ecosystem Nature” into “Sealed Building Biotopes” The consequences call for a new hygiene concept with updated microbial perspectives In the very short timeframe of the last two hundred years, industrial revolutions shifted our living and working spaces more and more into buildings, which became increasingly airtight in the last fifty years. Spending more than ninety percent of our lifetime in sealed rooms, often mechanically ventilated with filtered air, limited our interactions with the diversified microbiomes of nature. Our building habitats have become sealed biotopes. For hygienic reasons, natural and locally available mainly porous materials for construction, interior finish and accessories have increasingly been replaced by fabricated, easy to clean, nonporous materials. These materials lack any water storage capacity and are either wet (if undesired condensation occurs) or dry, independent of the relative humidity in the surrounding air. From the perspective of surface attached bacteria and mould, this results in a permanent, year round stress by deprivation of water and nutrients. This stress scenario, intended for hygienic reasons in order to efficiently prevent any proliferation of bacteria and mould turned out to have dangerous side effects on human beings. The provoked stress adaptations (morphologic, metabolic and genetic) of bacteria result in multi-resistance, including antibiotic resistance. This issue is becoming particularly evident in hospital environments where we had to recognize that the multi-resistant gram positive and negative bacteria causing hospital associated infections share the common feature of resistance to water and nutrient deprivation. Dryness and cleanliness as hygienic concepts need to be reconsidered in the light of updated knowledge of microbial stress reactions that combines culture based methods with stress testing and genome sequencing.

Keywords: infectious disease

MO-PL-D4-EAM: Use of Biomarkers to Assess Health Risks

MO-PL-D4-82

Non-targeted approaches to screen new markers of chemical exposure from human matrices: application and challenges regarding sample preparation and data processing

Pourchet M. [1], Cariou R. [1], Bichon E. [1], Hutinet S. [1], Le Bizec B. [1], Antignac J-P. [1]; [1]Laboratoire d'Etude des Résidus et Contaminants dans les Aliments (LABERCA), INRA, Nantes, France

Our environment is vector of human exposure to hundred thousands of chemicals. This opened view of the chemical space contaminating the environment-food-human continuum is today supported by advanced chemical profiling instrumentation opening the door to holistic sample characterisation. One particular issue in this context is related to the early warning detection of emerging chemicals. Non-targeted screening (NTS) approaches are new valuable strategies aiming to detect unknown markers of exposure without any a priori. These screening approaches are associated to important expectations and promises, but also major challenges for scientists, particularly concerning sample preparation and data processing. In the frame of the workpackage WP16 "emerging chemicals" of the Human Biomonitoring for European Union project (HBM4EU, EU H2020), the present work proposes a first focus on these two last components, in order to determine which are suitable for NTS. The halogenated compounds were the selected chemicals for this study, because of their persistence in the environment and their toxicological concern. The first step consisted in developing appropriated sample preparation protocols for various human matrices of interest, especially accessible matrices from mother-child cohorts, namely breast milk, meconium or placenta. The main lines of these investigations will be presented and discussed, with a particular – but not exclusive – focus on placenta analysis. The challenging sample preparation issues will be described, as well as the final analytical strategy which was a combination of a Bligh and Dyer extraction[1] followed by a LC-HESI-HRMS Q-Orbitrap analysis. The second step consisted in developing a user-friendly application, HaloSeeker[2], based on isotopic patterns and mass defect, to highlight chlorinated and brominated clusters, resulting in several chemical formulae assigned as a finalised output. This first proof of concept finally illustrates the real capability and complexity of these NTS approaches in support of human biomonitoring and environmental-health researches. [1]doi.org/10.1139/o59-099
[2]doi/10.1021/acs.analchem.8b05103

Keywords: analytical methods, biomarkers, biomonitoring, environmental health, prenatal

MO-PL-D4-83

Saliva Metabolome in Assessing Internal Exposure to Traffic Related Air Pollution

Liang D. [1], Chang C. [1], Li Z. [1], Vu B. [1], Golan R. [2], Sarnat S.E. [1], Yu T. [1], Greenwald R. [3], Jones D.P. [1], Chang H.H. [1], Sarnat J.A. [1]; [1]Emory University, Atlanta, GA USA, [2]Ben Gurion University of the Negev, Beersheba, Israel, [3]Georgia State University, Atlanta, GA, USA

Background. High-resolution metabolomics (HRM) is a sensitive tool for measuring complex environmental mixtures and biological responses. However, most HRM applications, to date, have examined blood metabolites, the collection of which is invasive, making it extremely difficult to perform longitudinal biosampling. Saliva, an oral fluid and natural filtrate of blood that contains omics-features worthy of interrogation, may serve as an alternative, non-invasive biospecimen to blood plasma for use in HRM application. Methods. We conducted the Saliva Metabolome in Traffic-related Health (SMITH) study to examine the feasibility of saliva as an alternative HRM biospecimen for use in detecting exposure to traffic-related air pollution (TRAP). We assessed exposures to several TRAP (e.g., black carbon, nitrogen oxides), and biomonitoring in a panel of 51 students living at various distance to a traffic hotspot. Monthly saliva and plasma samples were collected over four months and analyzed using HRM. We then conducted an untargeted Metabolome Wide Association Study (MWAS) workflow, including biological pathway analysis and chemical identification. Results. A total of 29,013 metabolic features were reliably extracted from 204 saliva samples and 20,766 metabolic features from 175 plasma samples. 9479 features were shared in both saliva and plasma samples with moderate-to-strong correlation ($\rho=0.55$, $p<0.0001$). Using an untargeted MWAS workflow, we detected eight metabolic pathways consistently associated with TRAP in both saliva and plasma samples, including pathways predominantly associated with xenobiotic-mediated oxidative stress and acute inflammatory response, including as leukotriene and vitamin E metabolism. Conclusions. With moderate-to-strong correlation and similar trends in pathway perturbation associated with TRAP in both biometrics, these results provide initial support for the use of saliva samples as a sensitive, and less invasive biometric for metabolomics analysis. Ongoing cluster analysis and chemical annotation will further clarify saliva metabolic perturbations associated with TRAP exposures to compare with metabolic changes observed in plasma metabolome.

Keywords: air, biomarkers, biomonitoring, environmental health, other (specify)

MO-PL-D4-84
Biomarker-Based Assessment of the Influence of Chemical Exposures on Chronic Kidney Disease

Brabec S.[1], Nguyen V.[1], Jolliet O.[1]; [1]University of Michigan, Ann Arbor, MI, USA

Background/Aim Chronic Kidney Disease (CKD) was the 9th leading cause of death in the United States in 2015. Previous studies have identified environmental toxicants such as Per- and polyfluoroalkyl substances (PFASs) and heavy metals to be associated with CKD but focused on only a handful of chemical classes corresponding with a few biomarkers of CKD. To better understand the influence of environmental toxicants on CKD, we implemented an overarching screening approach to systematically evaluate the associations between chemical biomarkers and biomarkers of CKD as well as identify populations at risk for CKD. Methods We used the National Health and Nutritional Examination Survey (NHANES) of 167 chemicals and 4 biomarkers of CKD (Cystatin, Glomerular Filtration Rate (GFR), Albumin, and Creatinine) measured in 18881 participants 1999-2002. For every pairwise combination of CKD and chemical biomarkers, we conducted a series of linear regression models to determine the association of one biomarker of CKD and chemical exposure adjusted for covariates such as age, race, gender, poverty income ratio, and cotinine (biomarker of smoking). Results For a 1 unit increase in chemical biomarker levels, we identified Heptachlorodibenzofuran and Oxychlordane decrease GFR by 29% and 28% respectively, and a few PFASs decreases albumin levels by 4-12%, implying an increased risk for CKD. Additionally, Non-Hispanic Blacks, Mexican Americans, & Other Hispanic Americans on average have Albumin levels that are higher compared to those of Non-Hispanic Whites by 57%, 10% and 19%, respectively, implying that they are at a higher risk for CKD. Conclusions Overall, we presented a framework that systematically identified populations susceptible to CKD and environmental toxicants that are either risk or protective factors for CKD. Ongoing work is understanding how these toxicants influence biological pathways associated with CKD to better elucidate the link between environmental exposures and kidney health.

Keywords: biomarkers, cumulative exposure, environmental health, health, chronic diseases, statistical methods

MO-PL-D4-85
Using Biomonitoring Data to Understand Multipollutant Exposure Burden in Canada

St-Amand A.[1], Willey J.[1], Pollock T.[1], Liang C.[1]; [1]University of Montreal, Montreal, Canada

Canadians are ubiquitously exposed to mixtures of chemicals that vary spatially and temporally, and that may have differential impacts on health depending on the exposure context. Moving from a chemical-by-chemical assessment paradigm to considering cumulative effects of multiple exposures will require further development of relevant conceptual approaches. Mixtures can be prioritised by considering the likelihood of co- or sequential-exposure, important exposure sources, and internal body burden, using biomonitoring data which incorporates relevant exposure routes.In the current study, representative Canadian biomonitoring data from the Canadian Health Measures Survey (CHMS) is used to characterize realistic cumulative exposures in the population. To do this a measure called Exposure Load (EL) is developed, which is based on determining whether each survey respondent is exposed or not exposed to a specific chemical group, based on a defined exposure threshold, and then summing counts to derive EL. The analysis uses 27 different chemical groups, based on cycles 3 and 4 of the CHMS. The Limit of Detection (LOD) is used as an exposure threshold. Results are stratified by age group, gender, chemical group or other parameters to explore differences in exposure burdens across the population.Preliminary results from the Exposure Load analysis provide an idea of the distribution of exposure burden across the Canadian population. Analysis by age, gender and other factors will help to identify subpopulations disproportionately exposed to chemicals, while analysis by chemical groups, such as metals or VOCs, will aim to identify whether certain chemical classes contribute disproportionately to higher exposure burdens.Moving toward characterizing the exposome, innovative thinking is needed for addressing multipollutant exposures in chemical risk assessment. The Exposure Load (EL) method focuses on characterizing exposure burdens in the Canadian population in an effort to understand chemical mixtures with the ultimate goal of advancing our understanding of population health impacts.

Keywords: VOCs, biomarkers, biomonitoring, cumulative exposure, environmental health

MO-PL-E4-CPC: Housing and Health

MO-PL-E4-86
Influence of house types on the prevalence of chronic respiratory diseases in Ho Chi Minh City

Tran N.T.[1], Tran B.N.[1], Tran T.M.H.[1], Tang K.H.[1], Ngo M.X.[1], Godin I.[2], Michel O.[2], Bouland C.[2]; [1]Ho Chi Minh City, Viet Nam, [2]Brussels, Belgium

Exposure to pollutants is related to house typology. Besides smoking, indoor air pollution is a significant risk factor of chronic respiratory diseases (CRD). The prevalence of CRD in different house types was studied in Ho Chi Minh City. 1561 people living in four house types, were enrolled. Those house types are apartment, rental house (a very small house with ground floor, toilet and main room, with or without windows, often built into row of rental houses), rural house (a one floor and separate house which is built in rural and suburban district by wood or semi-solid materials) and tube house (several floors, built close together, very small in width and deep in length). Information on respiratory health, lung function, residential characteristics, and indoor environmental sources was obtained by questionnaire and spirometry. The two main respiratory health outcomes studied are clinical CRD (CCRD) (symptom of CRD) and chronic obstructive respiratory diseases (CORD) (FEV1/FVC < 0.7). The prevalence of CCRD (24.3%) and CORD (5.3%) did not equally distribute among studied house types ($p<0.0001$, Chi-square). It is equal in tube house and apartment. Comparing to people living in apartment, those who live in rental house had a 46% higher risk of developing CCRD. The odds of CORD for people living in rental house and in rural house were 4.64 (1.97-10.5) folds and 2.99 (1.21-7.37) folds higher. Controlled factors were age, sex, in home time, smoking level, occupational risk, previous tuberculosis and presence of pet, rat, cockroach, flooded wall, biomass usage and air-conditioner. House type was associated with a likelihood of CCRD and CORD morbidity.

Keywords: respiratory health

MO-PL-E4-87

Explanations of differences between living room temperatures in England and Denmark

Bonderup S. [1], *Jensen O.* [1], *Gunnarsen L.* [1]; [1]*Danish Building Research Institute, Copenhagen, Denmark*

Indoor temperatures have an impact on both occupant health, wellbeing, and on the relative humidity of the indoor air. Room temperature is colloquially considered to be 20-22 degrees Celsius, and WHO has set 18 - 24 degrees Celsius as the range of safe and well balanced indoor temperatures. In northern Europe's temperate climate, central heating is necessary to obtain these indoor temperatures in the winter months. We know that relatively small differences in temperature can change the energy use for heating remarkably. Differences in usage patterns and temperature preferences can therefore affect both heating expenses and occupant health and wellbeing. We examine possible technical and socio-cultural answers to notable differences seen in temperature and heating pattern data in England and Denmark, where climate and building style are relatively comparable. The data used consists of measurements from two SBi projects in Denmark, and secondary data from two studies in England. The English CaRB HES and 4M studies report winter temperature measurements in a total of 697 households from a diverse sample of dwelling types, ages and tenures. The Danish data consists of measurements from 100 detached houses, and approximately 500 apartments in 3 different estates. Building physics aspects that can affect indoor temperature and heating pattern range from overall energy efficiency of the building shell, over ventilation, to the type of central heating system and how it is operated. The socio-cultural aspects are more ambiguous, for example, when low temperatures have different meanings, whether they are regarded as an expression of energy saving behaviour or an indicator of energy poverty. Getting a more nuanced view of indoor temperature and heating patterns can aid in understanding and developing the complex relationship between occupant comfort and health, building health, and sustainable use of energy and resources.

Keywords: behavior, built/indoor environment, thermal environment

MO-PL-E4-88

Prediction of wheezing at ages 3 and 7 and asthma at age 7 using quantified observational exposure data collected at age 1

Cox J. [1], *Ryan P.* [2], *Burkle J.* [2], *Jandarov R.* [1], *Zhu Z.* [1], *Mendell M.* [3], *Reponen T.* [1], [1]*University of Cincinnati, Cincinnati, OH, USA,* [2]*Cincinnati Children's Hospital Medical Center, Cincinnati, OH, USA,* [3]*Indoor Epidemiology, El Cerrito, CA, USA*

Although the association of observed dampness and mold with asthma is well established, most epidemiologic studies have used dichotomous metrics of observed mold that cannot assess exposure-response relationships. This study investigated specific exposure-response relationships within an ongoing birth cohort, the Cincinnati Childhood Allergy and Air Pollution Study (CCAAPS) (n=762). The primary exposure variable was the maximum damaged surface in a home from water damage or mold. Potential confounding variables were identified with directed acyclic graphs. Logistic regression models were used to estimate odds ratios (ORs) between year 1 exposure and year-3 wheeze, year-7 asthma, and three wheeze outcomes at year-7: early transient (ETW), late onset (LOW), and persistent (PW) wheeze. Analyses were adjusted for income, race, presence of cockroach antigen, and reported presence of mice or rats. Potential interactions of exposure variables with water damage history and moldy odor were assessed. The upper quartile of individuals with any damage from water or mold had increased odds for year-7 asthma (OR=1.9, $p<0.1$), year-3 wheeze (OR=2.0, $p=0.05$) and year-7 PW (OR=3.3, $p<0.01$). Several significant interactions ($p<0.05$) were identified. The upper quartile of individuals with any damage from water or mold, when water damage history was absent, had increased odds for year-3 wheeze and year-7 ETW, and when moldy smell was absent, had increased odds for year-7 PW. Our data suggested that moldy odor and water damage history may be more important when there is little to no visible mold or water damage than in cases of higher levels of observed dampness or mold.

Keywords: Indoor Air, Mold, Exposure, Wheeze, Asthma

MO-PL-E4-89

Influence of the Built Environment Factors on Mental Health of Occupants

Hoisington A. [1], Stearns-Yoder K. [2], Lowry C. [3], Postoloache T. [4], Brenner L. [2]; [1]Air Force Institute of Technology, Wright-Patterson, OH, USA, [2]Department of Veteran Affairs, Denver, CO, USA, [3]University of Colorado Boulder, Boulder, CO, [4]University of Maryland, Baltimore, MD, USA

Awareness of Veteran mental health conditions have been poignant since the start of the Global War on Terrorism. While research on how to best diagnose and assist Veterans with mental illness abounds, little is known about their residences and the related impacts on these illnesses. Growing research is indicating multiple ways in which the built environment affects occupant mental health. For these reasons, a joint research study into the built environment of Veteran homes and the associated mental health outcomes was undertaken. Currently there is not a set measurement for analyzing the built environment in terms of mental health outcomes. Therefore, the team developed a self-reported survey of residential quality. In this paper we present the survey and discuss preliminary results that were compared against five psychometrically sound measures of mental health conditions, to include depression, post-traumatic stress disorder, and insomnia. Although further analysis is forthcoming with additional survey respondents is planned, this analysis includes the first 92 Veterans. Preliminary results show that a lack of nature pictures or paintings in the home significantly increased odds of severe mental health symptoms across all measures. Additional aspects of the built environment, such as water damage to the home, poor air quality, and low ceiling height, were also associated with negative mental health symptoms validating current academic literature. Further studies between interdisciplinary teams of social scientists and building designers should focus on longitudinal effects of exposures to built environment factors and interventional studies to more completely understand how these factors influence mental health.

Keywords: health

MONDAY POSTER SESSION

MO-PO-01
Impact of construction stages on Indoor Air Quality

Gallon V.[1], Le Cann P.[1], Sanchez M.[1], Dematteo C.[1], Le Bot B.[1]; [1]Univ Rennes, Inserm, EHESP, Irset (Institut de recherche en santé, environnement et travail), Rennes, France, [2]Univ Rennes, Inserm, EHESP, Irset (Institut de recherche en santé, environnement et travail), Rennes, FranceAQC (Construction Quality Agence), Paris, France, [3]INDDIGO, Chambéry, France

People spend 85% of his time in closed environments and they are exposed to physical, chemical and microbiological agents which have or are suspected to have impacts on human health. These pollutants or contamination come from different sources of emission including building materials and indoor activities, but less information are available regarding contamination during construction process which could persist during the entirely life of building. This study aims to evaluate the emission of substances during different construction stages: volatile organic compounds, fungal contamination, particles and semi volatile organic compounds. It focuses on the impact of the different trades (painter, plasterer, electrician, mason...) considering differences in process (materials, products...) and implementation of ventilation equipment. The objectives are to identify the potential sources of pollution that could impact the indoor air quality at the building reception and thus contribute to the exposure of the future occupants. The results show that the future indoor air quality does not only depend on the emissions of each building product but also closely related to the whole implementation process. The results of mold spore measurements can reach 1400 CFU/m3, which is greater than the guide value proposed by Anses as abnormally high, i.e. 1000 CFU/m3. These molds, even hidden at the delivery phase, could reappear during the occupation of buildings or remain active behind a lining wall, and could be correlated with the onset of asthma in children. The most emitting tasks of volatile organic compounds correspond to the use of products considered as secondary or annexes which are not always the subject of prescription in the special technical clauses. Some phtalates, permethrin, hydrocarbons were found in significant concentrations upon the delivery of the building as well as Triclosan, suspected to be endocrine disruptor and yet prohibited in the treatment of materials and construction since 2014

Keywords: VOCs

MO-PO-02
Volatile organic compounds in school buildings with indoor air quality problems

Vornanen-Winqvist C.[1], Alapieti T.[1], Mikkola R.[1], Kurnitski J.[1], Salonen H.[1]; [1]Aalto University, Espoo, Finland

Indoor air quality (IAQ) has an important contribution to health. Thus, understanding the significance of different factors influencing IAQ and human comfort is of great importance. Volatile organic compounds (VOCs) are often measured as a significant part of IAQ investigations when IAQ is experienced unsatisfactory. The aim of our study was to measure VOC concentrations in four Finnish comprehensive school buildings built or renovated between 2005 and 2009. They were located in the Helsinki Metropolitan area and had mechanical supply and extract ventilation system. Surface materials were plastic floorings and painted walls. Classrooms were furnished with modern school furniture. Occupants had reported significant symptoms or discomfort but no specific cause had been found. VOC sampling (24 samples) and analysis were carried out in six classrooms per school according to the ISO standard 16000-6:2011. Air samples were taken from the central area of an empty, closed classroom in the main working zone at a height of 1.5m. TVOC (total volatile organic compounds) concentrations varied from <10 to 40 µg/m3 (averages per school 12, 9, 8 and 12 µg/m3, respectively). Single VOC concentrations varied from 2 µg/m3 and existing in more than one classroom of all schools were detected: 2-ethyl-1-hexanol, decanal, nonanal, decamethylcyclopentasiloxane, toluene and acetone. Typical sources of the detected VOCs are e.g. different cleaning agents and human cosmetics. In conclusion, measurements showed that the variety and quantity of VOCs was remarkably minor even in the indoor air of schools with reported prolonged IAQ problems.

Keywords: VOCs

MO-PO-03
Comparison between PMF and isotopic compositions on carbon and Pb sources investigation

Jung C-C. [1]*, Chou, C.-K.* [2]*, Huang Y-T.* [2]*, Chen M-J.* [2]*, Huang S-H.* [2]*, Huang W-J.* [2]*, Chang K-Y.* [2]*, Hsieh B-J.* [2]*, Tsai C-Y.* [2]*, Chang S-Y.* [3]*, Lee C-T.* [4]*;* [1]*Research Center for Environmental Trace Toxic Substances, National Cheng-Kung University, Tainan City, Taiwan,* [2]*Research Center for Environmental Changes, Academia Sinica, Taipei City, Taiwan,* [3]*Department of Public Health, Chung Shan Medical University, Taichung City, Taiwan,* [4]*Graduate Institute of Environmental Engineering, National Central University, Taoyuan City, Taiwan*

The objective of this study was to compare the difference between positive matrix factorization (PMF) modeling and isotope analyses to carbon and Pb sources in PM2.5. The PM2.5 samples in the atmosphere was collected in the November of 2016, February and August of 2017, and March and July of 2018 in south-central Taiwan (nine sampling sites). This study also collected PM2.5 samples from traffic emission, straw burning, oil burning plant, coal power plant, incinerator, and steel plant. 14C, Pb isotope signals (206Pb, 207Pb, and 208Pb), elemental metals, ions, organic carbon, and elemental carbon were analyzed for all samples. Our results revealed that the average PM2.5 concentrations (23.2 ± 13.2 μg/m3) was higher in the atmosphere than Taiwan's standard (annual average: 15 μg/m3). PMF modeling showed that both industrial emissions (mix) and biomass burning were major sources on carbon; industrial emissions (cement ash, oil refining, and steel plant) and traffic emissions were important contributors on Pb. 14C analyses indicated that the contributions of modern (49%) and fossil (51%) carbons were similar; Pb isotope ratios suggested that traffic and industrial emission (including steel plant, coal power plant, and incinerator) contributed about 72% of Pb concentration. In comparison, PMF modeling can quantize the contribution of pollution sources on carbon and 14C analyses provide the percentage of modern and fossil carbons. On the hand, there were large gaps in the Pb pollution source identification between PMF modeling and Pb isotope ratios analyses. In summary, PMF combined with 14C increase the benefits of carbon pollution source identification which further applied in exposure assessment or air quality management. However, it necessary to collect more information in investigating Pb pollution sources in the near future, including the Pb isotope signal from other pollution sources.

Keywords: air, analytical methods, metals, particles, radiation

MO-PO-04
On the aerosol pollution levels and its sources in some primary schools in Vilnius city

Prokopciuk N. [1]*, Franck U.* [2]*, Dudoitis V.* [1]*, Tarasiuk K.* [1]*, Juskiene I.* [3]*, Valiulis A.* [3]*, Cepuraite D.* [4]*, Staras K.* [4]*, Davuliene L.* [1]*, Ulevicius V.* [1]*,* [1]*SRI Center for Physical Sciences and Technology, Vilnius, Lithuania,* [2]*Helmholtz Centre for Environmental Research–UFZ, Leipzig, Germany,* [3]*Vilnius University Faculty of Medicine, Institute of Clinical Medicine, Vilnius, Lithuania,* [4]*Vilnius University Faculty of Medicine, Public Institution Vilnius Centro Outpatient Clinic, Vilnius, Lithuania*

The air quality in school environment is a significant cause for concern, because children spend more time at school than anywhere else except for home. Usually, this quality is assessed evaluating aerosol pollution levels. In the present work, aerosol particle number (PN) and mass concentrations (PM) were studied in eleven elementary schools during the 2017-2018 school year (from September to May) in Vilnius city with the aim to evaluate the main aerosol pollution sources and their levels. Expeditious information on the total aerosol particle concentration in the size range of 0.01 – 1.00 μm was estimated using a Condensation Particle Counter (CPC, TSI model 3007). Particle number and mass concentrations in the size range of 0.3 - 10 μm were measured using an Optical Particle Seizer (OPS, TSI model 3330). The main sources responsible for the elevated levels of indoor PN and PM aerosol concentrations were determined, as follows: local canteens in the absence of ventilation (particle concentrations up to 97500 part/cm3 (CPC)), the school-grader activity during the lesson breaks (275-586 μg/m3), soft furniture and carpets in the classrooms and corridors (~220 μg/m3) and in some cases (city centre) - elevated outdoor aerosol pollutions (in the range of 16800-18170 part/cm3). As well, significant aerosol indoor pollution levels may be related to the outdoor occasional pollution sources, as follows: constructional works in schools during lessons (in the range of 200 - 1000 μg/m3 (OPS)), scraping the exterior walls of buildings near schools (in the range of 1700 - 4600 μg/m3 (OPS)), use of petrol-powered trimmers during cutting the green plantings (22500 - 66400 part/cm3 (CPC)). In general, indoor aerosol particle number and mass concentrations in school classrooms varied in the range of 1100 - 31000 part/cm3 (CPC) and of 16-1350 μg/m3 (OPS), respectively.

Keywords: air, children, particles, ventilation

MO-PO-05
Transgenic pollen corn dispersion at Cordoba, Colombia

Mazo Castaño C. [1]*, Pinto Herrera L.C.* [1]*;* [1]*University of Los Andes, Bogota, Columbia*

Transgenic crops were created in order to supply the high food demand as a result of population growth. Through biotechnology, different ways have been developed to satisfy this need, among them, the creation of transgenic seeds, capable of being resistant to pests and offering greater yield and productivity. Since its first sale in the 90s, several countries in the world have planted these seeds with the aim of increasing the potential of their crops; however, different authors and organizations have created campaigns against their commercialization, since, according to them, they have impacts on the environment, health and the social sector. Based on this concern, several researches have been made with the intention to figure out the effects of their use, such as possible genetic contamination by altering the native ecosystems, the effects on the use of pesticides, the tearing of cultural traditions and other types of affectations. This research work seeks to study one of its characteristics and its possible problem: the dispersion of pollen from a batch of transgenic corn crops in the municipality of Tierralta, Córdoba. Several models of pollen dispersion were carried out with the Hysplit program, changing the sedimentation rate and the density of the particle, in order to conclude the deposition speed and the viability of the transgenic corn pollen. The results showed that the greatest dispersion occurs in the first months of the year, causing the pollen particles to move away up to 23 km from the source. This study has great relevance, it serves as a basis for future research on genetic contamination of traditional crops.

Keywords: air, analytical methods, community, exposure factors, food, geospatial analysis/GIS, soil

MO-PO-06
Volatile organics off-gassed from dry-cleaned fabrics

Chien Y-C. [1]*, Wang Z-C.* [2]*;* [1]*Undergraduate Program in Safety, Health and Environmental Engineering, Department of Safety, Health and Environmental Engineering, National Yunlin University of Science and Technology, Douliu, Taiwan,* [2]*Department of Safety, Health and Environmental Engineering, National Yunlin University of Science and Technology, Douliu, Taiwan*

Fabrics are known to absorb and re-emit organic compounds. Thus dry-cleaned clothes may release volatile organic compounds (VOC) into surrounding indoor environment, causing inhalation exposures to dry-cleaning workers and consumers. Petroleum-based solvents are commonly used in Taiwan as the dry-clean solvent. This study examines the residues and reemission of the VOC from dry-cleaned fabrics, aiming at finding the emission factors to improve the process for the industry, and to minimize VOC exposures for consumers. The emission kinetics and factors of VOC from dry-cleaned fabrics were tested in a controlled facility. The results indicate that re-emission of the VOC from dry-cleaned fabrics was affected by environmental factors, in addition to fabric type and washing parameters. Wool showed greater re-emission compared to cotton and silk under similar cleaning treatment. For wool, the first–order concentration decay constant (k) increased 67% when test temperature increased by 10°C (25°C to 35°C), while k increased 221% when air change rate increased from 0.5 h-1 to 2.0 h-1. Conversely, k reduced by 67% as the relatively humidity increased (60% to 80%). It was estimated that 3.65d is required for a dry-cleaned sweater to re-emit VOC completely under normal environmental condition (25°C/60%RH/0.5ACH). A newly dry-cleaned sweater will result in an increase of indoor total VOC to 20-60 µg/m3 in a typical room (80 m3) during a 12-h period.

Keywords:

MO-PO-07
DEHP transfer from PVC sheet to different kinds of particles.

Shinohara N. [1], *Uchino K.* [1]; [1]*National Institute of Advanced Industrial Science and Technology, Tsukuba, Ibaraki/Japan*

DEHP emission rates and transfer rates to house dust from PVC sheet were determined using PFS. The transfer rates from PVC sheet to house dust were 40- to 500-times higher than the emission rates from PVC sheet to the adsorbent through air layer. The transfer rates from PVC sheet to house dust were constant regardless of the thickness of air layer on the house dust. We will present on the transfer rate of DEHP from PVC sheet to house dust, soda–lime–silica glass particles, and polyethylene particles.

Keywords: SVOCs, air, behavior, particles, phthalates

MO-PO-08
The Impact of Cooking Pan Material on UFPs Emission

Amouei Torkmahalleh M., Nazarbayev University, Astana, Kazakhstan

Cooking is one of the major sources of indoor UFPs. Oil, meat, stove and pan are among the key sources contributing to cooking UFPs. However, the relative importance of these sources has not been yet investigated under a controlled experimental protocol. It was found in the literature that desorption of semivolatile organic compounds (SVOCs) from the heated metal surfaces including pans and burners contributes to UFPs emissions. Thus, the material coated on the surface of the pans may impact the amount of surface deposited SVOCs, and therefore the UFP emission rate. Among different cooking components, the emission rate of UFPs generated from a heated cooking pan itself has not been yet investigated. The aim of this study is to examine the impact of type of cooking pans including Teflon, ceramic, granitium and uncoated aluminum pans on UFP emission rates under controlled experimental conditions. Furthermore, this study investigates the relationship between the pan exposure time to indoor air prior to heating, and the UFPs emission rates during the heating. Currently, we have been monitoring the UFPs concentration using an ultrafine particle counter, Nanotracer (Aerasesne, NL) capable of monitoring total UFP concentrations from 10 to 300 nm up to 106 UFPs/cm3. The results so far indicated that un-coated aluminum pan produces significantly higher UFPs compared to other three pans. We are planning to study the effect of the exposure of pan to the indoor air for 23, 47, 72 and 144 hr on UFPs emission rates. Effective April 2019, our laboratory will be equipped with an 1nm SMPS (Model 3938EL57, TSI, USA) capable of monitoring size segregated particle number concentrations from 1.4 nm to 1µm up to 107 Particles/cm3. Thus, we will have the chance to extend our experiments using this SMPS and presente the results during the ISES 2019.

Keywords: SVOCs, air, built/indoor environment, nanoparticles, particles

MO-PO-09

Factors affecting the indoor concentration of ozone and ozone reaction products in school buildings and means to reduce exposure

Salonen H.[1], Salthammer T.[2], Morawska L.[3]; [1]Aalto University, Department of Civil Engineering, Espoo, Finland, [1]Fraunhofer WKI, Department of Material Analysis and Indoor Chemistry, Braunschweig, Germany, [3]Queensland University of Technology, International Laboratory for Air Quality and Health, Brisbane, Australia

Several epidemiological studies have reported that ozone and products of ozone reactions cause several acute and chronic health effects, and that even a slight amount of ozone is po-tentially deleterious to human health. Knowledge of the factors affecting the concentrations of ozone in different environments is important to define the means to reduce the exposure to ozone and to ozone reaction products. [1]The literature review was performed using Web of Science, SCOPUS, Google Scholar, and PubMed. Altogether 34 search terms and their combinations were used to identify journal papers and reports. This study revealed that the following factors affect the concentration of ozone and ozone reaction products indoors: 1) Outdoor ozone penetrat-ing into the indoor environment; 2) Housing characteristics (e.g. type of ventilation, filters, decorations and finishing materials); 3) Indoor chemistry (e.g. ozone reactions with cleaning chemicals); 4) Occupants (ozone reactions with skin, clothes and hair) and their activities (e.g. use of copy machines, laser printers, air cleaning devices, and window opening); 5) Sea-sonality, temperature and relative humidity. The means for reducing exposure to ozone and ozone reaction products indoors include e.g. filtering ozone from the air supply using suita-ble filters, limiting the use of cleaning chemicals and ozone emitting devices, and avoiding window openings when the concentration of ambient ozone is high. Although the ambient ozone concentration constitutes an important part of the total individual ozone exposure, reducing outdoor ozone concentration is difficult. Thus, means to reduce the concentration of indoor ozone and its oxidation products are crucial in limiting exposure.

Keywords: exposure factors

MO-PO-10 WITHDRAWN

MO-PO-11

Impact of household products on indoor air quality: definition of a simple and harmonized test protocol for volatile compounds emission

Nicolas M.[1], Karr G.[2], Maupetit F.[1]; [1]Scientific and Technical Centre for Building, Champs-sur-Marne, France, [2]French National Institute for Industrial Environment and Risks, Verneuil-en-Halatte, France

Consumer products in general and household products in particular are a significant source of indoor air pollutants. Exposure to these pollutants is important as it takes place in confined, poorly ventilated environments where people spend most of their time. The PEPS research project focused on studying the impact of household products on indoor air quality. A simple, adapted and reproducible test protocol for the characterization of volatile organic compounds (VOC) emitted by cleaning products is proposed. The developed methodology specifies: test conditions (type of test chamber, environmental parameters, application scenario of the product), the compounds of interest and the sampling conditions. In addition, it is representative of the use of cleaning products in real environments. Thus, definition of test conditions for the products and of performing the cleaning action during the test constitutes a key point of the work. Emissions of household products are evaluated under controlled conditions. A panel of 33 products, including 27 manufactured products and 6 home-made products, is constituted. Product composition is studied by collecting data provided by manufacturers and by the direct analysis of their content. A robustness study is conducted to define an harmonized and reproducible test protocol, which is then deployed to analyze the VOC emissions of the products. In addition to the development of this methodology, a field campaign is conducted in a realistic environment to collect knowledge on household products emissions in real conditions of use. This campaign is organized in the experimental house of the CSTB, MARIA, to evaluate, in real conditions, the emissions of 10 selected household products from the panel. The cleaning products are used in an experimental room according to realistic scenarios of use. VOC emissions are measured before, during, and after the use of the cleaning product showing the potential impact of those products on IAQ.

Keywords: water

MO-PO-12

Experiment for Reduction on NOx Concentration in Photocatalytic Paint with UV Lamps

Song Y.W. [1], Park J.C. [2], [1]Graduate School, Chung-Ang University, Seoul Korea, [2]School of Architecture and Building Science, Chung-Ang University, Seoul, Korea

This study conducted a Mock-up experiment using a UV lamp to check the effect of reducing nitrogen oxides, a fine dust precursor in a paint containing TiO2 photo catalyst. The Mock-up experiment was applied to paint containing TiO2 photo catalyst on the gypsum board, the representative indoor finish material, and then the UV lamp on/off and UV lamp light levels were changed to confirm the reduction of nitrogen oxide concentration. As a result, the effect of reducing the concentration of nitrogen oxides compared to the off state was obtained and the reduction of the concentration of nitrogen oxides was confirmed to change with the variation in the amount of light. However, there was no proportional change in the concentration of nitrogen oxides due to the increase in the amount of light.

Keywords:

MO-PO-13

Residential indoor exposure to semivolatile organic compounds and cumulative risks in France

Pelletier M. [1], Glorennec P. [1], Mandin C. [2], Le Bot B. [1], Ramalho O. [2], Mercier F. [1], Bonvallot B. [1]; [1]Univ Rennes, EHESP, Inserm, Irset (Institut de recherche en santé, environnement et travail), Rennes, France, [2]University of Paris-Est, Centre Scientifique et Technique du Bâtiment, Marne la Vallée, France

Nationwide measurements of SVOCs in PM and settled dust in housing were used as a basis for modelling exposure, for different age groups, to 32 SVOCs with two-dimensional Monte Carlo simulations. Estimates spanned orders of magnitude, with median exposure ranging from pg/kg-bw/d to µg/kg-bw/d. Inhalation, ingestion and dermal pathway contributed at var-ying proportions depending on compound, and air was the dominant medium for 28 of the 32 compounds (either by inhalation or dermal contact). Indoor exposure variance was mainly driven by indoor contamination variability, and secondarily by uncertainty in chemical pa-rameters. Subsequent risks were assessed chemical by chemical and cumulatively under the additivity hypothesis. Unacceptable risks were shown for a certain portion of the French population depending on compound, and mainly for children. Furthermore, it allowed the identification of the most contributive compounds for different health effects (reproductive, neurologic or immunologic) namely DBP, DEP, DEHP, BDE 47, lindane, and PCBs.

Keywords: SVOCs, aggregate exposure, built/indoor environment, cumulative exposure, risk assessment

MO-PO-14

SO2 and CHCO vertical column densities in major cities of Kazakhstan from 2005 to 2016

Amouei Torkmahalleh M. [1], *Darynova Z.* [1], *Sabyrzhan A.S.* [1], *Sagynov S.* [1], *Abdrakhmanov T.* [1]; [1]*Nazarbayev University, Astana, Kazakhstan*

Tropospheric SO2 and CH2O vertical column densities were analyzed for period 2005-2016 over the atmosphere of Kazakhstan using the Ozone Monitoring Instrument onboard the NASA Aura satellite. We found main hot spots, seasonal trends and overall trend over 11 years. Major hot spots in terms of SO2 relate to regions with high population and industrial objects. For example, the city Ekibastuz has the biggest thermal power plants in the country. Determinants of the hot spots are fuel sources and electricity usage. Power plants in Ekubastuz burned brown coal that contains 30% toxic minerals with long lifetimes. The annual average concentration of SO2 for this city was 2.5x10-5 kg/m2, whereas for the rest of the country it was 6 times lower. Other hotspots includes megacities such as Almaty and Astana experienced high concentrations of SO2 in 2008. One of the main reasons for that is the financial crisis in 2008, as the cities used cheap low-quality coal. Regarding CH2O VCD, one of the hot spots was the city Atyrau, the major oil producer of the country with two massive oil fields. The highest VCD of formaldehyde (9 x 1015 molecules/cm2) is appeared in the summertime due to secondary production in the presence of hydroxyl radicals as a result of photo-oxidation of VOCs emitted from industrial objects and vehicles. Moreover, stimulation of CH2O emissions by wildfires was observed in the capital city Astana in 2012, when the abnormal number (37) of forest area fires happened. Nevertheless, two significant steps were done for improvement of the air quality in 2013: the introduction of the emission trading system (ETS) and Euro-4 standard. It led to the negative trend in CH2O concentration between 2012 and 2015 in the two highly populated cities.

Keywords: SVOCs

MO-PO-15

Examine the Impacts of Human Activities on Indoor Exposure Profile Using a Low-cost Passive Air Pollutant Monitor

Lin E. [1], *Pollitt K.*; [1]*Yale University, New Haven, CT, USA*

In most developed countries, people spend 90% of their time indoors. Personal activity patterns will determine an individual's exposure profile. Cooking, cleaning and personal care products are major emission sources in the indoor environment. The HomeChem study was conducted in June 2018 in Austin, Texas to explore these indoor exposure sources. As part of this study, non-selective passive air samplers were deployed in the model house to profile organic contaminants associated with common indoor activities, including chrysene, Benzo[a]pyrene, fluoranthene. Sampling devices were placed in the kitchen over a 24-hour period and analyzed using thermal desorption gas chromatography time-of-flight mass spectrometry to quantify time-weighted average exposures to a panel of 36 volatile and semi-volatile organic compounds during typical daily activities. Detailed activity logs were recorded within 1 hour on individual experimental day. Experimental days were designed to imitate typical human activity patterns. Those experimental days can be categorized into four types of days, they are Friend Visit, Cleaning, Cooking and Thanksgiving Day. On each of the experimental days, three sampling replicates were conducted. Principal Component Analysis was applied to exposure results to examine relationships between experimental days and individual analytes. Comparisons were made between the length of activity and the exposures to demonstrate the does-response relationship. At the end of the study, key factors that determine the exposure profile were identified; does-response relationships between exposure and activity were established among EPA identified carcinogenic organic compounds.

Keywords: SVOCs, built/indoor environment, cumulative exposure, exposure factors, sampling methods

MO-PO-16

Evaluating vertical concentration variation of non-polar organic compounds for PM2.5 in Taipei metropolis

Tsai W.C. [1]*, Liao, H.T.* [2]*, Wu, C.F.* [1]*;* [1]*Institute of Environmental Health(IEH), National Taiwan University, Taipei, Taiwan,* [2]*Research Center for Environmental Changes, Academia Sinica, Taipei, Taiwan*

Numerous people live in multistory buildings in Taipei metropolis. However, vertical variation in exposure to air pollutants is not well characterized. This study was conducted in Taipei, Taiwan from September to October 2018. Three sampling sites were set up on balconies at three different floor levels (low-, mid-, and high-level sites) of one multistory building. Twenty-four-hour integrated PM2.5 samples were collected on quartz-fiber filters simultaneously at all sites twice a week, making a total of 36 samples. Non-polar organic compounds (NPOCs), including 24 polycyclic aromatic hydrocarbons (PAHs), 16 hopanes, 12 steranes, 23 n-alkanes, and 10 Iso-/anteiso-alkanes, were identified and quantified using Thermal Desorption Gas Chromatography/Mass Spectrometry (TD–GC/MS) method. At the sampling building, n-alkanes showed highest concentrations [10.12±0.96 ng/m3 (Mean±SD)] that accounted for 62% in the analyzed NPOCs, followed by PAHs (3.88±0.33 ng/m3, 24%), and Iso-/anteiso-alkanes (1.86±0.25 ng/m3, 11%). The Heneicosane (n-C21) to Tritriactotane (n-C33) were the most abundant components of n-alkanes, whereas the four- and five-ring elements were dominant in total PAHs. The average concentration of all analyzed species decreased as the height increased. However, there were no statistically significant differences ($p>0.05$) in the average concentration of the three floors. The preliminary results in this study reveal the vertical distribution of air pollutants in the Taipei metropolitan area. Nevertheless, long-term monitoring campaigns are needed to further evaluate the seasonal effects.

Keywords: SVOCs, air, environmental health, particles, sampling methods

MO-PO-17

PFAS and total fluorine levels in fire station dust

Young A. [1]*, Sparer-Fine E.* [1]*, Peaslee G.* [2]*, Sunderland E.* [3]*, Pickard H.* [3]*, Allen J.* [1]*;* [1]*Harvard T.H. Chan School of Public Health, Boston, MA, USA,* [2]*University of Notre Dame, Notre Dame, IN, USA,* [3]*Harvard John A. Paulson School of Engineering and Applied Sciences, Cambridge, MA, USA*

Per- and polyfluoroalkyl substances (PFAS) are a class of over 4,000 highly persistent, fluorinated chemicals used in consumer products, building materials, outdoor gear, and firefighting foams. Firefighters are particularly vulnerable to PFAS exposure due to inhalation of smoke during combustion of consumer products, use of firefighting foams, and contact with turnout gear (which may potentially be treated with PFAS for water resistance). Studies on firefighter exposures to PFAS have focused on exposures encountered while fighting fires, yet most career firefighters spend the majority of their shifts at the fire station itself. In addition, the limited studies available on firefighter PFAS exposure have only analyzed fewer than 15 specific PFAS of the thousands of possible regrettable substitutes to the two legacy PFAS phased out of firefighting foam. Total fluorine can provide a proxy for cumulative levels of fluorinated chemicals. In this study, we measured total fluorine levels with PIGE and analyzed for 25 PFAS compounds in dust samples from six room types (truck bays, turnout gear locker spaces, sleeping quarters, kitchens, living rooms, and gyms) in 15 fire stations in Massachusetts. Fluorine was detected in every dust sample collected, ranging from 7.33 ppm (kitchen) to 953 ppm (truck bay) with a median of 158 ppm. Our multilevel model showed that dust in the truck bays and gear locker spaces had significantly higher total fluorine levels compared to the reference living rooms ($p<0.01$). Our PFAS analysis will allow us to determine if the 25 targeted PFAS account for a significant fraction of the high total fluorine detected, to compare PFAS profiles by room type and corresponding potential sources, and to estimate health risks for firefighters. These results will help establish whether firefighting activities and turnout gear use are contributing sources of firefighter exposure to PFAS.

Keywords: SVOCs, building materials, built/indoor environment, occupational, workplace

MO-PO-18

Inhalation risk assessment of particulate phase polycyclic aromatic hydrocarbons (PAHs) at a naturally ventilated kerbside office building.

Jyethi D. [1], Khillare P. [1], [1]Indian Statistical institute, Colcata, India

The megacity of Delhi has a ring radial road system and bulk of the traffic is road based. Three Ring Roads were planned in the Delhi master plan of 1962, namely, the inner Ring Road, the Ring Road and the outer Ring Road. The Ring Road is now the major traffic artery. Naturally ventilated office buildings in close proximity of high traffic roads are common in developing countries. Sixteen particulate phase polycyclic aromatic hydrocarbons (PAHs) were analysed for weekly particulate samples for a year at a kerbside official building near a major traffic intersection on the Ring Road. Eight-stage Anderson cascade impactor, a simulation of human respiratory system, with a back up filter stage was also used to collect size fractionated particulate samples. High concentration of ∑16PAHs was observed (mean annual ∑16PAHs = 159.8±33.0 ng m-3, range 102.5 - 240.1 ng m-3). PAH isomer diagnostic ratios were used for preliminary source identification. Principal component analysis coupled with multiple linear regression apportioned the following sources: diesel emission- 32 %; natural gas combustion- 21 %, vehicle (gasoline)- 19 %, coal combustion- 15 %, wood/biomass combustion- 7% and oil combustion- 6%. Potentially high inhalation risk assessment was found out for the observed PAHs level at the office building by Incremental Lifetime Cancer Risk (ILCR) methodology.

Keywords: SVOCs, air, particles, risk assessment

MO-PO-19

Formaldehyde and TVOCs sensors evaluation for reliable Indoor Air Quality monitoring

Goletto V. [1], Mialon G. [2], Wang Y. [3], Lesieur I. [4]; [1]Saint-Gobain Research Paris, Aubervilliers, France, [2]Saint-Gobain Research Provence, Cavaillon, France, [3]Saint-Gobain Research North America, Northborough, USA, [4]Saint-Gobain ISOVER FRANCE, Rantigny Research Center, Rantigny, France

Volatile Organic Compounds (VOCs) are major indoor pollutants with multiple origins. Standard methods exist to measure them, but they can only be used by experts, and gives at best an average value of their concentrations. There is a need to monitor them continuously during several days, weeks or even months. Recently, portable devices become available as well as connected objects that can be used by the general public. There is a need to evaluate those devices and assess their limits. The work presented here will show the results of our experimental evaluations of commercial devices that claim to monitor formaldehyde and/or Total Volatile Organic Compounds (TVOCs). These measurements were done in controlled conditions (23°C, 50%RH) in small scale or full scale chambers (12m3 rooms) located in one of our research and devel-opment center in France and in the United States. Influence of humidity (from 50% to 65% RH) was also investigated for the most promising devices. Our objective is to assess their performance in controlled conditions in real rooms to conclude on their reliability and know their limits. The work presented here will detail our methodology and the results of our evaluation.

Keywords: VOCs, analytical methods, sensor technology

MO-PO-20

Disinfection by-products in the atmosphere of indoor swimming pools

Hyttinen M. [1], Ruokolainen J. [1], Pasanen P. [1]; [1]University of Eastern Finland, Kuopio, Finland

Disinfection chemicals are needed in swimming pools to control formation and spreading of pathogenic microorganisms between swimmers. Disinfection by-products (DBP) are formed when disinfectants such as chlorine or sodium hypochlorite have been introduced in to the water which contains organic or inorganic impurities. All the DBPs are considered to cause irritation symptoms in respiratory tract and eyes. In present study, indoor air quality of 32 Finnish swimming pool and spa areas were monitored. VOC-samples (including DBPs) were collected using Tenax TA and Chromosorb 106 adsorbent tubes and SKC AirChek 3000 pumps. VOC-samples were analyzed using TD-GC-MS device (Markes TD-100, Agilent 7890A, Agilent 5975C). Trichloroamines were collected to Na2CO3 + As2O3 impregnated filters. Analysis was made according to EN ISO 10304-1 standard. Trihalomethanes and trichloroamines were the most commonly found DBPs. Chloroform was the main trihalomethane and its concentrations varied from 6 to 270 µg/m3 (mean 61 µg/m3). Concentration of trichloroamines varied from 15 to 330 µg/m3 (mean 117 µg/m3). In addition, around 50 other disinfection by-products were found and their concentrations were typically below 5 µg/m3. Although, the concentration of DBPs were relatively low, previous studies suggest that mixtures of DBPs might have synergistic effect on the irritation of mucous membranes.

Keywords: VOCs

MO-PO-21
Exposure assessment of VOC emissions from polymer-based consumer products

Even M.[1], Girard M.[1], Rich A.[1], Roloff A.[1], Hutzler C.[1], Luch A.[1]; [1]German Federal Institute for Risk Assessment, Berlin, Germany

The emissions of volatile organic compounds (VOCs) from building materials have been studied in detail for some time. This resulted in the development of standardized analytical methods and health-related indoor air concentration guidelines. Polymer-based consumer products such as toys may also emit harmful substances that could be inhaled and potentially influence the occupant's health. However, emissions from such articles have not been systematically studied so far. After the identification of polymer-derived products with a significant release of VOCs and the characterization of their emission profiles, we provide a realistic evaluation of the inhalation exposure of consumers. First, the polymer type was identified by pyrolysis coupled online to a gas chromatograph with mass spectrometric detection (py-GC/MS). In parallel, a qualitative assessment of emitted VOCs from the same samples was performed using GC/MS in dynamic headspace mode. Subsequently, emissions of relevant samples were quantified in emission chambers with varying dimensions. To this end, the emitted VOCs were sampled on TENAX® TA tubes and analyzed after thermal desorption by GC/MS. Indoor air concentrations were derived and compared to the guideline values, considering different exposure scenarios. It was observed that soft plastics like polyvinyl chloride emit a broader range of VOCs with higher concentrations compared to harder polymers like polypropylene. VOC concentrations in the emission chambers decreased rapidly after loading. Therefore, the first hours after unpacking are of special relevance for the exposure assessment of such compounds. Corresponding indoor air concentrations (30 m3 room) of single target analytes were found to be at least 35-fold lower than thresholds provided by indoor air guidelines. However, when multiple samples or smaller breathing volumes are considered, the derived concentrations can exceed guidelines values. We emphasize that consumer products should be considered as an additional source of VOC emissions indoors beside well-known sources such as building materials.

Keywords: VOCs, analytical methods, children, consumer and personal care products, risk assessment

MO-PO-22

Outdoor and indoor air quality in schools in rural eastern Poland

Polednik B. [1], Dumala S. [1], Piotrowicz A. [1], Guz L. [1], Siuta-Olcha A. [1], Dudzinska M. [1]; [1]Lublin University of Technology, Lublin, Poland

Outdoor air pollutants are currently a serious issue due to their adverse effects on humans and the environment. Air quality studies in Poland carried out as part of the State Environmental Monitoring include only a limited number of measurement stations. In the case of many towns in eastern Poland, the nearest air quality monitoring stations are not located in their vicinity. As smog is often a local characteristic, information from such stations is not always accurate. Indoor and outdoor air quality measurements were carried out in a number of rural schools located approximately 20-30 km away from the border with Belarus and Ukraine. Continuous measurements of physical and chemical pollutants, including the concentrations of particles (PM10, PM2.5, PM1), CO2, volatile organic compounds (VOC), HCHO, NO2, SO2 and microbiological pollutants were performed. Measurements of air temperature, humidity and pressure as well as surveys concerning the perceived air quality were also carried out. The studies indicated that in the heating season significant amounts of outdoor air pollutants (smog) are infiltrated to the school premises and along with the pollutants generated in such premises they deteriorate the quality of indoor air in classrooms. The students' exposure increased in such way may, as demonstrated by numerous studies, have an adverse effect on their health and performance. Studies have also confirmed that the presence and activity of students contributes to a multiple increase of coarse aerosol particles and bioaerosols as well as CO2 concentrations in classrooms. In turn, during the students' absence (at the weekend) VOC and HCHO concentrations were increased in all monitored classrooms. The obtained number dependencies allow for determining individual actions for each school to effectively ventilate the classrooms

Keywords: VOCs

MO-PO-23

Headspace Analysis for Screening of Volatile Organic Compound Profiles of Electronic Juice Bulk Material

LeBouf R. [1], Burns D. [1], Ranpara A. [1], Attfield K. [1], Zwack L. [1], Stefaniak A. [1]; [1]National Institute for Occupational Safety and Health, Morgantown, WV, USA, [2]California Department of Public Health, Richmond, CA

Use of electronic nicotine delivery systems continues to gain popularity, and concern exists for potential health risks from inhalation of aerosol and vapor produced by these devices. We developed an analytical method that provides quantitative and qualitative chemical information for characterizing the volatile constituents of bulk electronic cigarette liquids (e-liquids) using a static headspace technique. We screened volatile organic compounds (VOCs) from a convenience sample of 146 e-liquids by equilibrating 1 g of each e-liquid in amber vials for 24 hours at room temperature. The headspace was transferred to an evacuated canister and quantitatively analyzed for 20 VOCs using a preconcentrator/gas chromatography/mass spectrometer system; additional compounds were tentatively identified using mass spectral matching. E-liquids were classified into flavor categories including Brown, Fruit, Hybrid Dairy, Menthol, Mint, None, Tobacco, and Other. The highest concentration of 2,3-butanedione was found in Brown flavor types, but also in Fruit, Hybrid Dairy, and Menthol types. Benzene was observed at concentrations that are concerning given the carcinogenicity of this compound (max 1.6 ppm in a Fruit type). Benzene was also found in all but the None flavor category. The proposed headspace analysis technique coupled with partition coefficients allows for a rapid and sensitive prediction of the volatile content in the liquid. The technique does not require onerous sample preparation, dilution with organic solvents, or sampling at elevated temperatures. Static headspace screening of e-liquids allows for identification of volatile chemical constituents, which is critical for identifying and controlling emission of potentially hazardous constituents in the workplace.

Keywords: VOCs, analytical methods

MO-PO-24

Derivation of EU-Lowest Concentration of Interest (LCI) of propylene glycol monomethyl ether for the health-based assessment of emission levels from construction products

Choi J. [1]*, Höhne D.* [1]*, Mune W.* [1]*, Rücker T.* [1]*;* [1]*Ramboll Environment & Health GmbH, Munich, Germany*

The health-based evaluation of chemical emissions from construction products is an integral part of the harmonisation framework for indoor products labelling schemes in the EU. One approach within this framework is the establishment of the "lowest concentration of interest" (LCI), which is a health-based reference concentration for inhalation exposure used to assess emissions after 28 days from a single product. These values are applied in product safety assessment with the goal to avoid health risks from long-term exposure of the general population. The EU-LCI derivation involves three main steps: compilation of toxicological data, data evaluation, and EU-LCI derivation. The EU-LCI derivation process for propylene glycol monomethyl ether (PGME, CAS 107-98-2) is presented here as an example. PGME is predominantly used as a solvent in chemical synthesis and manufacturing of paints and coatings with an annual volume in the EU summing up to 100000-1000000 tonnes. PGME exhibits low acute toxicity to experimental animals. It is not a skin sensitiser or irritant and is a mild eye irritant. Effects observed from repeated dose studies include mild central nervous system depression, liver changes and decrease in body weight gain, but they are often observed at the highest tested doses (e.g. ≥3000 ppm). There is little evidence to indicate that PGME is a reproductive toxicant, and PGME has not been shown to be genotoxic or carcinogenic. The derivation of the EU-LCI for PGME is based on the chronic 2-year inhalation study conducted in rodents. The increased incidence of eosinophilic foci in the livers of PGME-exposed male rats was considered as the most critical effect, and 300 ppm (no-observed-adverse-effect-concentration/NOAEC for this effect) was set as the point of departure for the derivation. Total assessment factor of 140 for exposure duration and inter-/intraspecies differences was applied, which resulted in a derived EU-LCI value of 7900 $\mu g/m^3$.

Keywords: VOCs, air, building materials, risk assessment

MO-PO-25

Estimation of life-time exposure to nitrogen dioxide and volatile organic compounds according to similar time-activity groups

Heo J. [1]*, Ryu H.* [1]*, Park J.* [1]*, Kim E.* [1]*, Choe Y.* [1]*, Cho M.* [1]*, Kim Y.* [1]*, Sung K.* [1]*, Yang W.* [1]*;* [1]*Daegu Catholic University, Gjongsanas, South Korea*

Personal or population exposure to hazardous air pollutants has been assessed by personal sampling. however, it has many problems such as cost and time. thus, exposure is assessed by the concentration of pollutants in microenvironments and the time-activity patterns. in addition, similar time activity groups(STG) were categorized through the time activity patterns. In this study, we examined the time activity patterns of the citizen in metropolitan in Korea and classify with STG. In addition, exposure assessment of the life cycle of nitrogen dioxide (NO2) and volatile organic compounds(VOCs) is performed, and the exposure characteristics of each substance are identified. The research subjects were designated aged 0 to 80 and the survey areas were limited to Seoul, South Korea. The research period was conducted on 1,000 people residing in Seoul for two months. The time activity patterns were conducted in the form of interview investigation by an individual visit through the time-use survey. The questionnaire was developed by dividing it into adults, teenagers and infants. In this study, the sampling of NO2 and VOCs was carried out using a passive sampler. Concentration analysis of NO2 was conducted using an ultraviolet spectrometer, and for VOCs , it was desorbed with Carbon disulfide(CS2) and analyzed using GC/MS. Of the 1,000 people surveyed for the time-activity pattern, 493 men and 507 women. On the basis of time-activity patterns, they were divided into infants, preschool child, elementary school students, middle and high school students, college students, office workers, production workers, sales and service workers, housewives, and senior citizens. Infants, housewives, and the elderly were spent more than 80% of their time at home. As a result of the analysis of NO2, housewives and sales worker showed a high concentration of exposure level each of 34.7 ppb, 34.9 ppb.

Keywords: VOCs, activity patterns, cumulative exposure, exposure factors, lifecycle analysis

MO-PO-26
High levels of benzene inside an office building due to a leakage of natural gas from a street pipeline

Johannesson S.[1], Almerud P.[1]; [1]Occupational and Environmental Medicine, University of Gothenburg, Gothenburg, Sweden

Background: Occupational and Environmental Medicine at the University Hospital in Gothenburg, Sweden, got an inquiry from the local Municipalities about an unpleasant smell in a multistore building, hosting offices and public areas such as library and a grocery store. The inquiry focused on the office for social services and the library managed by the local Municipalities. An air screening had shown elevated indoor levels of benzene. Methods: Air sampling of benzene was performed with Perkin-Elmer samplers (Carbopack X) at 20 ml/min during 5-6 hours. Samplers were placed in the library and social office, and where the smell was most intense (entrance hall, grocery store). In addition, sampling was performed in the parking garage below ground level. The week after, a leak in a pipeline for natural gas in the street outside the building was discovered. After the leak was repaired, the smell decreased considerably and a follow-up sampling was performed. The samples were analysed using GC-MS. Results: In the office and library, the concentrations of benzene were 21 and 66 µg/m3, respectively. Considerably higher concentrations were found in the entrance hall, 430 µg/m3, and in the grocery store, 230 µg/m3. The concentration in the parking garage was 30 µg/m3. The follow-up sampling showed benzene concentrations between 1-2 µg/m3. Along with the air samples, a sample from the natural gas was analysed. The concentration of benzene in the natural gas sample was 14 ppm (45 mg/m3), and the chromatograms from the gas and the air samples showed a similar pattern for some other hydrocarbons e.g. n-hexane. According to the gas supplier, the natural gas may contain 20-30 ppm benzene. Conclusion: A leak in an underground pipeline for natural gas outside the building caused elevated indoor levels of benzene. After the pipeline was sealed, indoor benzene concentrations were detected at background levels.

Keywords: VOCs, air

MO-PO-27
Concentration of Volatile Organic Compounds in Small Beauty Salons in a Predominantly Latino Community

Beamer P.[1], Wagoner R.[1], Lopez-Galvez N.[1], Lothrop N.[1], Sandoval F.[2], Parra, K.[1], Wertheim B.[1], Griffin S.[1], Cortez I.[2], Bell M.[1], Wolf A.M. [1]; [1]Mel and Enid Zuckerman College of Public Health, University of Arizona, Tucson, AZ, USA, [2]Sonora Environmental Research Institute, Tucson, AZMel and Enid Zuckerman College of Public Health, University of Arizona, Tucson, AZ, USA

Small businesses, like beauty salons, employ a high proportion of minority and low-wage workers, yet occupational health interventions seldom reach these workers due to physical, social, cultural, and economic barriers. Our objective was to characterize workers' exposure to VOCs in beauty salons in the marginalized, predominantly Latino area of southern Arizona. Businesses were recruited via community health workers. A photo-ionization detector, calibrated with isobutylene, was used to measure real-time VOC concentration in 20 second intervals over 40 work shifts across 10 beauty salons. Summa canisters were used as a stationary monitor to measure the concentration of 70 specific VOCs in workplace air. In half of the businesses studied, Summa canister measurements were taken on two separate days. Workers' activities, ventilation conditions, and other business characteristics were recorded. Chemical inventories were compiled at each business. Mixed models were used to estimate VOC exposure and variance components between and within shops and workers. Thirty-one specific VOCs were detected in workplace air across all 10 shops with a median of 8 chemicals detected per sampling event (range: 1-18). Different chemicals were detected on different days within the same shop. We inventoried 123 products across the shops and identified 231 specific chemicals. Mixed models showed that the highest VOC exposures occurred during hair styling and cutting, followed by hair processing. Peak exposures were most commonly associated with application of hair oil followed by blow drying. The concentrations reported in this study are higher than those reported in previously published studies. Variability was greatest within worker, followed by between-shop, -day, and –worker, respectively. Beauty salon workers have highly variable VOC exposures related to their activities and chosen products at their salon. Interventions aimed at reducing exposures for this at-risk population.

Keywords: VOCs

MO-PO-28
Indoor sinks of terpenes : experimental screening of various surfaces

Harb P. [1], Angulo Milhem S. [1], Brun R. [1], Verriele M. [1], Thevenet F. [1]; [1]IMT Lille Douai, Lille, France

Terpenes are ubiquitous VOCs of indoor environments. Among various sources, wood-based materials provide steady emission of terpenes while fragranced household products are transient but intense sources. Once emitted in the indoor environment, the fate of terpene is crucial since these VOCs easily react either in the gas phase or in the adsorbed phase with indoor oxidants and could lead to the formation of secondary organic aerosols and oxygenated VOCs. Noticeably, the reaction rate of terpenes are by one or two orders of magnitude higher in the adsorbed phase. To that regard, the determination of the partionning coefficient of terpenes on typical surfaces can help identifying the main surfaces acting as sinks of terpenes and subsequently potential surface sources of side-products. Limonene, evidenced as the major indoor terpene, has been used as a model terpenic VOC. The uptake of limonene has been determined, using Field and Laboratory Emission Cell, on ten typical and actual indoor surfaces (glass, paintings, wood and wood-based materials, clothes and gypsum boards). Experiments have been carried out under realistic indoor environmental conditons : T = 23°C, RH = 50% and limonene concentrations ranging from 50 to 800 ppb. The experimental protocol allows the determination of partitionning coefficients (K) as well as the evaluation of the reversibility of the uptake process. Results enable the precise ranking of the various surfaces according to their uptake ability of terpenes. They clearly evidence that specific materials such as gypsum boards and natural clothes can act as significant sinks of terpene and impact the gas phase budget of terpenes. The high uptake capacities of these materials suggest they are surface pools of terpenic VOCs and potential platforms for the heterogeneous oxidation of adsorbed terpenes.

Keywords: VOCs, air, building materials, consumer and personal care products, surfaces

MO-PO-29
Cleaning practices and chemicals: Effects on indoor microbiome and air quality

Alapieti T. [1], Täubel M. [2], Mikkola R. [1], Valkonen M. [2], Leppänen H. [2], Hyvärinen A. [2], Salonen H. [1]; [1]Aalto University, Department of Civil Engineering, Espoo, Finland, [2]Environmental Health Unit, National Institute for Health and Welfare, Kuopio, Finland

Chemicals resulting from the use of cleaning products has been suspected as one of the causes behind indoor air problems. Exposure to these substances can be harmful for occupants and may have selective impact on the indoor microbiome. Therefore, we studied the effects of cleaning practises and chemicals on indoor microbiome and air quality in two school buildings and one day care center. Our research included microbial assessment from settled dust and indoor air, and the chemical composition of indoor air was observed with chemical assessment of volatile organic compounds (VOCs) and carbonyl compounds. The field study consisted of three research periods: 1. "Normal" cleaning where practises and products were the ones that are used daily. 2. Cleaning intervention where only water and microfiber cloths were used in cleaning of the surfaces. 3. Returning to the original cleaning practises. We did not observe any consistent trend of the impact of cleaning intervention on microbial levels from settled dust. None of the microbial groups that were measured were consistently lower of higher during the cleaning intervention compared to baseline. However, levels of viable bacteria assessed from indoor air were consistently higher during the intervention in both schools and the day care center regardless of changes in the levels of outdoor microbes. Overall concentrations of chemical compounds were on a low level (<50 µg/m3 for VOCs, <21 µg/m3 for carbonyls) during the field study. Contrary to expectations, total concentrations of VOCs (TVOC) were slightly higher during the cleaning intervention. Nonetheless, we did not find any consistent changes in the individual compounds that could connect the increase of TVOC concentrations directly to the cleaning intervention.

Keywords: VOCs, built/indoor environment, microbial agents

MO-PO-30

Occupational exposures to volatile organic compounds among U.S. hairdressers serving ethnically diverse clienteles.

Kavi L. [1], Boyle M. [1], Lam A. [2], Pool W. [3], Wilson S. [1], Thomas S. [4], Rule A. [2], Quiros-Alcala L. [2]; [1]Maryland Institute for Applied Environmental Health, University of Maryland School of Public Health, College Park, MD, USA, [2]Department of Environmental Health and Engineering, Johns Hopkins Bloomberg School of Public Health, Baltimore, MD, USA, [3]Centro de Apoyo Familiar, Riverdale, MD, USA, [1]Center for Health Equity, University of Maryland School of Public Health, College Park, MD, USA

Background: Hairdressers are routinely exposed to numerous toxic chemicals, including volatile organic compounds (VOCs) from the hair products used. Of increasing concern are occupational exposures among hairdressers servicing clients of African and Latino descent given the services provided and the products used, which involve chemicals of concern, including formaldehyde, benzene, toluene and methylene chloride. However, occupational exposure data in this population of hairdressers is scarce. Objective: To measure occupational exposures to VOCs in salons predominantly serving clients of African and Latino descent. Methods: We conducted a pilot study in 6 salons in Maryland, where we collected 8-hour integrated VOC samples using coconut-shell charcoal tubes at a flow rate of 0.5 L/min using personal air sampling pumps. Each salon was sampled for the entire 8-hour work shift over two days. The charcoal sampling tubes were stored at –80oC until laboratory analysis using GC-MS. Questionnaires were also administered to capture data on worker demographics, workplace exposures, and salon characteristics. Results: Acetone, benzene, toluene, isopropyl alcohol, methylene chloride, chloroform, methyl chloroform, n-butyl acetate, ethylbenzene, and o,m,p-xylene were detected in 100% of the samples, while 1,1-dichloroethane was below the limit of detection for all samples. We found median (range) 8-hour time weighted average concentrations of 0.14 (0.06-1.33), 0.002(0.001-0.003), and 0.004 (0-0.007) 0.02(0.01-0.11) ppb, for acetone, benzene, isopropyl alcohol, and toluene, respectively. Methylene chloride exhibited a 100% breakthrough in all samples with values >0.06ppb while formaldehyde, which was detected in 75% of samples, exhibited 45% breakthrough (range: 0 to >0.04 ppb). Conclusions: Preliminary results indicate that hairdressers servicing an ethnically diverse clientele are exposed to several VOCs. Further studies are warranted to fully characterize these exposures and determine what mitigation strategies should be implemented to reduce occupational VOC exposures among hairdressers.

Keywords: VOCs, air, built/indoor environment, industrial hygiene, workplace

MO-PO-31

The effect of wind velocity on airborne release of ochratoxin A from cultivated Aspergillus niger

Lee S.A. [1], Chen J.C. [1]; [1]Department of Environmental Engineering and Science, Feng Chia University, Taichung, Taiwan, [2]Department of Environmental Engineering and Science, Feng Chia University, Taichung, Taiwan

Mycotoxins are secondary metabolites produced by fungi and might cause serious health problems in human. Among mycotoxins, ochratoxin A is carcinogenic, teratogenic, genotoxic, and immunosuppressive, and has been classified as "possibly carcinogenic to humans" (group 2B) by the International Agency for Research on Cancer (IARC). This study investigated the effect of wind velocity on airborne release of ochratoxin A from cultivated Aspergillus niger as well as particles concentrations and size distribution, viability and culturability of airborne fungi. Aspergillus niger was dispersed using our newly developed automated fungal particle aerosolizer (AFPA) and the air samples were collected using NIOSH two-stage cyclone bioaerosol samplers for subsequent microbial analysis. The particle concentrations and size distribution of airborne fungi were measured using an electrical low pressure impactor (ELPI). The results show that no ochratoxin A was found at wind velocity < 1.7 m/s and only particles of 11.6 m/s, we began to see airborne release of fungal spores and ochratoxin A, which mostly fell in the particle size >1.8μm. The concentrations of airborne fungi and ochratoxin A, viability, and culturability significantly increased with increasing wind velocity. This study shows that airborne fungi and ochratoxin A might not be a respiratory issue indoors in terms of wind velocity.

Keywords: air, built/indoor environment, dampness/mold, exposure factors, respiratory health

MO-PO-32

Applying Land Use Regression Models with a Mobile Platform in Cycle Network to Estimate PM2.5 Spatial Variation in Taipei, Taiwan

Wu C-F. [1], Wu T-G. [2], Chen B-H. [1], Chang J-C. [1], Chang C-C. [3]; [1]Institute of Occupational Medicine and Industrial Hygiene, College of Public Health, National Taiwan University, Taipei, Taiwan, [2]Institute of Environmental Health, College of Public Health, National Taiwan University, Taipei, Taiwan, [3]Innovation and Policy Center for Population Health and Sustainable Environment, College of Public Health, National Taiwan University, Taipei, Taiwan

Vehicle emission in urban areas is a major source of air pollutants. In recent years, biking has been promoted as a non-polluting transportation tool. However, bikers could receive high exposures to air pollutants if cycling routes are not chosen properly. Therefore, aiming to provide the bike commuters the healthier routes with lower PM2.5 exposure during the commuting periods, in this study PM2.5 was monitored by low cost sensor with mobile monitoring platforms (MMP) to build the land use regression (LUR) models in rush hours. A total of 12 routes were selected to construct the route networks for collecting PM2.5 measurements and land use characteristics in Taipei metropolitan area. The monitoring results showed that the average PM2.5 along the riding routes ranged between 4.7 and 73.0 μg/m3. The R2 of the LUR models with different spatial and temporal resolutions ranged between 0.76 and 0.83 with the leave one out cross validation (LOOCV) R2 ranging between 0.75 and 0.81. The major variables selected in the models included the industrial and business area, the road area and length, and the PM2.5 measurements from the air quality monitoring stations in the study area. Based on the exposure contour maps estimated from the LUR models, bikers could choose the low exposure routes for commuting.

Keywords: air, exposure models, geospatial analysis/GIS, particles, sensor technology

MO-PO-33
Microenvironmental PM2.5 Exposure in a Mixed Land-Use and Heavily Burdened Air Basin

Ivey C.[1], Do K.[1], Yu H.[2]; [1]University of California, Riverside, Riverside, CA, USA, [2]University of Central Florida, Orlando, FL

The South Coast Air Basin of California (USA) is well-known for its historically poor air quality. In recent years, the Basin has seen tremendous progress due to the implementation of effective emissions mitigation strategies. The Basin is still subject to poor air quality due to the expansive network of rail and interstate corridors, high volume of shipping activity in the ports of Los Angeles and Long Beach, and operations of several industrial point sources. Further, meteorological conditions in the Basin are conducive to photochemical smog formation due to landward sea breeze, the opposing mountain range, temperature inversions, and infrequent rainfall. The approximately 18 million residents of the South Coast Air Basin are subjected to high levels of primary and secondary particulate and ozone pollution as a result. In this work, we seek to determine the spatiotemporal variability in personal exposure to PM2.5 in the inland portion of the Basin, where emissions sources are commonly adjacent to residential areas and secondary pollutant formation is extensive. Further, few personal exposure studies have been conducted for the inland Basin compared to the neighboring, coastal counties of Orange and Los Angeles. In a pilot study, we measure daily PM2.5 exposure for 20 community participants from diverse backgrounds each for one week using advanced, real-time, wearable monitoring technology that samples every 15 seconds. Participants are also outfitted with fast-response GPS data loggers for precise microenvironment characterization. Results elucidate the mincroenvironments in the Inland Basin that pose the highest risks for PM2.5 exposure. We stratify results by median household income to investigate the relationship between socioeconomic status and exposure in this unique, mixed land-use area.

Keywords: air, environmental justice, particles, sampling methods

MO-PO-34
Study of a Saharan Dust Intrusion into the Colombian Atmosphere

Pinto Herrera, L.C.[1], Mendez Espinosa, J.F.[1], Belalcazar Ceron, L.C.[1]; [1]University of Los Andes, Bogota, Columbia

Sahara desert is one of the main sources of particulate matter (PM) in the world. This Particulate Matter has the potential to affect the air quality, thousands of kilometers away from the source. The dust intrusions in the Caribbean are well documented; however, there are no reports of this pollutant in Colombian large cities. In this article we analyze and establish the possible relation between the Sahara's dust emissions and an episode of elevated PM levels registered at the end of June 2014 in Bogota and Medellin. For this purposes we used data collected by the automatic air quality monitoring stations installed in these cities, the column mass density of the MERRA-2 project, and results from the NMMB/BSC-Dust model. We also implemented the HYSPLIT model to determine the origins and trajectories of air masses during the episode.The impact of this type of episodes affects vulnerables comunnities, this is assessed through incidence of respiratory illnesses and ecological studies.

Keywords: air, ecological exposure, emergency response, geospatial analysis/GIS, respiratory health, risk assessment, sensor technology, source control, statistical methods

MO-PO-35
Exposure to Airborne Particulate Matter in High Traffic Urban Locations

Parker M.[1], Hopke P.[1], Korfmacher K.[2]; [1]University of Rochester, Rochester, NY, USA, [2]Rochester Institute of Technology, Rochester, NY, USA

The availability of low-cost particulate matter sensors has opened opportunities for measuring particulate pollution concentrations at multiple locations over an extended time. Prior studies in Monroe County, NY have used sensors to examine the spatial and temporal variability of PM and ozone across a spatial scale of 10s of kilometers. In the current study, ten Purple Air PM monitors were deployed to examine potential "hot" spots where it was expected that higher volumes of traffic or higher frequency of higher emitting vehicles were likely to be present. These sites included the central municipal bus terminal, an inner city school serviced by busses and located next to the Rochester Inner Loop, and at sites adjacent to urban stretches of interstate highways. One unit was deployed next to an official USEPA monitoring station for calibration purposes. To be able to provide better estimates of exposure in the vicinity of these sources, data from these sensors are being analyzed using a Gaussian dispersion model, AERMOD, on small spatial scales and complex urban environments to estimate its ability to model the measurement values. The temporal patterns of these high traffic site PM values will be presented along with the comparisons with the model results.

Keywords: air, environmental health, particles, sensor technology

MO-PO-36
Addressing artifacts on black carbon data from microAeth® monitors (AethLabs AE-51) due to rapid temperature changes

Ross J.[1], Chillrud S.[1], Yang Q.[1], Jack D.[1], Smith C.[1], Vera J.[1]; Columbia University, New York, USA

Monitors for personal sampling can experience rapid changes in temperature and relative humidity as subjects move in/outdoors. Since outdoor exposures, though relatively brief, can be a significant source of exposure for black carbon (BC), it is necessary to ensure data quality is not impacted by changing conditions. Here we present data showing the impact of temperature changes on BC data collected by multiple AE-51s. Both in field and lab settings, we have consistently noted that rapid drops in ambient temperature of ~10-20°C, as would be experienced by a wearer going outside on a cool day, cause positive BC artifacts of up to several µg/m3, which are large enough to obscure the true BC reading and persist until the internal temperature of the units stabilizes. Similarly, negative artifacts are produced when a unit rapidly warms by going indoors. During rapid temperature change there are significant inverse relationships of the AE-51 optical signals to internal temperature, which are seen in all the reference detector data investigated (129 field and 14 lab deployments) and most sensing detector data under zero or HEPA flow (10 lab deployments). These relationships are linear in nature (R2 = 0.96-1.00). The impact of temperature on the reference detector in most units tested is about 1.5-4 times greater than on the sensing detector, depending upon the unit. The differing impact on the two detectors causes the BC artifact. We are investigating correction methods to remove these artifacts. Initial tests of a simple linear correction algorithm to remove the impact of temperature on selected field data with these artifacts generally made marked improvements. Two of the units tested in lab did not show a strong relationship between the sensing signal and T in lab tests; we are developing separate corrections for these units.

Keywords: air, analytical methods, particles, sampling methods, sensor technology

MO-PO-37
Organophosphate flame retardants: a review of indoor contamination and human exposure in Europe, and human health effects.

Chupeau Z.[1], Bonvallot N.[1], Mercier F.[1], Le Bot B.[1], Chevrier C.[1], Glorennec P.[1]; [1]Univ Rennes, EHESP, Inserm, Irset, Rennes, France

Organophosphorus Flame Retardants (OPFRs) are increasingly used as in replacement of Brominated flame retardants. The literature published from 2000 to 2018 investigating their presence in indoor environments and human exposures in Europe, as well as worldwide epidemiological evidence of adverse effects was reviewed. Considering studies with at least 10 samples, 30 OPFRs were detected in air and dust of homes, day care centers, offices and cars from 11 European countries. The most concentrated compounds in both air and dust were tris(2-

butoxyethyl)phosphate (TBOEP) , tris(2-chloroisopropyl)phosphate (TCIPP), tris(2,3-dichloropropyl)phosphate (TDCIPP) and tris(chloroethyl)phosphate (TCEP) .Concerning biological measurements of exposure , OPFRs or their metabolites were measured mainly in urine, but also in hair and breast milk. The most consistently detected in humans were TBOEP, TCIPP, TDCIPP, TCEP, tris(butyl)phosphate (TBP) and tris(phenyl)phosphate (TPHP), which appears consistent with air and dust concentrations except for TBP and TPHP. TBOEP and TPHP found in paints, lacquers and construction material were the most prevalent in general population. Concentrations of TDCIPP and TPHP were found to be higher in children than in adults. Regarding epidemiological evidence, the most studied OPFRs were TDCIPP, TPHP, mono-isopropylphenyl phenyl phosphate (IP-PPP) and TBP. The most consistent associations were found between TDCIPP and decrease of thyroid hormone levels, TDCIPP and decrease of sperm quality, IP-PPP and decrease of cognitive abilities, and TNBP and increased risk of allergy. Finally OPFRs are very prevalent in indoor environment and human populations. Given the data analyzed in this work, future epidemiological studies are required to focus on TBOEP, TCIPP, TPHP and IP-PPP, a metabolite of isopropylated triarylphosphate isomers (ITPs) used in Firemaster® 550.

Keywords: air, biomonitoring, built/indoor environment, epidemiology, flame retardants

MO-PO-38
Proof of principle testing for using small black carbon monitors for real-time PAH measurement

Jaramillo J. [1]*, Yan B.* [2]*, Ross J.* [2]*, Pitiranggon M.* [2]*, Yang Q.* [2]*, Blair J.* [3]*, Perzanowski M.* [4]*;* [1]*Hunter College, Manhattan, NY, USA,* [2]*Lamont-Doherty Earth Observatory at Columbia University, Palisades, NY, USA,* [3]*NYAethLabs, San Francisco, CA, USA,* [4]*Columbia University Mailman School of Public Health, New York, NY, USA*

Exposures to PAH compounds have been associated with various adverse health outcomes. Real-time PAH measurements would allow a better understanding of temporal and spatial variability in urban settings. Black carbon is a reliable tracer of incomplete combustion emissions, and optical methods exist for making real-time measurements. The purpose of this work is to explore whether BC data from residential microAeth® monitors (AethLabs, MA350) can be used to estimate real-time levels of high molecular weight PAHs. We collected PM2.5 onto Teflon filters at 4 LPM co-located with a MA350 which collected particles at two spots simultaneously at two flow rates (0.1 and 0.05 LPM) onto a Teflon tape roll. BC measurements made on the co-located Teflon filters agreed well with the mean BC levels from the MA350 (R2 = 0.90 and slope = 0.99). Mean BC concentrations from the MA350 were associated with levels of individual PAH compounds extracted from the Teflon filters such as B[k]F (R2 = 0.94), providing a basic premise for using the MA350 to estimate levels of PAH compounds. PAHs were measured with a highly sensitive fluorescent detector linked to LC/MS. Direct PAH measurements from the used MA350 tape were impacted by high blank levels, whereas initial experiments suggested low blank levels from unused MA350 tapes. Future work is needed to determine whether the high blank levels are due to tape-to-tape variability or contamination during sampling. If successful, site- and time-specific calibrations for MA350s could be developed for high molecular weight PAHs without the need for co-located filters, thus expanding the possibility for high-quality, real-time data for these pollutants.

Keywords: air, analytical methods, other (specify), particles, sampling methods

MO-PO-39
Preliminary Discussions on Effects of Ventilation Rates on Moist-related Issues for Residential and Office Buildings in China

Tang S.[1], Ye W.[1], Su X.[1], Zhang X.[1]; [1]Tongji University, Shanghai, China

Ventilation is the most common method to dilute indoor air pollutants in buildings. In practice, what the adequate ventilation rate for a specific building space should be is often of interest to many and applicable to debates. During the last ~200 years, the recommended ventilation rates are determined mostly based on bio-effluents (e.g., CO2, etc.) that may cause discomfort or odor, rather than concrete health-based evidences. At present, health requirements have raised more attention and in the Europe, 4 L/s has been proposed to be the minimum ventilation rate per person that prevent occupants from suffering unaccepted health consequences. In this paper, a preliminary discussion on methods to determine ventilation rate based on one aspect of the health requirements, i.e., humidity environment and habitability of microbes, is presented. Both residential (natural ventilation) and office (mechanical ventilation) settings have been included in the scenarios and meteorological data from 5 cities, that each has been selected to represent one climate zone in China. Two generally-agreed relative humidity levels, i.e., 1) 30%; and 2) 60% are adopted as the low and high thresholds, respectively, to, 1) reduce health symptoms, such as dry eye and irritation;) reduce the proliferation of house dust mites, as well as the emergence and development of molds, respectively. The results showed that, first, for offices, the mechanical ventilation strategies should be closely corresponding to the occupation patterns. Second, for residences, although mechanical ventilation is not widely used in China, an equivalent 4 L/s of natural ventilation may not be sufficient for regions that have higher outdoor humidity levels, suggesting that dehumidification is needed. Third, infiltration plays a surprising and sometimes unwanted role in controlling indoor humidity level. When building airtightness is not good, a revised ventilation rate based on climate conditions would be required.

Keywords: built/indoor environment, dampness/mold, environmental health, ventilation, workplace

MO-PO-40
Spatiotemporal distribution of particulate air pollution in a pub

Drahler, D.[1], Fishbain, B.[1]; [1]Dept. of Environmental, Water and Agricultural Engineering, Faculty of Civil & Environmental Engineering, Technion – Israeli Institute of Technology, Haifa, Israel

Environmental tobacco smoke (ETS) or passive smoke, is the side product of smoking cigarettes, cigars, pipes and other smoldering tobacco products (STP). The negative health outcomes of exposure to ETS are well-known and include: lung cancer, cardiovascular disease, stroke and asthma. In an attempt to control the non-smokers' exposure to ETS, governments around the world have restricted or prohibited smoking in public places, particularly in hospitalities, such as restaurants and pubs. A number of laws were updated through the years to contain some area outside of public places in the smoking ban, such that smoking is forbidden within a few meters of a pedestrian access to a building; The Israeli law has not been updated in this way. This study aims to investigate the spatiotemporal distribution of ETS in an Israeli pub, caused by smoking activity that is concentrated adjacently to the front door. The study was hosted in a local pub, at downtown Haifa, Israel. The estimation of ETS concentrations was done by measuring particle number concentration (PNC) as an indicator, using low cost sensors (Plantower PMS5003) and small single-board computers (Raspberry-Pi 3). At the indoor area of the pub, 8 sensors were placed throughout the pub, next to the potential sources and sinks of particles. Another sensor was placed outside of the pub, next to the front door, where most of the smoking activity was expected to take place. Preliminary results showed association between smoking activities that occurred outside the pub, and increased particulate concentrations that were measured in the indoor environment. The preliminary findings add to the importance of expanding the legislation dealing with smoking in and nearby public places. Furthermore, it seems that exposure assessments in a complex indoor environment with high spatiotemporal resolution, can be performed using low-cost PNC micro-sensing units.

Keywords: built/indoor environment, environmental policy, environmental regulation, particles, secondhand smoke

MO-PO-41
Indoor VOCs concentrations in a case-study of residential buildings

Rovelli S. [1], Campagnolo D. [1], Cattaneo A. [1], Fazio A. [1], Spinazzè A. [1], Borghi F. [1], Dossi C. [2], Cavallo D.M. [1]; [1]Department of Science and High Technology, University of Insubria, Como, Italy, [2]Department of Theoretical and Applied Sciences, University of Insubria, Varese, Italy

Indoor air quality (IAQ) plays an important role in the well-being of occupants and may affects their behavior and health. IAQ is strictly dependent on the occurrence of several chemicals, among which the volatile organic compounds (VOCs) represent one of the most important categories of pollutants. In the present study, 12 target VOCs (benzene, toluene, ethylbenzene, xylenes, ethylene glycolbutyl ether, 2-ethylhexanol, styrene, benzylether, α-pinene and d-limonene) were selected on the basis of their occurrence in typical indoor environments (e.g. residential buildings and offices) and their potential impact on human health and comfort. These compounds were subject to a solid adsorbent-based sampling and a fully quantitative determination by thermal desorption combined with gas chromatography and a flame ionization detector (TD-GC-FID). The TD-GC-FID method was optimized to obtain the separation of all the analytical peaks (including m- and p-xylene) and a satisfactory sensitivity, with low detection (between 0.14 and 0.31 ng) and quantification (between 0.47 and 1.02 ng) limits. The reproducibility of the whole procedure – from air sampling to thermal desorption and chromatographic quantification – was assessed with the analysis of four co-located tubes and results revealed a very low inter-tubes variability (relative standard deviations of parallel measurements <5%). The optimized protocol was subsequently used to quantify the indoor concentrations of the target VOCs in nine different homes during the dishwasher washing cycle (2-h sampling interval). The most abundant detected VOC in all dwellings was d-limonene (mean: 231 μg m–3; maximum: 611 μg m–3), whereas all the other compounds were found at concentration levels one or two orders of magnitude lower. In terms of health concern, the measured concentrations were always well below the safe levels established for the protection of the general population in living environments.

Keywords: built/indoor environment

MO-PO-42
Indoor air quality in salt rooms

Kakucs R. [1], Szigeti T. [1]; [1]National Public Health Center, Budapest, Hungary

Salt inhalation therapy is often used by patients with pulmonary diseases as alternative therapy to complement or replace conventional treatments. Although, the results of the available studies on the therapeutic effectivity of salt rooms are inconsistent, besides commercial salt rooms for public use, most of kindergartens and other institutions (e.g., hotels) operate salt chambers. There are no regulated standards for these establishments, therefore several technologies are used: halotherapy with dry finely-dispersed sodium chloride aerosols, vaporization of certain concentrations of sodium chloride solutions as well as small chambers with salt-brick wall panels. To investigate whether the indoor environment of salt rooms comply with the indoor air quality guidelines, the concentration of CO2, bacteria and fungi as well as sodium chloride aerosol concentrations in 30 salt rooms during salt inhalation therapy were measured. The concentration of CO2 in all cases exceeded 1000 ppm, and in most cases were higher than 1500 ppm. The concentrations of bacteria in the air of salt rooms were 10-300 times higher than those measured outdoors. The concentration of sodium chloride in the air varied between 2 and 33 mg m-3 in rooms when sodium chloride aerosolizers or nebulizers were applied, while it was lower than 0.5 mg m-3 in chambers with salt brick walls or pots with concentrated salt solutions. The results highlight the need of adequate ventilation of these establishments built for prevention and therapeutic purposes. Frequent cleaning of the rooms are also needed to reduce the microbiological contamination. On the other hand, it is advisable to keep the concentration of the salt aerosol proposed in the literature (5-16 mg m-3). Exceeding the recommended value might has a health hazard, while chambers having very low sodium chloride concentration might not be efficient for prevention or as complementary therapy.

Keywords: built/indoor environment, health, other/general, microbial agents, particles, ventilation

MO-PO-43
Effects of Particle Volatility to Particle Size Distribution of Exhaled Aerosol when Using New Nicotine Containing Devices

Meisutovic-Akhtarieva M. [1], Ciuzas D. [1], Tichonovas M. [1], Misevicius S. [1], Martuzevicius D. [1], Kauneliene V. [1]; [1]Kaunas University of Technology, Kaunas, Lithuania

The rapid spread of new nicotine containing devices as alternative to traditional cigarettes raises many discussions on the potential reduction of harm to both mainstream smoke users as well as bystanders. The exhaled particles following the use of such devices have been indicated as highly volatile by several chamber studies, thus indicating potentially lower impact to particulate matter concentrations indoors compared to exhaled aerosol of traditional cigarettes. This research analysed particle size distributions (PSDs) in exhaled breath using an electronic cigarette and electrically-heated tobacco product. The inhaled main stream aerosol has been exhaled to either non-heating conductive tube (length of 2 m) or heated tube (heater temperature 150 oC, exiting air flow temperature 105 oC). The PSDs were determined at 10 Hz temporal resolution using electrical low pressure impactor (ELPI+, Dekati, Inc., Finland). The exhaled aerosol was diluted 2 times with dry filtered air just before entering the impactor. The particle size distributions based on number concentration indicated bimodal distributions, having majority of particles in 10-20 nm range, and the second mode at 80-200 nm. As expected, exhaled particles proved to be more volatile than mainstream particles, especially in case of heated tobacco product. At the same time, the accumulation mode of exhaled particles was heavily affected by the heating, i.e., almost completely removed in the aerosol denuding system, indicating high volatility. At the same time, the highest mode at 10-20 nm did not seem to be impacted by the heating, indicating either the presence of non-volatile primary particles, or the sampling related artifacts.

Keywords: built/indoor environment, nanoparticles, other (specify), particles, secondhand smoke

MO-PO-44
Improvement of indoor air quality in school buildings in Lodz, Poland

Kozajda A. [1], Szigeti T. [2]; [1]Nofer Institute of Occupational Health, Lodz, Poland, [1]National Public Health Center, Budapest, Hungary

Thermo-modernization of buildings is still one of the major determinants of indoor air quality. Due to the airtightness of the new windows, the air exchange rate is generally lower in the Polish schools as the teachers and the pupils do not open the windows more frequently after the renovation. The poor ventilation is strongly correlated with the increase of the CO2 concentration in the indoor air. The health effects related to the exposure to high CO2 concentrations are associated with the occurrence of drowsiness and deficits of attention. Accordingly, the learning performance of the students and a the teachers' work efficiency might be lower in the classrooms characterized by high CO2 concentrations. However, the uncontrolled ventilation of classrooms may causes increased PM2.5 concentration in the indoor air when the PM2.5 concentration is high outdoors. An action plan has been developed to improve indoor air quality in the school buildings after thermo-modernization. The campaign has started on the 11th March 2019 in selected school building in Lodz. Four classrooms were instrumented with CO2 and PM2.5 monitors. Moreover, PM2.5 monitors were placed outdoors to have information on the outdoor PM2.5 concentration levels. The visual signals of high CO2 and PM2.5 concentration will support teachers to maintain proper air exchange rates in the classrooms. The expected results are: (i) improvement of indoor air quality in classrooms; (ii) raising the awareness on indoor air quality among teachers; (iii) providing support for authorities to conduct successful renovation works.

Keywords: built/indoor environment, children, environmental health, particles, ventilation

MO-PO-45

Evaluation of selected portable solutions for the assessment of indoor air quality

Rouget E. [1], Medan C. [1], Tharin M. [1], Glabasnia A. [1], Mitova M. [1], Goujon C. [1]; [1]Philip Morris International R&D, Philip Morris Products S.A., Neuchâtel, Switzerland,

Smoke-free products have the potential to reduce individual risk and population harm in comparison with smoking cigarettes. In order to assess the impact of smoke-free products on indoor air quality (IAQ), Philip Morris International (PMI) built an environmentally controlled exposure room and combined it with an analytical platform [1-2]. This platform consists of 14 validated and accredited analytical methods covering a total of 28 analytes and considers mainly offline measurements. Additionally, to enable the assessment of air quality in real-life environments, PMI is looking for possibilities to enlarge its analytical capabilities to allow for portable and online analyses of selected IAQ markers. A project was thus initiated to assess the performances of transportable indoor air analyzing solutions, including: 1. Selection of commercial solutions, considering suppliers' performance claims as well as target level concentrations.; 2.Definition and application of a fit-for-purpose protocol allowing comparison of online or near-real-time solution results with average results obtained from validated and accredited reference methods. In total, six technical proposals were investigated, aiming at the determination of carbon monoxide (CO), benzene, ethylbenzene, toluene, xylenes, and formaldehyde, among others. Online or near-real-time quantitative determination of low-content pollutants in indoor air remains a challenge. A portable solution was found for CO analysis. Near-real-time results obtained from portable gas chromatography-mass spectrometry analysis revealed opportunities to monitor the presence and abundance of volatiles even at low concentration levels (down to ppt) and enabled their identification. The trans-portable proton transfer reaction mass spectrometry demonstrated the ability to monitor a wide range of chemicals, including highly volatile compounds like formaldehyde. [1] Mottier N. et al, 2016, DOI: 10.1016/j.talanta.2016.05.022; [2] Mitova M.I. et al, 2016, DOI: 10.1016/j.yrtph.2016.06.005; [3] Gómez Lueso M. et al, 2018, DOI: 10.1016/j.chroma.2018.10.037

Keywords: analytical methods, built/indoor environment, sensor technology

MO-PO-46

Evaluation of Porous Tube-type Passive Samplers for Volatile Organic Compounds according to the ISO 16107

Oishi J. [1], Nishio H., Wang Q., Tokumura M., Miyake Y., Amagai T., Fukushima Y., Suzuki Y., Enomoto T.; [1]University of Shizuoka, Japan

Volatile Organic Compounds (VOCs) are widely used as adhesives, solvents, and paints. Some VOCs cause adverse effects to human health depending on the exposure amount of VOCs. It is necessary to measure the concentrations of VOCs in the work environment according to Industrial Safety and Health Act in Japan. The active sampling method is commonly used to measure VOC concentrations in the work environment. However, the active sampling method requires a sampling pump, electrical power, and so on. Although the passive sampling method is useful for measuring VOCs, the accuracy of this method was often insufficient since sampling rates (SRs) of VOCs with passive samplers were affected by various environmental factors. In this study, we have developed a chamber to evaluate the performance of porous tube-type passive samplers under various environmental conditions according to the ISO16107 method. The chamber developed in this study was capable of controlling the temperature, humidity, wind speed, and VOC concentrations in the chamber by real-time monitoring. Our results showed that the sampling rates of eighteen VOCs targeted in this study were almost constant under the different conditions. In addition, we investigated the relationship between the environmental factors and physico-chemical properties of target chemicals. It was suggested that the validated passive sampling methods enabled us to measure the accurate concentrations of VOCs under the various conditions operated in this study.

Keywords: VOCs

MO-PO-47

Low-cost, high-resolution sensor networks for air quality monitoring in logisitically difficult environments: Ghana case study

Hodoli C.G.; Cranfield University, Cranfield, England

Pollution is a major cause of death in both low and middle income countries. In 2012 alone, exposure to polluted soil, water and air contributed to an estimated 8.4 million deaths in these countries.1 Low-cost feasible new approaches is required for air quality monitoring in these regions. Ghana, among others, is an emerging economic powerhouse in Sub-Saharan African (SSA) region featuring rapid industrialization and urbanization associated with a high demand for energy, food and water that has repercussions on local air quality. Appropriate air quality management plan is required to inform and guide environmental policy and pollution mitigations measures. This study seeks to demonstrate that low-cost miniaturised environmental sensing technologies can be used to develop cheaper and regionally specific methodologies for air quality monitoring in the SSA region by providing fast, detailed, accurate and near real-time data. In this work, a deployment in Ghana will be used as a regional case study. Currently, the use of multiple low-cost sensors within a network is being validated for the routine analysis of selected key atmospheric species, including NO, NO2, CO, O3, and particulates matter (PM1, PM2.5 and PM10). Other key environmental parameters, such as relative humidity and temperature, are also measured by the sensor nodes and will be used, along with auxiliary data, to establish the influence of meteorology on the observed air quality. The analysis of results and validation of these deployments will allow the implementation of appropriate air quality networks initially in Ghana, and later across the wider SSA region.

Keywords: air, children, climate change, ecological exposure, environmental health, environmental justice, environmental policy, environmental regulation, epidemiology, respiratory health, sensor technology

MO-PO-48

Characteristics of Particular Matters by Locations at Hospital Using Internet of Things (IoTs) for Indoor Air Monitoring

Jeong N-N.[1], Lee H.[1], Sohn J-R.[2]; [1]Public Health Science, Korea University, Seoul, [2]School of Health and Environmental Science, Korea University, Seoul

On spring in Korea, people are suffering from high concentrations of particulate matters(PM) and spend much more times indoors for avoiding exposure to high level of outdoor PM. For those reasons, the interest in the monitoring with air quality sensors based on the internet of Things (IoT) and air purification has been increased constantly. Though IoT sensors for air monitoring come into the market and have been improved, a controversy exists regarding the accuracy of them. Thus, the aim of this study is to evaluate application of IoT sensors to indoors for monitoring changing patterns of contaminants. A general hospital was chosen as the target place because many patients visit and are hospitalized. By analyzing real-time data on PM with IoT device, the concentration of PM10 (particulate matter less than 10 ㎛ in diameter) and PM2.5 (particulate matter less than 2.5 ㎛ in diameter) has different characteristics according to detecting location. The lobby and an outpatient space, which are located near the entrance, were high concentrations of PM2.5 and the lobby and an outpatient space, which a lot of patients get treatment in, were high PM10 levels. The results showed that indoor air quality has the characteristics according to locations, pollutants and number of people in that space.

Keywords:

MO-PO-49

Estimation of ventilation rate in Korean residential house using IoT sensor, time-activity patterns and CO2 generation rate from breathing

Choe Y.[1], Ryu H.[1], Heo J.[1], Kim E.[1], Park J.[1], Cho M.[1], Kim Y.[1], Yang W.[1]; [1]Department of Occupational Health, Daegu Catholic University, Gyeongsan, Republic of Korea

The purpose of this study was to estimate the ventilation rate of residential house in Korea using indoor and outdoor concentrations of carbon dioxide (CO2) and CO2 generation rates using breathing rate and time-activity patterns of Korean adults. Indoor and outdoor CO2 concentrations, indoor CO2 generation rate by breathing, volume of the house were applied to mass balance model for residential indoor environments. Indoor CO2 generation rate was calculated using time-activity patterns, the average number of people per household from the Korean National Statistical Office and the amount of CO2 generation by breathing. The amount of CO2 generation rates by breathing were measured by activity loads for 194 Korean adults. The average volume of residential house was provided by the Korean National Statistical Office, indoor CO2 concentrations were measured by IoT sensors, and outdoor CO2 concentrations provided by the Korea Meteorological Administration. The average number of people were 2.53 per household and Koreans spend 61.0% of their day at home. The average volume of Korean residential house was 172.6 m3. The CO2 generation rate from breathing was 13.9±5.3 L/h during sleep and 15.1±5.7 L/h in a sedentary state. Indoor and outdoor CO2 concentrations were 849 ppm and 407 ppm, respectively. The ventilation rate in Korean residential houses calculated by the mass balance model were 42.1 m3/h and 0.71 air change per hour. The estimated ventilation rate tended to increase with an increase in the number of occupants. Since the IoT sensor devices were used to collect data, sustainable data could be collected to estimate the ventilation rate of Korean residential houses, which enables further studies such as on changes in the ventilation rate by season resulting from the activities of occupants. The results of this study could be used as a basis for exposure and risk assessment modeling.

Keywords: activity patterns, air, environmental health, exposure factors, ventilation

MO-PO-50

Estimating Personal Exposures from a Multi-Hazard Sensor Network

Koehler K.[1], Zuidema C.[2], Stebounova L.[3], Sousan S.[3], Gray A.[3], Stroh O.[3], Thomas G.[3], Peters T.[3]; [1]Johns Hopkins School of Public Health, Baltimore, MD, USA, [2]University of Washington, Seattle, WA, USA, [3]University of Iowa, Iowa City, IA, USA

Occupational exposure assessment is almost exclusively accomplished with personal sampling. However, personal sampling can be burdensome and suffers from low sample sizes, resulting in inadequately characterized workplace exposures. Sensor networks offer the opportunity to measure occupational hazards with a high degree of spatiotemporal resolution. Here, we demonstrate an approach to estimate personal exposure using hazard data from a sensor network. We developed a multi-hazard monitor, constructed with low-cost sensors for particulate matter (PM), carbon monoxide (CO), oxidizing gases (OX) and noise (using a sensor developed in-house), and deployed a 40-node network in a heavy-vehicle manufacturing facility. During typical production periods, one-hr mean hazard levels ± standard deviation across all monitors for PM, CO, OX and noise were 0.62 ±

0.2 mg/m3, 7 ± 2 ppm, 155 ± 58 ppb, and 82 ± 1 dBA respectively. Next, we simulated stationary and mobile employees that work at the study site. Network-derived exposure estimates compared favorably to measurements taken with a suite of reference direct-reading instruments (DRIs) deployed to mimic personal sampling but varied by hazard and type of employee. The median magnitude of the percent bias between network-derived exposure estimates and DRI measurements for mobile employees was 33% for PM, 18% for CO, 119% for O3, and 3% for noise. Correlation between network-derived exposure estimates and DRI measurements ranged from 0.39 (noise for mobile employees) to 0.75 (noise for stationary employees). Despite the error observed estimating personal exposure to occupational hazards it holds promise as an additional tool to be used with traditional personal sampling due to the ability to frequently and easily collect exposure information on many employees.

Keywords: industrial hygiene, occupational, particles, sampling methods

MO-PO-51

Land Use Regression Models of Traffic-Related SVOC Pollutants in an Urban Area with Elevated Prevalence of Pediatric Asthma

Esenther S.[1], Lin E.[1], Pollitt K.;[1]Yale University, New Haven, CT, USA

While land use regression models have been commonly used to model inorganic gaseous traffic emission marker pollutants such as nitrogen dioxide (NO2), the difficulty of deploying equipment to measure SVOCs across a study area has limited the number of LUR models of SVOCs associated with traffic emissions. Deployment of the PDMS sorbent material mounted in PTFE housing case has enabled the measurement of PAHs at 40 sites across the greater Springfield, MA area (~185 km2). The objective of this study was to assess concentrations and determinants of PAHs traffic emission pollutants in Springfield. The PDMS bars were deployed for five consecutive weekdays (Monday to Friday) at a sites across the urban and suburban area under summer conditions. PDMS sorbent material was thermally desorbed and analysed by gas chromatography time-of-flight mass spectrometry to determine time-integrated exposure measures. Detectable concentrations were measured for six PAHs (isophorone, benzene,1-bromo-4-phenoxy, benzene, hexachloro, pyrene, Benzo[b]fluoranthene, Benzo[k]fluoranthene) at 20 or more of the sample locations. LUR models will be developed using built environment and land use characteristics and compared with NO2 LUR model of the area using ambient exposure data collected through deployment of 40 Ogawa passive sampler badges. Future work will extrapolate these models to estimate exposures for children with asthma.

Keywords: SVOCs, air, children, community, geospatial analysis/GIS

MO-PO-52

Association between manipulation activities of asbestos containing brake products and respiratory abnormalities among brake mechanics.

Giraldo M.[1], Cely-García M.[1], Ramos-Bonilla J.[1], Durán M.[2], Torres-Duque C.[3], González-García M.[3], Pérez C.[4], Parada P.[3]; [1]Department of Civil and Environmental Engineering, School of Engineering, Universidad de Los Andes, Bogotá, Colombia, [2]Medical Department, Fundación Neumológica Colombiana, Bogotá, Colombia, [3]Research Department, Fundación Neumológica Colombiana, Bogotá, Colombia, [4]Radiology Department, Fundación Cardioinfantil - Instituto de Cardiología, Bogotá, Colombia

Introduction In Colombia, asbestos containing brake products (ACBP), which include brake linings and pads, are still distributed separated from the support. Riveters are a specific group of brake mechanics responsible for attaching ACBP to the supports, which require different manipulation activities including drilling, countersinking, riveting, beveling, grinding, and cutting, which release asbestos fibers. Thus, riveters can be exposed to high asbestos concentrations. A cohort of 56 riveters from Bogotá, Colombia have been studied to determine variables from this group of workers that could be associated with respiratory health abnormalities. Methods A cohort of 56 riveters has been created. All workers had a comprehensive respiratory health evaluation. Workers answered a standardized occupational questionnaire. Daily asbestos exposure of each worker was estimated applying models previously validated models that only require as input variable the number of ACBP manipulated in a work shift. Cumulative asbestos exposure was calculated from work shift asbestos exposure estimations. Logistic regressions models were used to determine the association between respiratory health abnormalities and different variables related with labor history and smoking status of brake mechanics. Results Sixteen mechanics (28.57%) had pulmonary abnormalities that can result from asbestos exposure. Cumulative lifetime asbestos exposure for the 56 riveters ranged between 0.122-19.064 f/cc-year. Logistic regression models showed a marginal significant association between respiratory abnormalities and cumulative asbestos exposure (OR=1.139,p= 0.078), adjusted for the number of shops where the worker was employed (OR=0.509, p= 0.108), if the riveter had been continuously working in this occupation (OR=1.363, p= 0.698), and if the riveter was a smoker (OR = 0.478, p=0.276). Conclusion The results suggest that each additional f/cc-year cumulative fiber asbestos exposure was associated with a marginal significant increase in the odds of respiratory abnormalities. Each additional shop where the mechanic worked had a marginal protective effect on respiratory abnormalities.

Keywords: cumulative exposure, exposure models, respiratory health, statistical methods, workplace

MO-PO-53

Multinomial logistic model for prediction of respiratory illness among household women exposed to cooking fumes in poor ventilation settings

Naseem F. [1], *Adnan M.* [2], *Rashid A.* [3]; [1]*Government Postgraduate College for Women, Satellite Town, Rawalpindi, Pakistan,* [2]*Department of Geography, Federal Postgraduate College (Men) H-8, Islamabad, Pakistan,* [3]*EcoHealth Research Lab, Department of Environmental Sciences, Arid Agriculture University, Rawalpindi, Pakistan*

Poor households in developing countries mostly rely on solid fuels for energy requirements. Additionally, small homes coupled with poor ventilation have been associated with several respiratory infection illnesses. Use of biomass for heating and cooking is highly prevalent in Pakistan. We conducted a case–control study among poor urban population in Rawalpindi city to predict the respiratory illness risk in women exposed to cooking fumes. Multinomial logistic model was developed using cases with respiratory illness and working women controls were enrolled from women cohort of housewives aged between 21–45 years. Using self-structured questionnaire, information about demographic characteristics, including ventilation, cooking characteristics and heating appliances and fuels were obtained. Active monthly respiratory illness monitoring was pursed and population-attributable risk (PAR) was calculated for type of fuel used. A total of 219 women (127 cases and 92 controls) were investigated. With reference to use of clean fuel (gas and/or electricity) for cooking, increased in respiratory illness was associated with solid fuel or biomass stoves used [odds ratio (OR) = 2.81; 95% CI: 1.26, 5.31], kerosene stoves (OR = 2.07; 95% CI: 1.33, 4.98). Heating carried out through coal, wood, or kerosene was significantly associated with high respiratory illness (OR = 3.28; 95% CI: 1.95, 7.48), among housewives compared to working ladies who use no heating or rely on electricity or gas for heating. PARs for respiratory illness were 21.5% and 16.9% for poor ventilation and biomass use as fuel type, respectively. Study concludes that biomass use in households as fuel, is a significant risk factor for respiratory illness especially among housewives who spend most of the time at home. We provide novel finding that poor ventilation in small homes and biomass use for cooking may constitute high risk for housewives than working ladies to develop respiratory illness.

Keywords: ventilation

MO-PO-54

A framework for estimating the US mortality burden of fine particulate matter exposure attributable to indoor and outdoor microenvironments

Azimi P.; Illinois Institute of Technology, Chicago, IL, USA

Exposure to fine particulate matter (PM2.5) is associated with increased mortality. Although epidemiology studies typically use outdoor PM2.5 concentrations as surrogates for exposure, the majority of PM2.5 exposure in the US occurs in microenvironments other than outdoors. We develop a framework for estimating the total US mortality burden attributable to exposure to PM2.5 of both indoor and outdoor origin in the primary non-smoking microenvironments in which people spend most of their time. The framework utilizes an exposure-response function combined with adjusted mortality effect estimates that account for underlying exposures to PM2.5 of outdoor origin that likely occurred in the original epidemiology populations from which effect estimates are derived. We demonstrate the framework using several different scenarios to estimate the potential magnitude and bounds of the US mortality burden attributable to total PM2.5 exposure across all non-smoking environments under a variety of assumptions. Our best estimates of the US mortality burden associated with total PM2.5 exposure in the year 2012 range from ~230,000 to ~300,000 deaths. Indoor exposure to PM2.5 of outdoor origin is typically the largest total exposure, accounting for ~40–60% of total mortality, followed by residential exposure to indoor PM2.5 sources, which also drives the majority of variability in each scenario.

Keywords: built/indoor environment, environmental health, exposure models, health, chronic diseases, particles

MO-PO-55

Agricultural Pesticide Exposure and Congenital Abnormalities (CA) in Mexico: A Systematic Review

Trejo B.; Boston University, Boston, MA, USA

Background: Pesticide exposure has been associated with congenital abnormalities (CA), which are a public health concern in Mexico. Objective: This systematic review identified literature that addressed the relationship between parental exposure to agricultural pesticides via agricultural work and CA among infants of Mexican parents. The systematic literature review sought to communicate the findings of the association, identify strengths and limitations, and make recommendations for further research. Methods: Articles were identified using PubMed with secondary searches of Scopus, PubMed Central, ProQuest Environmental Science, Academic Search Complete, Popline and hand search of references of the identified articles. Results: Seven studies were identified that investigated the association between agricultural pesticide exposure and CA in Mexico. Six studies observed a positive association; one study found no association. Five studies investigated exposure to specific pesticide(s) and two investigated general pesticide exposure. Five studies relied on self-reporting for ascertainment of exposure and/or outcome. Two studies used more objective methods, such as biomarkers and clinical diagnosis for both the exposure and outcome of interest. Five studies investigated specific CA; two studies investigated CA generally. Conclusion: The reviewed literature supported the positive association between pesticides and CA while demonstrating a need for improved exposure and outcome ascertainment.

Keywords: pesticides

MO-PO-56

Cell-on-Particles Nanofiber Based Human Bronchial Epithelial Model for Testing Cytotoxicity of Aerosol Particles

Krugly E.[1], Martuzevičius D[1]., Aldonyte R.[2]; [1]Kaunas University of Technology, Kaunas, Lithuania, [2]Centre for Innovative Medicine, Vilnius, Lithuania

Exposure to fine aerosol particles is among the most important current health issues. Exposure studies associate fine aerosol particles with the chronic respiratory and cardiovascular diseases, along with other pollutants. Part of the inhaled fin aerosol particles can penetrate alveolar barrier and due to interaction with endothelial cells can be transferred into the blood. Various direct and indirect methods are being used for evaluation of the toxicity of pollutants. Among those, the most relevant method is in vitro lung model. We have developed a novel substrate to be used as in vitro lung model providing an opportunity for cells to directly interact with fine aerosol particles. The substrate of the model is composed of natural hydrophilic polymer converted to nanofibres by solvent electrospinning, followed by chemical-physical treatment to achieve sufficient surface parameters and favorable environment for cells. Sampling of fine aerosol particles are performed by pumping air sample through an aerodynamic separator which screens out coarse aerosol. Fine particles are collected directly on designed substrate acting as a filter membrane. After sampling, the substrate is placed into holder and immersed chamber with cell growth media, followed by seeding of selected type of respiratory tissue cell on top of substrate. Such in-vitro model allows to analyze toxicity of collected fine particles using various types of cells.

Keywords: exposure models

MO-PO-57

Estimation of Potential Dermal Exposure to Organophosphorus Flame Retardants via Direct Contact with Products Using Silicone

Shindo M.[1], Tokumura M.[1], Wang Q.[1], Miyake Y.[1], Amagai T.[1], Makino M.[1]; [1]University of Shizuoka, Shizuoka, Japan

Organophosphorus compounds are presented in indoor products at high concentration, and some of them have toxic consequences like carcinogens. Although the dominant exposure route to organophosphorus compounds is considered as the dust ingestion, recent researches suggest that the dermal exposure rate via direct contact with products could be expected to be higher than those via other exposure routes. As conventional devices to estimate the potential dermal exposure rate, human skin and EpiSkin epi dermal model have been used. However, they have disadvantages (e.g., ethical problems, expensive, and hard-to-use). Therefore, they seem to be not suitable for testing numerous products. In this study, to develop the simple and cost effective devise which can estimate a potential dermal exposure to organophosphorus flame retardants via direct contact with products, we carried out the migration test of phosphorus compounds from a product to the devise made of silicone. The amount of the phosphorus compounds in the silicone increased with the contact time. A higher phosphorus compound concentration in the product was likely to result in a faster migration rate. Phosphorus compounds with a lower n-octanol/water partition coefficient (log Kow) were likely to show a faster migration rate. The average migration rates of phosphorus compounds for 48 hours from the product to the silicone were in the range of 0.07–2.2 ng/cm2/h. We compared the migration rates of phosphorus compounds from the products to the devises (silicone and EpiSkin).

Keywords: SVOCs, flame retardants, risk assessment

MO-PO-58

A Study on the Aggregate Exposure Evaluation of phthalates through Comparison of Exposure Estimation Methods

Junghyun P., Lee, J. [1]; [1]*EH R&C, Seo-gu, Incheon/South Korea*

Exposure to chemical from various residential environmental media needs to be evaluated comprehensively. Since exposure dose in quantitative risk assessment is critical part to calculate possible adverse health effects of the human, realistic estimates of aggregate exposure are needed. Estimation of exposure dose is generally used by deterministic and probabilistic methods. Recently, a method of personal-based assessment has also been proposed. In this study, all three methods were applied to calculate aggregate exposure to DEHP, DBP and BBP in food, consumer products. DEHP, DBP, BBP have been used in various products as plasticizers, and are detected in residential environmental media. To propose an evaluation method that reflect realistic exposure levels, the phthalates concentrations in urine for general population and the total exposure derived by each method were compared. phthalates concentrations in the residential exposure media (food, consumer products) applied from domestic monitoring data. Personal intake amount for food was from the Korea National Health and Nutrition Examination Survey (KNHANES) data from 2015 to 2016. Individual multi-use patterns of consumer products were applied domestic consumer survey data from the Safety and biocide management technology development business.

Keywords: aggregate exposure

MO-PO-59

Aerosolization of silver nanoparticles during simulated use of a commercially available spray disinfectant product advertised to contain colloidal silver

Stefaniak A. [1], *O'Boyle S.* [2], *Bowers L.* [1], *Knepp A.* [1], *Navratilova J.* [3], *Al-Abedd Souhail R.* [4], *Potter P.* [5], *Gitipourf A.* [6], *Radwan I.* [7], *Nelson C.* [3], *Bradham K.D.* [3], *Rogers Kim R.* [3]; [1]*National Institute for Occupational Safety and Health, Morgantown, WV, USA,* [2]*Messiah College, Mechanicsburg, PA, USA,* [3]*U.S. Environmental Protection Agency, Research Triangle Park, NC, USA,* [4]*U.S. Environmental Protection Agency, Cincinnati, OH, USA,* [5]*Oak Ridge Institute for Science and Education, Cincinnati, OH, USA,* [6]*Pegasus Technical Services, Inc., Cincinnati, OH, USA,* [7]*University of Cincinnati, Cincinnati, OH, USA*

Silver nanoparticles (all three dimensions 1 – 100 nm) are being exploited in diverse products such as spray surface cleaners used in homes and workplaces to take advantage of their potent antimicrobial properties. The use of these products may generate aerosols, and inhalation of silver nanoparticles causes chronic lung inflammation and liver abnormalities, cytotoxicity, and generation of reactive oxygen species in vitro and in vivo. Twenty-two products purported to contain silver were characterized to determine whether they actually contained silver and if it was present as nanoscale particles. One product, a sanitizer spray containing 6.1 mg/L of silver nanoparticles (mode = 20±13 nm by nanoparticle tracking analysis) was selected for testing in a 12.85 m3 chamber. Air was monitored using two aerodynamic particle sizers (range: <0.52 to 20 µm), one positioned near the spray bottle nozzle and the other 25 cm away near a tile surface. Aerosol was also collected onto filters for off-line analysis using scanning electron microscopy. Three different bottles of the spray product were tested. MPPD2 software was used to model respiratory deposition of aerosol. During spraying, aerosol concentration increased near the spray bottle nozzle (breathing zone) and near the tile surface. The geometric mean (GM) size and geometric standard deviation (GSD) of the aerosol droplets was 0.87 µm (1.67) and 1.03 µm (1.79) at the nozzle and tile, respectively. Electron microscopy analysis identified nanoparticles in the dried aerosol droplets. Based on modeling for the breathing zone location, particle deposition fractions were 0.59, 0.03, and 0.10 for the head, tracheobronchiolar, and alveolar regions, respectively. A commercially available antibacterial spray product released aerosol containing nanoparticles during simulated use. These particles have potential to deposit throughout the respiratory tract, including in the nasal passages, which could lead to translocation to other organs, including the brain.

Keywords: consumer and personal care products, green buildings, industrial hygiene, nanoparticles

MO-PO-60 MOVED TO MO-PL-D2-38

MO-PO-61

Using a passive sampler to measure the dermal migration of di(2-ethylhexyl)phthalate from polyvinylchloride flooring at different skin condition

Kim D.Y. [1], Kwon, J.H.[1]; [1]Division of Environmental Science and Ecological Engineering, Korea University, Seoul, Korea

Di(2-ethylhexyl)phthalate (DEHP) is still widely used as plasticizer in polyvinylchloride (PVC) products, although it is known to be toxic to human. One of the exposure pathways of DEHP from PVC products is the dermal migration. Generally, the dermal migration is measured from the transferred mass of a chemical per contact area and time. The dermal migration can occur through direct contact between human skin and PVC products or deposition of gaseous DEHP to human skin. In the case of the direct contact between skin and PVC products, the presence of sebum should be considered because sebum has high affinity to DEHP and might affect to the dermal migration rate. In this study, the dermal migration of DEHP from a PVC coated heating pad in direct contact condition with and without artificial sebum between a polydimethylsiloxane (PDMS) passive sampler and the heating pad were measured. The pre-cleaned PDMS sampler was attached on the heating pad or over filled artificial sebum in the gap between the sampler and the heating pad using custom made glassware. Those samples were stored at constant temperature 60 °C to simulated heating condition, taken after pre-designated sampling time, and DEHP was extracted using n-hexane and was analyzed with GC-MS. The dermal migration rates of DEHP from PVC with and without artificial sebum were 0.0378 and 0.00122 ug cm^-2 min^-1 respectively, showing that artificial sebum accelerates the process by a factor of 30. Artificial sebum might be absorbed into PVC and PDMS, mimicking the real skin condition better than dry skin condition, without artificial sebum. Considering the covering matrices such as sebum is needed when assessing the exposure to DEHP through direct contact.

Keywords: SVOCs, phthalates, sampling methods

MO-PO-62

Results from the Norwegian human biomonitoring study in the EuroMix project: Exposure to the pesticides boscalid and imazalil from the diet in Norway

Husøy T. [1], Sonnet F. [1], Sakhi A.K. [2], Thomsen C. [2], Dirven H. [1], [1]Department of Toxicology and Risk, Norwegian Institute of Public Health, [2]Department of Environmental Exposure and Epidemiology, Norwegian Institute of Public Health

Background: The fungicides boscalid and imazalil were among the most frequently detected pesticides in the residues monitoring programs 2013-2017 in Norway. The aim of the present study was to estimate the daily intake of these two pesticides and compare with measured concentrations in 24 h urine samples. Methods: A human biomonitoring study was performed to study the exposure to chemicals present in food and personal care products (PCPs). In two 24 h periods two-three weeks apart, 144 participants (100 women and 44 men) kept detailed weighted food diaries and PCP diaries and collected all urine excreted. Individual-specific consumption data from both 24 h periods were used to estimate boscalid and imazalil exposure deterministically. A sensitive ultra-performance liquid chromatography coupled to tandem mass spectrometry (UPLC-MS-MS) method was developed to measure the boscalid metabolite 2-chloro-N-(4'-chloro-5-hydroxybiphenyl-2-yl)nicotinamide (M510F01) and imazalil in the 24 h urine pools collected at Day 1. Results: Overall, the estimated dietary exposure of boscalid and imazalil was comparable between males and females. In the lower bound exposure scenarios, the estimated dietary exposure of boscalid ranged from 0-0.9 µg/kg bw/day and the estimated exposure of imazalil ranged from 0-0.81 µg/kg bw/day. In 99 % of samples M510F01was detected in concentrations from 0.04-15.03 ng/ml. There was a statistically significant difference between genders ($P<0.0001$) with a median concentration of

0.98 ng/ml for females, and 0.46 ng/ml for males. The lower bound exposure estimate of boscalid was in the same range as the measured concentrations of M510F01 in urine. The major food items contributing to boscalide exposure were berries and carrots. Imazalil was detected in 1 % of the samples, but the low detection rate may be due to considerable metabolism before excretion. Conclusion: Widespread human exposure to the fungicide boscalid as measured by one of its metabolites in urine samples was observed.

Keywords: pesticides, boscalid, imazalil, biomonitoring

MO-PO-63
Synthetic Musks by DLLME and APGC-MS/MS in human serum

Bélanger P.[1], Laflamme O. [1], [1]Centre de toxicologie du Québec, Québec, Canada

Galaxolide (HHCB) and Tonalide (AHTN) are odoriferous synthetic compounds part of the musk family. These compounds are widely used in personal care products (perfume, deodorant, body cream, detergent, etc.). These polycyclic musks are very lipophilic and are also found in the aquatic environment. The sources of exposure in humans are varied: personal care products, food, water, air, etc. (1) Since these compounds accumulate in fats, the half-life are a few months in humans (2). They are suspected endocrine disruptors (3). Few human toxicity studies are available at this time. In contrast, several studies demonstrate toxicity in various living organisms (4,5,6). Since these compounds are semi-volatile, they are found in the air so they become a significant source of contamination when preparing samples during the analysis. During the development of the method, we observed that the sources of contamination were multiple (solvent, GC Septum, pipette tips, glass tube, evaporation, etc ...) So the reduction of the preparation steps is essential to reduce interference and obtain the lowest detection limits. The DLLME (dispersive liquid-liquid microextraction) makes it possible to minimize the steps of preparation and thus to answer the needs of the analysis of the musks in human serum. The APGC-MS/MS technique provides excellent sensitivity. Unlike in GC-EI-MS/MS (the most common technique for these compounds), in APGC ionization source we observe the absence of fragmentation in the source which allows to obtain mainly the molecular ion for the two compounds which allows to be more sensitive and more specific. Detectable levels are observed in the population of the Quebec area.

Keywords: Musk, Galaxolide, Tonalide, DLLME, APGC-MS/MS

MO-PO-64 MOVED TO TH-PL-E1-27

MO-PO-65
Nonlinear model analyses of urinary metabolite levels of organophosphate pesticides corresponding to cholinesterase activity inhibition in rats

Sato H. [1], Ito Y. [1], Bundo I. [1], Hanai C. [1], Nishimura M. [1], Ueyama J. [1], Kamijima M. [1]; [1]Nagoya City University Graduate School of Medical Sciences, Nagoya, Aichi/Japan

Background and aim Urinary dialkylphosphates (DAPs) are generally used for the exposure assessment of organophosphate pesticides (OPs) whereas they are common metabolites of OPs and not specific indices for each agent. This study aimed to clarify the relationship between cholinesterase (ChE) activity inhibition as the neurotoxicological endpoint of OPs and the amount of DAP excreted into urine in rats exposed to dichlorvos (DDVP) and fenitrothion (FNT), of which the magnitudes of acute toxicity are different. Methods A total of 40 male Wistar rats (11-week-old) was divided into four groups for each OP and orally administered 0 (control), 0.2 (low), 1.0 (middle) or 5.0 (high) mg/kg/day of the respective OP for 10 days with a 2-day interruption after the first 5 days. The urine was collected 6, 12 and 24 hours after the final administration, and then the blood and brain were taken. The ChE activities of the brain, red blood cell (RBC) and plasma were analyzed by Ellman method. DAPs were measured by using gas chromatography-mass spectrometry. Statistical analyses and nonlinear model estimation were conducted with drc R package. A lower confidence limit of excretion corresponding to 20% inhibition of each ChE activity (ELL20) was estimated. Results & Discussion The ELL20 (nmol/24-hour urine) of DDVP and FNT were 1156 and 261 for the brain, 481 and 262 for RBC, and 735 and 315 for plasma, respectively. The ratio of DAPs excreted in 24-hour urine for DDVP and FNT was 99% and 52% for low, 63% and 29% for middle, and 40% and 16% for high group, respectively. The differences in ELL20 between two OPs were considered to be attributable to a possibly slower metabolism of FNT than DDVP, and/or to the production of other metabolites than DAPs.

Keywords: wildlife

MO-PO-66

Exposure to triclosan is a significant predictor of urinary 2,4-dichlorophenol in Polish population: experimental and epidemiological evidence

Wielgomas B.[1], Klimowska A.[1], Dabrowska P.[1], Dziomba S.[1], Rodzaj W.[1], Jurewicz J.[2], Radwan P.[1], Radwan M.[1], Hanke W.[2]; [1]Medical University of Gdansk, Gdans, [2]Nofer Institute of Occupational Medicine, Lodz, Poland, [3]Gameta Hospital, Lodz, Poland

In numerous human biomonitoring studies urinary 2,4-dichlorophenol (24DCP) is considered as a biomarker of exposure to 2,4-dichlorophenol itself and few organochlorine pesticides. Other sources of urinary 24DCP are not studied at sufficient level. We noticed a significant correlation between urinary concentrations of 24DCP and triclosan (TCL) in two observational studies performed recently in Poland. TCL is an antibacterial used in several personal care products as well as other consumer goods. Older studies suggested possible metabolic pathway of TCL leading to formation of 24DCP. To verify the hypothesis that exposure to triclosan can affect urinary concentration of 24DCP we studied: metabolism of TCL in vitro using rat and human liver microsomes as well as a kinetics of elimination of TCL and 24DCP in 3 volunteers following controlled oral exposure to a single dose of triclosan. Determination of TCL and 24DCP in both cases was performed by isotope dilution gas chromatography and ion-trap tandem mass spectrometry following enzymatic hydrolysis, liquid-liquid extraction and trimethylsilylation. Rat and human liver microsomes were shown to generate 24DCP from TCL under standard conditions. In human volunteers' study, single oral exposure to triclosan at a dose about 40 ?g/kg bw resulted in significant increase of urinary 24DCP in all study participants and its elimination followed the kinetic profile of triclosan. The study documents that TCL is metabolized in humans to 24DCP and both are eliminated with urine. These findings clearly emphasize the need for considering triclosan exposure as a significant source of 2,4-dichlorophenol.

Keywords: biomarkers, biomonitoring, consumer and personal care products, exposure factors

MO-PO-67

Endocrine disrupting chemicals in follicular fluid – methodological aspects of sample preparation before GC-MS/MS analysis

Klimowska A.[1], Wielgomas B.[1]; [1]Medical University of Gdańsk, Department of Toxicology, Gdańsk, Poland

Follicular fluid forms micro-environment for oocyte growth. Maturation of ovum strictly depends on hormones concentration. Environmental exposure to endocrine disruptors (EDCs) might be responsible for the hormonal imbalance. In this work we focused on analyzing bisphenol A and its analogues, parabens, triclosan, oxybenzone, and metabolites of selected pesticides in follicular fluid using GC-MS/MS method. To facilitate the measurement of both free and total concentrations of analyzed xenobiotics, samples have to be analyzed twice: with and without enzymatic hydrolysis. Several sample preparation techniques were evaluated for their suitability in extraction and cleanup. Liquid-liquid extraction (LLE), solid phase extraction (SPE) as well as QuECHERS were compared. Acetonitrile, perchloric and m-phosphoric acid, and zinc sulfate were tested as deproteinization reagents. LLE was performed using mixture of n-hexane and MTBE, whereas SPE was carried out with Bond Elut Plexa columns. Sample preparation appeared to be crucial for determination of: bisphenol S, 3,5,6-trichloro-2-pyridinol, and methyl paraben. SPE was the most promising extraction technique for the isolation of analytes from hydrolyzed and non-hydrolyzed follicular fluid while low recovery of bisphenol S was observed for LLE and QuECHERS. Protein removal step was required before any type of extraction. Satisfactory recovery for most analytes including bispenol S was obtained when sample was deproteinized with acetonitrile then diluted with formic acid and extracted by SPE. Other protocols failed to provide acceptable results due to the incomplete protein removal leading to SPE column clogging or stable emulsion formation during LLE. Follicular fluid is still a matrix rarely used in human biomonitoring studies and thus may generate some analytical challenges. In this work we optimized sample preparation step enabling GC-MS/MS analysis of several EDCs in follicular fluid.

Keywords: analytical methods, biomonitoring, environmental health

MO-PO-68

Effects of Biophilic Interventions in Office on Stress Reaction and Cognitive Function: A Randomized Crossover Study in Virtual Reality

Yin J.[1], Arfaei N.[1], MacNaughton P.[1], Catalano P.[2,3], Allen, J.[1], Spengler J.[1]; [1]Department of Environmental Health, Harvard T.H. Chan School of Public Health, Boston, MA, USA, [2]Department of Data Sciences, Dana-Farber Cancer Institute, Boston, MA, USA, [3]Department of Biostatistics, Harvard T.H. Chan School of Public Health, Boston, MA, USA

Biophilia hypothesis suggests humans have an innate connection to nature. Yet we currently live in a world that is rapidly urbanizing with people spending more time indoors than ever before. We simulated one open and one enclosed office space in virtual reality (VR) and quantified the acute effects on stress reaction and cognitive function in 30 participants who each visited twice to experience three biophilic environments (natural elements, natural analogs, and combo) compared to a non-biophilic environment in both spaces, with the order randomized. While in VR, we measured blood pressure, heart rate variability and skin conductance level, and administered cognitive tests to measure their reaction time and creativity. These three biophilic interventions affect lower stress levels, higher creativity scores (0.25, 0.15, and 0.20 standard deviation increase in creativity task scores), yet longer reaction times (5.5%, 1.0% and 2.8% increases respectively) on focused tasks, comparing to the non-biophilic environment. Natural elements intervention had more prominent effects on stress reduction and cognitive function than the natural analogues intervention. In addition, effects of virtual exposure to biophilic elements on reduction of stress level and improvement of cognitive function were not linear, with optimal effects seen when the percentage of time participants looked at biophilic elements was around 50%. Biophilic interventions could help reduce stress and improve creativity; however, the effects may be different based on the workspace type (open vs. enclosed). The effects are related to both the type and amount of biophilic elements, with natural elements and 50% space saturation being optimal for stress reduction and improved creativity. This research is essential to help developers, architects and designers know what are the appropriate biophilic design interventions to use, and how these vary from one workspace type to another.

Keywords: biomonitoring, built/indoor environment, environmental health, productivity, workplace

MO-PO-69

A high resolution mass spectrometry-based metabolomics strategy for comprehensively screening biomarkers of di-(2-propylheptyl) phthalate exposure and their applications

Hsu J. [1]*, Shih C-L.* [2]*, Wong H.L.* [2]*, Liao P-C.* [2]*;* [1]*National Health Research Institutes, Miaoli County, Taiwan,* [2]*National Cheng Kung University, Tainan, Taiwan*

Di-(2-propylheptyl) phthalate (DPHP) is an alternative plasticizer used to replace other phthalates currently being scrutinized, and its use and production volumes are increasing. This study aimed to develop a high-resolution mass spectrometry (HRMS)-based metabolomics strategy for comprehensively screening urinary biomarkers of DPHP exposure and filtering out potentially useful DPHP exposure markers for human exposure assessments. This strategy included three stages: screening of biomarkers, verification of dose-response relationships in laboratory animals, and application in human subjects. The multivariate data analysis method known as orthogonal partial least-squares discriminant analysis (OPLS-DA) was used to screen and find meaningful signals in the MS dataset (18,146 signals) generated from urine samples collected from DPHP-administered rats. A total of 6 to 372 MS signals were filtered out by the OPLS-DA method with different criteria. Furthermore, 31 MS signals were verified as exposure marker candidates by assessing the dose-response relationships in an animal feeding study. Ten of these 31 exposure marker candidates were ultimately detected in more than 30% of human subjects (N=116). However, three previously reported oxidized DPHP metabolites (mono-(2-propyl-6-oxoheptyl phthalate, mono-(2-propyl-6-hydroxyheptyl) phthalate, and mono-(2-propyl-6-carboxyhexyl) phthalate) were detected in fewer than 30% of human subjects. Thus, the use of metabolites as exposure markers could underestimate DPHP exposure. After chemical structure identification, a new metabolite of DPHP, mono-(2-propyl-7-dihydroxy-heptyl) phthalate, was suggested as a DPHP exposure marker for human exposure assessments.

Keywords: analytical methods, biomarkers, phthalates

MO-PO-71

Breath Volatile Organic Compounds Profile After Short Exposure to High Concentrations of Indoor Fragrance Chemicals

Vicent-Claramunt A., Center for Physical Sciences and Technology, Vilnius, Lithuania

Nowadays people spend most of their time indoors; sleeping at home, working in the offices, eating in canteens or restaurants, or spending free time in cinemas, shopping centers and other stores. It is in those places where we are exposed to a high content of volatile organic compounds (VOCs). In this work, we focused on places where the concentration is especially high like cosmetic stores, drug stores and bathrooms. Fragranced products emit dozens of different volatile compounds. As some VOCs might be harmful for human health, the aim of the work was to monitor how people respond to a short exposure to environments with very high concentration of fragrances. Air samples were taken from the selected places, to benchmark the background compounds. Samples were collected using thermal desorption (TD) tubes and purged into a gas chromatography with mass spectrometer (GC-MS). Then, breath samples from participants were collected before and after being exposed to places with high fragrance content. Optimization of system parameters was performed to get good compound separation and identification.The results shows the high concentration of several VOCs in a study place. If people are exposed, some of them are incorporated into respiratory system and their variation in breath could tell us how those compounds behave inside our body, some will be exhaled, others might be retained and different metabolites can be formed. In the next stage we will look into the possible pathways of these VOCs through the body, as some of them might be very dangerous for our health. Knowing the effect of those compounds inside our body can give us an understanding of the toxicity levels we are exposed to in our every day life.

Keywords: VOCs, air, biomarkers, consumer and personal care products, environmental health, respiratory health, workplace

MO-PO-72

Serum concentrations of polybrominated biphenyls (PBBs) and the other persistent organic pollutants (POPs) among the Michigan PBB cohort

Chang C.J. [1]*, Terrell M.* [1]*L, Marcus M.* [1]*, Marder M.E.* [1]*, Panuwet P.* [1]*, Ryan, P.B.* [1]*, Barr, D.B.* [1]*;* [1]*Department of Environmental Health, Rollins School of Public Health, Emory University, Atlanta, GA, USA*

Polybrominated biphenyls (PBBs) were highly produced and widely used compounds added as flame retardants in various consumer products. Reported health effects associated with exposure to PBB include alterations to timing of pubertal development, adverse reproductive outcomes, thyroid dysfunction and certain cancers. From 1973 to 1974 widespread PBB contamination occurred in Michigan when the commercial PBB mixture was accidentally added to livestock feed instead of a nutritional supplement. Consequently, people in Michigan were exposed to PBBs via consuming contaminated meat, egg, and dairy products. The Michigan PBB Registry was established by the Michigan Department of Health to monitor exposure and health effects among Michigan residents. PBB stores in adipose tissues and is only slowly eliminated. Additionally, PBBs can be transferred from a mother to her fetus through the placenta and to her developing child through breastfeeding. Due to the long biological half-lives of PBBs, as well as the concerns from the exposed community, we carried out biomonitoring measurements from 2012 – 2015. Because of their similar structures, this study measured serum PBB, polychlorinated biphenyl (PCBs), and polybrominated diphenyl ether (PBDEs) levels 40 years after the industrial incident. We found that the serum PBB-153 among the original highly-exposed groups (chemical workers, the family of chemical workers, and individuals who lived on or received food from the contaminated farms) remains significantly higher than other Michigan residents. Overall, the serum PBB-153 concentrations [geometric mean 25.69 ng/g lipid (95% CI 22.01-29.98 ng/g lipid)] are significantly higher than 2003-2004 NHANES data [geometric mean 2.29 ng/g lipid (95% CI 1.82-2.87 ng/g lipid)], and those who were born after the contamination incident (after 1973) also have higher serum PBB levels when compared with similar age groups in the US population, which suggests transgenerational exposure and/or continued environmental exposures following the contamination period.

Keywords: biomonitoring

MO-PO-73

Age, gender and lifestyles in difference urinary levels of cadmium, lead, cobalt, copper in Taiwanese – Taiwan Environmental Surveys for Toxicants 2013-2016.

Liao K-W. [1]*, Chen P-C.* [2]*, Lin C-C.* [2]*, Tsai M-S.* [2]*, Chung C-J.* [3]*, Chen J-K.* [4]*, Ling M-P.* [5]*, Huang H-B.* [6]*, Huang, P-C.* [1]*;* [1]*National Institute of Environmental Health Sciences, National Health Research Institutes. Miaoli, Taiwan,* [2]*Institute of Environmental and Occupational Health Sciences, National Taiwan University College of Public Health, Taipei, Taiwan,* [3]*Department of Health Risk Management, College of Public Health, China Medical University, Taichung, Taiwan,* [4]*Institute of Biomedical Engineering & Nanomedicine, National Health Research Institutes, Miaoli, Taiwan,* [5]*Department of Food Science, National Taiwan Ocean University, Keelung, Taiwan,* [6]*School of Public Health, National Defence Medical Center, Taipei, Taiwan*

Background Our previous study revealed that Taiwanese exposure to higher levels of certain hazardous metals than the other developed countries. However, the information of difference levels of cadmium (Cd), lead (Pb), cobalt (Co) and copper (Cu) by demographic characteristics and lifestyles are limited. The aim of this study was to determine the association between demographic characteristics, lifestyles and the urinary levels of Cd, Pb, Co, Cu in the general Taiwanese. Methods We randomly recruited 1992 participants age above 7 years old from Taiwan Environmental Surveys for Toxicants during 2013-2016. The levels of Cd, Pb, Co and Cu in the participants' urine samples were analyzed by ICP-MS. A questionnaire survey was applied to obtain individual's demographic characteristics and lifestyles. Results The median (range) levels of urinary Cd, Pb, Co and Cu in participants were 0.61 (ND-8.44), 0.80 (ND-66.0), 0.30 (ND-23.22) and 10.66 (ND-706.13) µg/L, respectively. We found that the urinary Pb and Cu levels of men were 1.2-fold significantly higher than in women ($P<0.01$), whereas, the urinary

Co levels in women were 1.7-fold significantly higher than in men (P<0.05). Besides, the urinary Co level in youths were 2.1-fold significantly higher than in adults (P<0.01). The major component of metal exposure in Taiwanese were Cd and Pb, regardless gender or age, using principal component and cluster analysis. Participants who self-reported betel quid chewing, smoking and passive smoking, respectively, had significantly higher median levels of urinary Cd and Pb than those without (P<0.01). Besides, participants with betel quid chewing and smoking had significantly higher 1.3 to 1.9-fold median levels of urinary Cd (1.0 v.s 0.6 μg/L), Pb (1.5 v.s 0.8 μg/L) and Cu (13.6 v.s 10.5 μg/L) than those without (P<0.05). We concluded that the urinary levels of certain metals exposure in the general Taiwanese varied by age, gender and different lifestyles.

Keywords: biomonitoring, environmental health, metals

MO-PO-74

Differences in human subjective responses depending on chemical concentration levels in indoor air

Nakaoka H. [1]*, Mori C.* [1]*, Todaka E.* [1]*;* [1]*Chiba University, Chiba, Japan*

Background: Indoor air environment is a key element for a healthy life because almost all individuals spend the majority of their times in indoor. Health effects related to indoor air pollutants have long been an inconclusive issue. Because even in the same environment, some people show symptoms and others do not. It may largely depend on individual sensitivity. Therefore, we tried to clarify the relationship between indoor air environment and its health effects on people who are more sensitve and less sensitive to chemicals Methods: Two laboratory houses (LH-A and LH-B) were built on the campus of Chiba University, Japan, in November 2017 for this study. The interior and exterior of these two LHs looked almost the same except the chemical concentrations in indoor air due to the difference of building and interior materials. From 2017 to 2018, 148 healthy volunteers evaluated the air quality of the LHs subjectively. They were requested to complete self-reported questionnaires on Sick Building Syndrome (SBS) symptoms, odour and so on during their stay at the LHs for around 90 min. Before every evaluation tests, indoor air samples were collected from the LHs, and they were analyzed. The relationship between these data of participants and indoor air quality were analyzed statistically. Results: The mean concentrations of the sum of VOCs (ΣVOCs) in LHs A and B were 5563 and 294 μg/m3, respectively. In the subjective evaluation, they showed the differences in complaint of SBS symptoms, odour intensity, their comfort, and relaxation degree in each of the two LHs. Conclusions: In this study, it was indicated that levels of ΣVOCs and the odour strength could affect human's quality of life such as SBS symptoms, indoor comfort and their degree of relaxation through subjective data.

Keywords: VOCs

MO-PO-75

A study on exposure levels and influencing factors of endocrine disruptors in Korean adults - The 3rd cycle of Korean National Environmental Health Survey (KoNEHS)

Kim S.Y. [1]*, Joo Y.K.* [1]*, Yoo J-Y.* [1]*, Lee C-W.* [1]*, Yu, S-D.* [1]*;* [1]*National Institute of Environmental Research, Incheon, South Korea*

Bisphenol A and phthalate metabolites are endocrine disruptors that enter the human body and interfere with normal endocrine functions (e.g. reproductive and developmental processes). This study aims to investigate exposure levels and effects of bisphenol A and phthalate metabolites, based on Korean National Environmental Health Survey (KoNEHS, 2015-2017). In this study, those taking obesity and/or metabolic syndrome medicine were excluded from the 3,787 subjects (233 districts) in 3rd KoNEHS. Urinary Bisphenol A and phthalates were analyzed by LC-MS/MS. DEHP metabolites were calculated by summing MEOHP and MEHHP. Body mass index(BMI) was divided into four groups :

low body weight(25.0). Statistical analysis was performed using GM, 95% confidence interval. Significance was determined by kruskal-wallis test. The Bisphenol A level was 1.18 μg/L on average, while it was 1.33 μg/L for males, higher than 1.04 μg/L for females. Bisphenol A concentrations showed a significant increasing trend($p<0.0001$) with BMI: 1.01, 1.17, and 1.37 μg/L for the normal, pre-obesity, and obesity groups, respectively. This trend was more apparent in females. DEHP metabolites had a similar relation with BMI(21.2, 24.0, and 25.6 μg/L for the normal, pre-obesity, and obesity groups, respectively) with the average concentration of 23.5 μg/L. Consumption frequency of canned beverage, plastic-bottled beverages, and frequency of plastic container use were considered, but did not show statistically significant relationship with these chemicals in this research. This study examined exposure levels of bisphenol A and DEHP metabolites, and confirmed their associations with BMI. The results can be used as basic data to understand health effects of endocrine disruptors.

Keywords: biomonitoring

MO-PO-76
Levels of metals in blood among Swedish schoolchildren and adolescents

Almerud P.[1], Lignell S.[2], Andersson E.[1], Lundh T.[3], Sallsten G.[1], Lindroos A.K.[2]; [1]Occupational and Environmental Medicine, University of Gothenburg, Gothenburg, Sweden, [2]Swedish Food Agency, Uppsala, Sweden, [3]Occupational and Environmental Medicine, Lund University, Lund, Sweden

Background: Cadmium (Cd), mercury (Hg) and lead (Pb) are toxic heavy metals that almost all humans are exposed to. Diet is the main source of these metals, in addition to smoking. Knowledge about sources, exposure and body burden of these metals are essential for preventing negative health effects. The aim of this study was to examine the blood levels of Cd, Hg and Pb in relation to diet and background factors, such as gender, age, and sociodemographic background. Methods: The study is based on Riksmaten adolescents 2016-17, a nationally representative cross-sectional school-based dietary survey in grades 5, 8 and 11 (mean age 12, 15 and 18 years) (n=3099). Diet was reported in a web-based 24-hour recall and information on eating habits, health, socioeconomic background etc. was collected through online questionnaires. Height and weight were measured in all participants, while blood and urine samples were collected in a subgroup (n=1105). Cd, Hg and Pb were measured in whole blood with inductively couple plasma mass spectrometry equipped with collision cell. Results: The median levels (5-95th percentiles) for all subjects were Cd: 0.12 (0.06-0.35) μg/L, Hg 0.72 (0.14-2.1) μg/L, and Pb: 7.1 (3.7-16) μg/L. Cadmium levels in blood were higher among girls than boys, whereas boys had higher blood levels of Hg and Pb. Cd levels increased by age, both among girls and boys. An increase in blood levels of Hg by age was seen among boys, whereas the levels decreased by age among girls. Conclusions: Blood levels of Cd, Hg and Pb differed by sex and age. Using the detailed information on diet and blood levels from over 1000 adolescents, we will further investigate associations between diet and blood levels, and how these differ by age and sex.

Keywords: biomarkers, biomonitoring, children, food, metals

MO-PO-77
Systematic Identification of Allostatic Load Components Associated with Mortality

Nguyen V.[1,2], Chung J.[1], Pho N.[2], Patel C.[1], Colacino J.[3], Jolliet O.[3]; [1]Harvard University, Boston, MA, USA, [2]University of Michigan, Ann Arbor, MI, USA, [3]University of Washington, Seattle, WA, USA

Physiological stress response can be characterized by measurements of health biomarkers, which represent various biological pathways including neuroendocrine, cardiovascular and metabolic system. These components can be summed as binary variables into a uniformly weighted sum known as the allostatic load (AL) to represent the aggregated burden of chronic stress. AL score components vary widely between studies. Further, many calculations of AL do not quantify biomarker measurement deviance from clinical thresholds as a measure of risk.

To generate a more robust AL score, here we identify components significantly associated with mortality to develop a weighting method based on an individual's biomarker distance from a clinical threshold. We used the National Health and Nutrition Examination Survey of biomarker measurements for 41 components in 62,160 participants 1999-2010. We used the mortality data linked with the National Death Index to ascertain mortality status and time to death. Ultimately, we plan to validate our weighted AL scores by evaluating the prediction performance on mortality, so AL components were excluded here based low overlap with mortality data. We conducted survey weighted cox proportional-hazards regression models with mortality status and time to death as the outcome variables and the AL component as the main predictor while adjusting for age, sex, race/ethnicity, smoking, and socioeconomic status. We identified 22 AL components that are significantly associated with mortality including creatinine, glycohemoglobin, and C-reactive proteins that increase the risk of death in the study period by 33%, 17%, and 14%, respectively, for a one-unit increase, while albumin decreases the risk by 68%. Here, we present a framework to systematically identified AL components that are associated with mortality for calculating the AL score. Using the remaining AL components, we are developing methods to incorporate weights based on distance from clinical thresholds to calculate a weighted AL score.

Keywords: biomarkers

MO-PO-78
Exposure Characterization of Drinking Water Haloacetic Acids in Human Urine

Pervez S. [1], Ashby J. [1], Kimura S. [2], Richardson S. [3]; Indiana University, Bloomington, IN, USA, [2]University of Calgary, Calgary, Canada, [3]University of South Carolina, Columbia, SC

Disinfected drinking water is the primary source of haloacetic acids (HAAs) which are considered mutagenic, cytotoxic, genotoxic, teratogenic, and carcinogenic. Despite adverse health risk of HAAs, their exposure levels and variability in humans are not well-known. Therefore, we measured HAAs in cross-sectional and longitudinal urine specimens from 17 individuals. Five regulated HAAs-mono, di, and trichloroacetic acid (MCAA, DCAA, and TCAA) and mono- and dibromoacetic acid (MBAA and DBAA)-and one unregulated HAA-bromochloroacetic acid (BCAA)-were measured. Measured levels of DCAA, MBAA, DBAA, and BCAA were always below the limits of detection (LOD). MCAA and TCAA were frequently detected, with higher interindividual variability of MCAA than TCAA. 24-hours repeat measures of urinary specimens showed peaked MCAA levels in post-shower specimens, while TCAA levels remain unchanged. Correlation between urinary MCAA and TCAA was moderate but statistically significant. High urinary MCAA in post-shower specimens suggest MCAA captures short-term exposure via dermal and/or inhalation, while urinary TCAA captures long-term exposure via ingestion. Although these findings are preliminary, they will inform future exposure and epidemiologic investigations. Further research is needed to investigate the toxicokinetic mechanisms of MCAA and TCAA for half-lives and route specific contribution.

Keywords: biomarkers, biomonitoring, cumulative exposure, environmental health, epidemiology, exposure factors, water

MO-PO-79
Hypertonic Saline Nasal Lavage Intervention in Dairy Workers

Erlandson G. [1], Magzamen S. [1], Carter E. [1], Sharp J. [1], Reynolds, S. [1], Schaeffer J; [1]Colorado State University, Fort Collins, CO, USA

Livestock workers experience an increased burden of bioaerosol-induced respiratory disease. We have previously demonstrated that dairy operations generate bioaerosols that span the inhalable size fraction. These aerosols contained complex bacterial communities and inflammagens. Because those particles with an aerodynamic diameter between 10-100 μm are known to deposit in the nasopharyngeal region, we believe that exposure health outcomes in the nose are equally important and need consideration. We conducted a pilot study to evaluate the effectiveness of a hypertonic saline nasal lavage in reducing inflammatory responses in dairy workers. We recruited a total of ten workers from a high-volume operation in the High Plains region. We collected inhalable personal breathing zone (PBZ) samples (SKC button) and pre-/post-shift nasal lavage samples from each participant over five consecutive days. The treatment group (n=5) received hypertonic saline (HTS) while the control group (n=5) received normotonic saline. The PBZ samples were analyzed for endotoxin using the Pyrogene assay (Lonza) and gravimetrically for particulate. Pro-inflammatory cytokines (i.e., for IL-8, IL-6, and TNF-⍰) were measured from nasal lavage samples using a multiplex assay (Meso Scale Discovery). Inhalable dust concentrations ranged from 0.15mg/m3 to 1.9mg/m3. Cytokine concentrations were analyzed with a hierarchical linear effects model. Significant attenuation of TNF-a, IL-6, and IL-8 was observed among workers receiving HTS compared to those in the control group (α = 0.05). For TNF-a, there was a significant day by group interaction and the expression of TNF-a in the treatment group was significantly lower than expression in the control group at the end of the study ($p < 0.001$). There was also an overall significant treatment effect for IL6 and IL8 cytokines. Funding for this research is provided by the National Institute for Occupational Safety and Health U01 grant #: 1U01OH010840.

Keywords: biomarkers, microbial agents, occupational, respiratory health

MO-PO-80

Association of exposure to metals and polycyclic aromatic hydrocarbons with expression of Receptor of Advanced Glycation End Products in welding workers

Chen W-L. [1]*, Chou C-C.* [1]*, Chuang H-C.* [1]*, Lin G-J.* [1]*, Pan C-H.* [1]*, Wang C-C.* [1]*, Lai C-H.* [1]*;* [1]*National Defense Medical Center, Taipei, Taiwan*

Shipyard welders will expose to fine particulate, aerosols, dust, metal fumes, polycyclic aromatic hydrocarbons (PAHs) and other substances during the work process. Modifiable environmental factors including particulate matter and its components drive endogenous formation Advanced glycation end-products (AGEs) overwhelm innate defenses of antioxidant and activate pro-inflammatory and pro-oxidant receptors for advanced glycation end-products (RAGE) associated cardiovascular effects have not been explored. This study was in a cross-sectional design. Study subjects consisted of 53 welders and 29 office workers from a shipyard of northern Taiwan. Participants were asked to wear PM2.5 samplers during their working hours on the first weekday (Monday) and collected urine and fasting blood samples on the next day (Tuesday) morning. Monocytic receptor of advanced glycation end products expression was assessed flow cytometry. AGEs, IL-6, TNF-α were analyzed by ELISA kits. Urinary metal concentration was analyzed by ICP- MS. Statistical analysis was conducted using SPSS 22.0. It showed Log(Ni) (β=0.329, 95% CI 0.031 to 0.172, p=0.006), or Log(Fe) (β = 0.258, 95% CI 0.044 to 0.556, p = 0.022) was positively associated with Log(AGEs) adjusting for creatinine, age, smoking status, and work group. Log(Cr) (β = 0.236, 95% CI 0.018 to 0.337, p = 0.03), Log(Ni) (β = 0.225, 95% CI 0.006 to 0.133, p=0.032), or Log(Cd) (β=0.309, 95% CI 0.006 to 0.139, p=0.028) was positively associated with the Log(RAGE), respectively. Log(RAGE) was found statistically significant with IL-6 (β = 0.294, 95% CI 0.083 to 0.732, p = 0.014). The possible mechanism is that the metal or PAHs adsorbed on the fine particles from welding fume might cause the increasing AGEs concentration, and then react with the receptor to produce the inflammatory reaction. We conclude that exposure to welding fume particles is associated with increasing AGEs levels and RAGE expression.

Keywords: occupational

MO-PO-81

Risk of metabolic syndrome and the level of the AGEs and RAGE in welding workers- a cross-sectional study

Lan Y.W.[1], Lai C-H.[1]; [1]School of Public Health, National Defense Medical Center, Taipei, Taiwan

Introduction: Metabolic syndrome is commonly recognized as a risk for cardiovascular disease or diabetes. Previous studies were shown that welding exposure may be a risk for metabolic syndrome. Advanced glycation end products (AGEs) is also a risk for chronic disease such as cardiovascular disease or diabetes. AGEs which binds to receptor for AGE (RAGE), and then resulting in pro-inflammatory reaction to lead to chronic disease. Previous studies were shown that welding exposure may influence the level of AGE and RAGE. Objectives: Investigating the association of the risk of metabolic syndrome and the level of the AGEs and RAGE in shipyard in Taiwan. Methods: The subjects included 65 welding workers, and 17 administration workers. According to American Heart Association, the definition of the metabolic syndrome indicators of abdominal obesity, high triglycerides, low high-density lipoprotein cholesterol, high blood pressure, and high fasting glucose. All risk components of metabolic syndrome were measured by physical examination, and the level of AGEs and RAGE were analyzed by flow cytometry. We use SPSS version 22 to analyzed the data. Continuous variables were reported by mean± SD and categorical variables were reported by n (%). The linear regression of risk component of metabolic syndrome were calculated by logistic regression. Results: The mean of the level of AGE in shipyard is 16.04±10.87 ng/mL. The level of AGE is positively associated with the level of RAGE. The level of the AGE is significantly associated with fasting glucose in shipyard workers.($p<0.05$) after adjustment of cooking methods in high temperature (e.g. barbecue), alcohol drinking, areca using, the experience of welding exposure, and the medical history of hypertension, stroke, and diabetes. Conclusion: The risk component of metabolic syndrome is associated with the level of AGE and RAGE.

Keywords: workplace

MO-PO-82

Association between chronotype and AGEs: a cross-sectional study

Hsueh Y-W.[1], Lai C-H.[1]; [1]School of Public Health, National Defense Medical Center

BACKGROUND: Security works 24 hours a day. Shift work can cause a significant disruption of circadian rhythms of biologic functions, more over lead to a long term disturbances of sleep, chronic fatigue syndromes, cardiovascular, and even cause to the cancer. Thus, the aim of this study is to investigate the relationship between chronotype and 10-year cardiovascular disease risk in shift-worker. METHODS: This was a cross-sectional study. Data were collected, cardiovascular disease was measured by Framingham Risk Score (FRS)and chronotype was measured by Morningness-Eveningness Questionnaire (MEQ). AGEs were measured by skin autofluorescence. There were 100 subjects including 63 day-shift workers and 37 night-shift workers. RESULTS: In the Spearman correlation coefficients, the Framingham score was significantly correlated with the chronotype ($p=0.006$). Framingham score and AGEs were significantly correlated ($p<0.001$). After adjusting for age, gender, and work type, the chronotype was not significantly associated with Advanced Glycation End Products from logistic regression analyses. Further large studies in longitudinal design are needed to explore the relationship between chronotype and 10-year cardiovascular disease risk in shift-worker

Keywords: workplace

MO-PO-83

Urinary metabolites of solvent exposure among workers in electronic factories in Batam, Indonesia

Liu S.[1], Hasan Z.[1]; School of Health Sciences, Purdue University, West Lafayette, IN, USA

Numerous chemicals have been used in the manufacturing processes in the electronics industry. As a result, workers have been exposed to a variety of hazardous chemicals, including organic solvents. Concerns about the possibility of excess cancer risks among workers in semiconductor manufacturing have been expressed for many years; however, results from several large industry-based epidemiological studies were inconsistent, partially due to the lack of quantitative exposure measurement for individual workers. Although a few studies reported that the exposures of individual chemicals in the electronics industry were much lower than occupational exposure limits, epidemiological studies conducted in this industry have shown that long-term exposure to low levels of chemicals may still adversely affect such health outcomes as spontaneous abortion, irregular menstrual cycles, endocrine disruption and kidney dysfunction. Additionally, due to rapid changes in this industry, manufacturing processes and their associated hazards may change quickly every few years. Moreover, most electronics manufacturing factories are located in developing countries with less industrial hygiene support and resources. All these factors make frequent hazard assessments necessary, but difficult to conduct. We are currently conducting a biomonitoring pilot project in Batam, Indonesia to investigate if urinary metabolites of solvents known to be used in factories in the area would be higher in exposed workers than in controls. We will recruit 45 exposed and 10 un-exposed workers and collect urine samples before and after the workweek in April 2019. Urine samples will be analyzed for metabolites of selected solvents and for untargeted metabolomics profiling. Analyses are anticipated to be finished in July 2019. The results of this work can help us identify biomarkers to be used to evaluate the exposure and potentially the effectiveness of control strategies, as well as to study potential adverse health effects related to the exposure.

Keywords: workplace

MO-PO-84 WITHDRAWN

MO-PO-85

Addressing Water Quality Alongside Community Health Workers in a Farm Worker Community

Wagoner R.[1], Verhougstraete M.[1]; [1]University of Arizona, Tuscon, AZ, USA

The community of Winchester Heights, Arizona is made up of approximately 400 migrant farm workers and has been identified as a community near the US-Mexico border lacking accessible potable water, adequate sewer services, and other infrastructural issues. Previous work has illustrated community concern over environmental health issues, including water quality. The objectives of this project were to increase community capacity to address environmental health concerns through the establishment of a community health worker (CHW) environmental health (EH) program, measure fecal indicators and metal concentrations in community water samples, and identify community water usage patterns and barriers to improvement. Five community members were recruited as CHWs and trained in environmental health principles, human subject research, water collection, sample processing, and result-reporting methods. The trained CHRs collected tap water samples from 43 households and administered questionnaires to 43 adults. Morning and evening tap water samples were analyzed for E.coli using compartmental bag test as well as membrane filtration followed by incubation on modified mTEC media. Water samples were analyzed for lead and arsenic via inductively-coupled plasma mass spectrometry. Water samples from all households were negative for E.coli. The concentration of lead and arsenic in all water samples were below the EPA Maximum Contaminant Level (MCL) for both contaminants. The majority of respondents used bottled water as their primary drinking water source and listed lack of regular testing and acceptable consumer confidence reports as barriers to improvement. This project demonstrates the utility of CHW programs in addressing community environmental health concerns along the US-Mexico border region. The CHWs recruited participants, conducted water sampling, and reported of results to their own community, which are transferrable skills that can be used to address other community issues. Furthermore, the project can be used as a model in other underserved border communities.

Keywords: community, environmental health, environmental justice, metals, microbial agents, susceptible/vulnerable, water

MO-PO-86

The Synthesis of Traditional Technologies and Modern Green Technologies in Natural Ventilation for the Renovation of Historical Vernacular Settlements: A Case Study of Sandaoguai Vernacular Dwellings in Chongqing, China

Liu K. [1], Zhao Y. [1], Huang J. [1], Zhang, H. [1]; [1]Tsinghua University, Beijing

It contributes more to the preservation of traditional cultures and the adaptability to environment and climate to renovate vernacular architecture with green technologies rather than build a new one especially in the villages of western China, which have large numbers of vernacular dwellings as well as increasing demands for modern living conditions. However, current standards relative to green buildings lack in special design guidelines or technology diagrams to instruct architects in the improvement of the green performance and the regional adaptability of vernacular architecture in western China. Therefore, the research takes the renovation of Sandaoguai, which is known as A Chinese Famous Historical and Cultural Street in Chongqing in HSCW Zone, as an example to show the effectiveness made by means of the synthesis of traditional technologies in typical vernacular dwellings and modern general green technologies in the aspect of natural ventilation, in order to provide design guidelines and technology diagrams on the improvement of the green performance and the regional adaptability of vernacular architecture as well as to provide design methods and practice cases on the modernization of vernacular architecture and the localization of modern general green technologies.

Keywords: Traditional Technologies, Green Technologies, Vernacular Dwellings, Natural Ventilation, Adaptability

MO-PO-87 MOVED TO MO-PL-E4-88

MO-PO-88 MOVED TO TH-PL-E2-46

MO-PO-89

Estimates of the Prevalence, Intensity and the number of Workers Exposed to Cigarette Smoking across industries and occupations in South Korea

Jung H.[1], Koh D-H. [1], Choi S.[2], Park J-H.[3], Kim, H-C.[4], Lee, S-G.[5], Park D-U.[6]; [1]Catholic Kwandong University, International St. Mary's Hospital, Incheon, South Korea, [2]Daegu Catholic University, Daegu, South Korea, [3]Dongguk University, Seoul, South Korea, [4]Inha University Hospital, Incheon, South Korea, [5]Korea Occupational Safety and Health Agency, Ulsan, South Korea, [6]Korea National Open University, Seoul, South Korea

The use of tobacco is a major cause of preventable death worldwide. It affects the incidence and prognosis of various illnesses such as lung cancer, respiratory diseases, and cardiovascular diseases. As an effort to reduce carcinogen-induced cancer, we aimed to estimate the prevalence, intensity and the number of workers exposed to smoking, which would be specific to the occupational and industrial circumstances in Korea. We used the

Korean Working Condition Survey (KWCS) and Korea Census data. Smoking prevalence and intensity were estimated using KWCS data. The number of smokers was estimated by multiplying smoking prevalence with the number of workers in the occupation or industry. This results were estimated for major, sub-major, and minor group of occupation and industry. Given that the total labor force in 2010 was 22,290,398 workers, 32.47% (Male: female= 6,788,502: 502,331) of the workers were estimated to be current smokers. The average smoking prevalence in male workers (52.66%) was about 10 times higher than that in female workers (5.24%). The average amount of smoking in male workers was 15.42 cigarettes/day, and that of female workers was 11.29 cigarettes/day. For occupations, in total workers, "Craft and Related Trades Workers" showed the highest smoking prevalence (52.24%). "Managers" showed the largest amount of smoking (16.63 cigarettes/day) and "Equipment, Machine Operating and Assembling Workers" showed the largest number of estimated smokers (1,368,726 workers), respectively. For industries, in total workers, "Mining and quarrying" showed the highest smoking prevalence (69.27%). "Construction" showed the largest amount of smoking (17.16 cigarettes/day) and "Manufacturing" showed the largest number of estimated smokers (1,629,893 workers), respectively. We examined high-risk groups to smoking according to occupation and industry. Our results may help setting properties for prevention-related activities and assessing the impact of changing regulations. And, it can be used for epidemiological studies controlling for the effect of smoking.

Keywords: environmental health, exposure factors, industrial hygiene, occupational, workplace

MO-PO-90

Residential indoor environment and self-reported health in Northeast China

Chen Y. [1], Chen B. [1]; [1]School of Civil Engineering, Dalian University of Technology, China, Dalian, LiaoNing Province/China

The indoor environment quality of residential buildings has a significant impact on the occupants' health and wellbeing. To determine associations between characteristics of residential indoor environment and health, A Pan-Northeast China survey has been undertaken in 1836 residential buildings in 2014. The results show that there were multiple risk factors in residential buildings: thermal problems in summer, noise exposure in the living room; air too dry, thermal problems in summer in the bedroom; odor, mold growth in the kitchen; thermal problems in winter , odor in the wash room, dressing room and bathroom; thermal problems in winter, odor in the toilet; in the entrance. Moreover, there were significant associations between indoor environment factors (thermal environment, lighting environment, acoustic environment, indoor air quality, safety, and hygiene) and self-reported health (subjective health, specific symptoms and diseases). Furthermore, overall indoor environment was also significantly associated with self-reported health. The Pan-Northeast China survey provides a useful database to identify risk factors related to indoor environmental quality and occupants' health, which is helpful for improving knowledge of the impacts of existing indoor environment conditions on health.

Keywords: acoustics/noise, health, chronic diseases, health, other/general, health, symptoms, light/lighting, thermal environment

MO-PO-91 MOVED TO TH-PL-C3-62

MO-PO-92

Perception analysis for health risk attributed to air pollution and policy interventions for preventing community exposure in urban environment

Rashid A.[1], Uddin I.[1], Malik A.[2], Ullah A.[1]; [1]Department of Environmental Sciences, Arid Agriculture University, Rawalpindi, Pakistan, [2]Civil Aviation Authority, Islamabad International Airport, Islamabad, Pakistan

Global disease burden due to adverse air quality in urban areas is on rise which needs exposure assessment of communities to air pollutants in relation to potential health risks. Policy interventions without considering public perception about existing impacts of adverse air quality would less likely result in health promotion. Therefore, we analyzed nine potential risk perception factors which includes: controllability, familiarity, equity, understanding, uncertainty, dread, trust in government policies, personal stake, and victim identity in a community health survey (n=396) conducted in Rawalpindi city of Pakistan. Major health risk concerns expressed by community were related to air pollution. For moderate to extreme category, health risk perception showed respiratory health (41%), lifestyle disruption (36%), other physical health (22%), nuisance effect (14%), and psychosocial effects (9%) as top concerns. Multivariate regression model for neighbourhood air pollution concern was significant as depicted by odds ratio (OR). Highest impact on health was perceived for particulate matter (OR=4.18; 95% CI=1.58–7.08), followed by vehicular emission in primary neighbourhood (OR=2.44; 95% CI=1.26–5.94) and atmospheric smog (OR=1.89; 95% CI=1.02–4.11). Results of principal component analysis (PCA) show that controllability, trust in local government policies and equity were strongly correlated with first PCA axis (variance 56.3%) whereas with the second PCA axis, personal stake and uncertainty were significantly correlated. Trend observed through PCA results indicate. Our findings were further confirmed by hospitalization rate for respiratory illnesses recorded for local population. We conclude that exposure prevention to air pollutants require public perception as first step to focus for establishing a intervention point where urban community involvement can play a significant role in health risk minimization.

Keywords: community

MO-PO-93

The effects of indoor air quality on sleep quality: a review

Liao C. [1], Laverge J. [1]; [1]Ghent University, Ghent, Belgium

Sleep quality is quite important for health and daily performance. Indoor air quality (IAQ) is related to sleep quality. Although some people have a relaxed mood, and sleep in the bedrooms with low noise, moderate light and appropriate temperature, they cannot sleep well because of poor air quality. The aim of this study is to conduct a review of subjective and objective assessments of sleep, IAQ measured in bedrooms and the effects of IAQ on sleep quality. A search was conducted in Web Of Science using keywords of "indoor air quality" and "sleep". Subsequently, we screened the publications by review the abstracts. Generally, EEG, PSQI, actigraphy, questionnaire were used for sleep measurements. Lower O2 levels (less or equal to 16%) reduced slow wave sleep (SWS) and increased awakening. Exposure to cooking fumes extended sleep latency and reduced sleep quality. People in dormitory with no ventilation compared to ventilated rooms had lower sleep efficiency. Exposure to cigarette smoke or room scents was associated with self-rated poor sleep. However, a study found no changes in sleep quality using personalized ventilation device. Good IAQ improves sleep quality. But the definitions of good IAQ and sleep quality were still not clear in previous studies, because the lowest sufficient ventilation rate can be figured out with those clear definition. Also, the methods to change and measure IAQ varies a lot. From the review, we can conclude that there is a lack of studies showing direct effects of IAQ on sleep and that a number of methodological questions remain to be addressed in the design of such studies.

Keywords: health

MO=PO-94

Recovery of viable Staphylococcus aureus bioaerosols

Chang C-W.[1], Lin M-H.[1], Horng Y-J.[1]; [1]National Taiwan University, Taipei, Taiwan

Staphylococcus aureus aerosols are present in facilities (e.g., animal barns and hospitals) and may threaten human health. Air monitoring by appropriate sampling techniques is required for exposure assessment. This study thus aims to assess the recovery efficiency (R) of liquid-based samplers for viable S. aureus based on cellular membrane integrity. By using propidium monoazide and qPCR, the R of six sampling methods, i.e. two widely-used samplers (all-glass impinger-30 (AGI-30) and BioSampler) filled with three collection liquids (deionized water (DW), Tween mixture (TM) and phosphate-buffered saline (PBS)), were determined at five sampling times (3, 6, 15, 30 and 60 min). Effects of liquid refilling every 15 min during 60-min sampling were also evaluated. Irrespective of sampler type, PBS presented the lowest cell recovery, statistically different from DW ($P<0.05$). Liquid replenishment improved cell recovery in DW by AGI-30 (P0.05). In contrast, samplings for 15 min or greater resulted in decreased recoveries in PBS or TM by both samplers compared to 3-min sampling. Overall, BioSampler recovered S. aureus greater than AGI-30 ($P<0.001$). In particular, BioSampler with DW presented great R independent of effects of sampling time and liquid refilling, supporting it as an appropriate sampling method. This study elucidates the factors affecting the collection of S. aureus aerosols, which helps professionals possess greater certainty of successful capture of S. aureus and collect reliable data.

Keywords: sampling methods

TUESDAY SESSIONS

Tuesday, August 20th

TU-PL-A1-HER: Biomonitoring of SVOCs

TU-PL-A1-01

Short-term temporal variability in urinary biomarkers of phosphate flame retardants and plasticizers (PFRs) and alternative plasticizers (APs)

Bastiaensen M.[1], Malarvannan G.[1], Gys C.[1], Covaci A.[1]; [1]Toxicological Center, University of Antwerp, Antewrp, Belgium

Emerging chemicals such as phosphate flame retardants and plasticizers (PFRs) and alternative plasticizers (APs) have recently been introduced as replacements to more persistent chemicals and are therefore increasingly studied in human biomonitoring studies. Many of these chemicals have a shorter half-life than the ones they are replacing, which means that urinary metabolite concentrations likely only reflect short-term exposure. However, it is not clear whether biomarkers levels of these chemicals are stable or fluctuating over time. This is of critical importance for biomonitoring studies that generally use only one spot urine sample per participant. We aimed to investigate the intra- and inter-day variability of PFR and AP metabolite concentrations over a 5-day period and to compare the reliability of spot urine samples versus 24 hours urine to predict general exposure. In this study, ten healthy volunteers of at least 18 years old provided every urine sample they produced for 5 consecutive days, followed by one day of 24-hour urine collection. PFR and AP metabolite concentrations were quantified by validated analytical protocols based on solid-phase extraction (SPE) and liquid-chromatography tandem mass-spectrometry (LC-MS/MS). Spearman's rank correlation was used to relate metabolite concentrations. Intraclass correlation coefficients were calculated to assess temporal variability. Of the 14 PFR metabolites measured, eight were found in more than half of the samples. Of the 11 AP metabolites measured, five had a detection frequency above 50%. Median concentrations were comparable to the levels reported in our pilot studies. Moderate to strong correlations were observed between metabolites coming from the same parent compound. Intra class coefficients were of moderate strength for the majority of the investigated compounds (i.e. 0.4 – 0.6), however some metabolites had much higher values (>0.6) which showed that these concentrations varied little across time. Correlation strength did not improve significantly after adjustment for SG or Cr.

Keywords: analytical methods, biomarkers, biomonitoring, flame retardants, phthalates

TU-PL-A1-02

Children's and Adolescents' Exposure to Glyphosate – Results of the German Environmental Survey, GerES 2014-2017

Conrad A.[1], Schmied-Tobies M.[1], Hoppe H-W.[1], Rucic E.[1], Kolossa-Gehring, M.[1], [1]German Environment Agency (Umweltbundesamt), Berlin/Dessau-Rosslau, Germany

Background and Methods: Glyphosate is one of the most used substances for weed control. The internal exposure of children and adolescents (3 to 17 years) to glyphosate and its metabolite aminomethylphosphonic acid (AMPA) were investigated in the fifth cycle of the German Environmental Survey (GerES V) carried out from 2014 to 2017. Glyphosate and AMPA were quantified in 2,144 morning urine samples by GC-MS/MS (limit of quantification (LOQ) = 0.1 µg/L). Standardized interviews i. a. on exposure-relevant behaviors were carried out with participants or their parents. GerES V was jointly performed with Wave 2 of the German Health Interview and Examination Survey for Children and Adolescents (KiGGS W 2) of the Robert Koch Institute (RKI). Hence, sociodemographics could be included in statistical analyses. Results: 52% of glyphosate and 46 % of AMPA levels were at or above LOQ. The overall geometric mean concentrations for both analytes resulted in 0.1 µg/L. The maximum level for glyphosate was 11.1 µg/L (AMPA: 13.4 µg/L). Glyphosate and AMPA levels differed significantly by age group.

Vegetarian diet seems not to be associated with exposure differences. Glyphosate and AMPA were significantly correlated (rank correlation = 0.56). Discussion: The estimated daily glyphosate intake based on the highest observed concentration is well below the acceptable daily intake of 0.5 mg/kg/d. As glyphosate is rated as probable carcinogen, detailed knowledge on the human exposure is warranted. Multivariate analysis of GerES V data will support the further elucidation of exposure sources and routes for the young generation in Germany. A combined consideration of glyphosate and AMPA in exposure assessment should be explored further. Acknowledgements: We thank all participants and the RKI for making GerES V possible. Funding by the German Ministries for the Environment, Nature Conservation and Nuclear Safety (BMU) and of Education and Research (BMBF) is gratefully acknowledged.

Keywords: biomonitoring, children, environmental health, epidemiology, risk assessment

TU-PL-A1-03

Long-lasting Legacies: Organochlorine Compounds in Children and Adolescents – Results of the German Environmental Survey, GerES 2014-2017

Conrad A.[1], Rucic E,[1], Murawski A.[1], Kolossa-Gehring M.[1], Bandow, N.[1], Sawal, G.[1]; [1]German Environment Agency (Umweltbundesamt), Berlin/Dessau-Rosslau, Germany

Background and Methods: A number of organochlorine compounds (OC) have been strictly regulated in Germany already decades ago. For providing data on the current internal exposure of children and adolescents (3 to 17 years), OC were investigated in the German Environmental Survey (GerES V) carried out from 2014 to 2017. GerES V was jointly performed with Wave 2 of the German Health Interview and Examination Survey for Children and Adolescents (KiGGS W2) of the Robert Koch Institute (RKI). Among other OC, dichlorodiphenyldichloroethylene (DDE, dichlorodiphenyltrichloroethane (DDT) metabolite) and polychlorinated biphenyls (PCB118, PCB153, and PCB180) were quantified in 1,135 blood plasma samples by GC-MS/MS. Sociodemographics and data on health-relevant behaviors were used in statistical analyses. Results: The overall geometric mean (GM) concentration of DDE was 0.158 µg/L. GMs increased with socioeconomic status and migrational background. For the sum of the three higher-chlorinated PCBs (PCB-Sum) GM was 0.303 µg/L with higher GMs in males. PCB-Sum GMs increased with socio-economic status but decreased with migrational background. DDE and PCB-Sum was higher in former East and West Germany, respectively. For both parameters, GM where higher in participants that had been breastfed. Discussion: Decades after national regulation and being included in the Stockholm Convention of 2004, health-relevant OC are today still ubiquitous in Germany's population. Current regional internal exposure differences can be explained by the country's separation until 1990 with different OC usage patterns. GerES V yields detailed information on sources and routes that can explain current difference between sub-populations and support further exposure reduction. Moreover, the study data provides sound empirical evidence of the long-lasting impact of persistent chemicals. Acknowledgements: We thank all participants and the RKI for making GerES V possible. Funding by the German Ministries for the Environment, Nature Conservation and Nuclear Safety (BMU) and of Education and Research (BMBF) is gratefully acknowledged.

Keywords: biomonitoring, environmental health, epidemiology, pesticides, risk assessment

TU-PL-A1-04
Human toxicokinetic profiles after the dermal exposure to methyl-, ethyl-, and propyl paraben in human.

Shin M.Y.[1], Choi J.W.[1], Lee S.[2], Kim S.[2]; [1]Department of Environmental Health Sciences, Graduate School of Public Health, Seoul National University, Seoul, South Korea, [2]Public Health, Seoul National University, Seoul, South Korea

Parabens are commonly used preservatives in cosmetics, but there have been concerns on the association between their exposure and adverse outcomes. Although humans are exposed to various parabens through food intake and dermal exposure, the information on the relative contribution to internal dose via each exposure pathway are limited. This study investigated urinary excretion of methyl- (MtP), ethyl- (EtP), and propyl paraben (PrP) after topical application in human volunteers and compares with the results from oral exposure study to find out the contributions. We determined the pharmacokinetic characteristics of MtP, EtP, and PrP in humans following a single dermal application of 6.5 g of cream containing deuterium labeled MtP, EtP, and PrP (approx. 1% each). Blood and urine were collected at several intervals over the course of 48 hours to measure levels of the parabens and their metabolites in five male volunteers. The unconjugated parent compounds (free parabens), glucuronide and sulfate conjugates were measured using HPLC-MS/MS. The fraction of dermally exposed dose excreted into urine (Fue) based on the applied amounts were ranged from 0.13 ~ 0.15 % for free parabens, 1.38 ~ 1.88 % for total parabens (sum of free and conjugated parabens). Meanwhile, the Fue calculated from orally administered PrP experiment in our previous study (male, n = 7) were lower than those in free PrP (0.04 %) and much higher than those in total PrP (8.5%) compared to the data from dermal exposure. Depending on the exposure pathways of parabens, the quantitative difference on excretion rates were observed. Especially, the difference of conjugated metabolites by exposure pathways was obviously larger than the other metabolite. The present study can be utilized to predict the contribution of parabens to the human body by exposure pathways. Eventually, it can provide important information on making strategies to reduce paraben exposures.

Keywords: exposure models, risk assessment

TU-PL-A1-05
Monitoring exposure levels to bisphenols in Japanese schoolchildren

Gys C.[1], Ait Bamai Y.[2], Araki A.[2], Caballero-Casero N.[1], Kishi, R.[2], Covaci A.[1]; [1]University of Antwerp, Antwerp, Belgium, [2]Hokkaido University, Sapporo, Japan

Bisphenol analogues such as bisphenol F (BPF) and bisphenol S (BPS) are currently replacing bisphenol A (BPA) in a number of industrial and consumer materials, such as epoxy resins and thermal paper. Although BPA is still the most prevalent bisphenol in human and environmental matrices, its analogues have also been detected frequently, in the environment and food products. Nevertheless, data regarding the human exposure levels of BPA analogues are scarce, in particular in children. To address this data gap, we assessed the exposure to 11 bisphenols through the analysis of urine samples from 200 children from Sapporo, Japan. Demographic variables and secular trends were studied in relation to urinary levels. This study was part of a prospective birth cohort (Hokkaido Study). Morning void urine samples from 7-year-old children were collected from 2012 to 2017 and analyzed for bisphenols using solid-phase extraction and GC-MS/MS. Demographic characteristics of the study population were acquired through a questionnaire. Multiple linear regression was used to evaluate secular exposure trends and possible associations between bisphenol concentrations and demographic aspects. Nine out of 11 measured bisphenol analogues were detected, indicating that exposure to these chemicals is ubiquitous, even for young children. Detection frequencies ranged from 2.7 % for bisphenol E (BPE) to as high as 94.6 % for both BPA and BPF. BPS (64.9 %) and bisphenol B (BPB, 48.6 %) were detected frequently as well. As expected, the highest median concentration was observed for BPA (1.5 ng/mL). The median concentration for BPS was 0.07

ng/mL and 0.11 ng/mL for BPF. Multivariate regression models were used to investigate the relation between demographic and urinary concentrations of bisphenol analogues and to determine secular trends. This is the first time exposure levels to bisphenols are assessed in Japanese children.

Keywords: analytical methods, biomonitoring, children

TU-PL-A1-06

Perfluorinated alkylated substances (PFAS) in Children and Adolescents – Results of the German Environmental Survey, GerES 2014-2017

Conrad A.[1], Rucic E.[1], Murawski A.[1], Lange R.[1], Kolossa-Gehring M.[1], Duffek A.[1], Wellmitz J.[1]; [1]German Environment Agency (Umweltbundesamt), Berlin/Dessau-Rosslau, Germany

Background/Methods: The current extent of German children's and adolescents' internal exposure to perfluorinated alkylated substances (PFAS) was unknown. To close this knowledge gap, 12 PFAS were investigated in German Environmental Survey, carried out from 2014 to 2017 (GerES V). The survey was jointly performed with Wave 2 of the German Health Interview and Examination Survey for Children and Adolescents (KiGGS W2) of the Robert Koch Institute (RKI) In a GerES V sub-sample of 1,109 children and adolescents (3-17 years), PFAS have been analyzed in blood plasma by LC-MS/MS. Limits of quantification (LOQ) ranged from 0.25 to 1 µg/L. Standardized interviews i. a. on exposure relevant behaviors and living conditions were carried out with participants or their parents. Sociodemographical data was used in statistical analyses. Results: For PFOA, PFOS and PFHxS 74% - 100% of measurements were at or above LOQ. For the other PFAS this fraction is 10% or less. Overall geometric mean (GM) concentrations were PFOA: 1.12 µg/L, PFOS: 2.49 µg/L, and PFHxS: 0.36 µg/L. For all three compounds, GMs are significantly higher in male participants and increase with socio-economic status and breastfeeding. For PFOA and PFOS, the current health-based guidance value (HBM-I) is exceeded by 21% and 7% of the participants, respectively. Discussion: GerES V provides population-representative data on German children's and adolescents' PFAS exposure. For a considerable fraction of the young generation, PFOS and PFOA exposure raises concern, as health effects can – according to current knowledge – not be ruled out with sufficient certainty. Further data analysis will provide an empirical basis for exposure reduction strategies and targeted risk communication. Acknowledgements: We thank all participants and the RKI for making GerES V possible. Funding by the German Ministries for the Environment, Nature Conservation and Nuclear Safety (BMU) and of Education and Research (BMBF) is gratefully acknowledged.

Keywords: biomonitoring, children, environmental health, epidemiology, risk assessment

TU-SY-B1: Meet-in-metabolite analysis (MIMA) for bridging the external exposures and biological response

TU-SY-B1-07

Meet-in-metabolite analysis bridging the internal dose of exposure and adverse outcome within a molecular epidemiology framework

Shen H.; Institute of Urban Environment, Chinese Academy of Sciences, Xiamen, China

In analytical chemistry, biomonitoring is the measurement of the body burden of toxic chemical compounds, elements, or their metabolites, in biological substances. To extend the traditional concept from measurement of the external exposure related internal residuals to biological response with emphasis of metabolism alteration is the key step forward to bridge the external exposure and the adverse outcome via the metabolism implied pathways. Combinations of the data-driven or hypothesis-free metabolomics with the traditional biomonitoring and the epidemiological study (such as case-control design), the endogenous metabolic molecular events (MMEs) that related to biomonitoring data and/or the adverse outcome can be screened, in which the internal exposure dose (IED) and adverse outcome (AO) can be used as the classifiers (i.e., latent variables) in partial least square discriminant analysis (PLS-DA), respectively. With considering the formative measurement model characterization, the causative linkages between AO and AO-related MMEs will be addressed firstly by PLS-DA. However, the relationships between the variable of IED and IED-related MMEs need to be confirmed by using the reflective measurement model, then the causative linkage of IED to the IED-related MMEs can be constructed (i.e., the relationships should be annotated as the latent variable of IED can be reflected by IED-related EMBs). In further, when observed that IED can associate AO in traditional epidemiology, there must be certain molecular pathways of linking IED to AO, and the analysis from IED-side and OA-side would meet-in-middle of these pathways that implied by the relevant MMEs. Finally, the meet-in-metabolite analysis (MIMA) is suggested because body fluids of blood and urine are the most common substances for human biomonitoring, while the other OMICS data would be applaudable either in together with metabolomic data. The strategy would useful for uncovering the black-box of epidemiology.

Keywords: biomarkers, biomonitoring

TU-SY-B1-08

Human Placenta - A Perfect Matrix to Connect the Biomarkers of Environmental Exposure to Biomarkers of Effect

De Angelis M.[1], Li Z-M.[1,2], Pauluschke-Fröhlich J.[3], Fritsche A.[4,5], Michalke B.[1], Benker B.[1], Bao Q.[7], Henkelmann B.[1], Shen H.[8], Schramm K-W.[1,9]; [1]Helmholtz Zentrum München-German Research Center for Environmental Health (GmbH), Molecular Exposomics, Neuherberg, Germany, [2]School of Life Sciences Weihenstephan (Nutrition), Technische Universität München, Freising, Germany, [3]Department of Obstetrics and Gynecology, University Hospital, Eberhard Karls University, Tuebingen, Germany, [4]Division of Endocrinology, Diabetology, Nephrology, Angiology and Clinical Chemistry, Eberhardt Karls University, Tuebingen, Germany; [5]Institute for Diabetes Research and Metabolic Diseases of the Helmholtz Center Munich at the University of Tuebingen (IDM), Tuebingen, Germany, [6]Helmholtz Zentrum München-German Research Center for Environmental Health (GmbH), Research Unit Analytical BioGeoChemistry, Neuherberg, Germany, [7]Ningbo Entry-Exit Inspection and Quarantine Bureau Ningbo, Zhejiang, China, [8]Key Lab of Urban Environment and Health, Institute of Urban Environment, Chinese Academy of Sciences, Xiamen, China, [9]Department für Biowissenschaftliche Grundlagen, Technische Universität München, Freising, Germany

The exposome of real organisms is defined as the measure of all the exposures of an individual in a lifetime and how those exposures relate to disease. Some exposure starts already in early life at prenatal conditions and leads to the onset of diseases in later life. To study prenatal exposure, the placenta is becoming a key tissue. In fact, due to its size, it is possible to analyze both the exogenous and the endogenous chemical substances. In addition, it is known, that placenta play a key role in the metabolism and detoxification of several drugs, compounds in diet, as well as environmental contaminants. Therefore, human placenta can also be used to identify possible biomarkers of environmental exposure and to link them to biomarkers of possible adverse outcomes. In several recent papers, to find the potential bridge between external exposure and biological response, researchers have reported the "meet-in-metabolite" method of analysis. Large part of these studies were conducted using human fluids such as urine or blood but we have supposed that placenta, for its characteristics, could also be a good matrix for this approach. For this purpose, we have analyzed in the human placenta some important biomarkers such as thyroid hormones and catecholamines. We have investigated the distribution of some metals, as well as the distribution of some persistent organic pollutants (e.g., polychlorinated dibenzo-p-dioxins and dibenzofurans, polychlorinated biphenyls, and polybrominated diphenyl ethers). Finally, we have studied if those molecules were homogenously distributed in the tissue. The results suggested that placenta could be a good matrix to correlate biomarkers of exposure to biomarkers of effects, although the sampling mode should be taken into consideration to avoid misleading results.

Keywords: biomarkers

TU-SY-B1-09

Associations between maternal complex environmental exposure and the risk of fetal neural tube defects in north China

Wang B.

The relationship between the levels of various environmental chemicals and the high prevalence of neural tube defects (NTDs) in north China remains unclear. Our study aimed to classify the associations of maternal exposures to polycyclic aromatic hydrocarbons (PAHs), passive smoking, and trace metal elements with NTD risk by a case-control study with large population size. It was concluded that maternal exposure to either PAHs or passive smoking was not a significant risk factor for NTDs, whereas the maternal intake of certain trace metal elements may be important risk factors contributing to the high NTD prevalence in north China.

Keywords: environmental health

TU-SY-B1-10

The modification of PM2.5 exposure to human health: an example of the meet-in-metabolite analysis

Huang Q.; Institute of Urban Environment, Chinese Academy of Sciences, Xiamen, China

Exposure to airborne fine particulate matter (PM2.5) has been associated with a variety of adverse health outcomes. To investigate the linkages between PM2.5 exposure and the health outcomes including chronic obstructive pulmonary disease (COPD), the meet-in-middle strategy based on urinary metabolic biomarkers were used. A cross-sectional study was designed to illustrate the mentioned linkages. Indoor PM2.5 and its element components were assessed in 41 Chinese elderly participants including COPD patients and their healthy spouses. Metabolic biomarkers involved in PM2.5 exposure and COPD were identified by using urinary metabolomics. The associations between PM2.5- and COPD-related biomarkers were investigated by statistics and metabolic pathway analysis. The results indicated the modification of PM2.5 to COPD via both modes of action of lowering participants' antioxidation capacity and decreasing their lung energy generation.

Keywords: air, biomarkers, environmental health, epidemiology, health, chronic diseases

TU-SY-B1-11

Multiple elements related to metabolic biomarkers on the context of gestational diabetes mellitus in meconium

Wu Y.[1], Shen H.[1]; [1]Institute of Urban Environment, Chinese Academy of Sciences, Xiamen, China

The classic epidemiologic approach has been referred to as the "black-box epidemiology" and can't elucidate these elements' biological roles on the damaged fetus developmental microenvironment. We aimed to characterize the associations of multiple elements with gestational diabetes mellitus (GDM) for illustrating their effects on the interrupted in-uterus microenvironment by conducting a case-control study (n=137 cases; n=197 controls). Twenty-one meconium elements were characterized using ICP-MS/OES. For shifting the present paradigm from a black-box approach to a molecular way, GDM-related metabolic biomarkers were identified in our previous metabolome report. Based on the meet-in-middle strategy, the relations among the elements, metabolic biomarkers and GDM incidence were assessed by using redundancy analysis and correlation-adjusted correlation; mediation analysis was further used to test the hypothesis that metabolic biomarkers mediate the associations of elements with GDM. As, Cd, Ba, and Ca significantly contributed to the variation of GDM-related metabolic biomarkers. Additionally, the associations of Cd, Ba, Ca and As with GDM were mediated by the metabolic biomarkers which majorly involved in the lipid metabolism and Adenosine/L-Arginine/Nitric Oxide (ALANO) pathways.

Keywords: biomarkers, children, environmental health, metals

TU-SY-C1: Exposure to SVOCs in the Indoor Environment – State-of-the-art Review and the Need for Consensus on a Suite of Exposure Models

TU-SY-C1-12
A fundamental SVOC emissions and transport model

Xu Y.; The University of Texas at Austin, Austin, TX, USA

A mechanistic model that considers particle dynamics and their effects on surface emissions and sorptions was developed to predict the fate and transport of phthalates in indoor environments. A controlled case study was conducted in a test house to evaluate the model. The model-predicted evolving concentrations of benzyl butyl phthalate in indoor air and settled dust and on interior surfaces are in good agreement with measurements. Sensitivity analysis was performed to quantify the effects of parameter uncertainties on model predictions. The model was then applied to a typical residential environment to investigate the fate of di-2-ethylhexyl phthalate (DEHP) and the factors that affect its transport. The predicted steady-state DEHP concentrations were 0.14 μg/m3 in indoor air and ranged from 80–46000 μg/g in settled dust on various surfaces, which are generally consistent with the measurements of previous studies in homes in different countries. An increase in the mass concentration of indoor particles may significantly enhance DEHP emission and its concentrations in air and on surfaces, whereas increasing ventilation has only a limited effect in reducing DEHP in indoor air. The influence of cleaning activities on reducing DEHP concentration in indoor air and on interior surfaces was quantified, and the results showed that DEHP exposure can be reduced by frequent and effective cleaning activities and the removal of existing sources, although it may take a relatively long period of time for the levels to drop significantly. Finally, the model was adjusted to identify the relative contribution of gaseous sorption and particulate-bound deposition to the overall uptake of semi-volatile organic compounds (SVOCs) by indoor surfaces as functions of time and the octanol–air partition coefficients (Koa) of the chemical. This model can be modified to predict the fate and transport of SVOCs and serves to improve our understanding of human exposure to indoor SVOCs.

Keywords: SVOCs, air, exposure models

TU-SY-C1-13
A critical comparison of Junge/Pankow and Goss/Schwarzenbach theory

Salthammer T.; Fraunhofer WKI, Braunschweig, Germany

Understanding the partitioning of semi volatile organic compounds (SVOCs) between gas phase and particle phase is essential for exposure analysis and risk assessment in the indoor environment. Numerous attempts have been made to calculate gas/particle partitioning coefficients Kip. Single-parameter adsorption and absorption models, which relate Kip to the vapor pressure Ps (according to Junge and Pankow) or the octanol/air distribution coefficient KOA (according to Harner and Bidleman) are usually applied. However, the fact that single-parameter models do not take into account the specific intermolecular interactions that govern every sorption process and depend on the respective chemical structure of the molecules and the sorbing phase should be regarded as disadvantageous. In this work we used poly-parameter Linear Free Energy Relationships (pp-LFER) (according to

Goss and Schwarzenbach) to describe the partitioning behavior of 14 SVOCs with high relevance for the indoor environment. The pp-LFER concept is based on Abraham descriptors and considers the size of the compound, interaction abilities like van der Waals, H-accepting (e-donating) and H-donating (e-accepting) as well as a dipolarity/po¬la¬riza¬bility parameter, which describes polar interactions that are not covered by the other parame¬ters. Aerosol descriptors were taken from studies in ambient air. Therefore it was also necessary to evaluate the data set for indoor applications. Under the assumption that absorption into a water-insoluble organic matter (WIOM) phase is the dominating mechanism, the temperature dependence of the partitioning process could be described by the van't Hoff equation and the enthalpy for the distribution between air and WIOM. The results were compared with experimental and calculated Kip values from single-parameter models and the suitability of the pp-LFER method for indoor applications was critically discussed. It is shown that the pp-LFER method is a strong alternative to single-parameter approaches and gives reliable coefficients for gas/particle distribution in the indoor environment.

Keywords: SVOCs, air, particles, phthalates

TU-SY-C1-14
From air through skin to blood: dermal uptake of SVOCs commonly found in indoor air

Weschler C.J.[1], Beko G.[2], Morrison G.[3]; [1]Rutgers University, NJ, USA, [2]Technical University of Demark, Kongens Lyngby, Denmark, [3]The University of North Carolina at Chapel Hill, Chapel Hill, NC, USA

During the past decade researchers have made significant advances in understanding the uptake of higher molecular weight organic pollutants via dermal absorption. Theories have been developed to predict their uptake from air, through bare skin, under steady-state and dynamic conditions. Further theories have been developed to estimate dermal uptake of organics from clothing. For SVOCs with the right combination of physical-chemical properties, exposure via dermal absorption is predicted to be larger than exposure via inhalation. Circumstantial evidence for dermal uptake of pollutants from the environment comes from an increasing number of reported correlations between specific chemicals in skinwipes and these same chemicals (or their metabolites) in blood and urine. This presentation will review key ideas that have taken us to this point. It also serves as an introduction to two talks that immediately follow: Gabriel Bekö summarizing specific experiments that have demonstrated substantial direct dermal uptake from air for several SVOCs, and Glenn Morrison discussing clothing as either an inhibitor or amplifier of SVOC uptake via the dermal pathway. Taken together, this research has alerted us to the fact that dermal intake of certain common indoor SVOCs (e.g., phthalates, parabens, alkylphenols, benzophenones, synthetic musks, chlorpyrifos) can be an important exposure pathway.

Keywords: pesticides

TU-SY-C1-15
Experimental verification of the dermal uptake of indoor chemicals from air

Bekö G.[1], Weschler C.J.[2], Morrison G.[3]; [1]Technical University of Demark, Kongens Lyngby, Denmark, [2]Rutgers University, NJ, USA, [3]The University of North Carolina at Chapel Hill, Chapel Hill, NC, USA

Exposure of building occupants to manufactured chemicals in the indoor environment can occur via non-dietary ingestion (e.g. indoor dust), inhalation and dermal absorption including dermal uptake directly from air. Dermal uptake of semivolatile organic compounds (SVOC) received limited attention compared to volatile organic compounds and therefore this exposure pathway is often neglected in SVOC exposure assessments. Modeling efforts indicate that direct uptake of certain SVOCs from air may occur. Experimental verification of this hypothesis is emerging. Using human subject experiments under controlled laboratory conditions coupled with biomonitoring, the fraction of the total SVOC intake attributable to a specific exposure pathway can be isolated.

One such study has demonstrated that for human participants following a 6-hr dermal-only exposure to elevated gas-phase concentrations of DEP and DnBP, the levels of their metabolites in urine samples collected over the next two days were roughly half those measured in urine samples following a 6-hr dermal plus inhalation exposure. Meaningful dermal uptake from air was demonstrated in a similar fashion for nicotine. These studies indicate that dermal uptake of certain common manmade chemicals present in indoor air can be comparable to or larger than the corresponding intake from inhalation. Substantial interpersonal differences in uptake and excretion pattern were observed. Skin washing following exposure may decrease exposure, but quantifying its effect warrants further experimental work. The talk will summarize the findings of these human subject experiments that uniquely probed the dermal uptake of SVOCs directly from air.

Keywords: SVOCs, biomonitoring, built/indoor environment, phthalates

TU-SY-C1-16
Clothing and exposure to SVOCs indoors

Morrison G.; The University of North Carolina at Chapel Hill, Chapel Hill, NC, USA

Clothing has recently been identified as an important mediator of exposure to indoor pollutants. Clothing can be protective, by acting as a physical barrier and also by sorbing or reacting with airborne chemicals, especially SVOCs. However, clothing and other textiles (e.g. bedding) can accumulate chemicals and particles and contribute to the overall body burden via dermal uptake, inhalation and ingestion. Controlled human participant experiments have demonstrated these effects for nicotine, phthalates and oxybenzone. Mechanisms controlling this exposure have been hypothesized, and models have been developed that predict chemical uptake that is consistent with the existing (but limited) experimental results from human subjects. Population exposure assessment using such models should be approached cautiously because the underlying transport and transformation mechanisms have not been firmly established, or rates quantified. Equally important and uncertain is the role of behavior. The frequency and manner of laundering, the frequency of wear, the storage environment and many other consumer-specific phenomena and behaviors all are predicted to strongly impact exposure.

Keywords: SVOCs, behavior, built/indoor environment, consumer and personal care products, exposure models

TU-SY-C1-17
On the accuracy of SVOC exposure assessment: how good is acceptable for different purposes?

Zhang Y.; School of Environment, Tsinghua University, Beijing, China

Exposure assessment is key to estimating health risk and formulating the level of control. Mechanistic models are especially useful. In contrast to empirical models, mechanistic models are based on well-recognized, physical/chemical mechanisms that can be relatively easily generalized and can also vary in their complexity based on the needs of the assessment. However, for different purposes and under different conditions, the models tend to be different. The talk will show the difference between rough and detailed exposure analysis. The effectiveness and limitations of various models will be described using the following typical SVOC exposure examples: (1) estimating indoor particle phase SVOC concentrations from known indoor gas phase SVOC concentrations; (2) estimating dermal exposure using various models. Each model has its advantages and limitations. Detailed models can reduce the relative errors by an order of magnitude but require more detailed information. How to develop or use the models strongly depends on the purpose of the analysis. References [1] Cao JP, Mo JH, Sun ZW, Zhang YP, Indoor particle age, a new concept for improving the accuracy of estimating indoor airborne SVOC concentrations, and applications, Building and Environment 2018, 136, 88-97. [2] Cao JP, Zhang X, Zhang YP, Predicting dermal exposure to gas-phase semi-volatile organic compounds (SVOCs): improved description of SVOC mass transfer between clothing and skin surface, Environ. Sci. Technol. 2018, 52, 4676–4683. [3] Tunga Salthammer, Yinping Zhang, Jinhan Mo, Holger M. Koch, Charles J. Weschler, Assessing human exposure to organic pollutants in the indoor environment, Angewandte Chemie-International Edition, 2018, 57, 12228-12263.

Keywords: SVOCs

TU-PL-D1-EAM: Assessing Exposures for Susceptible and Vulnerable Population I

TU-PL-D1-18

Modelling residents' exposure to pesticides from boom sprayer applications

Figueiredo D.[1], Duyzer J.[2], Jacobs C.[3], Holterman H.[4], van de Zande J.[4], van den Berg F.[3], Huss A.[1], Vermeulen R.[1,5]; [1]Institute for Risk Assessment Sciences - UU, Utrecht, The Netherlands, [2]TNO Urban Environment and Safety, Utrecht, The Netherlands, [3]Wageningen Environmental Research - WUR, Wageningen, The Netherlands, [4]Wageningen Plant Research - WUR, Wageningen, The Netherlands, [5]Julius Centrum - UMC, Utrecht, The Netherlands

Pesticides can be transported from the site of application to residential homes via different routes (e.g. spray drift, volatilization, dust) and eventually lead to exposure of residents, raising concern regarding pesticide exposure. We built a model framework to assess exposure of residents living near fields where pesticides were applied. A framework, consisting of 5 existing models, was developed to assess residential pesticide exposure. The models were run independently and connected via their inputs and outputs at an hourly timescale and high spatial resolution. Models included descriptions of spray drift, volatilization from vegetation, dispersion through air, exchange between outdoor and indoor air and air to dust. Fourteen bulb field applications, comprising 12 different pesticides and weather conditions, were modelled and results were compared with measurements. Model verification indicated suitability of the developed framework to estimate residential exposure levels at high temporal scale. Most of the steps in the model framework resulted in estimates that ranged a factor of 10 from measured levels. Large differences for some steps could be explained by uncertainties in input data and modeling concepts. For some steps the modeling concepts are simplistic, e.g. exposure to dust. Estimates of concentrations in dust differed between factors of 10 to 50 from measured levels. Integrated exposure calculations showed that volatilization from the field after spraying, spray drift in case the wind direction is towards the homes and pesticides in house dust may be important routes of residents' exposure to pesticides. Our developed framework covers the major processes needed to describe residents' exposure to pesticides as result of spraying on nearby fields. Based on the model testing in this project, the framework shows promise for predicting residents' exposure to pesticides. Further development of modeling steps is required to improve the overall performance of the model framework.

Keywords: exposure models, other (specify), pesticides

TU-PL-D1-19

Comparison of low-tier modelled occupational exposure estimates with measured data for petroleum substances' REACH dossiers

Urbanus J.[1], Hahn S.[2], Basinas I.[3], Galea K.[3], van Tongeren, M.[4], Trantallidi M.[5]; [1]Shell (Shell Health - Risk Science Team), Brussels, Belgium, [2]Fraunhofer Institute for Toxicology and Experimental Medicine, Hanover, Germany, [3]Institute of Occupational Medicine, Edinburgh, United Kingdom, [4]School of Health Sciences, Faculty of Biology, Medicine and Health, The University of Manchester, Manchester, United Kingdom, [6]CONCAWE, Brussels, Belgium

For the exposure assessment in the 2010 REACH dossiers of petroleum substances, Concawe used the Tier 1 exposure model ECETOC TRA v.2. In order to account for the more complex petroleum substances and the exposure situations, several modifications of the TRA were developed, including an approach to estimate liquid aerosol, along with some risk management measures describing the use and handling of petroleum substances in use in Europe. Afterwards, a project was started to compare these modelled estimates with measurement data sets. Data were collected from the database built for the Evaluation of Tier 1 Exposure Assessment Models (E-team) project (www.eteam-project.eu) by IOM and ITEM, from Concawe reports and via literature search. These included kerosines (mainly jet fuel), heavy fuel oils, two naphtha groups, gas oils (mainly diesel fuels) and other lubricant base oils. The quantity and quality of data sets varied between the groups and there were still a number of scenarios which could not be evaluated due to lack of suitable data. The comparison exercise showed some discrepancies depending on substance group and the specific scenarios. Although some underestimations were observed, there were also cases where clear overestimations were observed and thus, a further refinement with higher Tier tools should be possible. There were situations where the comparison of measurements and estimates suggested reasonable results and a controlled risk, whereas for other situations a clear need for new measurement data emerged. A particular problem was the lack of high quality aerosol data for semi-volatile petroleum substances. The effort and time dedicated in this project to collect and collate available exposure data into the estimates needed for chemical safety assessment across substance life-cycles would have been prohibitive in the run-up to the first REACH registration deadline, whilst still leaving important data gaps.

Keywords: workplace

TU-PL-D1-20

Where are the landfilled sites? Methodological approaches to investigate the presence of asbestos in the soils of Sibaté (Cundinamarca, Colombia).

Lysaniuk B.[1], Cely Garcia M.F.[2], Ramos Bonilla J.P.[2]; [1]IRD (MàD par le CNRS), UMR PRODIG, Bogota, Colombia, [2]Universidad de Los Andes, Facultad de Ingeniería, Departamento de Ingeniería Civil y Ambiental, Bogota, Colombia

Introduction: Sibaté is a small town (38,000 inhabitants) located 25 kilometers South-West of Colombia's capital, Bogotá. Residents from Sibate report that during the 1970s and 1980s, parcels were filled with asbestos-containing materials (ACM), to push out the heavily polluted waters of a reservoir from the urban area and create flat spaces to promote urbanization. A sampling campaign consisting of four exploratory soil perforations confirmed the presence of an underground friable asbestos layer. Considering the risk of asbestos exposure resulting from excavations that may disturb the asbestos layer, we propose methodological approaches to identify potential landfilled zones without conducting excavations. Data and Methods: Two approaches were combined to establish the potential location of landfilled zones: 1 – Development of social cartography consisting of the elaboration of maps by inhabitants of Sibaté in which landfills zones were located, 2 – Identification of both land use and topographic changes over time analyzing topographic maps from different years (1950, 1978, 1989) combined with a Digital Elevation Model. Both approaches were integrated using Geographic Information Systems to determine the potential boundaries of the landfilled zones that may contain asbestos. Results: The social cartography method allowed delimitating those areas with the highest frequency of being identified as landfilled zones by inhabitants of Sibaté. Crossing these reported landfilled zones with the areas identified using pre-existing geographic information resulted in a strong coincidence between "probable" zones – defined by land-use and topographic changes - and "frequently reported zones" identified by the inhabitants. Discussion: This method allows for a cross-validation process of social cartography and diachronic land-use analysis. Although the resulting maps could be used as an initial step for risk management purposes, extreme caution should be exercise when using these results because of the underlying uncertainty. Thus, these boundaries do not establish asbestos contaminated and asbestos free zones.

Keywords: built/indoor environment, environmental health, risk assessment, soil

TU-PL-D1-21
Childrens' Particulate Matter Exposure Characterization as Part of the New Hampshire Birth Cohort Study

Thornburg J.[1], McCombs M.[1], Cho S-H.[1], Siripanchignon N.[1], Butler E.[2], Karagas M.[2]; [1]RTI International, Research Triangle Park, NC, USA, [2]Dartmouth College, Hanover, NH, USA

The New Hampshire Birth Cohort Study (NHBCS) is an ongoing, prospective study of rural pregnant women and their children, and a pediatric cohort for the NIH-funded Environmental influences on Child Health Outcomes (ECHO) program. The primary aim of the NHBCS is to investigate the effects of environmental exposures, including use of wood stoves, on fetal growth and childhood development. A subset of children 3 to 5 years of age, nearly 25% of whom live in a home with a wood stove as the primary heating source, participated in a personal monitoring study of PM2.5 exposures. This study is the first to report personal PM2.5 exposure using a wearable sensor on children younger than 5 years old, not proximity measurements. Three days' worth of physical activity and personal air monitoring were collected using the MicroPEM™ (RTI International, Research Triangle Park, NC) for PM2.5. To date, 272 children have been enrolled in the study to wear the MicroPEM™ for 72 hours. Exposure data were collected successfully from 244 participants who had more than 48 hours of participation. Participants wore the MicroPEM™ an average of 70% of their time awake. Across all participants, real-time filter corrected nephelometer PM2.5 concentrations showed a mean of 23.1 µg/m3 (standard deviation, SD = 136). Each filter was analyzed for speciation of environmental tobacco smoke (ETS), black carbon (BC), and brown carbon (Br). Average values were 4.2 µg/m3 ETS (SD = 6.1), 0.5 µg/m3 BC (SD = 3.1), and 0.9 µg/m3 BrC (SD = 1.0). Personal PM2.5 data showed higher values for winter-spring months versus other months of the year. Personal sensors can be worn by young children to measure their air pollution exposure and advance environmental epidemiological research. The data shows seasonal differences in PM2.5 as well as higher levels of ETS exposure over BC and BrC.

Keywords: air, children, epidemiology, exposure factors, particles

TU-PL-D1-22
The concerning young age of diagnosis in mesothelioma cases from Sibaté, Colombia

Ramos-Bonilla J.P.[1], Cely-García M.F.[1], Lysaniuk B,[2], Comba P.[3], Terracini B.[4], Pasetto R.[3], Marsili D.[3], Ascoli V.[5], Mazzeo A.[6], Magnani C.[7], López Panqueva R.[8], Baldión M.[8], Cañón D.[8], García-Herreros L.[9], Pinzón B.[10], Giraldo M.[1], Hernández L.J.[11]; [1]Department of Civil and Environmental Engineering, School of Engineering, Universidad de los Andes, Bogotá, Colombia, [2]IRD, UMR Prodig, Paris, France, [3]Department of Environment and Health, Istituto Superiore di Sanità, Rome, Italy, [4]Biostatistics, University of Turin (Now Retired), Turin, Italy, [5]Department of Radiological, Oncological and Anatomo-Pathological Sciences, Sapienza University, Rome, Italy, [6]Department of History and Cultures, University of Bologna, Bologna, Italy, [7]Dipartimento di Medicina Traslazionale - Università del Piemonte Orientale, Novara, Italy, [8]Department of Pathology and Laboratories, Fundación Santa Fe de Bogotá, Bogotá, Colombia, [9]Thoracic Surgery Section, Department of Surgery, Fundación Santa Fe de Bogotá, Bogotá, Colombia, [10]Diagnostic Imaging Department, Fundación Santa Fe de Bogotá, Bogotá, Colombia, [11]School of Medicine, Universidad de los Andes, Bogotá, Colombia

Introduction: In 1942 the first asbestos cement-facility of Colombia initiated operations in Sibaté, Colombia (Current population 38.000). Inhabitants of Sibaté report that large quantities of asbestos containing materials (ACM) were disposed in the urban area of Sibaté. The ACM was later used to build landfills. Recent evidence suggests the existence of a mesothelioma cluster in Sibaté, with a concerning number of cases diagnosed at a young age (range 38 – 58 y.o.). This study proposes potential explanations for the young age of mesothelioma diagnosis. Methods: Self-reported mesothelioma cases were identified using a socio-economic survey (n=355). The survey was applied, door-to-door, in an area that covers 84% of potential landfilled zones, and 44% of the urban area of Sibaté. Medical records of 17 self-reported cases were analyzed by six physicians. A participatory workshop was conducted with 11 residents of Sibaté to determine the history of both ACM disposal and landfilled zones. Results: Fifteen of the 17 self-reported cases were classified as "certain malignant pleural mesothelioma". Subsequent analysis was focused on 13 cases diagnosed over the last 10 years. Based on the information collected in the participatory workshop, it was determined that the disposal of ACM in the urban area of Sibaté occurred between ~1975 to ~1985. Construction of the landfilled zones occurred between ~1985 to ~1990. Eleven of the 13 mesothelioma cases lived in Sibaté between the years 1975 and 1990. Only one case had certain occupational exposure. All other cases had non-professional exposures, household exposures, and/or environmental exposures. Conclusions: Based on the latency period of pleural mesothelioma, the unrestricted access to the disposed ACM and the construction of the landfills may have been a major source of asbestos exposure for 11 cases. Asbestos exposure may have occurred during childhood for 10 of the 13 cases analyzed (<14 years-old).

Keywords: children, cumulative exposure, respiratory health, risk assessment, sampling methods

TU-PL-D1-23

A comprehensive analysis of racial disparities in chemical exposure biomarkers in US women

Nguyen V.[1], Kahana A.[1], Heidt J.[1], Polemi K.[1], Kvasnicka J. [1], Jolliet O. [1], Colacino J. [1]; [1]University of Michigan, Ann Arbor, MI, USA

African American women are 2 to 3 times more likely to be diagnosed with triple negative breast cancer, compared to European American women. The biological mechanisms for this disparity are not well understood but are likely due to an interaction between genes and environmental exposures. To better understand environmental factors which may impact triple negative breast cancer risk, we implemented an untargeted approach to systematically compare chemical exposure differences by race. We used the National Health and Nutrition Examination Survey (NHANES) of 148 chemical biomarker concentrations measured in 38,080 female participants 1999-2014. We conducted a series of multiple survey weighted regression models with chemical biomarker levels as the outcome variable and race/ethnicity as the main predictor while adjusting for age, cotinine levels (biomarker of smoking), poverty-income ratio (surrogate variable of socioeconomic status), and cycles (years) in which the participants were sampled. To understand chemical exposures by age and race, we conducted stratified analyses by age groups (0-11, 12-25, 26-50, and 51-up). Across age groups, African American women had significantly higher biomarker concentrations of many chemicals, including methyl paraben (2.39-fold), propyl paraben (2.09-fold), mono-ethyl phthalate (1.74-fold), 1,4-Dichlorobenzene (2.35), and 2,5-dichlorophenol (4.56-fold), relative to non-Hispanic white women. For the youngest age group (0-11 years), the chemical exposure disparities between African American and non-Hispanic white women are further exaggerated for parabens and 1-4 dichlorobenzene. Overall, we identified substantial differences in chemical body burden by race in a representative sample of US women. Many of the chemicals with exposure disparities are found in cosmetics, pesticides, and other consumer products. Ongoing work is determining whether these toxicants impact biological pathways associated with breast cancer to better understand the link between the environment and triple negative breast cancer disparities.

Keywords: chemical prioritization

TU-PL-E1-SEM: Improving Indoor Air Quality in Buildings

TU-PL-E1-24
Smart homes and the control of indoor air quality

Schieweck A.[1], Uhde E.[1], Salthammer T.[1], Salthammer L.[2], Morawska, L.[3], Mazaheri M.[3], Kumar P.[4]; [1]Fraunhofer WKI, Braunschweig, Germany, [2]Leuphana University, Lüneburg, Germany, [3]Queensland University of Technology, Brisbane, Australia, [1]University of Surrey, Guildford, United Kingdom

Global climate change will go along with changes in climatic conditions such as temperature, water balance and direct sunshine. As this will lead to a worsening of air quality, we have to look intensively at the matter of how we can create healthy living environments in areas with extreme air conditions. In order to serve primarily the efficient use of energy and the optimization of ventilation technology connected with new ways of constructing buildings (low energy and passive houses), "smart home technology" was introduced also in private homes. Facing demographic change, higher demands on a comfortable life and advancing mechanization of everyday life, sensor technology is increasingly implemented in order to create acceptable and improved living conditions. Thus, the term "smart home" is also linked today with the networking of home automation systems, home appliances and communications and entertainment electronics. In general, low-energy houses, which as a rule have a system of artificial ventilation, require special technologies to achieve good indoor air quality. Using modern sensor technology, it is possible to monitor not only the climatic parameters, but also the concentrations of air polluting substances, such e.g. carbon dioxide, sum parameters of volatile organic compounds (VOCs) and particles and record them in Home Energy Management Systems (HEMS). Indoor air quality and air hygiene are taken now as important aspects of smart home technology. Nevertheless, living in a smart home often puts demands on the occupants who are required to change some of their living habits. Due to the significant impact of smart home technology on everyday life in the near future, the authors have summarized the actual state-of-the-art of housing technology on indoor air quality, individual thermal comfort and living behaviour for the temperate climate zone. The main findings will be presented at the conference.

Keywords: VOCs, air, climate change, sensor technology, thermal environment

TU-PL-E1-25
Stinky Butts: Airborne emission rates from non-smoldering cigarette butts

Poppendieck D.[1], Gong M,[1]; [1]National Institute of Standards and Technology, Gaithersburg, MD, USA

The global disposal of five trillion cigarette butts every year is a major environmental challenge. Although there is a reasonable body of knowledge on aqueous emissions from cigarette butts, the environmental impact of cigarette butts due to airborne emissions is not well studied. This research quantified airborne cigarette butt emission rates for the first time. More specifically, this research measured the chemical emission rates into air from cigarette butts in a walk-in chamber. Consistent cigarette butts were generated using a custom-built smoking apparatus. Eight target chemicals were quantified in the experiment, including nicotine. Three-hour air samples were collected on sorption tubes from the chamber up to 24 h after 18 cigarette butts were placed in the walk-in chamber. Samples were analyzed using thermal desorption (TD) gas-chromatograph-mass spectrometry (GC-MS). The average emitted mass of the target chemicals over the first 24 h ranged from 2 ⍰g/butt to 98 ⍰g/butt. The 24 h emitted mass for nicotine from a cigarette butt (62 ng/butt) in this study could be up to 25 % of the literature reported emitted mass from a burning cigarette (241 ng/cigarette to 3370 ng/cigarette). Hence, the long-term emitted nicotine mass from a butt could be comparable to the nicotine mass emitted from mainstream and side stream smoke of a cigarette if the cigarette butt were left longer than 24 h in a room or at higher temperature. Cigarette butts are likely to be sources of nicotine for prolonged periods (greater than 24 h), which may be a concern for butts disposed in indoor spaces like cars. The emission rates and relevant risk to receptors will be dependent upon environmental factors including temperature, relative humidity, and water saturation. Airborne cigarette butt emissions warrant further study.

Keywords: VOCs, built/indoor environment, exposure factors, secondhand smoke

TU-PL-E1-26
Stinky Butts: Environmental influences on airborne chemical emissions from cigarette butts

Poppendieck D.[1], Gong M.[1]; [1]National Institute of Standards and Technology, Gaithersburg, MD, USA

The global disposal of five trillion cigarette butts every year incurs a major environmental challenge. Although there is a reasonable body of knowledge on aqueous emissions from cigarette butts, the environmental impact of cigarette butts due to airborne emissions is not well studied. This research studied the characteristics of airborne emissions systematically for the first time. More specifically, this research determined the relative distributions of the target chemicals in freshly smoked butts to understand which part of butts the chemicals were emitted from, examined the influence of various environmental factors (i.e., temperature, relative humidity, ultraviolet radiation, water saturation and air change rate) on airborne emissions from cigarette butts, and compared the cigarette butt airborne emissions outdoors in summer and winter seasons. Most of the target chemicals released from the butt were the result of the burning process (including both combustion and pyrolysis) and were primarily emitted from the burned filter. Results from the conditioning experiments showed cigarette butts to be significant sources for six of the eight target chemicals for the first 24 hours after smoking. Emissions of heavier chemicals like nicotine continued at elevated levels up to 144 hours after smoking. Increased temperature and relative humidity increased airborne chemical emission rates from cigarette butts in both chamber and outdoor experiments. During butt saturation in a rain event, chemicals with high water solubility and water-to-air partitioning coefficients (Kwa), like nicotine, may migrate into the surrounding environment via aqueous rather than airborne routes. Ultraviolet radiation and air change rate had minimal influence on the rate of emitted target chemicals under the conditions tested. This work demonstrates that cigarette butts may be a significant source of the target chemicals to the environment and the source emission rate will be dependent upon environmental factors including temperature, relative humidity, and water saturation.

Keywords: VOCs, analytical methods, built/indoor environment, secondhand smoke, source control

TU-PL-E1-27
Measuring Dynamic Behaviour of Volatile Organic Compound Levels in a Zero-Energy Building

De Jonge K. [1], Laverge, J. [1]; [1]Ghent University, Research Group of Building Physics, Ghent, Belgium

With increasing building airtightness, the design of an adequate ventilation system increases in importance. The first generation of ventilation systems, based on continuous supply of the nominal airflow rate, are now being replaced by Demand Controlled Ventilation (DCV). These systems, often H2O and/or CO2 controlled, do not take into account the emissions of Volatile Organic Compounds (VOCs) to the indoor environment. Previous research, using airflow simulation software, shows an important need to integrate the dynamic characteristics of the VOC emissions, especially when the ventilation system airflow rates are controlled dynamically. In this study, measurements are done aiming to support the conclusions made from results obtained through simulations. A small, airtight, zero energy building that has been designed as the Belgian contribution to the international solar decathlon competition (2011) was rebuilt afterwards in Ostend, Belgium. This building is used as test facility for the measurements. Both indoor parameters (room temperature, humidity, formaldehyde levels, 'TVOC' levels, system temperatures, airflow rates) and outdoor parameters (outside temperature, humidity, outdoor formaldehyde levels, solar irradiance, wind speed and direction) are monitored in an attempt to measure all possible influences on the indoor VOC level. With this set-up, different loading scenarios (difference in amount of source material) as well as different ventilation control scenarios are replicated to obtain measurements representative for a continuous airflow and a demand controlled, variable airflow rate, ventilation system. The results show an important influence of environmental parameters on the indoor VOC levels, with VOC concentrations exceeding health guidelines. It is therefore important to design DCV systems and controls taking into account possible elevated VOC levels and while doing so, incorporate the dynamic behaviour (influence of temperature and humidity) of the VOC emissions.

Keywords: VOCs, air, built/indoor environment, ventilation

TU-PL-E1-28
Improving the indoor air quality by using a surface emissions trap for exposure reduction

Larsson L.[1], Mattsson J.[2]; [1]Lund University, Lund, Sweden, [2]cTrap Ltd, Lund, Sweden,

Unsatisfactory indoor air quality (IAQ) may be due to emissions of volatile organic compounds (VOCs) from building construction parts leading to adverse health symptoms. We describe the use of a surface emissions trap (cTrap) to prevent such emissions from reaching the individuals that reside in the building. The cTrap is a flexible four layer laminate with two protective sheets of nonwoven polyester fabric surrounding an adsorption layer and a hydrophilic polymer sheet. The device is applied directly at the source of the emissions (floor, walls, ceiling, over cavities etc) using adhesive tape or a staple gun. A room in a ground apartment of a building erected on a concrete slab in the 1970:s, with a PVC flooring and a long history of health complaints, was studied. The PVC flooring was removed and the cTrap cloth (cTrap Ltd, Lund, Sweden) was attached over the floor. The tenants reported already the day after the cTrap installation that symptoms had disappeared. The air concentrations of 2-ethylhexanol were measured two months before and two months after the cTrap installation, and decreased from 12 ug/m3 to 1.4 ug/m3 in the air of the studied room and from 63 ug/m3 to 1.5 ug/m3 according to the directed floor emission measurements. The cTrap has been shown to stop and bind adsorb VOC such as odours, alcohols, aldehydes including formaldehyde, chloroanisols, terpenes, PAH, aromates, sulfur compounds etc efficiently. cTrap is gas tight, at the same time moisture passes through the cloth virtually unhindered (Z = 200 s/m), thus there is no risk for vapour condensation which might lead to mould growth. The cTrap does not contain any chemicals, has no measurable self emissions, and provides new possibilities for health-focussed remediation e.g. of water-damaged buildings.

Keywords: VOCs, air cleaning, built/indoor environment, environmental health, exposure factors

TU-PL-E1-29
Residential balanced ventilation and tested impact on indoor pressure and air quality

Ouazia B.[1], Aubin D.G.[1], Won D.[1], Yang W.[1], So S.[1], Arsenault C.[1]; [1]National Research Council Canada, Ottawa, Canada

Ventilation is a key component in an energy efficient and healthy home. Ventilation standards and building codes do not attempt to address delivery of outdoor air to each space or circulation (distribution) of ventilation air and they don't differentiate or give credit to the best ventilation strategy based on the IAQ related benefits. Theis study focused on the in-situ impacts of two ventilation systems (balanced energy recovery ventilation system and single-spot exhaust-only ventilation system) in a side-by-side setting (twin research houses) with various indoor mixing scenarios (no mixing, partial mixing and continuous mixing), including the impacts of indoor pressure, IAQ and whole house ventilation effectiveness. Exhaust-only ventilation showed lower whole- house air change rate, lower uniformity of air exchange rate between indoor spaces, higher concentration of formaldehyde and VOCs and higher whole house energy consumption than did the balanced energy recovery ventilation system. The differential envelope pressure data showed that, compared to the exhaust-only system, balanced ventilation systems achieved a positive indoor pressure (1 to 3 Pa) and the exhaust ventilation systems depressurized the house by -1 to -2.5 Pa. Compared to the exhaust-only ventilation system and for all mixing scenarios, balanced ventilation system reduced formaldehyde by 37% to 70% in winter and by 6% to 34% in summer, reduced concentration of VOCs; α-pinene (α-Pin) by 26% to 89% and Toluene-d8 (dTol) by 8% to 76% in winter and VOCs; alpha-Pinene (α-Pin) by 87% to 339%, Toluene-d8 (dTol) by 45% to 88%, Heptane (Hep) by 40% to 91% and Benzaldehyde (BenzA) by 35% to 406% in summer. The PFT data analysis showed that balanced ventilation system improved the whole house air change rate by 6% to 34%. Balanced (ERV) ventilation confirmed the energy saving benefit by reducing the average weekly heating/cooling and ventilation energy consumption by up to 8.5%.

Keywords: VOCs

TU-PL-A2-EHR: Exposure to Fine and Ultrafine Particles II

TU-PL-A2-30
Particle concentrations in shoeless schools

Pasanen P.[1], Leppänen M.[1]; [1]University of Eastern Finland, Kuopio, Finland

Airborne particles may have a substantial effect on indoor air quality in schools. In addition to other particle sources, outdoor dust and soil are commonly transported indoors on people's shoes. These soil-originated particles can be resuspended into indoor air, and be further inhaled by pupils and teachers. Recently, many Finnish schools have applied a so-called shoeless course, when shoes are taken off right when entering the building and they are kept on shoe shelves near the entrance. The aim of this study was to investigate how the shoeless course affects the particle concentrations in schools. Particle mass (particle size range 0.1-15 μm) and number concentrations (particle size range 0.02-1 μm) were measured on-line in seven shoeless schools and ten traditional schools, where shoes are worn until pupils' coat racks. Measurements were conducted in fall, during wet weather conditions, and they lasted for one school day in each school. Results showed that particle mass concentration varied remarkably during the day, and the highest concentrations were measured from the corridors full with pupils wearing shoes. Average PM2.5 and PM10 concentrations were 1.7 and 2.0 times higher in traditional schools (25.9 and 55.5 μg/m3) compared to shoeless schools (15.0 and 27.1 μg/m3), respectively. The shoeless course did not affect on the particle number concentrations. The number concentrations were over two-fold in corridors compared to classrooms, and the highest concentrations were measured in diners or hallways in contact with diners (up to 98 843 particles/cm3). In conclusion, this study showed that by taking the shoes off when entering the school building you can enhance the indoor air quality.

Keywords: particles

TU-PL-A2-31
Assessing lower airway exposure to monodispersed micron-sized fluorescent particles by an in-vitro respiratory system

Duan M.[1], Liu L.[1], WangY.[2], Da G.[3], Géhin E.[3]; [1]School of Architecture, Tsinghua University, Beijing, China, [2]State Key laboratory of Green Building in Western China, Xi'an University of Architecture and Technology, Xi'an, China, [3]Université Paris-Est, CERTES, Créteil, France

Millions of people died of COPD and pneumoconiosis per year. Lower airway exposure to micron-sized particles is considered as the most threatening factors. Many efforts have been dedicated to quantify the concentration of particles in human breathing zone or inside the human respiratory system. But the discrepancy between different diameter particles from breathing zone into lower airway is still unknown. That's exactly an important information for assessing the harm of particles to human health. We developed an in-vitro respiratory system with a CT-imaging-based airway replica and a conjunct breathing machine. The airway replica contains face, nasal cavity, throat, trachea, the first five stages of bronchi and lung volumes. It is 3D-printed based on the CT scanning from a healthy volunteer. Monodispersed fluorescent particles with the aerodynamic diameters dae 0.6-5 μm were generated by TSI VOAG 3450® and monitored by TSI APS 3321®. The lower airway exposure was measured as the total accumulative particles that enters the lung volumes during the sampling period. SKC BioSampler® was used for collecting the fluorescent particles in the lung volumes. The quantity of the fluorescent strength was measured by Horiba Fluoro Max-4®. Particle concentration of breathing zone was evaluated through numerical simulation based on the experimental set-up. Results shown more than 30 % of 0.6 μm, 1 μm and 2 μm particles in human breathing zone can reach lower airway. The percentage becomes much smaller for larger particles. (20.4 ± 8.2)% for 3 μm particles and (5.1 ± 0.2)% for 5 μm particles are able to reach the lower airway. Our results are close to but still different with the classic ICRP66 model, which is based on anatomy replica measurements. This low-cost in-vitro system inspires a new method to access personal exposure to indoor particulate contaminants, especially for people with respiratory preconditions.

Keywords: cumulative exposure, environmental health, exposure models, particles, respiratory health

TU-PL-A2-32
Characterization of Personal Fine and Ultrafine Particle Exposures in Children with Asthma

Koehler K.[1], Majd E.[1], Zamora M.[1], Hansel N.[1], McCormack M.[1], Green T.[1], Meade L.; [1]Johns Hopkins School of Public Health, Baltimore, MD, USA

Fine particulate matter air pollution (particles with diameter less than 2.5 μm, PM2.5) is known to exacerbate asthma symptoms in children; however, recent toxicological evidence suggests that respiratory health effects may be more strongly related to ultrafine particles (UFP; particles with diameter less than ~ 0.1 μm). Due to complexities of exposure assessment and the lack of a regulatory monitoring network for ultrafine particles (UFP), the relationship between UFP and PM2.5 exposures and their impacts respiratory health remains unclear, especially for susceptible populations. We measured personal exposures to PM2.5 and UFP for 25 children aged 8-16 with asthma on 4 consecutive days. Exposure measurements were paired with a GPS receiver allowing for apportionment of exposure to different microenvironments (ME; home, school, transit). Exposures to both particle size fractions was highly variable and short-duration, high-intensity peak exposures were found to contribute substantially to cumulative exposures. We found that peaks contributed to 25% of cumulative exposure at home, on average, and high maximum concentrations were observed in all MEs for both PM2.5 (up to 1021.5 μg/m3) and UFP (up to 1488 μm2/cm3). We also found that while UFP concentrations increased due to activities such as cleaning and burning candles at home, PM2.5 concentrations did not follow the same trend in some cases, leading to poor correlations between PM¬2.5 and UFP exposures during these events. Our findings suggest that PM personal exposure assessment should be conducted at high spatiotemporal resolution in order to appropriately characterize the short-duration, high PM concentrations, and to provide insight as to the associated activities and sources. Furthermore, the distinct trends in UFP and PM2.5 concentrations associated with different activities suggests that UFP is an exposure that is independent of larger PM and should be monitored in personal exposure assessment campaigns.

Keywords: particles

TU-PL-A2-33

Home Environmental Intervention to Reduce Exposure Level to PM2.5 for Children with Asthma using Clustered data of ESCORT Child Asthma Panel Study, South Korea

Kim S.[1], Seeun O.[1], Lim D.[1]; [1]Inha University, Incon, South Korea

BACKGROUND: Children with asthma in inner-city communities may be particularly vulnerable to adverse effects of indoor air quality because they spend most of their time in indoor, yet little is known about the effect of environmental interventions in the Korean home setting. OBJECTIVE: We piloted the feasibility and effectiveness of a home-based air cleaner intervention to reduce particulate matter ≤ 2.5 μm (PM2.5) in homes of children with asthma. METHODS: In this pilot randomized controlled trial, with a case-crossover design, we assessed the effect of air cleaners on indoor PM2.5 concentrations at 30 homes (15 control, 15 intervention) in the metropolitan city of Incheon, South Korea. We obtained 867 episodes from children with asthma (15 control, 15 intervention) aged 7 to 12 years. Daily home air pollutant measurements and peak expiratory flow rate were completed over 6 weeks (3 weeks before and another 3 weeks twice after randomization). Indoor air quality were compared between control and interferon group and using mixed models the association of fluctuations of daily indoor air quality and PEFR. RESULTS: Indoor PM2.5 concentration levels (median (Q1~Q3) for overall intervening period was 8.8(3.2~12.8) ug/m3 for intervention group and 15.8(8.5~20.5) for case group. Unit increase in concentration of PM2.5 significantly reduced the PEFR by 0.8 l/min (95% confidence interval [CI]: 0.1–1.5) after adjusting for indoor CO2, temperature, relative humidity, sex, age and body mass index (BMI). In this study, we found that fluctuation of indoor PM2.5 were also associated with PEFR and a home-based air cleaner intervention was effective to reduce indoor PM2.5 levels.

Keywords: environmental health

TU-PL-A2-34

Seasonal difference of indoor-outdoor relationship of PM2.5 in various microenvironments

Guak S.[1], Lee K.[1]; [1]Seoul National University

Background/Aim: Exposure to fine particle (PM2.5) has been associated with increased morbidity and mortality in numerous epidemiologic studies. Since people spend most of their time indoors, outdoor concentrations are not sufficient to assess personal exposure to PM2.5. Indoor-outdoor relationship could provide valuable information of emission and exchange of air pollution. The purpose of this study was to determine I/O ratio of PM2.5 in various microenvironments in two seasons. Methods: Field technicians measured daily personal exposure to PM2.5 for 50 person-day both in summer and winter of 2017. PM2.5 concentrations were continuously measured using MicroPEM (RTI incorporated, USA). Microenvironments were classified into six categories including home, workplace/school, restaurant, other locations, private transportation, and public transportation. Outdoor concentrations were obtained from ambient monitoring stations in Seoul (AirKorea). I/O ratio of PM2.5 was calculated by microenvironmental concentration and outdoor concentration. Results: Average daily personal PM2.5 exposures were 20.8±30.7μg/m3 in summer and 22.2±35.3μg/m3 in winter, while average outdoor concentrations were 22.2±12.3μg/m3 in summer and 30.1±20.7μg/m3 in winter. The highest PM2.5 I/O ratio in summer and winter were 2.5±3.5 and 3.8±8.3 in restaurant, respectively. The lowest PM2.5 I/O ratio in summer and winter were 0.8±0.7 in workplace/school and 0.6±0.5 in home, respectively. Other indoor locations, restaurant, and public transportation had PM2.5 I/O ratio higher than 1. Home, workplace/school, and private transportation had PM2.5 I/O ratio lower than 1. Conclusions: I/O ratio of PM2.5 was different by microenvironments. I/O ratio of PM2.5 in microenvironments was not changed by seasons. Indoor-outdoor relationship could indicate characteristics of the microenvironments.

Keywords: particles

TU-PL-A2-35

NanoOffice – nanoparticles in new and renovated office buildings

Orru H.[1], Hagenbjörk A.[1,2]; [1]Umea University, Umea, Sweden; [2]University of Tartu, Tartu, Estonia

More and more people work in offices with complicated ventilation and heating systems. Sometimes, people feel uncomfortable or sick in a certain building and one possible reasons could be exposure to nanoparticles (<100 nm). Nanoparticles can be penetrated either from outdoors or emitted by heating and ventilation systems, by office equipment and from used materials, and chemicals. The current study aims to determine the levels nanoparticles (both indoors and outdoors), temperature, humidity and CO2 levels in indoor environments. Also establish potential the links between indoor, and outdoor particle levels and characteristics of the buildings. The concentrations of nanoparticles were measured for one week in twelve offices in Umea (Northern-Sweden) using SMPS 3938 for indoor and DiscMini for outdoor measurements. Indoor temperature, humidity, CO2 levels and ventilation air flows were measured with AMI310. In the measurements we could see very large difference in nanoparticle number concentration, varying from just few particles cm–3 to more than 100,000 cm–3. The levels were highest in the offices close to bus terminal and lowest in offices near a park. Also the very strong temporal effect appeared as levels were highest during the day (often 50,000-100,000 cm–3), whereas less than 100 cm–3 nanoparticles were present during the night. Also there appears to be very high infiltration rate of outdoor nanoparticles into indoor air (indoor/outdoor differences usually less than 20% in polluted area). The reason for very high infiltration rate could be very small size of the particles (average diameter around 20 nm among outdoor particles). The effect of ventilations systems appeared as well, but it is seems to be somewhat smaller compared to outdoor air infiltration effect. The levels of nanoparticles were relatively lower in energy efficient offices with modern ventilation, where air flows are regulated with sensors, being relatively lower based on our measurements.

Keywords: nanoparticles

TU-SY-B2: The European Human Biomonitoring Initiative HBM4EU: Current status and latest results

TU-SY-B2-36

Science and policy for a healthy future - HBM4EU, the European Human Biomonitoring Initiative – latest results

Kolossa-Gehring M.[1], Kobosil N.[1], Lange R.[1], Pack K.[1], Reiber L.[1], [1]German Environment Agency, Berlin, Germany

Pollutants like plasticisers, glyphosate or mercury are spread throughout the environment. A problematic issue, as exposure to these chemicals can have negative effects on human health and well-being. Human Biomonitoring allows for the measurement of internal human exposure to these chemicals by analysing preferably human blood or urine, but also hair and tissue. Since 2017, the European Human Biomonitoring Initiative (HBM4EU), coordinated by the German Environment Agency (Umweltbundesamt), follows an innovative approach to, amongst other aims, gather comparable exposure data for Europe as a whole. Within five years, HBM4EU aims to establish a Europe-wide Human Biomonitoring that provides results tailored to feed directly into policy making in the areas of health, environment and chemicals to protect human health more effectively.

Keywords: biomarkers, environmental health, environmental regulation, epidemiology, health, other/general

TU-SY-B2-37

Collection and use of EU wide human biomonitoring data in HBM4EU

Govarts E., Remy S., Gilles L., Rambaud L., Vogel N., Kolossa M., Montazeri P., Vrijheid M., Schoeters G.

Human biomonitoring studies in Europe are fragmented and disperse. HBM4EU brings together datasets on human biomonitoring from different European studies in the IPCHEM platform hosted by JRC and makes them accessible for policy makers and the research community respecting ethical and data protection constraints. Different strategies are being developed for using and analysing the data at the European level. We will present the different approaches that are used to evaluate and analyse the datasets for deriving European exposure distributions and reference values, for analysing geographical and time trends and for identifying determinants that influence levels of exposure biomarkers. Some examples of these approaches will be presented.

Keywords: biomarkers, biomonitoring, epidemiology, flame retardants, metals, phthalates

TU-SY-B2-38

Building up a complete QA/QC scenario for the analytical phase in HBM4EU

Esteban López M.[1], Göen.T.[2], Mol H.[3], Koch H.M.[4], Thomsen C.[5], Vorkamp K.[6], Hajslova J.[7], Nuebler S.[2], Zarrabi K.[2], Lancova D.[7], Antignac J.P.[8], Vaccher V.[8], Bury D.[4], Kasper-Sonnenberg M.[4], Småstuen Haug L.[5], Ramos J.J.[1], Pedraza S.[1], Bartomé M.[1], Castaño A.[1]; [1]CNSA, Instituto de Salud Carlos III, Majadahonda, Madrid, Spain, [2]Institute and Outpatient Clinic of Occupational, Social and Environmental Medicine of the University of Erlangen, Erlangen-Nuremberg, Germany, [3]Wageningen University & Research, RIKILT, Wageningen, Netherlands, [4]Institute for Prevention and Occupational Medicine of the German Social Accident Insurance, Institute of the Ruhr, Bochum, Germany, [5]Norwegian Institute of Public Health, Oslo, Norway, [6]Aarhus University, Aarhua, Denmark, [7]University of Chemistry and Technology, Prague, Czechia, [8]Oniris, INRA, Nantes, France

Quality assurance and quality control (QA/QC) are essential in analytical chemistry. Different internal and external measures are routinely applied in analytical laboratories. QA/QC has an important role in human biomonitoring since it ensures that observed differences are the result of variations in exposure (diet, life styles, habits, etc.) and not because of contamination of samples or irregularities during the analytical phase. Laboratories from different countries and with different background and experience participate in HBM4EU and so, the establishment of a strict QA/QC measures is required. The QA/QC in HBM4EU involves several activities with the aim to ensure the quality and comparability of the analytical results within the participating laboratories by applying a step-by-step approach. Firstly the biomarkers, matrices and analytical methods for each group of substances are prioritised (phthalates, DINCH, per- and poly-fluorinated alkyl substances, polyaromatic hydrocarbons, brominated flame retardants, phosphorus flame retardants, cadmium, chromium and aromatic amines). In parallel, laboratories with experience in chemical analysis in human samples are identified as candidate laboratories to perform the analysis after the successful participation in the ICI/EQUAS scheme. Laboratories with experience in organising ICI/EQUAS are also identified and selected to support in the implementation of the scheme that includes more than 70 parameters in different human matrices. Until now 2 rounds (one ICI and one EQUAS) have been completed showing different results depending on the substances group. The third round started in May 2019. First results and conclusions of the QA/QC scheme related to the 1st list priority substances will be presented.

Keywords: biomonitoring

TU-SY-B2-39

How to close the gap between human biomonitoring (HBM) data and policy related questions?

Buekers J.[1], Lobo Vicente J.[2], Kubelt J.[3], Bessems J.[1], Schoeters G.[1]; [1]VITO (Flemish Institute for Technological Research), Moi, Belgium, [2]EEA (European Environment Agency), Copenhagen, Denmark, [3]UBA (German Environment Agency), Berlin, Germany

In ancient times, cavemen indicated the presence of deer and mammoths with paintings. Indicators have thus existed for a long time and still play a crucial role in today's society. Our society is complex and we are continuously exposed to chemicals via different routes and from different sources. HBM data reflecting internal concentrations give a clear picture of the integrated exposure. The internal exposure may vary across e.g. age categories, gender, geographical locations, socio-economic status (SES; measured via education, income) and time. Furthermore, not all chemicals are harmless and environmental health risks need to be minimized. With the results of HBM, policy questions arise or they did ex ante. Is there spatial difference in exposure? Is there inequity in exposure by groups of different SES? Are there groups of people at (higher) health risk for exposure to chemicals (than others)? Here, the concept of indicator comes in. Indicators try to bridge the gap between the HBM data we gather and the questions that exist. Within the HBM4EU project, indicators are developed to present HBM data in an understandable way (https://www.mdpi.com/1660-4601/15/10/2085). Several selection criteria were developed, if and when HBM data need to be translated into HBM indicators: e.g. indicator's relevance for policy, society, health and quality of the HBM data. In the HBM4EU project, indicators were already developed for bisphenol A (BPA), per- and polyfluoroalkyl substances (PFASs) and phthalates (PHT) each answering different policy questions. Questions related to the disproportional affected social disadvantaged and vulnerable groups by chemical exposure remain an important topic on the WHO and EU environmental research agenda. Also here indicators are being developed under HBM4EU, providing information to help policymakers making right decisions to minimize exposure as well as exposure differences and raising awareness among citizens on how to make smart choices in relation to chemicals.

Keywords: aggregate exposure, biomarkers, biomonitoring, environmental health, environmental policy, susceptible/vulnerable

TU-SY-B2-40

Correlation patterns between multiple chemicals measured through human biomonitoring to describe changes over time

Ottenbros I.[1,2], Govarts E.[3], Vogel N.[4], Reiber L.[4], Conrad A.[4], Vlaanderen J.[2], Lebret E.[1], Schoeters G.[3], Vermeulen R.[2]; [1]National Institute for Public Health and the Environment (RIVM), Bilthoven, the Netherlands, [2]Institute for Risk Assessment Sciences (IRAS), Utrecht University, Utrecht, the Netherlands, [3]Flemish Institute for Technological Research (VITO), Health, Mol, Belgium, [4]German Environment Agency (UBA), Berlin, Germany

While mixture toxicology and mixture risk assessment have been on the research and policy agenda for decades, assessment of the actual exposure of the population to mixtures of chemicals has received rather little attention. Within the HBM4EU project, individual human biomonitoring (HBM) data of multiple chemicals was analysed to better understand the patterns in correlation and network structures over time. These analyses are informative in understanding the common exposure patterns over time to combinations of chemicals. Three Flemish birth cohorts (FLEHSI-II-III) were included; recruited in ~2003, 2008 and 2013 from the general population. Heavy metals (Cd, Pb), p,p'-DDE, HCB and PCBs (PCB138, PCB153, PCB180) were measured in cord blood at all three points in time. Additionally, in 2008 and 2013 the heavy metals (As, Cu, Mn, Tl, MeHg (in hair)) and PFASs (PFOA, PFOS) were measured. Besides the Flemish data, preliminary comparative results of the cross-sectional population-representative German Environmental Survey for Children and Adolescents (GerES V) will be presented. At first, correlation structures within and across the datasets were assessed visually with Circos plots.

Next, partial correlations were calculated to measure strength and direction between two chemicals whilst controlling for the effect of the other chemicals in the data. Based on the partial correlations, the best undirected graph networks were estimated using sparsity parameters. 'Undirected graphs' is an approach to describe the conditional independence among many variables. Within the networks, groups of closely related chemicals were identified (communities), which can provide insight into the existence of groupings. Changes over time were assessed with differential network analysis, assessing the similarities and differences between network structures and the communities. Here, differences are changes in correlation strength and direction between chemicals.

Keywords: biomarkers, biomonitoring, chemical prioritization, statistical methods

TU-SY-B2-41

A novel combined text mining and systems biology approach linking priority chemicals to Adverse Outcome Pathways: Bisphenol S as a case study

Barouki R.[1], Carvailo J-C.[1], Coumoul X.[1], Audouze K.[1]; [1]Inserm, Université Paris Descartes, Paris, France

Integrated exposure to chemicals of concern can be assessed through human biomonitoring (HBM). The aim of HBM4EU is to develop HBM surveys at the EU level and to support policy decision making by gathering evidence linking chemicals to health outcomes. Available toxicity data can be optimally interpreted if they are integrated using computational approaches such as systems biology and the Adverse Outcome Pathway (AOP) framework. We have developed a new approach combining a text mining procedure and integrative systems biology to comprehensively analyze the scientific literature and to enrich AOPs related to environmental stressors. This approach was applied to Bisphenol S (BPS), a substitute for Bisphenol A. First, to identify relevant associations between BPS and different AOP components (Molecular Initiating Events, Key Events, Adverse Outcomes), a list of abstracts was screened using the developed text-mining tool 'AOP-helpFinder' which calculates scores based on the graph theory to prioritize the findings. Then, to fill gaps between BPS and AOP components, a systems biology approach was used to integrate information from the AOP-Wiki and ToxCast databases, followed by manual curation of the relevant publications. Links between BPS and 48 AOP key events were identified and scored via 32 references. The main outcomes were related to reproductive health, endocrine and metabolic disruption, and obesity. We then explicitly analyzed co-occurrence of the terms BPS and obesity by manual curation of the full text of the selected publications. Several molecular and cellular pathways were identified linking BPS and obesity. By analyzing dispersed information from the literature and databases, our novel approach can identify links between chemicals and AOP key events. The findings associating BPS and obesity illustrate the use of computational tools as a complement to human biomonitoring and highlight the relevance of the approach to decision makers assessing substitutes for toxic chemicals.

Keywords: environmental health

TU-SY-C2-42

Characterizing SVOCs in real indoor environments and implications for exposure assessment

Mandin C.; Scientific and Technical Center for Building (CSTB), Paris, France

Estimating human exposure to semi-volatile organic compounds (SVOCs) implies the use of environmental concentrations that are already available through measurements or the validation of models that can be used to predict concentrations, a process which is also based on available measurements. However, SVOCs include numerous groups of chemicals with a large range in physicochemical properties, which makes it very challenging to measure them reliably in indoor air, settled dust, or on surfaces. In addition, the complexity of the indoor environment is such that it is essentially impossible to capture the range of likely exposures via measurements. We therefore need to rely on models to estimate exposure, but the models themselves need to be developed in such a way as to capture the major features of the indoor environment. There is thus a need to consider the development of models in close association with real world measurements of indoor environments. This presentation addresses several key issues of measuring SVOCs in real indoor environments from the perspective of human exposure assessment. One major issue is the level of representativeness of sampled media relative to the media actually causing exposure. Neither indoor air nor dust sampling and analyses are standardized, leading to different sample collection methods and sample treatments, for example varying selections of dust size fractions. Spatial and temporal variabilities should also be considered to correctly assess exposure. Major studies performed worldwide to measure SVOCs in dwellings, schools, and offices are presented and discussed to describe these problems in detail. The potential of SVOC exposure models to address some of the problems is highlighted.

Keywords: SVOCs, air, built/indoor environment, chemical prioritization, sampling methods

TU-SY-C2-43

Key drivers for science to address exposures to SVOCs

Cohen Hubal E.; United States Environmental Protection Agency, Research Traingle Park, NC, USA

Large numbers of SVOCs are on the market and present in consumer products and articles. Potential health effects from SVOCs in products are associated with disruption to complex biological processes. Yet, many of these chemicals have not been thoroughly evaluated for potential risks to human health, particularly when considering the consequences of use over the chemical/product life cycle. Globally, chemical management decisions are being made at the national, regional, and local levels. Policy makers and consumers are demanding robust information on chemicals in products and are driving large retailers and industry to make changes. Tools for evaluating chemical substitutions and product alternatives are evolving to meet the demand for action. However, current approaches for characterizing exposures across the chemical/product life cycle are data and resource-intensive. In addition, important information gaps remain that are related to emission and transport parameters for the full range of commercial chemicals in many commonly used products. To build strength in current exposure models used by agencies around that world as well as emerging structure-based models for estimating critical emission and transport parameters, experimental data are necessary to cover additional chemical space and product types. Specifically, data on emissions from articles of legacy and substituted SVOCs are required. Associated mechanistic modeling approaches and tools for estimating emission, transport, and exposure should be extended to incorporate considerations of product and article categories. The resulting advanced predictive exposure tools deployed for a variety of risk-based evaluations will ultimately be used to increase the pace, transparency, and defensibility of chemical safety assessments. Addressing these priority needs for information and methods will facilitate better informed, timelier decisions about SVOCs. The views expressed in this presentation are those of the authors and do not necessarily represent the views or policies of the U.S. Environmental Protection Agency.

Keywords: SVOCs

TU-SY-C2-44
A modeling framework for SVOC emission, transport, and exposure

Eichler C.[1], Cao J.[2], Little C.[1]; [1]Virginia Tech, Blacksburg, VA, USA, [2]Sun Yat-sen University, Guangzhou, China

This talk will review emissions from the different product categories including solid products (small or large), liquid products and sprayed products. These categories cover most consumer products and building materials containing SVOCs. We present a consensus-based framework of mechanistic emission, transport, and exposure models for SVOCs within a larger framework of chemical risk assessment. Mechanistic models suitable to rapidly assess risks posed by SVOCs in consumer products must be able to 1) support the development of predictive structure-based relationships such as quantitative structure-activity relationships (QSARs) for the estimation of emission and transport parameters, 2) evaluate and build confidence in exposure estimates, and 3) inform risk-based chemical management decisions. Specifically, the objective is the development of a screening-level exposure modeling framework that incorporates a mechanistic understanding of emission, transport and exposure to SVOCs indoors. This will be achieved by deriving mechanistic product emission categories as the central elements of emission modeling. Any product or consumer article can be categorized as belonging to one of the basic emission categories. From there, emission models relevant for this category can be applied. Model parameters are predicted based on available data and QSAR models. Subsequently, SVOC transport into different environmental compartments (gas phase, airborne particles, aerosols, settled dust, and sink surfaces) and partitioning among compartments is modeled, resulting in a realistic description of the fate of the chemicals. Based on this, SVOC concentrations available for exposure are estimated, giving insights into the most relevant exposure pathways associated with a specific SVOC-product combination, and based on their respective product emission category.

Keywords: SVOCs, built/indoor environment, consumer and personal care products, exposure models, risk assessment

TU-SY-C2-45
A suite of SVOC exposure models – estimating parameters and illustrative results

Cao J.[1], Eichler C.[2], Little J.[2]; [1]Sun Yat-sen University, Guangzhou, China, [1]Virginia Tech, Blacksburg, VA, USA

This talk will review the consensus models used to predict exposure resulting from the various product categories and the subsequent contributions to exposure, as well as illustrative results. Building on the consensus-based framework previously introduced, a detailed mechanistic description of SVOC emission, transport, partitioning, and fate is presented for different product categories and environmental compartments, integrating the knowledge and understanding that has been achieved over the past decade. Emphasis is given to those models which will serve as the basis for an international consensus. The resulting suite of mechanistic exposure models can be used to support the development of predictive relationships for the estimation of emission and transport parameters, evaluate and build confidence in rapid, high-throughput exposure estimates, and inform risk-based chemical management decisions. Additionally, important gaps in the current understanding will be identified as waypoints for further research. To illustrate the framework, several examples will be presented and discussed for each product category (small and large solid products, liquid products, and sprayed products). The importance of each exposure pathway (ingestion of dust, inhalation, dermal uptake from air, and dermal uptake by direct contact) will be ranked based on the resulting exposure for each source category and pathway. Important parameters that significantly influence human exposure to SVOCs indoors will be identified by sensitivity analysis and a consistent set of methods for measuring or estimating the model parameters will be reviewed. Finally, the effectiveness of several common control strategies, such as source removal, ventilation, vacuum cleaning, and application of air purifiers, on the exposure to SVOCs from each product emission category will be evaluated.

Keywords: SVOCs, built/indoor environment, exposure models, particles, surfaces

TU-SY-C2-46

Exposure to SVOCs in the indoor environment – the need for consensus on a suite of exposure models

Little J.[1], Cao J.[2], Eichler C.[1]; [1]Virginia Tech, Blacksburg, VA, USA, [1]Sun Yat-sen University, Guangzhou, China

The talk will describe the process by which an international consensus was established for the suite of exposure models, summarizing the outcomes of the consensus building process, and next steps for implementation. SVOC exposure modeling has made significant progress in the past decade. However, many research groups are working in isolation, often without a shared knowledge base. This impedes further development of useful exposure modeling and subsequent chemical risk assessment and management. Acknowledging this need, a group of researchers in the field of SVOC emission, transport, and exposure modeling established contact with experts and stakeholders to engage in a project with the goal of identifying existing models that can be considered “Consensus Models”. The suite of models include those that have been found to describe emission, transport, partitioning, and exposure in a consistent, realistic, and process-based fashion, using parameters that can either be measured or estimated based on physicochemical relationships. To strengthen the endeavor, a “Consensus Protocol” was developed. The protocol identified key personnel, highlighted the proposed strategy, and encouraged the participation of a wide range of researchers. In May 2019, a core group of experts participated in a workshop at Tsinghua University in Beijing, China, to reach a preliminary agreement regarding which of the models qualify as Consensus Models. The resulting modeling framework and the proposed models have been described in the preceding talks. This presentation highlights research needs and next steps to improve the widespread adoption and routine application of the Consensus Models.

Keywords: SVOCs, built/indoor environment, chemical prioritization, exposure models, risk assessment

TU-PL-D2-EAM: Indoor Air Exposures

TU-PL-D2-47

MALDI-TOF mass spectrometry – a quicker tool to identify dampess related actinomycetes?

Pessi A-M.[1], Harju I.[2], Kankaanpää.S.[1], Saarto.S.[1], Hakanen A.[2]; [1]Biodiversity Unit of the University of Turku, Turku, Finland, [2]Clinical Microbiology Laboratory, Turku University Hospital, Turku, Finland

Thousands of microbial samples are collected yearly from Finnish buildings suspected of dampness problems, either from building materials, surfaces or indoor air. The present laboratory techniques rely mostly on the cultivation method, and on the identification of colonies with light microscopy. The interpretation of results are based on the microbial concentrations and the composition of microbiota. The list of microbial ‘signature taxa’ of moisture damage in Finland consists of some 25 fungal taxa, and actinomycetes-type bacteria. The group of Actinomycetes is large and polyphyletic, and there is a clear need to determine more accurately, which subtaxa arereally connected to dampness and/or potential health problems and which are not. Matrix-assisted laser desorption-ionization time - of - flight mass spectrometry (MALDI-TOF MS) is a proteomic based technique for identification microbes. This technique is widely in use in the identification of pathogenic microbes in hospital settings. Our project aims at testing the applicability of MALDI-TOF in identification of environmental microbes, sampled from buildings with dampness problems. In the first phase of the project we have developed an extraction procedure for aerobic Actinomycete isolates and started to build a non-clinical MALDI-TOF library for these environmental Actinobacteria. The potential use of MALDI-TOF in indoor investigations would have several assets. From cultivated samples, idenfication procedure is fast, only some minutes, simple and cost-effective. The technique brings new possibilities to species level identification, previously needed DNA sequencing or chemotaxonomical methods.

Keywords: analytical methods, built/indoor environment, dampness/mold

TU-PL-D2-48
Differences Between Indoor and Outdoor Air Composition at Nighttime Using an Indoor Chemistry Model

Ward C.[1], Cheng P.[1], Jarrett J.[1], Jones R.[1]; [1]Centers for Disease Control and Prevention, Atlanta, GA, USA

Much of human exposure to air pollution takes place while sleeping indoors at night, when indoor and ambient air compositions are expected to be most alike, since they are not affected by photochemistry. We explored the effects of emissions, temperature, and ozone on nighttime dissimilarity between indoor and ambient air in a modeling study using the USEPA Simplified Indoor Air Chemistry Simulator (SIACS). METHODS: The concentrations of 74 chemical species were modeled for a 140 m2 box (representing a residential space) to equilibrium conditions, using ambient concentrations and indoor emission rates of volatile organic compounds (VOCs) from the literature. The model scenarios varied ambient temperature and ozone concentrations (10 to 50 ppb), with steady indoor temperature at 22 ^{0}C. We used Euclidean distance as a metric of dissimilarity between indoor and ambient air, normalizing all concentrations to fixed ambient levels. RESULTS: Ambient and indoor air composition were most different when ambient and indoor temperatures were similar, due to the smaller air exchange rate (AER). Lower outdoor temperatures (0^{0}C) resulted in greater differences between indoor and outdoor composition than higher outdoor temperatures (40^{0}C), if AER was held constant. A non-linear interaction between ambient ozone concentration and temperature was observed. Peroxidation byproducts, predominantly organic hydroperoxides and peroxy-carboxylic acids, accounted for most of the dissimilarity in composition between indoor and ambient air. The selective removal of indoor VOC emissions decreased the dissimilarity, particularly for species that readily react with ozone (e.g. terpenes). CONCLUSIONS: Reactive oxidation byproducts, despite very low concentrations, deserve further attention, as they represent the chemical peculiarities of indoor air even more than VOC emissions. Understanding their interaction with indoor surfaces (including biological surfaces) may help identify any significance for health or material wear. Reducing VOC emissions may mitigate exposure to reactive pollutants even at night.

Keywords: VOCs, air, built/indoor environment, exposure models

TU-PL-D2-49
Development of a simple tool for guiding mold inspection and remediation in U.S. homes

Hegarty B.[1], Shaughnessy R.[2], Haverinen-Shaughnessy U.[2], Peccia J.[1]; [1]Yale University, New Haven, CN, USA, [2]The University of Tulsa, Tulsa, OK, USA

The goal of this research is to provide a technically defensible, economically-practical tool for defining the dampness-associated fungal contribution to a building's fungal ecology. We hypothesize that the fundamental data contained in home fungal community DNA sequence analyses can be leveraged to build databases and indices to accurately guide home mold inspection and restoration. Specific objectives include the following: (1) Conduct an extensive, nation-wide field campaign to sample fungi in water damaged and non-water damaged homes; and (2) Leverage this nationwide fungal ecology data to produce indices that quantitatively define the contributions of dampness-associated fungi in homes across the U.S. The sampling plan involves 12 representative metropolitan areas, a total of 120 homes, and 780 samples. Three categories of samples have been collected including: (1) settled dust from homes with no history of water damage or visible mold ("no mold" home dust), (2) settled dust from homes with documented water damage and visible mold ("mold" home dust), and (3) direct surface samples from building materials with visible fungal growth ("direct mold"). Preliminary results from fungal sequencing has found compositional differences between the ecological profiles of the fungal samples as a function of sample type. There exist a significant difference in richness between direct mold, mold dust, and no mold dust samples. The direct mold samples have a lower richness than the dust samples from moldy or non-moldy homes. Geography, defined as specific climate regions, also drives some of the changes in fungal community structure. Taking these relationships into account, comparisons between mold, no mold, and direct mold indicate even larger differences, suggesting the important role of geography in obscuring the differences between mold and no mold samples.

Keywords: building materials, built/indoor environment, dampness/mold, exposure factors, microbial agents

TU-PL-D2-50
Influence of housing typology on energy retrofit efficiency and indoor air quality

Milando C.W.[1], Dols W.S.[2], Levy J.[1], Fabian M.P.[1]; [1]Department of Environmental Health, Boston University, Boston, MA, USA, [2]National Institute of Standards and Technologies (NIST), Gaithersburg, MD, USA

Background Residential energy-efficient retrofits recommended in city-scale climate action plans can substantially lower residential energy consumption, improve ambient air quality, and potentially mitigate climate impacts. However, weatherization without additional ventilation may result in worsened indoor air quality, and this tradeoff may vary by housing typology. Such dynamics have not typically been explored in climate action planning. In this study, we examine how energy efficiency retrofits affect energy use and air pollutant exposure across common Massachusetts housing structures. Methods We modeled annual and seasonal changes in energy consumption and indoor air quality (PM2.5) in common housing templates (n = 10) following various retrofits (e.g., weatherization) using coupled EnergyPlus and CONTAM simulation models. Templates included single-family structures and apartment buildings, each with a variety of foundations and heating systems commonly found in Boston, MA. Results Air quality and energy impacts of energy-efficiency retrofits varied greatly between housing templates. Weatherizing without ventilation improvements increased indoor PM2.5 by an average factor of 2.6 and 3.7 for apartments and single-family houses, but average modeled concentrations of PM2.5 in weatherized apartments were twice as high as those in single-family houses (152 and 75 µg/m3). The same retrofit resulted in larger decreases in heating energy use in single family houses than apartments (by factors of 2.9 and 2.3 respectively), while the total annual energy usage for gas heating in weatherized single family houses was 3.5 times higher (28,215 kBtu) than for apartments (8,129 kBtu). Conclusions Overall, our results indicate that city-scale planning of energy efficiency retrofit programs requires tailoring to city-specific housing typologies. Our modeling framework allows for quantification of various co-benefits and potential hazards related to intervention implementation in specific housing types. Such tailoring may enable more specified retrofit programming in future plans, e.g., those attempting to mitigate city-specific climate effects in coming decades.

Keywords: air, built/indoor environment, environmental health, exposure models, risk assessment

TU-PL-D2-51
An agent-based model to quantify human exposure variability in residential indoors

Micolier A.[1], Taillandier F.[2], Sonnemann G.[1], Jolliet O.[3]; [1]University of Bordeaux, Bordeaux, France, [2]IRSTEA, Aix-en-Provence, France, [3]University of Michigan, Ann Arbor, MI, USA

Low-energy buildings are confined spaces in which a wide variety of chemicals expose the occupants to health hazards through various exposure pathways that affect chemicals fate. Besides, chemicals residence-time is affected by physical removal from occupants activities such as windows opening. Occupants adaptive actions in response to thermal incomfort influence the thermal behavior of low-energy buildings particularly sensitive to variable internal gains while chemicals release from building materials is temperature-driven. The combined effect of environmental and human factors should be considered to model holisticaly the dynamics of pollutants fate indoors. Therefore, there is a need for a model that addresses chemical fate together with heat transfer and human profiles heterogeneity. We developed a model of chemical and heat balance in which mass transfer parameters evolve at each time step as a function of the temperature. Since Artificial Intelligence techniques are promising in describing human dynamics, we developed an agent-based model simulating occupants' behavior in residential dwellings. Both models are coupled in order to quantify the impact of the occupants' behaviour on the chemicals fate as well as the occupants' exposure to the resulting indoor chemicals. A global sensitivity analysis on the coupled mass and heat transfer model shows that 17, 19 and 29% of the variance of the product intake fraction (PiF) for DEHP in a vinyl flooring the can be attributed to the floor's thickness, room dimension and number of persons present respectively, while for ethylbenzene in the same conditions, 31, 7 and 17% of the variance of the PiF can be attributed to the floor's thickness, air renewal rate and inhalation rate. At the time of the conference, K-mean clustering algorithm will be applied to identify several households' archetypes representative of the exposure variability. This study offers a better understanding of exposure patterns for healthy energy-efficient buildings.

Keywords: activity patterns, behavior, exposure models, thermal environment

TU-PL-D2-52
Mitigating radon in existing housing in Canada: a cost effectiveness analysis

Gaskin J.[1], Whyte J.[2], Coyle D.[3]; [1]Radiation Protection Bureau, Health Canada, Ottawa, Canada, [2]Construction Research Centre, National Research Council Canada, Ottawa, Canada, [3]School of Epidemiology and Public Health, University of Ottawa, Ottawa, Canada

The percentage of Canadians in 2012 living in homes with radon concentrations above the Radon Guidelines threshold of 200 Bq/m3 was estimated to be 6.9%, but homeowners are responsible for all the mitigation costs and compliance is voluntary. A cost-utility analysis is conducted to determine the incremental cost effectiveness ratios (ICERs) for the gradual implementation of radon screening and mitigation of existing housing in Canada. The structural uncertainty resulting from the choice of model structure is assessed by comparing the results from a Markov cohort model with those from a discrete event simulation. A period life-table analysis is based on data derived from two recent Canadian radon surveys, the BEIR VI exposure-age-concentration model for attributable risk of lung cancer mortality, and Canadian mortality and quality of life data for the 2012 population. A societal perspective, a lifetime horizon and a discount rate of 1.5% are adopted. At a radon mitigation threshold of 100 Bq/m3, the discounted ICER for current rates of screening and mitigation of existing housing compared to no specific radon measures in Canada is $58,866/QALY using a Markov cohort model and $59,556/QALY using discrete event simulation. The interaction of the built and social environment is important to represent when modeling interventions to reduce radon and associated lung cancer. The sensitivity of the ICER estimates to a mitigation tax credit incentive and to the housing stock renewal rate is investigated because renovations rates have recently become higher than new housing starts in Canada as a result of changing housing and employment policies over the last two decades. The mitigation of radon in existing housing in Canada is cost effective at higher radon mitigation thresholds and the cost effectiveness would improve under a theoretical incentive program for home owners to increase the rates of radon testing and mitigation.

Keywords: risk assessment

TU-PL-E2-SSS: Smart Sensing Strategies of Pollutants

TU-PL-E2-53
Smart building performance monitoring with IEQ sensor networks

Boerstra A.[1], Raue A.[1], Cheng L.[2]; [1]Binnenmilieu, Haga, South Holland, [2]Pure Living China, China

The recent development of affordable and quite accurate CO2, particle and e.g. VOC sensors has led to a growing interest in continuous indoor climate monitoring. Not just amongst scientists and engineers but also amongst building owners, developers and e.g. architects interested in boosting our buildings' health and comfort qualities. The introduction of advanced Indoor Environmental Quality (IEQ) sensor networks would allow us to better analyse indoor exposure both in new and existing buildings. However, to successfully deploy IEQ sensor networks inside one needs more than just accurate sensors. More important is to develop an overall view on what to measure, where, how, how often, etcetera. Apart from that, a general methodology has to be developed that allows to analyse and present the enormous amount of IEQ data that will be gathered in a way that building users and decision makers can relate to. This position paper addresses a couple of fundamental questions that need to be answered to assure the successful introduction of IEQ sensor networks in buildings at a larger scale. The paper presents a first attempt to answer these questions, based upon a review of the literature and the authors' experience with several kinds of IEQ sensor networks. Recommendation are illustrated with practical examples from some ongoing pilot projects. Some relevant standards and platforms that have been developed lately (e.g. RESET) are described. The results presented in this paper can be used to further develop IEQ sensor networks both for academic and more practical purposes, e.g. the application of sensor networks in the context of PPP/DBFMO contracts (performance verification).

Keywords: sensor technology

TU-PL-E2-54
Sensor Technologies for Indoor Air Quality Management and Monitoring

Puglisi D.[1]; [1]Linköping University, Linköping, Sweden

It has recently been reported by the World Health Organization that three billion people use polluting technologies and fuels for cooking, heating, and lighting, releasing dangerous smoke and air pollutants, and that over 90 % of the world's population lives in places where air pollutant levels exceed the recommended threshold limits. This causes eight million of premature deaths per year, of which 3.8 million due to indoor air pollution. The development of appropriate technologies for indoor air quality management and monitoring is a global drive to tackle this problem and achieve real and measurable air emission reduction, together with the design of effective air quality control strategies, policies and investments supporting cleaner transport, energy-efficient buildings, power generation, industry, and better waste management. But what does appropriate technologies mean? This contribution aims to identify the current trends in indoor air quality management and monitoring, and to highlight advantages as well as limitations of air quality technologies with main focus on sensor technologies. Novel, lower-cost, and easy-to-use sensor technologies are poised to support a paradigm shift by allowing ubiquitous air pollution monitoring with high spatio-temporal resolution available at every person's fingertips. However, they must be thoroughly evaluated to ensure that their performance specifications (precision, accuracy, sensitivity, interferences,...) meet designated monitoring objectives. Air pollution sensors provide citizens and communities with unprecedented opportunities to monitor the local air quality that can directly impact their daily lives, and facilitate the development of community-based strategies in line with the growing concept of citizen science, believed to play a key-role for effective reduction of air pollution emissions and protection of human health.

Keywords: VOCs, air, analytical methods, built/indoor environment, community, sensor technology

TU-PL-E2-55

Assessment of real-time contextual information at the workplace to support new generation miniature sensor data of airborne substances

Goede H.[1], Timmermans R.[1], Le Feber M.[1] Kuijpers E.[1], Pronk A.[1]; [1]Erasmus University Rotterdam, Rotterdam, Netherlands

Lightweight miniature sensors are entering the market with the potential to transform exposure assessment technologies in the occupational setting. In particular sensors that can detect airborne particles and specific (chemical) substances are being developed for this purpose. As a result, a transition from time-integrated measurement techniques to real-time sensor technologies are expected in the future. These sensors offer the possibility to assess individual worker exposure to substances in real-time with a higher sampling resolution in time. However, measurement data obtained from sensor data is particularly useful if we have sufficient contextual information of the when, where and why exposures occur, preferably in real-time. To inform sensor data, new technologies are therefore required to replace previously used logbooks, questionnaires and observations typically applied in traditional time-integrated sampling techniques. In this study we explored various technologies that can be applied to obtain contextual information. Based on a literature and internet review, various technologies suitable for registering contextual information was identified and evaluated, including different types of location tracking devices, video monitoring and other 'smart sensors' such as proximity sensors. An overview of these technologies are presented, their suitability for different workplace scenario's, together with the advantages and disadvantages of implementation in the occupational (indoor and outdoor) setting. The added value of these technologies are evaluated for the purposes of real-time exposure and risk assessment, for example by synchronizing real-time exposure concentrations with data such as active emission sources, worker distance from sources, use of tools or control measures, worker movement and posture, noise levels and heart rate. An increased use of such technologies is expected in industry to provide real-time feedback to workers and to improve working conditions and reduce exposures.

Keywords: sampling methods

TU-PL-E2-56

Opportunities and baries for microsensor sytems for air quality: practical experiences from Oslo

Bartonova A.[1], Castell N., Schneider P., Dauge F.R., Grossberndt S., Lepioufle J-M., Liu H-Y., Fredriksen, M.F., Kaastad B-A.; [1]NILU Norwegian Institute for Air Research, Kjeller, Norway

Monitoring air quality has decades long tradition, and has undergone much technological developments. Reference monitoring methods are today supplemented by technologies from satellites to micro sensors. Modelling has become ubiquitous and there are many more actors who can measure, including the responsible national agencies, local communities and private individuals. Undoubtedly, information on air quality has been improved, but there is also an increase in information noise. In the last decade, we have been conducting research on low-cost sensor systems for monitoring criteria air pollutants. Addressing the individual sensors' dependency on their physical and chemical environment and considering not one unit but networks of sensor system devices have been two areas of research. We have established several networks of low-cost sensor systems monitoring NO2 and particulate matter, each with several months operation. We were able to use such networks to improve air quality information, and our experiences have also contributed to development of calibration methods. Together with various stakeholders, we are gaining knowledge on the extent to which such networks are useful for different purposes, ranging from awareness raising, education, knowledge co-creation, to providing the responsible authorities and communities with information on air quality. Barriers to a more ubiquitous use of micro sensor systems in practice lie at this time mostly in the applied technologies. Information on performance, precision and accuracy of the available units is often not sufficient and makes it difficult to judge if they are fit for the purpose. Connectivity of the units is often a challenge – even if a number of technical solutions exist, linking a number of units in one system is difficult. And not least, making use of the data is a barrier in itself, one that is subject to quickly developing research. Successful applicatoin examples are beginning to appear, here demonstrated for Oslo.

Keywords: air, community, exposure models, particles, sensor technology

TU-PL-E2-57

Short-Term Measurements vs. Long-Term Continuous Monitoring of Indoor Environmental Quality

Clinchard S.[1], della Vecchia S.[1], Aller R.[1], Novotny T.[1], Haverinen-Shaughnessy U.[1]; [1]720°, Helsinki, Finland

Assessment of indoor environmental quality (IEQ) is changing from intermittent, short-term measurements to continuous, long-term monitoring, using sensors to monitor various IEQ parameters. Cloud technology enables real-time monitoring, and continuous data provide opportunities to detect spatial and temporal trends. This presentation focuses on demonstrating the results from long-term (several months to years) monitoring as compared to short term (a few days or weeks) measurements. Data on IAQ parameters, including T, RH, CO2, VOCs, and PM2.5, have been continuously collected (15 s. resolution) from >100 office buildings and also some schools. With respect to each monitored parameters, an 8-hour (corresponding to typically occupied hours) mean or median values were calculated. A total of 41 rooms (sensors) from 12 buildings had more than 1095 datapoints, indicating over three years of continuous monitoring. Statistical analyses were done with IBM SPSS Statistics version 24. In addition, times series analyses have been conducted using Python version 3.7.1. An obvious benefit of long-term monitoring is the more comprehensive picture of IEQ over time. Whereas short-term measurements represent the conditions only during the time of the measurement, long-term monitoring can result in frequency of noteworthy events over time or percentage of time during which the levels fulfill recommendations. These types of results can be used, for example, in optimizing ventilation rates based on CO2 concentrations. Several months of monitoring allow detection of trends, e.g. deteriorating or improving conditions over time. Such results are often seen in VOCs concentrations, related to material emissions during and/or after refurbishment. Monitoring over 12 months provides opportunity to detect seasonality, which are commonly seen for thermal parameters and PM. Finally, understanding trends and seasonality opens to time series modeling and forecasting that could help minimizing the delays in responding to potential IEQ issues and improving indoor environments proactively.

Keywords: built/indoor environment, productivity, sampling methods, sensor technology, statistical methods

TU-PL-E2-58

How many patrons in the university library ? CO2 sensors as tools to determine the occupancy of a whole building.

Hanoune B.; Univ. Lille, CNRS, Physicochimie des Processus de Combustion et de l'Atmosphère, Lillie, France

The Lilliad learning center of the University of Lille is a multi-use facility including the science library, a food court, a conference hall, and a technological showcase. It is located in an architecturally complex building from the 1960', renovated in 2017. It consists mostly of an open hall from the ground to the third floor, with one large reading room at each level, and about 40 small working spaces. The conference hall and technological showcase are separated from the main hall, but share the same entrance, which is equipped with electronic people counters, however with very limited reliability. The library receives daily a maximum of ~4000 patrons. Sensor nodes developed at the University of Lille were installed in December 2018 in 4 different places of the Lilliad learning center, one in the food court, one in a 4-person working space, and two in the main hall, on the second and third floors. These nodes contain sensors for CO, CO2, VOC, NOx, O3, for size-resolved number concentrations of particles (0.3-10 µm), as well as comfort and physical parameters: temperature, relative humidity, pressure, light, sound, motion. The aim was to investigate indoor air quality inside the different environments of the building, and to estimate the real occupancy. To our knowledge this is the first time that CO2 sensing is used for a whole building. We show that in most part of the buildings the air quality is within the recommendation levels, except in the food court where the ventilation is not sufficient. From the CO2 sensors located in the main hall, we were able to determine retrospectively the total occupancy of the building of any day, with a ~10% error. Further analysis is underway to assess the occupancy in real time.

Keywords: air, built/indoor environment, sensor technology, ventilation

TU-PO-01
How Human Presence and Activities of Daily Living Influence Indoor Air Quality

Bielik N.[1], Rotach M., Mitova M., Cluse C., Goujon C., Rouget E.; [1]Philip Morris International, Neuchâtel, Switzerland

To assess the impact on indoor air quality (IAQ) of products with the potential to reduce individual risk and population harm in comparison to smoking cigarettes, an environmentally controlled exposure room was built in Philip Morris International Research & Development facility. An analytical platform, including 11 online and offline methods, was developed, validated, and subsequently applied for several studies involving volunteer panelists. It has long been known that the presence of human beings and their daily activities in an enclosed environment can significantly modify IAQ by releasing several hundreds of volatile organic compounds from breath and skin. We evaluated the impact of panelists present in the exposure room for two hours and also of different common human activities (drinking wine, using cosmetics, and cooking) during so-called "activity sessions." Each activity was assessed separately using our analytical platform and exposure room under controlled environmental conditions. One method in particular, which is part of the analytical platform, measures the total volatile organic compounds (TVOC) yield and main contributors within a given range of volatility present in air. Its measurement relies on the use of gas chromatographic instrumentation equipped with parallel mass spectrometry and flame ionization detectors, allowing simultaneous quantification and identification of compounds present in the room atmosphere. The results obtained by TVOC analysis showed that prolonged human presence as well as indoor activities led to a deterioration of IAQ. Moreover, it was shown that TVOC patterns would vary qualitatively and quantitatively depending on the activities. Few constituents were specific for a type of experiment (e.g., fusel alcohols for the wine session), while most had just different patterns of increase (e.g., methylcyclosiloxanes had highest concentration in the cosmetic session). These learnings were applied to optimize the design for IAQ studies.

Keywords: VOCs, air, analytical methods, built/indoor environment, consumer and personal care products

TU-PO-02
Particulate science dynamics in indoor micro-environments of Agra, India

Rohra H.[1], Taneja A.[1]; Dr. B.R. Ambedkar University, Agra, India

Improved understanding of particle size distribution of residential air particles is valuable owing to a larger fraction of time spent in homes. Size segregated aerosol measurements was conducted using a multi staged cascade impactor in residential microenvironments of Northern India. Particles with diameter<2.5µm contributed about 64-80% of PM10 in which over 29% of mass was shared by ultrafine PM0.25. Increased significant inter-correlation among adjacent sized PM fractions was further noted. The accumulation mode dominated the lognormal particle size distribution during the complete monitoring campaign whereas those of elemental constituents reflected the distinct emission sources. These findings in future will be used to classify the impactor aerosol samples for toxicological studies and would provide source information to state agencies responsible for development of effective pollution control strategies.

Keywords: metals

TU-PO-03

Source characterization and apportionment of river pollution in central Taiwan

Lin[1], Wang[1], Huang[2], Wang[3], Wang[3], Hsu[3], Wu[1]; [1]National Taiwan University, Taipei, Taiwan, [2]Academia Sinica, Taipei, Taiwan, [3]Environmental Analysis Laboratory EPA, Taoyuan, Taiwan

The systematic forensic technology and pollution source tracking model have been widely used in analyzing sources of air pollution. However, these methods were not commonly used to trace sources of water pollution. In this study, receptor modeling was applied to investigate the spatial and temporal distributions of water pollution sources in Wu River in Taiwan by the chemical mass balance (CMB) model. Six types of industrial manufacturing plants are located along the river, including brewing industry (BI), agricultural and environmental pesticide manufacturing (AEP), leather manufacturing (LM), chemical manufacturing (CM), wastewater treatment (WT) and electroplating industry (EI). Two monitoring sites at a distance of 7.3 km were selected as the as the receptor sites. Two water samples at each site were taken in July and September, respectively. In addition, wastewater samples including the effluent and process wastewater at 24 industrial plants near the Xinan Bridge and Neixin Bridge were collected to represent the potential source profiles. The sample collected were analyzed for fifty-five heavy metals were analyzed by ICP-MS. The main characteristic species were identified for different types of industry, e.g. Mn and Sn in CM, Fe in EI-process, and Al in LM. The CMB modeling results showed that in July the main sources at Xinan Bridge and Neixin Bridge were CM and EI process wastewater with contribution rates of 49% and 47%, respectively. The sources at these two sites of the samples collected in September were both LM (66% and 85%). As the results showed, the sources of pollution in rivers exhibited temporal and spatial variation.

Keywords: environmental health, metals, statistical methods, water

TU-PO-04

Fugitive Emissions from a Breath Actuated Jet Nebuliser and a Vibrating Mesh Nebuliser for a Paediatric Patient

O'Toole C.[1], McGrath J.[1], Bennett G.[2], Joyce M.[2], MacLoughlin R.[2], Byrne M.[1]; [1]NUI Galway, Galway, Ireland, [2]Aerogen, Galway, Ireland

Introduction Recent studies have identified the potential for fugitive emissions to be released during nebulisation treatments. Here we quantify fugitive emissions from a commercially available breath actuated compressor driven jet nebuliser (JN) designed to reduce emissions, and a vibrating mesh nebuliser (VMN) and valved spacer. Methodology A breath actuated jet nebuliser with a mouthpiece (Aeroeclipse2, Monaghan Medical, Canada) was operated at 8 L/min. Two modes on the nebuliser were tested during simulated paediatric breathing (Vt 250 mL, breath rate 25 BPM, I:E ratio 1:2); breath-actuated nebulisation (BAN) and continuous nebulisation (CN) (n = 5). A vibrating mesh nebuliser (VMN) (Aerogen Solo with Aerogen Ultra, Aerogen, Galway, Ireland) was used with a mouthpiece in CN mode only. A standard prescription dose of 3 mL of 2 mg/mL salbutamol was nebulised. An Aerodynamic Particle Sizer (TSI Inc., US) placed at 0.8 m (arm's length) from the nebuliser was used to characterise the fugitive aerosol concentration (mg drug/m3) over a 45 minute period. Results Fugitive emissions were released during nebulisation treatment for both nebuliser types and both modes. CN mode on the jet nebuliser had a higher fugitive aerosol concentration of 0.061 ± 0.048 mg/m3 compared to the BAN and the VMN (0.044 ± 0.030 mg/m3 and 0.011 ± 0.006 mg/m3). MMAD averaged 1.21 ± 0.05 for JN CN, 1.22 ± 0.04 for JN BAN and 1.20 ± 0.12 µm for VMN. Conclusion The results confirm that fugitive emissions are released into the surrounding environment during respiratory therapy using both CN and BAN mode on the jet nebuliser as well as CN mode for the vibrating mesh nebuliser. However, across all tests, the vibrating mesh nebuliser in combination with a valved spacer recorded the lowest secondary emissions.

Keywords: air, built/indoor environment, children, cumulative exposure, environmental health, exposure factors, health, chronic diseases, health, other/general, particles, respiratory health

TU-PO-05
Local Determination of the Building Envelope Air Leakage

Bjoerling M.[1], Kumlin.A.[2], Carlsson.P.[3]; [1]University of Gävle, Gävle, Sweden, [2]Anders Kumlin AB, Stockholm, Sweden, [3]Bengt Dahlgren AB, Stockholm, Sweden

A method to determine local air leakages in the climactic barrier of existing buildings has been devised and tested. In particular we could measure that tracers could enter the office indoor air from the construction frame. Therefore pollutants originating from the construction frame need to be removed. Alternatively, the tightness of the climactic barrier may be improved in the offices showing the highest apparent emission rates that probably are due to high leakage rates. The method is also more generally applicable to convert concentration to emission rates. Emission rates are more indicative of the location of pollution sources and may therefore be useful in investigations of polluted buildings.

Keywords: air, built/indoor environment, sampling methods, source control, ventilation

TU-PO-06
Control of airborne microbial contamination in hospital wards using a highly antimicrobial nano-silver/ TiO2-chitosan composite

Yu K-P.[1], Chen Y-C.[1], Liao C-H.[1], Shen W-T.[1], Wu Y-C.[1], Su C. [1]; [1]National Yang-Ming University, Taipei, Taiwan (ROC)

In this study, a novel antimicrobial composite of silver nanoparticles (AgNPs), titanium dioxide (TiO2) and chitosan (CS) was preparedvia photochemical deposition of AgNPs on a CS-TiO2 matrix (AgNPs@CS-TiO2). Electron microscopy images showed that the AgNPs@CS-TiO2 microspheres had diameters of approximately 1.7 mm, and the AgNPs were well-dispersed on the TiO2-CS matrix with diameters ranging from 6.69 to 8.84 nm. X-ray photoelectron spectra (XPS) indicated that the most of the AgNPs were reduced to metallic Ag. Fourier transform infrared (FTIR) spectroscopy confirmed that some AgNPs formed a chelate with CS through coordination of Ag^+ with the chitosan amide II groups. The zone of inhibitionof AgNPs@CS-TiO2 against bacteria (E. coli and S. epidermidis) and fungi (A. niger and P. spinulosum) was 6.72–11.08 mm and 5.45–5.77 mm, respectively, and the lowest minimum (critical) concentrations of AgNPs required to inhibit the growth of bacteria and fungi were 4.3 and 73.2 μg, respectively. As an air cleaner, the removal efficiency of an AgNPs@TiO2-CS bed filter for bioaerosols increased with packing depth, and over 68% of bioaerosols could be removed when the packing depth was 6 cm. However, the optimal filter quality occurred when the packing depth was between 2 and 4 cm. When the AgNPs@TiO2-CS bed filter was installed in the ventilation system of two hospital wards, as much as 88% of bacteria and 97% of fungi was removed within 30 minutes. Consequently, this novel AgNPs@TiO2-CS composite has potential applications in antimicrobial air purification.

Keywords: nanotechnology

TU-PO-07
The road to a non-toxic environment in day care centers

Langer S.[1], de Wit C.[2], Fäldt J.[3]; [1]IVL Swedish Environmental Research Institute, Göteborg, Sweden, [2]Stockholm University, Dept. of Environmental Science and Analytical Chemistry, Stockholm, Sweden, [3]Environment and Health Administration, City of Stockholm, Sweden

Hazardous substances in materials and products and their emissions to air and dust in the indoor environment can lead to human exposure which in turn can contribute to negative health effects. Children are exposed more than adults. Most pre-school age children spend much of their time in day care centers. It is therefore of particular importance to minimize children's exposure by minimizing the load of the hazardous chemicals in the indoor environment of day care centers. One way to affect the occurrence and levels of the hazardous substances is to take measures to eliminate their presence in indoor building materials. In this study, a day care center in Stockholm was evaluated with respect to the levels of hazardous substances before, one month after and one year after renovation that was carried out in accordance with the property owner's environmental management routines, which include chemical requirements for construction products. We report here the presence of volatile organic compounds (VOC) in air as well as the occurrence of semivolatile organic compounds (SVOC) in air and dust samples. The relevant SVOCs were banned and alternative phthalates, alternative plasticizers and organophosphate esters. We will present the levels found in our study, compare these with relevant international studies and evaluate the effect of the renovation directly after and one year after renovation.

Keywords: SVOCs, VOCs, children, flame retardants, phthalates

TU-PO-08
Hepatic lipidosis after inhalation exposure to 2-ethyl-1-hexanol in mice

Wakayama T.[1], Ito Y.[1], Miyake M.[1], Sakai K.[1], Oya N.[1], Sato H., Ohno H.[2], Kamijima M.[1]; [1]Department of Occupational and Environmental Health, Nagoya City University Graduate School of Medical Sciences, Nagoya, Japan, [2]Department of Environmental Health, Nagoya City Public Health Research Institute, Nagoya, Japan

Introduction: In recent years, 2-ethyl-1-hexanol (2EH) has been postulated to cause "sick building syndrome". This indoor air pollutant is emitted by hydrolysis of plasticizers in polyvinyl chloride flooring materials. 2EH is a known metabolite of the plasticizer di(2-ethylhexyl) phthalate (DEHP), a hepatic carcinogen in rodents. Orally ingested 2EH was reported to increase relative liver weights and to affect hepatic lipid metabolism in rodents. Since effects of inhaled 2EH on the liver remain unknown, this study aimed to reveal the hepatic histopathological and biochemical changes after 2EH inhalation exposure. Methods: This study was carried out in accordance with the Guide for Animal Experimentation of Nagoya City University. After 1-week acclimation, male 10-week-old ICR mice were exposed to 0 (fresh air), 20, 60, or 150 ppm 2EH for 8 hours per day, 5 days per week for 3 months using a whole-body inhalation exposure chamber system. At the end of exposure, mice in each group were divided into two groups. The half of the mice were perfused and fixed with 4% paraformaldehyde and the liver was stained with hematoxylin and eosin (HE) or oil red O. In the remaining mice, lipids level in the liver and plasma were measured. Results and discussion: HE staining showed dose-dependent vacuolization, but no staining was observed with Oil Red O. Inhalation exposure to 2EH significantly increased the levels of hepatic triglyceride and phospholipid, while it did not affect cholesterol levels. Microsomal triglyceride transfer protein was increased dose-dependently, but no significant differences in the plasma lipid levels including triglyceride were observed. Peroxisome proliferator activated receptor (PPAR) α related proteins were not changed by 2EH exposure, although 2-ethyl hexanoic acid, one of 2EH metabolites, is known to act as a weak agonist of PPARα. Conclusion: 2EH inhalation exposure in mice for 3 months increases hepatic phospholipid levels.

Keywords: VOCs, air, built/indoor environment, health, other/general

TU-PO-09
Water intake patterns and sociodemographic determinants of Chinese children: Results from the First National Population-based Survey

Guo Q.[1], Wang B.[2], Cao S.[1], Jia C.[3], Zhao L.[4], Zhang Q.[4], Yu D.[4], Zhao X.[4], Citrine A.E-M.[1], Ma G.[5], Duan X.[1]; [1]School of Energy and Environmental Engineering, University of Science and Technology, Beijing, China, [2]State Key Laboratory of Environmental Criteria and Risk Assessment, Chinese Research Academy of Environmental Sciences, Beijing, China, [3]School of Public Health, University of Memphis, TN, USA, [4]National Institute for Nutrition and Health, Chinese Center for Disease Control and Prevention, Beijing, China, [5]Department of Nutrition and Food Hygiene, School of Public Health, Peking University, Beijing, China

Accurate data on water and beverage intakes are essential for assessing hydration adequacy and setting proper guidelines. The objective of this study is to identify the patterns and sociodemographic determinants of water intake and to assess the intake adequacy for children in China. The study team recruited 41,439 children aged 6-17 years using a multi-stage cluster random sampling method. Daily water and beverage intakes were investigated with standard questionnaires and measuring containers in face-to-face interviews. Each participant was assigned an adjustment weight to obtain a nationally representative sample. Sociodemographic factors influencing water intake were identified using multi-variable regressions. Water intake adequacy was evaluated by comparing with the recommended water intake (RWI). The mean ± standard deviation of total water intake (TWI) was 1,603±731 mL/d for boys and 1,487±661 mL/d for girls. Plain water, food moisture, and other beverages contributed 51%, 20%, and 29% of the TWI. Multivariable analyses showed that TWI of children increased with age, in urban areas and day schools, and with parents' economic and educational levels. The majority (82%) of children had TWI not meeting the corresponding RWI, and the percentage increased with age except for 14-17y boys. The study indicateed that plain water is still the major source of daily water intake by children in China. Unfortunately, the majority of children do not have sufficient water intake, which warrants future actions and guidelines targeting adequate hydration.

Keywords: activity patterns

TU-PO-10
Novel air filtration materials via combined melt-solution electrospinning

Buivydiene D[1]., Krugly E. [1], Todea A. [1], Asbach C. [1], Kliucininkas L. [1], Martuzevicius D.; [1]Kaunas University of Technology, Kaunas, Lithuania

Hierarchically structured nanofibers are gaining attention in the fields of energy and the environment, particularly for air filter and fluid membrane applications. Non-woven fibrous media have been used extensively in air filtration applications and nanofibrous air filters have previously shown grate results for high efficiency air filtration. However, the durability is still questionable because nanofibers form high packing density filtersfeaturing high pressure drop even with the presence of slip flow phenomena. Mixing micro and nano sized structures has been proven to increase the performance of the air filter by allowing a higher surface area to trap more particles and introducing the presence of slip flow at the same time. Synthesis of dual-sized fibrous material using the method of combined melt-solution electrospinning is reported here for the first time. The formation of well-mixed nano- and submicron-size Polyamide (PA 12 and PA 6/6) composites was accomplished simultaniously utilizing melt and sollution electrospinning via custom made electrospinning apparatus. By manipulating the process conditions and the ratios between melt and solution electrospinning, different structures of polimeric mats were achieved. Air filtration parameters were tested and with the particle size of 300 nm filtration efficiency reached up to 89.9 % and pressure drop was 55 Pa. In comparison, uniform nano-fibrous mat formed using only solution electrospinning had higher filtration efficiency of 99.5 % but presure drop was almost 3 times higher at 161 Pa.

Keywords: air, air cleaning, nanoparticles, nanotechnology, particles

TU-PO-11
Eco-Efficiency of Air Cleaning Technologies

Stasiulaitiene I.[1], Ciuzas D.[1], Tichonovas M.[1], Krugly E.[1], Martuzevicius D.[1]; [1]Kaunas University of Technology, Department of Environmental Technology, Kaunas, Lithuania

Occupants' comfort and health is negatively affected by poor indoor air quality. Achieving good indoor air quality is a difficult engineering task due to numerous pollutants and pollutions sources. Air cleaning technologies serve as a primary technological option capable in removing pollutants from both indoor and ventilation (outdoor air).

While air cleaning removes air pollutants, it also increases the energy consumption and operating costs of a building. From the perspective of sustainable building concept, air cleaners should also follow the principles of eco-design and resource efficiency. These aspects should be evaluated by eco-efficiency method, combining both environmental burden and economic benefits through the whole air cleaner lifecycle. This research aimed in the eco-efficiency assessment of a complex air cleaning technology (combination of low temperature plasma, photocatalysis, electrostatic precipitation, and catalytic filtration), and compared against a filtration/adsorption system. Eco-efficiency was performed according to the standardized rules set in ISO 14045:2012. The system boundary was set to the processes of the chosen air filtration technologies. Functional unit was set to the 876 Km3 of treated air, which represents a half year operation at an hourly flowrate of 200 m3/h of an average single family house air handling unit. The results obtained revealed complex air cleaning system may be advantageous with respect to material consumption, but emerges as less friendly from the energy consumption perspective. The air cleaning must be including to ecoefficiency calculations of the entire ventilation system of low energy buildings.

Keywords: VOCs, air, air cleaning, environmental policy, lifecycle analysis, sustainability

TU-PO-12
Engineered wood residues as a boiler fuel: potential impacts on air quality

Kaunelienė V.[1], Buinevičius K.[2], Jančiauskas A.[2], Arūnienė K.[3], Martuzevičius D.[1]; [1]Kaunas University of Technology, Department of Environmental Technology, Kaunas, Lithuania, [2]Depatment of Thermal and Nuclear Energy, Kaunas University of Technology, Kaunas, Lithuania, [3]Institute of Environmental Engineering, Kaunas University of Technology, Kaunas, Lithuania

Solid fuel burning during heating season is among major pollution sources in northern and central Europe significantly affecting outdoor air quality and subsequently impacting indoor air. Biomass serves a primary solid fuel in Lithuania, which is comparatively friendly fuel considering its lifecycle. At the same time, furniture industries produce residue containing biomass bound with synthetic resins. Fuels derived from such residues are increasingly used in both small scale and industrial scale boilers. We have simulated the incineration of several types of engineered wood in domestic and industrial boilers under controlled combustion conditions. Wood pellets and wood chips were incinerated as a reference. The real time concentration of the main combustion parameters – NO/NO2, CO, CO2, SO2, H2O, O2 – as well as specific organic species in the flue gas was measured by FTIR spectrometer DX4000 (Gasmet). Size resolved particulate matter concentrations were investigated by the real time measurements using the electrical low pressure impactor ELPI+ and PM10 impactor (Dekati). The incineration of most engineered wood types under optimal combustion conditions in the industrial boiler resulted in similar emission levels compared to emission from wood chips, however some types of engineered wood produced significantly higher levels of nitrogen oxides, while under conditions of incomplete combustion in several cases unacceptably high levels of carbon monoxide were observed. Incineration of engineered wood in the domestic boiler resulted in significantly higher emissions, especially of carbon monoxide, unsaturated hydrocarbons, and total hydrocarbons illustrating that incomplete combustion prevailed. Thus the incineration of engineered wood residues for energy production is acceptable only in industrial boilers under well controlled conditions, while incineration of engineered wood residues in domestic boilers is strictly not recommended in order to minimize the probability of the exposure of population to the harmful products of incomplete combustion.

Keywords: source control

TU-PO-13
The Finnish way to minimize emissions and odours from the ventilation

Kukkonen E; Helsinski, Finland

Different contaminants from the ventilation systems have been one of the common reasons for bad indoor air. The ventilation system will not operate as hoped. In Finland FISIAQ and staff from different universities together with the people in practice developed a new system to avoid these problems. The system with higher requirements for the systems in practice has been in use already years and now a new version has been introduced year 2018. The cleanliness of the ventilation system has been warranted by 3 actions: 1. The ducts and other parts must be fabricated so, that the emissions and odors are minimized. 2. All parts of the ventilation system must be stored and protected in a way that the cleanliness is permanent also on site. 3. The cleanliness of the completed ventilation system must be checked, measuring the dust on the surfaces or visual control methods. The cleanliness of the components will be verified by The Cleanliness Classification of Air-handling Components M 1, which is an essential part of Classification of Indoor Environment, which has been developed by FISIAQ for with better indoor conditions. To achieve the highest class, S 1, of indoor environment also the used ventilation system must be especially clean and no source of impurities. More than 400 different components have already got the classification to the best class M 1. The other actions have also been effective. To check the cleanliness of the ventilation system in practice, on site, also new methods were developed to measure the dust and other impurities of the ducts and components. Also visual control methods had been developed. Requirements were presented also to the clean and moisture proof stocking of prefabricated parts and ducts on site.

Keywords:

TU-PO-14
Exposure to greenspace, indoor PM2.5 and noise: A case study of four European cities

Mueller W.[1], Steinle S.[1], Pärkkä J.[2], Parmes E.[2], Liedes H.[2], Kuijpers E.[3], Sarigiannis D.[4], Chapizanis D.[4], Maggos T.[5], Stamatelopoulou M.[5], Wilkinson P.[6], Milner J.[6], Vardoulakis S.[1], Loh M.[1]; [1]Institute of Occupational Medicine, Edinburgh, UK, [2] VTT Technical Research Centre of Finland, Espoo, Finland, [3] TNO, The Hague, Netherlands, [4] Aristotle University of Thessaloniki, Greece, [5]National Centre for Scientific Research 'Demokritos', Athens, Greece, [6]London School of Hygiene & Tropical Medicine, UK

Background: Exposure to urban greenspace may affect health via a complex set of pathways, including reducing levels of particulate matter (PM) and noise. The HEALS study piloted an extensive exposure protocol in four European cities (Edinburgh, UK; Utrecht, Netherlands; Athens and Thessaloniki, Greece), aimed at direct gathering of exposure data, and indirect modelling based on collected exposure factor data fused with publically available environmental data. Data collected include monitoring of participant location, time activity, and indoor home environment. Method: We include three metrics of greenspace (Normalised Difference Vegetation Index [NDVI], tree canopy density, and proximity to green land use) at 50 m residential buffers and track objective usage of such spaces through GPS data. NDVI values are generated for summer and during HEALS monitoring periods. We use personal GPS data collected from the MOVES mobile phone application and Fitibit devices to assess active, e.g., walking, and passive, e.g., driving, transport. Indoor PM2.5 and noise levels. In addition, subjective noise annoyance data were collected from households. Available ground-level air quality monitoring and modelled data were combined to assign outdoor PM2.5 levels. Random-effects regression models were used for indoor PM2.5 and noise, and a logistic regression model was developed for noise annoyance. Results: Initial results suggest a 0.9 μg/m3 (95% CI 1.8 to 0.6 μg/m3) ($p = 0.035$) decrease in indoor PM2.5 concentrations per 0.1-unit increase in surrounding NDVI (summer) levels after adjustment for outdoor PM2.5, city, and season. Although this was the only greenspace metric with a statistically significant association, all greenspace coefficients were in the negative direction. Conclusions: Our analysis suggests an inverse association between specific greenspace exposure metrics and indoor PM2.5 levels, with no apparent effects on objective or subjective noise indicators. We will test the robustness of these findings, including the use of different residential buffer sizes.

Keywords:

TU-PO-15

Contributions of exposure-related behaviors and dietary factors to organophosphate pesticide exposure in Japanese toddlers

Oya N.[1], Ito Y.[1], Ebara T.[1], Kato S.[1,2], Hioki K.[1], Aoi A.[3], Ueyama J.[3], Oguri T.[1,4], Shoji N.[1,5], Sugiura-Ogasawara M.[6], Saitoh S.[7], Kamijima M.[1]; [1]Department of Occupational and Environmental Health, Nagoya City University Graduate School of Medical Sciences, Nagoya, Japan, [2]Department of Pediatrics and Neonatology, Nagoya City University Graduate School of Medical Sciences, Nagoya, Japan, [3]Department of Pathophysiological Laboratory Sciences, Field of Radiological and Medical Laboratory Sciences, Nagoya University Graduate School of Medicine, Nagoya, Japan [4]National Institute of Advanced Industrial Science and Technology (AIST), [5] Asahi University, School of Health Sciences, Mizuho, Gifu Prefecture, Japan, [6]Department of Obstetrics and Gynecology, Nagoya City University Graduate School of Medical Sciences, Nagoya, Japan, [7]Department of Pediatrics and Neonatology, Nagoya City University Graduate School of Medical Sciences, Nagoya, Japan

Several epidemiological studies suggested that exposure to organophosphate (OP) pesticides could elicit adverse health effects on children. The objectives in this study were to reveal the OP exposure levels in Japanese toddlers and to explore the possible contributions of their exposure-related behaviors and their mothers' considerations of food selection and preparation to the exposure levels, considering one-day intake of foodstuffs. We recruited children participating in Japan Environment and Children's Study at Aichi regional center, and collected used disposable diapers from 1037 children (16–23 months old, 531 boys and 506 girls) from June 2015 to August 2016. Six urinary dialkylphosphates (DAPs) were measured using urine extracted from diapers by ultra-performance liquid chromatography with tandem mass spectrometry. The information about exposure-related behaviors, considerations of food selection and preparation, and foodstuffs were obtained from questionnaires. The geometric mean of ΣDAP (total DAPs) concentrations and amounts excreted onto diaper during a night were 120 nmol/L and 13.4 nmol. In binary logistic regression analyses, use of an insecticide, an herbicide, an insect repellent spray, and a fragrance or a deodorant increased odds ratios (ORs) of the exposure levels (top 5% of urinary DAP concentrations or amounts) (ORs: 2.1–2.6). Receiver-operating characteristics curve analysis revealed that discriminatory power to determine whether or not ΣDAP concentrations and the amounts exceeded their 95th percentile values was lower for the respective propensity scores for the exposure-related behaviors (area under the curve 0.72 and 0.74) and the considerations of food selection and preparation (0.60 and 0.67) than that for the foodstuffs taken on the survey day (0.89 and 0.91). We revealed the exposure levels of OP pesticides in Japanese toddlers. The contributions of the exposure-related behaviors and the considerations of food selection and preparation to the exposure levels were likely much smaller than that of foodstuffs.

Keywords: biomonitoring, children, epidemiology, exposure factors, pesticides

TU-PO-16

Estimated Daily Dose and Cumulative Risk of Phthalates Exposure in the General Children and Adolescents: Results from Taiwan Environmental Survey for Toxicants (TEST) 2013-16

Huang P.C.[1], Liao K.W.[1], Lin C.C.[2], Tsai M.S.[2], Chen P.C.[2]; [1]National Institute of Environmental Health Sciences, National Health Research Institutes, Miaoli, Taiwan, [2]Institute of Environmental and Occupational Health Sciences, National Taiwan University College of Public Health, Taipei, Taiwan

Introduction. In 2011, some Taiwanese children exposed to phthalate-tainted products due to a phthalate contamination episode. However, no information is available to evaluate the exposure doses and risk of phthalates in Taiwanese children and adolescent after the episode. We aim to investigate the daily exposure dose and cumulative risk of phthalates in Taiwanese youth using a biomonitoring approach. Materials and Method. We conducted a national-wide human biomonitoring study, the Taiwan Environmental Survey for Toxicants (TESTs), to systematically assess the exposure doses and cumulative risk of phthalates in Taiwanese children and adolescents. A total of around 300 participants (7-18-year-old) were recruited during 2013 to 2016. First morning urine sample was collected and analyzed for 11 phthalate metabolites using LC-MS/MS. We estimated the daily intake doses of 6 phthalates and the cumulative risk for reproductive and hepatic hazard in our participants. Results and Discussion. Median and range of estimated daily dose of DiBP, DnBP and DEHP in children were 0.75 (<0.01~33.6), 0.83 (<0.01~29.5) and 3.2 (0.73-23.6) ?g/kg/day, respectively. Median levels of 7 urinary phthalate metabolites (included DBP, DiBP and DEHP metabolites) in children were significantly higher than those in adolescent, whereas no significant difference was found between boys and girls. We observed that median levels of phthalate metabolites in children and adolescent in Taiwan were around 1.5-4 fold higher than those from NAHANES 2011-12, except for MEP and MBzP. About 1.45% of our participant had cumulative HI above 1 for reproductive hazard. Our data indicated that part of Taiwanese children were still exposed to high level of DnBP, DiBP and DEHP.

Keywords: biomonitoring, children, cumulative exposure, phthalates, risk assessment

TU-PO-17

Comparison between methods for analysis of urinary metabolites of insecticides using gas chromatography-mass spectrometry and high performance liquid chromatography-mass spectrometry

Simaremare S.R., Hung C-C.[1], Hsieh C-J., Yiin L-M.; [1]Department of Medical Science,Tzu Chi University, Hualien, Taiwan (ROC)

Urinary metabolites are commonly used for biomonitoring of exposure to insecticides. Analysis using high performance liquid chromatography-mass spectrometry (LCMS) has been popularly applied and considered effective with excellent sensitivities, albeit costly in analysis and maintenance. We have developed a rapid, sensitive, and cost-effective extraction assay method to analyze urinary metabolites of insecticides using gas chromatography-mass spectrometry (GCMS) combined with liquid-liquid extraction (LLE), and successfully confirmed the alleged performance and effectiveness. In this study we analyzed urine samples using the developed GCMS and regular LCMS assay methods, and compared the results to show the excellence of this GCMS method as alternative. The urine samples were prepared by spiking insecticide metabolites to urine that was confirmed to be blank in advance. Thirteen urinary metabolites of the common commercial insecticides, including carbamate, organophosphates and pyrethroids, were selected for urine spiking. A total of 20 pairs of spiked urine samples were prepared with the concentrations being set at 10 μg/L, which were normally close to the median concentrations derived from genuine urine. Urine samples were first processed with acidic hydrolysis for de-conjugation, and then with LLE; the extract-containing organic layers were dried using an auto-programmed concentrator. Residues of the samples were added with acetonitrile to be ready for analysis on LCMS; as for GCMS analysis, derivatization by N-tert-butyldimethylsilyl-N-methyltrifluoroacetamide was conducted prior to the final analysis on instrumentation. Results derived from both methods indicated consistency between the two methods, showing approximately 10% error within the expected concentration (10 μg/L). Comparisons between data from GCMS and LCMS also demonstrated an excellent outcome of relative errors less than 8.6%. In summary, this GCMS method is not only as good as LCMS in performance, but is advantageous in terms of cost effectiveness.

Keywords: biomonitoring, pesticides

TU-PO-18
Comparison between methods for analysis of insecticides in human blood using gas chromatography-mass spectrometry and high performance liquid chromatography-mass spectrometry

Hung C.C. [1], Simaremare S.R., Hsieh C-J., Yiin L-M.; [1]Department of Medical Science Tzu Chi University, Hualien, Taiwan (ROC)

For biomonitoring of exposure to insecticides, blood sample analysis is an effective tool to assess the instant exposure without concern with metabolism in the body. Analysis using high performance liquid chromatography-mass spectrometry (LCMS) has been popularly applied and considered superior to others because of the excellent sensitivity; nevertheless, use of LCMS is never cost-effective. We have developed a rapid, sensitive, and cost-effective whole blood extraction assay method to analyze insecticides in blood using gas chromatography-mass spectrometry (GCMS). In this study we analyzed blood samples using the developed GCMS and regular LCMS assay methods, and compared the results side by side to show the excellence of this GCMS method as alternative. The blood samples were prepared by spiking 9 insecticides, including carbamate, organophosphates and pyrethroids, to blood that was confirmed to be blank in advance. A total of 20 pairs of spiked blood samples were prepared with the concentration being set at 10 μg/L for each compound. Pretreatment of blood samples started with extraction of whole blood and serum for analysis on GCMS and LCMS, respectively; the organic extracts were dried and reconstituted with solvent prior to the respective analyses. Results derived from both methods indicated consistency between the two methods, showing less than 8% error within the expected concentration (10 μg/L). Comparisons between data from GCMS and LCMS also demonstrated an excellent outcome of relative errors less than 8.2%, confirming that this GCMS method was as good as LCMS in performance, and was advantageous in terms of cost effectiveness.

Keywords: biomonitoring, pesticides

TU-PO-19
Imidacloprid Exposure in Grape Workers

Lopez-Galvez N. [1], Wagoner R. [1], Canales R. [1], de Zapien J. [1], Rosales C. [1], Beamer P. [1]; [1]University of Arizona, Tuscon, AZ, USA

Imidacloprid is a neonicotinoid insecticide commonly injected through agricultural drip irrigation systems to reduce the population of vine mealybugs in grape farms. In the present study, we evaluated the exposure to imidacloprid in 20 grape workers two weeks after imidacloprid injection in the irrigation system. We administered a questionnaire on work activities, exposure characteristics, and socio-demographics. We collected personal air and hand wipe samples to determine possible inhalation and dermal exposure. We collected urine samples to quantify metabolite concentrations and we measured hand surface temperatures. Most participants migrated from the state of Chiapas, Mexico, all participants had less than high school education and about half identified an indigenous language as their first language. Imidacloprid was not detected in air samples, but it was detected in 85% of the hand wipe samples (GM: 0.41 μg/wipe). The majority of participants (75%) had detectable concentrations of imidacloprid as a parent compound in their urine (GM: 0.15 μg/g), and almost all of the participants (95%) had detectable concentrations of 5-hydroxy-Imidacloprid, a urine metabolite, (GM: 1.79 μg/g). There was a significant correlation ($p < 0.001$) of the imidacloprid concentration found in hand wipes with the concentrations found in urine (rho: 0.67, and 0.79 for imidacloprid and 5-hydroxy-Imidacloprid). Hands' temperature was significantly correlated ($p < 0.05$) with the pesticide concentration found on hands (rho: 0.70), and the urine biomarkers (rho: 0.68, and 0.60 for imidacloprid and 5-hydroxy-Imidacloprid). No levels of exposure via the inhalation route and higher levels via dermal route were found. Our results suggested that hand wipe can be use as less intrusive measurement to assess imidacloprid exposure in farm workers. Imidacloprid urine and hand wipes concentration were significantly higher during the summer than during the spring season. Further research needs to be conducted on microclimatic conditions to evaluate and reduce dermal pesticide exposure.

Keywords: biomarkers, biomonitoring, cumulative exposure, environmental health, exposure factors, occupational

TU-PO-20
Occupational exposures to phthalates among U.S. hairdressers primarily serving an ethnically diverse clientele.

Quiros-Alcala L.[1], Boyle M.[2], Zhu L.[2], Sapkota A.[2], Rule A.M.[1], Kavi L .[2], Pool W.[3], Thomas S.B.[4]; [1]Department of Environmental Health and Engineering, Johns Hopkins Bloomberg School of Public Health, Baltimore, MD, USA, [2]Maryland Institute for Applied Environmental Health, University of Maryland School of Public Health, College Park, MD, USA, [3]Centro de Apoyo Familiar, Riverdale, MD, USA, [4]Maryland Center for Health Equity, University of Maryland School of Public Health, College Park, MD, USA

Background: Phthalates are endocrine disrupting compounds present in many personal care products (PCPs), including hair and skin care products, which have been linked to adverse health effects. Hairdressers may experience elevated exposures to phthalates from routine use of PCPs on themselves and their clients. Still, no studies have assessed phthalate exposures among hairdressers via biomonitoring. Additionally, hairdressers primarily serving clients of African and Latino descent may experience elevated exposures due to the unique products used. Objective: To assess exposures to phthalates among female hairdressers primarily serving clients of African and Latino descent. Methods: We recruited 20 female hairdressers from salons in Maryland and quantified urinary concentrations of 9 phthalate biomarkers [Mono-ethyl phthalate, MEP; Mono (3-carboxypropyl) phthalate, MCPP; Mono(2-ethylhexyl) phthalate; MEHP; Mono(2-ethyl-5-hydroxyhexyl) phthalate, MEHHP; Mono(2-ethyl-5-oxohexyl) phthalate, MEOHP; Mono(2-ethyl-5-carboxypentyl) phthalate, MECPP); Mono-n-butyl phthalate, MBP; Mono-isobutyl phthalate, MiBP; and Monobenzyl phthalate, MBzP)] using liquid chromatography tandem mass spectrometry (LC-MS/MS) based methods. Results: Phthalate metabolites were detected in 77-100% of urine samples. Median (p25, p75) concentrations (ng/mL) were: MEP: 300.8 (79.2, 339.4); MBP: 14.8 (2.8, 25.8), MiBP: 8.8 (1.6, 11.2); MEHP: 4.0 (2.3, 5.3); MEHHP: 6.6 (3.1, 13.2); MEOHP: 4.6 (1.7, 7.5); MECPP: 5.3 (4.0, 10.4); MBzP: 1.2 (0.4, 2.6); and MCPP: 0.9 (0.2, 1.5). Compared to women from the U.S. general population, median concentrations among hairdressers were 1.2-7.3 times higher for several phthalate metabolites (MEP, MBP, MEHP, MEHHP, MEOHP). Notably, median concentrations for MEP, a biomarker for Diethyl-phthalate, which is found in many hair and skin care products, was 7.3 times higher in hairdressers compared to women from the U.S. general population. Conclusions: Hairdressers servicing an ethnically diverse clientele may experience elevated phthalate exposures. Larger studies are needed to characterize phthalate exposures among hairdressers and develop exposure mitigation efforts to reduce the risk of potential long-term impacts.

Keywords: biomonitoring, consumer and personal care products, industrial hygiene, phthalates, workplace

TU-PO-21
Insecticide use on pets is a strong predictor of pyrethroid exposure in an urban population of young adult men in Poland

Rodzaj W. [1], Wileńska M. [1], Dziewirska E. [2], Klimowska A. [1], Jurewicz J. [2], Radwan M. [3], Radwan P. [4], Hanke W. [2], Wielgomas B. [1]; [1]Department of Toxicology, Faculty of Pharmacy, Medical University of Gdańsk, Gdańsk, Poland, [2]Department of Environmental Epidemiology, Nofer Institute of Occupational Medicine, Łódź, Poland, [3]Faculty of Health Sciences, State University of Applied Sciences, Plock, Poland, [4]Department of Gynecology and Reproduction, "Gameta" Hospital, Rzgów, Poland

Pyrethroids are synthetic insecticides widely used in agriculture, households and veterinary medicine to kill and control pests. It is well documented in the literature that exposure to these pesticides may pose a threat to human health. A deleterious effect on male fertility is one of concerning issues. The aim of the study was to determine levels of exposure to pyrethroids through measurements of five pyrethroid metabolites in urine: 3-phenoxybenzoic acid (3PBA), cis-2,2-dimethyl-3-(2-chloro-3,3,3-trifluoro-1-propenyl)-cyclopropanecarboxylic acid (BIF), cis/trans-3-(2,2-dichlorovinyl)-2,2-dimethylcyclopropane-1-carboxylic acids (cis/trans-DCCA), and cis-(2,2-dibromovinyl)-2,2-dimethylcyclopropane-1-carboxylic acid (DBCA) in the population of young men living in urban area of Poland (n=242) and to investigate the relationship between dietary, lifestyle and sociodemographic factors and urinary levels of these biomarkers in enrolled group. The study cohort was recruited in Lodz, central Poland, and in nearby areas. Along with spot urine sample, the participants provided a completed questionnaire regarding selected exposure factors. Concentrations of pyrethroid metabolites in urine were measured using a validated GC-MS method and adjusted for specific gravity (SG). The limit of detection (LOD) for all analytes was 0.1 ng/mL. BIF, cis-DCCA and DBCA were detected in less than 50% of the samples, whereas 3PBA and trans-DCCA in 67% and 78%, respectively. The geometric mean (GM) concentrations (SG-adjusted) for 3PBA and trans-DCCA were 0.229 and 0.274, respectively. These two metabolites were also positively correlated ($r = 0.4236$, $p < 0.001$), but the strongest correlation was noticed between cis- and trans-DCCA ($r = 0.7828$, $p < 0.001$). Although analysis of dietary habits, sampling time and sociodemographic data failed to identify significant predictors of pyrethroid exposure, we found that insecticide use on pets was the strongest predictor ($p < 0.01$) of 3PBA concentration. In conclusion, a magnitude of pyrethroid exposure in studied population was assessed and factors contributing to this phenomenon were identified but require further investigation.

Keywords: biomarkers, biomonitoring, pesticides

TU-PO-22
GIS analysis on relationship between traffic density parameters and near-road air pollutants

Xue J.[1], Liu S.[1], Chen F.[1]; [1]United States Environmental Protection Agency, Research Triangle Park, NC, USA

Traffic-related air pollution has become an important issue in human health as more people moving to near-road areas and more vehicles passing through various road ways. Chronic and cumulative exposure to near-road air pollutants has become a focus of human health risk studies. Because vehicle release is an important source of near-road air pollutants we focus on traffic density - a metric integrating major roads and traffic count –as a potential index for near-road air pollution. Large air quality survey (AQS) data on Ozone (O3), carbon monoxide (CO), nitrogen oxides (NOx), Sulfur dioxide (SO2), Lead (Pb), and particular matter (PM2.5 and PM10) were retrieved from US Environmental Protection Agency (EPA)'s Aerometric Information Retrieval System (AIRS). Their daily-averaged concentrations were used to quantify the correlation with Major-Road Density (MRD) and All-Traffic Density (ATD). CO and NOx were found as air pollutants with larger correlation coefficients in the range of 0.22 to 0.47, while O3, SO2, Pb, and PMs showed correlation coefficients less than 0.19. CO and NOx's correlation with traffic density index was higher in the peak traffic time than the low traffic time. Mean correlation coefficients between NOx and MRD fall in the range of 0.36 and 0.50 within a distance of 50- to 300-meters off the road. Mean correlation coefficients between NOx and ATD fall in the range of 0.16 and 0.41 within a distance of 50- to 300-meters off the road. These correlations are similar between spring and fall and remain constant between 2011 and 2012. Our study shows that near-road air pollutants contributed primarily by vehicle release are closely correlated with major road density and traffic density. Thus, these traffic density metrics may be useful indicators of near-road vehicle-released air pollutants for cumulative health risk studies. Such indicators can be used to inform community-based transportation decisions.

Keywords: air, exposure models

TU-PO-23
Factors affecting the indoor concentration of nitrogen dioxide in school and office environments

Salthammer T.[1], Salonen H.[2], Morawska L.[3]; [1]Fraunhofer WKI, Braunschweig, Germany, [2]Aalto University, Aalto, Finland, [3]Queensland University of Technology (QUT), Brisbane, Australia

Potential health hazards caused by nitrogen oxides (NOx), especially nitrogen dioxide (NO2), are being intensively discussed, particularly as regards the release of NOx from diesel engines. Moreover, several epidemiological studies conducted in residential environments have examined the link between NO2 exposure and respiratory diseases. The World Health Organization (WHO) has published guidelines for indoor areas of 40 $\mu g/m^3$ (annual) and 200 $\mu g/m^3$ (1 h). The new NO2 guideline values derived for indoor areas in Germany are 0.25 mg/m^3 (1h, precautionary) and 0.08 mg/m^3 (1 h, health hazard). However, NO2 pollution of the indoor and outdoor atmosphere can only be considered and evaluated under regional aspects. As NO2 is a traffic- and industry-related pollutant, emissions are generally higher in urban than in rural areas and generally higher in outdoor than indoor if there are no specific indoor sources available. In this case I/O ratios strongly depend on the air exchange rate and on the NO2 removal efficiency. Various building and indoor environment characteristics, such as type of ventilation, air-tightness of the envelope, furnishing and surface characteristics of building, location of the building (urban versus suburban and proximity to traffic routes), as well as occupants behavior such as window opening, have been significantly associated with indoor NO2 levels. NO2 is effectively degraded on surfaces. Taking into account the materials used in indoor areas, degradation rates between 1.04 1/h and 1.45 1/h are common. Typical primary sources of indoor NO2 in residential buildings, are gas-fueled cooking, heating appliances, candles, and ethanol fireplaces. Our study focuses on the situation in schools and offices, where the information about global and local exposures to NO2 is limited. Data were extracted from the available literature and evaluated by means of existing criteria.

Keywords: air, environmental regulation, other (specify), risk assessment, ventilation

TU-PO-24
A Pilot Study Evaluating Indoor Environment Quality in Energy Efficient Irish Dwellings: The VALIDate Project

McGrath J.[1], Connolly A.[1], Byrne M.[1]; [1]National University of Ireland, Galway, Ireland

Previous studies evaluating energy-efficient dwellings have highlighted the practical challenges of maintaining good indoor environmental quality. Measures that increase air tightness have been shown to have both positive and negative influences on thermal comfort, and thus on health and well-being. Currently, there is a knowledge gap regarding the effectiveness of ventilation required to maintain a healthy and comfortable environment within energy efficient residential dwellings. Recent advancements in the development of low-cost sensors offer the capability for remotely transmitting information means that there is now a unique opportunity to gather large-scale monitoring data without the traditional constrictions. The current longitudinal study uses low-cost sensors to continuously monitor the thermal comfort and indoor environmental quality in energy efficient Irish residential dwellings. Sensors are placed in four rooms per dwelling; two habitable rooms (living room and bedroom) and two wet rooms (kitchen and bathroom) due to varying ventilation characteristics. Environmental (temperature, humidity) and air quality data (CO2, radon and VOCs) is remotely collected. The advantages of continuous monitoring in four separate rooms within a dwelling identifies temporal and spatial variations in indoor environmental quality accounting for factors including occupants' activities, housing characteristics, seasonal variations, and the effectiveness and durability of the ventilation system. In addition, continuous monitoring reduces the uncertainties associated with short-term monitoring, and reliable air pollution exposure estimates can be made. Detailed contextual information is also collected to accompany sensor data, which documents the presence of ventilation and extraction systems and their accompanying switches. The initial data from the first ten homes monitored will be presented at the conference. These results will include the thermal comfort survey data and the analysis of the indoor environmental quality, including pollutant concentration data. This will provide an accurate representation of the effectiveness of whole-house ventilation as opposed to snapshots of individual rooms and periods.

Keywords: air, built/indoor environment, radiation, thermal environment, ventilation

TU-PO-25

Improved methods of assessing the impact of air pollution on public health.

Morhulova V. [1]*, Petrosian A.* [1]*, Maremukha T.* [1]*;* [1]*State Institution – "O.M. Marzeiev Institute for Public Health of the National Academy of Medical Science of Ukraine", Kyiv, Ukraine*

Air pollution is a major threat to health, resulting in seven million premature deaths around the world according to WHO in 2018. The quantitative determination of the air pollutants dose requires the use of regression analysis based on remote sensing data, digital relief models, meteorological data, and infrastructure of study area data. Aim. Provide a comparative analysis of the tools for assessing the air quality to scientific justify using of tools for different scenarios of distribution and impact of pollution, development of the foundations for the introduction of new tools for assessing the impact of air pollution on public health. Methods. Mathematical modeling methods (TM5-FASST, ISC-AERMOD View average concentrations calculation of air pollutants and Transboundary transfers); statistical methods; mapping (ArcGis 10.2, ArcMap Esri); dispersion and regression analysis; assessing the impact of air pollution on public health (AirQ+ and Air Quality Index). Results. Creation algorithms of work with software complexes of air pollution modeling and impact on the public health. Using the prognostic models of the influence the pollutants concentrations on the studied areas with consideration their peculiarities. Conclusions. Based on the analysis of field data and modeling will be deepened theoretical understanding of the air pollution influence mechanisms on public health, will be developed the scientific basis for the use of advanced technologies in air quality management, will be researched the influence of territorial features of air pollution on public health.

Keywords: air, environmental health, exposure models, geospatial analysis/GIS

TU-PO-26

Indoor air quality at nail salons in Taipei

Tsai S.W. [1]*, Lin Y.R.* [1]*, Lee, Y.C.* [1]*, Liu Y.C.* [1]*;* [1]*National Taiwan University, Taipei, Taiwan*

In nail salons, a wide variety of nail products are constantly used. Hence, different airborne hazardous chemicals may be released during the process of nail treatment. The occupation as a nail salon worker has been associated with various health effects. To assess the indoor air quality in nail salons as well as the associated health risks for nail salon workers in Taipei, the walk-through surveys were performed in nail salons. The airborne concentrations of volatile organic compounds, formaldehyde, and phthalates were also determined. Including charcoal, XAD-2, and OVS tubes were used for the air sampling, and the flow rates were about 50ml/min, 100ml/min, and 1000ml/min, respectively. After 5-hour sampling, the samplers were desorbed by carbon disulfide or toluene, followed by the analysis with GC/MS. In addition, salon workers were asked to complete a questionnaire to collect the information regarding work conditions and related symptoms. It was found that most of the nail salons were in enclosed space, and the concentrations of CO_2 were relatively high. Besides, when certain products were applied, the concentration of TVOCs rose instantaneously. In the part of active air sampling, 11 hazardous chemicals were detected in different salons. Among them, formaldehyde was detected in all nail salons, and the concentrations were even 2.7-4.2 times higher than PEL at some salons. The findings indicate that the possible exposures of formaldehyde for workers in nail salons deserve further attentions. To lower the levels of exposures, it is suggested that the nail salons should improve the ventilation efficiency and increase the air exchange rates. In addition, wearing proper personal protective equipment is also important to lower the possible health risk. Besides, using products which contain formaldehyde or formaldehyde releasers should be avoided. In addition, it is suggested that the concentrations of airborne hazardous chemicals should be monitored periodically.

Keywords: VOCs, air, built/indoor environment, consumer and personal care products, cumulative exposure, environmental health

TU-PO-27
Characteristics of indoor air quality and source Identifications of Pollutants in schools of Taiwan

Lee C.W.[1], Huang Y.C.[1], Lin, G.Y.[1], Hsu P.C.[1], Su H.J.[2], Chien Y.C.[3]; [1]National Kaohsiung University of Science and Technology, Kaohsiung, Taiwan, [2]National Cheng Kung University, Tainan, Taiwan, [3]National Yunlin University of Science and Technology, Yunlin, Taiwan

Inhalation exposure to indoor pollutants in schools is a health concern for children. The main purposes of this study were to measure pollutant levels (volatile organic compounds (VOCs), carbonyls, CO, CO2, O3, PM10 and PM2.5) in indoor/outdoor air at schools and to evaluate the contributions of indoor and outdoor sources to indoor air concentrations using mass balance model and receptor model. In the first stage of the study, fifty schools were selected for short-term air monitoring according to population density and industrial development in the study areas. Results reveal that the indoor air qualities in natural ventilated classrooms were mostly affected by the penetration of outdoor pollutants. Air pollutants such as VOCs and carbonyls emitted from indoor sources would accumulate in the seldom used classrooms. The indoor VOCs levels for schools located near industrial areas were higher than that of the schools located in the densely populated areas. Six schools were selected for more detailed evaluation. The results indicated that the school TVOCs levels in summer and in fall were similar, while PM2.5 levels in fall were usually higher than that in summer. Indoor PM2.5 and O3 were lower than that in outdoors, which suggest the indoor PM2.5 and O3 were mainly contributed by outdoor sources. Higher CO2 levels were found in fall season because the windows were closed due to the high PM2.5 outside. Receptor models were used to identify and quantify the major sources contributing to VOCs and PM2.5 indoors. Results from receptor model revealed that indoor VOCs were mainly contributed by vehicles exhaust and indoor sources, while major indoor PM2.5 sources were crustal materials and vehicles exhaust.

Keywords: VOCs

TU-PO-28
A Study on the Concentration Characteristics of Indoor and Outdoor VOCs in Some Houses, Ulsan

Lim H.B.[2], Park S.H.[1], Yoon D.K.[1], Lee H.W.[1], Hong H.J.[1], Yoon D.K.[1], Lee C.M.[1,2]; [1]Institute of Risk Assessment, Seokyeong University Seoul, South Korea, [2]Department of Chemical & Biological Engineering, Seokyeong University, Seoul, South Korea

This research was performed to provide basic data on health risk assessment of VOCs exposure of residents in Ulsan industrial complex area. VOCs was measured in according to the protocol of the indoor air quality management law setting by the Korea Ministry of Environment(KMOE) and the measuring instrument is located in the living room that represent the IAQ of the house and in the veranda that is affected by the outdoor air. The results of the study showed that the indoor concentrations of VOCs, benzene and Toluene were higher than outdoors and the outdoor concentrations of ethylbenzenes, m,p-Xylene and o-Xylene were higher than indoors. These results show that the sources of Benzene and Toluene are present in indoor and that Ethylbenzene and

Xylene are outdoor. The results of the study showed that indoor and outdoor VOCs concentration didn't exceed indoor air quality maintenance standards by KMOE's law and additional research is needed to improve the limitations of measurements.

Keywords:

TU-PO-29 MOVED TO TH-PL-C3-67

TU-PO-30 MOVED TO TH-PL-B3-59

TU-PO-31 MOVED TO TH-PL-A3-53

TU-PO-32
Inhalation exposure to fine particles and particle size distributions among rural Chinese

Shen G.; Peking University, Peking, China

Air pollution is one leading environmental risk factor in China causing millions of premature deaths every year. It has been documented that adverse health impacts of particulate matter are not solely associated with the mass but factors like particle size and toxic chemical components. In rural China, severe indoor as well as ambient air pollution has been identified due to high dependence on traditional solid fuels including coals and biomass fuels in daily lives particularly for space heating. In this study multiple-site field measurements have been carried to investigate daily inhalation exposure to particles in rural China, by using personal carried samplers. Results showed that daily inhalation exposure to fine PM2.5 varied in nearly two orders of magnitude ranging from 10 to over 1200 μg/m3. The overall median and geometric means were around 130 μg/m3 that was several times of WHO guidelines for indoor air quality. Primarily fuels are one main cause causing seasonal difference, as well as variations among rural residents. The study also evaluated size distributions and presence of submicron particles in inhaled particles, which found similar size distributions among the residents using clean and polluting fuels, but significant differences between winter and summer periods. Results from this study fill in important data gaps in inhalation exposure among rural Chinese, and provided strong supports for clean intervention and risk assessment on air pollution in rural China and controls on traditional solid fuel use.

Keywords: particles

TU-PO-33
Experimental investigation on the relative importance of mucosal exposure and inhalation exposure to airborne particles

Duan M.[1], Liu Li.[1], Da G.[2], Wang Yi.[3], Géhin E.[2]; [1]School of Architecture, Tsinghua University, Beijing, China, [2]Université Paris-Est, CERTES, Créteil, France, [3]State Key Laboratory of Green Building in Western China, Xi'an University of Architecture and Technology, Xi'an, China

Occupational health and safety have long been a global concern, especially for developing countries. Airborne particles are considered as the main threatening factors. The harm of airborne particles come from inhalation exposure and mucosal exposure, separately. Identifying relative importance of these two exposure risks provide possibility to realize better individual protection. There are seldom progress on simultaneous assessment of the particles deposited on facial mucosa and the particles inhaled into the lower airway. We developed an in-vitro respiratory system from a healthy volunteer by CT-imaging and 3D printing, and the replica of facial details was added for this study. Monodispersed florescent particles with the aerodynamic diameters dae of 0.6-5 μm were generated by TSI VOAG 3450® and monitored by TSI APS 3321®. The lower airway exposure was measured as the total accumulative particles that enters the lung volumes during the sampling period. SKC BioSampler® was used for collecting the suspended fluorescent particles in the lung volumes, while preset foils were used to sample the particles deposited on the surface of the lung volumes and face mucosas. Results shown that 0.6-2.0 μm particles that reach the lower airway can be 10 times as many as the total particles deposited on eyes and lips during the same exposure period. 5 μm particles, on the contrary, mainly cause mucosal exposure, 90% of which were deposited on the lips. The results prepare a solid input for further dose-response measurements with in-vitro cells.

Keywords: cumulative exposure, exposure models, occupational, particles, respiratory health

TU-PO-34
Exposure Assessment of Particulate Matters with Smart Spatial Interpolation Based on a Low-Cost Location-Aware Sensing System

Chen P-C.; National Cheng Kung University, Taiwan, China

Increasing studies are finding how the incidences of certain health problems are associated with the exact exposure to the air pollutants of concerns. The majority of the studies target chronical disease for which the long-term exposure has less requirement of precise estimation of exposure in a day or an hour. For the acute diseases, the studies assessing the exposure based on the official air quality monitoring stations might face a common issue when many cases are far away from the closed air stations. The epidemiological analysis based on spatial interpolation could lead to biased or uncertain conclusions. The spatial-temporal variation of the concentrations of air pollutants might be higher in the area of varying weather condition and uneven terrain of the land surface. Our objective is to build a database and model for smart interpolation for exposure assessment. The database encompasses both hourly-based data of 11 air pollutants from 76 air quality stations for 11 years and real-time data PM10 and PM2.5 from more than two thousands of low-cost sensors for two years. The modeling of site-specific exposure uses inversed distance weighted interpolation among the data from a set of representative air quality stations, which are generated through a machine learning on the massive data from the Location-aware Sensing System. We found the developed methods can generate a better exposure database by selecting suitable sites for spatial interpolation smartly with the consideration of changing clusters of air quality zones due to the change of local weather and terrain conditions. The developed database will support further analysis of the influences of air pollutants on related acute health effects.

Keywords: geospatial analysis/GIS, particles, sensor technology

TU-PO-35
Inter and intra-variability of personal exposure to air pollution in different microenvironments

Hossain M.[1], Che W.[1], Lau A.[1]; [1]The Hong Kong University of Science and Technology, Hong Kong, China

Ambient concentrations of air pollution measured at fixed-site monitor are typically used as a surrogate for exposure in estimating health effects in epidemiological studies. However, people spend more than 85 percent of their time in indoor microenvironments which may have different exposure concentrations than ambient levels. To better estimate of health risk, it is important to know the variability of personal exposure to air pollution in different microenvironments. This study was carried out to investigate the inter and intra-variability of personal exposure to air pollution in different microenvironments including home indoor, office indoor, others indoor, outdoor and transportations (e.g., MTR, bus, minibus and car/taxi). Personal measurements with time-location information were performed to determine the concentrations of particles (PM2.5, PM10, PM1) and gaseous pollutants (NO2, O3, CO2, CO) using portable air sampler (PAS). The results revealed the significant differences of the concentrations and exposure among microenvironments visited by a participant due to the variation in local emission sources. The study also demonstrated the substantial variations in daily personal exposure patterns among the participants owing to the difference of the concentrations and time spending by the participants in different microenvironments. Understanding the variability of the personal exposure to air pollution among the microenvironments will help in reducing exposure errors in health estimates and can help to mitigate the air pollution exposure for individuals and at population level. Further study is needed to evaluate the health risk of air pollution considering personal exposure variability among microenvironments.

Keywords: activity patterns, aggregate exposure, air, built/indoor environment

TU-PO-36
STUDY OF AIR POLLUTION BY PARTICULATE MATTER (PM10 AND PM2.5) IN KYIV, UKRAINE

Turos O.[1], Maremukha A., Petrosian A.[1], Brezitska N.[1], Mykhina L.,[1] Ananyeva O.[1]; [1]State Institution "O.M. Marzeiev Institute for Public Health of the National Academy of Medical Science of Ukraine", Kyiv, Ukraine

Background/Aims. According to available estimates, particulate matter is responsible for about 3 % of cardiopulmonary diseases and 5% of lung cancer mortality. The purpose of the work is an assessment of the air pollution level by particulate matter (PM10 and PM2.5) in Kyiv, Ukraine. Methods. Instrumental measurements of PM10 and PM2.5 concentrations in the air were carried out from January to September 2018. The stationary monitoring station equipped with APDA-371 (HORIBA) analyzers, which provide automatic continuous measurement of concentrations with hourly averaging of data. The obtained data was processed using the software package Statistica, results were compared with the limit values recommended by WHO and Directive 2008/50/EC. Results. Analysis of data obtained in 2018 (January - September) allowed to conclude that in the levels of air pollution of PM10 and PM2.5 in Kiev exceed the limits recommended by WHO or Directive 2008/50 / EC: average concentrations of PM10 were 36 ± 0.4 mg / m3, PM2.5 - 21 ± 0.2 mg / m3; the number of days with excess of the average daily concentration was 43 days for PM10 - 19 days (19%), for PM2.5 - 61 days (25%); maximum monthly concentrations of PM2.5 were observed during the heating season - from January to March and were at the level of 26-27 mg / m3. According to the results of the correlation analysis, a significant correlation was found between the concentrations of PM2.5 and PM10 ($r = 0.79$, $p < 0.005$); PM2.5 = 0.66 * PM10. Conclusion. The results indicate the need to establish standards in Ukraine for PM10 and PM2.5 and the inclusion of particulate matter in the mandatory monitoring program of air quality.

Keywords: particles

TU-PO-37
Risk assessment by exposure to nitrogen dioxide and ozone during lifetime

Park J.[1], Ryu H.[1], Heo J.[1], Kim E.[1], Choe Y.[1], Cho M.[1], Kim Y.[2], Yang W[1]; [1]Department of Occupational Health, Daegu Catholic University, Gyeongsan, Republic of Korea, [2]EH-Service Center, Daegu Catholic University, Gyeongsan, Republic of Korea

Although the risk assessments for air pollutants such as nitrogen dioxide (NO2) and ozone (O3) has been extensively studied, most of them were limited mainly to the indoor environment. Therefore, integrated risk assessment considering exposure in all microenvironments including outdoor is needed. The purpose of this study was to assess risk by population group according to time-activity patterns, and the lifetime risk of Korean. In this study, we calculated time-weighted average concentration for preschool child, student, housewife, worker and the old using residential time and concentrations of indoor (house, workplace or school, others), outdoor and transport by meta-analysis and data offered from the Korean National Statistical Office. The analyzed data were compared with measured data for validation. The risk for NO2 and O3 were assessed by hazard quotient using reference concentration, 30 ppb and 60 ppb respectively. The risk assessments were conducted by 1,000,000 times of Monte-Carlo simulation for probabilistic analysis. Preschool child, student, housewife, worker and the old spent 91.9%, 86.0%, 79.8%, 82.2% and 77.3% of their day at house and workplace or school respectively. The R2 value of meta-analyzed and measured NO2 concentration by population group was 0.7793. The risk assessment for the lifetime showed 28.4% and 33.8% of hazard quotient of NO2 for worker and housewife exceed 1, and more than 99% of hazard quotients of O3 were less than 1. The risk of NO2 and O3 by sub-population groups and for the lifetime of housewife and worker were assessed. The risk of NO2 was higher than O3 and showed different risk by sub-population groups. Since both NO2 and O3 showed a higher risk in the housewife than the worker. This study can be used as a basis of exposure and risk assessment for the lifetime to NO2 and O3.

Keywords: activity patterns, aggregate exposure, environmental health, exposure models, risk assessment

TU-PO-38
Exposure to traffic-related pollution on a road in Lublin, Poland

Polednik B., Dudzinska M.[1], Siuta-Olcha A.[1], Piotrowicz A.[1], Guz L.[1], Staszowska A.[1]; [1]Lublin University of Technology Faculty of Environmental Engineering, Lublin, Poland

Traffic-related emissions, apart from emissions from fuel combustion for heating purposes, significantly deteriorate air quality in cities. The above mainly concerns areas located close to busy traffic routes. According to epidemiological studies, traffic-related emissions have an adverse health effect. This specifically affects commuters (drivers and car passengers), pedestrians as well as inhabitants of houses located near roads. The aim of this study was to determine the variations of particle number and mass concentrations along a busy road in Lublin, Poland and their impact on the particle exposure for commuters and pedestrians and also for inhabitants of houses located near roads. Mobile and fixed-site measurements were performed in the summer (June) with a focus on peak and off-peak traffic hours and road sections with low and high traffic intensity. The obtained results indicated that the highest concentrations of traffic-related pollution were measured near 4-way traffic intersections during peak hours. During peak hours, the average number concentration of ultrafine particles (PN0.1) in the road section near 4-way traffic intersections amounted to 17.2 ±13×103 pt/cm3 and was about 2 times higher than during off-peak hours. The average mass concentration of fine particles (PM2.5) was also approximately twice as high than in off-peak hours and amounted to 21.8 ±9.4 µg/m3. Similar relations were found for other measured aerosol particles as well as with respect to particle exposure. The estimated particle doses inhaled by commuters and pedestrians within an hour near 4-way TIs in peak hours amounted to 4.3 ±3.3×109 pt/h (PN0.1) and 2.9 ±1.4 µg/h (PM2.5). While estimating the doses inhaled by the inhabitants of premises located near the road one should take into account different fractional penetration of particles from outdoors to indoors and other important factors. Howsoever assessed, their values are relatively high and may have adverse health effects.

Keywords: particles

TU-PO-39
A study on the Evaluation and Validation of Plasma Sterilizing Dust Collector Unit

Jeon H. [1],Sohn, J. [2], Seo J. [1], Choi J[1]; [1]Naver Corporation, Seongnam, South Korea, [2]Korea University, Seoul, Korea

With the continuous progress of urbanization and the improvement of economic standards, modern people spend most of their day indoors as well as changes in the lifestyle of modern people. According to a U.S. EPA report, modern people have high indoor activity times, with an estimated 90 percent of the time spent indoors. Interest in indoor air quality has been extended to the problem of indoor air pollution in office spaces. Accordingly, air purifiers and dust collectors that can clean up office space's indoor air have been developed and distributed continuously, and a dust sterilizer that introduces plasma has been also developed. Since the impact of outdoor air penetration in Pohang Steel Mill in Korea is expected to be significant due to the characteristics of the workshop, it is deemed necessary to verify the performance through field application evaluation to assess the performance change according to the location characteristics. For further evaluating and analyzing the efficiency reduction caused by the number and activity of the office workers, an on-site application assessment is also necessary to scientifically demonstrate the performance of Plasma Sterilizing Dust Collector Unit through verification of performance in actual buildings. Accurate assessment will be carried out, considering performance of actors in the office such as commuting/retiring, lunchtime, and other events, analysis of the reduction efficiency of fine dust by time. The purpose of this research is to evaluate the performance of Plasma Sterilizing Dust Collector Unit installed directly in buildings and verify their efficiency in the field, as the office space at the center of Pohang Steel Mill has many limitations, including the wide range of places to apply general air `purifiers, and the fact that indoor air quality needs certain improvement due to the characteristics of the place.

Keywords: activity patterns, air, air cleaning, built/indoor environment

TU-PO-40

Simplified method for oral bioaccessibility of metal(loid)s in soils and alternative to the Unified Barge Method.

Le Bot B.[1], Glorennec P.[1], Waterlot C.[2], Douay F.[2], Pelfrêne A.[2]; [1]Univ Rennes, Inserm, EHESP, Irset (Institut de recherche en santé, environnement et travail) - UMR, Rennes, France, [2]LGCgE-ISA, Yncréa Hauts-de-France, Lille , France

Incidental ingestion of soil can expose people to toxic metal(loid)s, especially on contaminated sites. Oral bioaccessibility, i.e. the fraction of a soil contaminant that is solubilised in the digestive fluids after ingestion, is recommended to refine exposure and risk assessments. Reference method is the Unified Barge Method (UBM) validated for As, Cd and Pb against in vivo measurement. However, UBM is a complex method that implies high technical skills and needs numerous reagents. The objective of the present study is to test different chemical extractions using acetic acid (0.11 M), citric acid (0.11 M), EDTA (0.16 M) and hydrochloric acid (0.65%) to mimic the bioaccessible fraction of As, Cd and Pb in 209 soil samples with a wide range of physicochemical parameters and metal(loid)s' concentrations. The underlying hypothesis is that pH of the simulated gastric step is the limiting condition that drives the bioaccessiblity of the soil contaminants on which the UBM validation study focused on (i.e. As, Cd and Pb). Linear regression was used to determine the most relevant single-extraction method(s) capable of solubilizing As, Cd and Pb similar as those measured in both gastric and gastrointestinal phases defined by UBM. Significant relationships were obtained between As, Cd and Pb extracted in gastric and gastrointestinal phases and using the single-extraction methods. The results showed that the dilute hydrochloric acid test could be used as a suitable proxy to assess the gastric bioaccessibility of As, Cd and Pb.

Keywords: exposure factors

TU-PO-41 WITHDRAWN

TU-PO-42

Exposure assessment of VOC emissions from Halloween masks

Even M. [1], Roloff A.[1], Hutzler C. [1], Haffke H. [2], Luch A. [1]; [1]German Federal Institute for Risk Assessment (BfR), Department of Chemical and Product Safety, Berlin, Germany, [2]Chemical and Veterinary Investigation Office Ostwestfalen Lippe (CVUA OWL), Detmold, Germany

Latex face masks that are frequently used at Halloween or Carnival are often associated with unpleasant odors and may even cause headache and nausea upon extended wearing. These effects can be induced by the emissions of volatile organic compounds (VOCs), which are inhaled and can potentially influence consumer's health. However, the exposure scenario for face masks is much more complex compared to other products. The emission source and the inhalation via nose are in close proximity within an almost enclosed space. Naphthalene was detected at 63 mg/kg in a toluene extract of a donkey face mask. Subsequently, emission tests were carried out in 114 ml microchambers at 23°C, 50% relative humidity and an air change rate of 45 h-1. Emitted VOCs were sampled on TENAX® TA tubes and analyzed after thermal desorption by GC/MS. The mask emitted 317 mg/m2 naphthalene over the first 5 hours, 781 mg/m2 over 24 hours and 1.19 g/m2 over 3 days. Ethylbenzene, xylenes and butylhydroxytoluene were also detected and quantified at 103, 260 and 423 mg/m2 over the first 5 hours respectively. Qualitative analysis of emissions of VOCs from 11 further latex masks was performed using GC/MS in dynamic headspace mode. Decane, xylenes and cyclohexanone were detected with high signals. To provide a realistic exposure assessment, additional surface emission tests with an innovative setup need to be carried out. The air change rates should be adapted to simulate the air exchange and breathing patterns under real use conditions of face masks. Different exposure scenarios that include higher temperature, increased humidity or longer exposure times should be considered.

Keywords: Halloween masks; VOCs; Emissions; Inhalation exposure

TU-PO-43

Monitoring the efficiency of a portable air cleaner for formaldehyde removal with a Cavity Ring-Down Spectrometer

Hofmann M.[1], Bonomaully J. [2], Knap H.[2], Russell H. [2], Johnson M. [2,3]; [1]Picarro Inc., Willemsplein 2, 's-Hertogenbosch, The Netherland, [2]Airlabs, Lersø Parkallé 107, Copenhagen, Denmark, [3]University of Copenhagen, Universitetsparken 5, Copenhagen, Denmark

Permanent formaldehyde-based adhesive used e.g. in plywood, composite wood product and carpeting can affect indoor air quality. The air quality inside vehicle cabins is not only affected by outgassing of fabrics inside the vehicle but also by the intake of polluted air caused by dense traffic in the surroundings. Airlabs developed a portable air cleaner, the airbubbl, to remove particulate matter, NO2 and other pollutants from air so that the air quality inside vehicle cabins can be improved efficiently. A new filter has been designed to also remove formaldehyde. Here, we will present the performance of the new formaldehyde filter, also in comparison to a range of different adsorbents and catalysts. The performance of the different filters for formaldehyde removal was tested using a Picarro G2307 Cavity Ring-Down Spectrometer. The Picarro G2307 analyzer can measure continuously and directly in air the formaldehyde concentration at a ppt level of precision. A key feature of the analyzer is the high stability and low calibration frequency. The analyzer also measures with a high-precision methane and water vapor concentrations allowing to use methane as a surrogate gas to validate the formaldehyde calibration. The different filter materials were tested in a 2m3 chamber with varying formaldehyde, NO2 and humidity levels. Variations in the humidity level and NO2 did not affect the formaldehyde removal rate and the high-precision measurements with the G2307 allowed to clearly distinguish formaldehyde removal rates for different filters. We also tested the long-term performance of the filters on a smaller scale, using filter material in a 16mm test column. These tests also showed clear differences in removal efficiencies for the different adsorbents and catalysts. The high time resolution and response time of the Picarro G2307 allowed to easily test the different filter materials under various conditions.

Keywords: formaldehyde, air cleaner, filter, CRDS, Cavity Ring-Down Spectrometer

TU-PO-44
High Density Air Quality Monitoring Sensor Network for Urban Hotspot Identification and Diagnostics

Sun L.[1], Wei P.[1], He J.[1], Zhang K.[1], Westerdahl D.[1], Ning Z.[1]; [1]Division of Environment and Sustainability, The Hong Kong University of Science and Technology, Hong Kong SAR, China

A public transport interchange (PTI) is a public place where buses, taxis, mini-buses or coaches pass or terminate, and where passengers queue for transport. Most of the PTIs (65 in total) in Hong Kong are located under residential or commercial complexes, and surrounded and enclosed by structural columns or walls. Many of these sites have active ventilation systems. Air pollutants may accumulate to high concentrations in such confined environments if the nature and mechanical ventilations are insufficient. From 2014 to 2018, Hong Kong Transport Department has received a total of 111 complaint cases concerning PTI's air quality, involving 37 PTIs and performed surveys at single sites of selected PTIs that showed exceedances (with a maximal value found of 0.65 mg/m3) of NO2's 1-hour concentration limit (0.30 mg/m3). However, such a survey could not provide sufficient information on the overall pollution distribution in the whole PTI nor could it identify potential hot spots in the PTIs. We deployed a sensor-based network of monitoring platforms to more fully characterize the air pollution and its distribution inside the PTIs, and to determine the potential hot spots inside the PTIs and analyze their formation causes. Gas pollutants (NO, NO2, CO and O3), wind speed and direction were simultaneously measured for 4 working days in 8 selected PTIs. Limit exceedances of NO2 (with a maximal value found of 1.10 mg/m3) were observed at several sites in the PTIs investigated, which mainly located in the middle of the PTIs or close to the bus stops occupied by older buses. The integrated impacts from large bus volume, insufficient ventilation, and improper bus idling operation were found contributing to the exceedances. This study confirmed that some PTI practices contributed to unacceptable air quality conditions. Mitigation measures were proposed and further studies are underway to evaluate their effectiveness.

Keywords: public transport interchange, sensor network monitoring, bus activity, ventilation

TU-PO-45
Seasonal spatial and temporal variability of Black Carbon in an elementary school catchment area of Milan, Italy

Boniardi L. [1], Dons E. [2], Campo L. [3], Van Poppel M. [4], Int Panis L. [2], Fustinoni, S. [5]; [1]Università degli Studi di Milano, Milano, Italy, [2]Hasselt University; Flemish Institute for Technological Research (VITO), Hasselt, Belgium, [3]Fondazione IRCCS Ca' Granda Ospedale Maggiore Policlinico, U.O.S Tossicologia, Milano, Italy, [4]Flemish Institute for Technological Research (VITO), Moi, Belgium, [5]Università degli Studi di Milano; Fondazione IRCCS Ca' Granda Ospedale Maggiore Policlinico, U.O.S Tossicologia, Milano, Italy

Introduction. The city of Milan is a high-traffic area with an estimated population of about 3.2 million inhabitants. The health impact of traffic-related air pollution is reported to be larger for the more susceptible population groups, such as children. Morning rush hour (MRH) is the most congested period and it is also the time during which schoolchildren commute to school and are directly exposed to traffic pollution. The aim of this study was to analyse the temporal behaviour and the determinants of Black Carbon (BC) during warm and cold seasons with a focus on MRH in an elementary school catchment area. Methods. Eight micro-aethalometers measured BC concentrations on 35 sites classified as street (S), urban traffic (UT) or urban background (UB) according to the influence of traffic sources. The monitoring campaigns lasted from May to June 2017, and from January to February 2018. Every site was monitored for one week during each season, while one reference site, located in the school courtyard, sampled for the entire period of both monitoring campaigns. Four seasonal Land Use Regression (LUR) models were developed. Results. In both seasons, MRH BC average was higher than the daily average: 2.6±0.7 µg/m3 (mean±SD) versus 1.5±0.4 µg/m3, and 4.3±0.7 µg/m3 versus 3.5±0.7 µg/m3, during the warm and cold season respectively. The highest spatial contrast was observed during the warm MRH with 3.7±0.5 µg/m3 for S-sites and 1.9±0.3 µg/m3 for UB-sites. LUR seasonal models explained from 56% to 65% of the BC variability and MRH-focused models performed better than models for daily average. Conclusion: MRH is the most critical time window for air pollution in Milan. The four developed LUR models show that traffic sources play a major role in explaining BC concentrations. Selecting low-pollution routes could lead to an important reduction in the exposure of school-aged children.

Keywords: air, children, geospatial analysis/GIS, particles, statistical methods

TU-PO-46
Characterizing peak exposure of secondhand smoke with a real-time PM2.5 monitor

Zhang T.[1], Yan B.[1], Ross J.[1], Yang Q.[1], Perera F.[1], Ji J.[1], Spira A.[1], Breysse P.[1], Miller R.[1], Chillrud S.[1]; [1]Key Laboratory of Surficial Geochemistry, Ministry of Education, Nanjing University, Nanjing, China, [2]Lamont Doherty Earth Observatory of Columbia University, Palisades, NY, USA, [3]Department of Environmental Health Sciences, Mailman School of Public Health, Columbia University, New York, NY, USA, [4]Boston Medical Center Cancer Center, Boston University School of Medicine, Boston, MA, USA, [5]Bloomberg School of Public Health, Johns Hopkins University, Baltimore, MD, USA, [6]Division of Pulmonary, Allergy and Critical Care of Medicine, Department of Medicine, College of Physicians and Surgeons, Columbia University, New York, NY, USA

Peak exposure to pollutants can trigger adverse acute effects given the relatively high levels of exposure in a short period of time. Secondhand cigarette smoke (SHS) leads to peak exposure if the smoking occurs in a confined space. Multiple adverse health effects are associated with SHS, the majority of which occurs in the residential environment. The aim of this study was to explore whether peaks in real-time PM2.5 data can estimate SHS exposure. A total of 41 children (21 with and 20 without SHS exposure based on self-reported questionnaire responses) from 28 families in New York City (NY, USA) were recruited for this study. Both real-time personal and residential PM2.5 monitoring were performed for five consecutive days using MicroPEM portable PM2.5 sensors (RTI International, USA). A threshold detection method based on a cumulative distribution function was developed to identify peaks. When children were home, the accumulated peak area (APA) for peak exposures was 297 hour*µg/m3 for children from smoking families, six times that of those from non-smoking families (50 hour*µg/m3). PM mass concentrations for SHS exposed and unexposed children were 24 µg/m3 and 15 µg/m3, respectively. The average SHS exposure duration was 1.2 hr per day. This represents ~5% of total exposure time, but ~13% of children's total PM2.5 exposure. SHS-derived PM2.5 was equivalent to an additional 2.6 µg/m3 per day. This study demonstrated the feasibility of peak analysis for estimating SHS exposure.

Keywords: air, particles, secondhand smoke

TU-PO-47
Potential contribution of dermal pathways to nicotine exposure in children residing in smoking households

Cooper E.[1], Blanco M.[1], Duncan B.[1], Shaffer R.[1], Kissel J.[1]; [1]University College London, London, United Kingdom

Cohabitation with adult smokers is associated with elevated urinary cotinine (a commonly reported biomarker for active smoking), in children. Inhalation of second hand (i.e., sidestream) smoke is an obvious exposure pathway of concern. An important research question is whether processes other than direct inhalation may be responsible for non-negligible contributions to observed urinary cotinine in children cohabiting with smokers. Increasing attention has been given to dermal exposure either directly to vapor phase nicotine, nicotine residues in fabrics (e.g. clothing and bedding), and nicotine on building material and furnishing surfaces. Theoretical predictions (Weschler & Nazaroff, 2012) of non-negligible dermal absorption of semi-volatile organic compounds (SVOCs) from the vapor phase have been confirmed in a series of chamber studies involving exposure of human volunteers to phthalates (Weschler et al., 2015), benzophenone (Morrison et al., 2017), and nicotine (Beko et al., 2017a,b). These studies have employed urinary biomarkers as indicators of exposure and have included some examination of clothing effects. In this work, Monte Carlo simulations of potential child nicotine exposures have been conducted using conventional exposure factors and apparent environment-to-skin mass transfer coefficients derived from the above chamber studies. These results are compared to results of studies of smoking households from which air nicotine and child urinary cotinine measurements are available.

Keywords: SVOCs, built/indoor environment, exposure factors, exposure models, secondhand smoke

TU-PO-48
Quantitative Risk Assessment: Uranium exposure from inhalation

Carl A.[1], Harris R[1], Alshammari M[1], O'Rourke, M.K.[1]; [1]University of Arizona, Arizona, USA

Uranium 238 (238U), a natural occurring radioactive metal is associated with kidney diseases, respiratory diseases/cancer, and bone cancer. Uranium is found in soil, water, and air. Information regarding 238U inhalation on Hopi lands is limited. We performed a quantitative risk assessment on the relative cancer risk from inhaling 238U. We measured particulate matter (PM2.5) in Northeastern Arizona on Hopi land using a personal DataRAM (pDR-1500). Placement occurred inside homes in the summer and winter seasons, each assessment ran for approximately twenty-four hours. The pDR-1500 contained filters that were analyzed for metal concentrations using inductively coupled plasma mass spectrometry (ICP-MS). Participant exposures were calculated using 238U data and personal questionnaires. Average duration was substituted for missing duration values. We obtained carcinogenic cancer risks by multiplying the exposure concentration and inhalation unit risk value obtained from the Agency for Toxic Substances and Disease Registry (ATSDR). Results: Data were collected in a total of 17 homes in the summer and 7 homes in the winter. Participants lived at home for an average of 24 years (sd ±22). The average concentration of 238U in the summer was 0.94 μg/m3 (sd ±0.39) and 0.82 μg/m3 (sd ±0.47) in the winter. The average exposure concentration in the summer was 0.24 μg/g (sd ±0.23) and 0.05 μg/g in the winter (sd ±0.04). The cancer risk factor was 7.3*10-5 and 1.5*10-5 for summer and winter seasons respectively. Preliminary data indicates that on Hopi lands the risk of respiratory diseases and cancer based on 238U exposure exceeded the acceptable risk limit of 10-6 carcinogens per lifetime.

Keywords:

TU-PO-49
Comparing wood and LPG cook-stove bioaerosol and particulate matter exposures in Tamil Nadu, India

Stapleton E.M.[1], Kizhakke Puliyakote A.S.[4], Shanmugam K.[5], Kesavan K.[5], Sirajunnisa A.[5], Geetha K.[5], Kumar D.[6], Metwali N.[3], Jeronimo M.5, Brauer M.[7], Thorne P.S.[3], Peters T.M.[3], Hoffman E.A.[2], Comellas A.P[1]; [1]University of Iowa, Department of Internal Medicine, Division of Pulmonary, Critical Care and Occupational Medicine, Iowa City, IA, USA, [2]University of Iowa, Department of Radiology, Iowa City, IA, USA, [3]University of Iowa, Department of Occupational and Environmental Health, Iowa City, IA, USA, [4]University of California San Diego, Department of Radiology, San Diego, CA, USA, [5]Periyar Maniammai Institute of Science and Technology, Thanjavur, India, [6]Kings of College of Engineering, Thanjavur, India, [7]The University of British Columbia, School of Population and Public Health, Vancouver, CA, USA

Background: Household air pollution causes approximately 0.5 million deaths in India annually, most occurring in people <70 y/o. Sixty-six percent of the population uses solid-biofuels to cook, increasing the risk of developing COPD, chronic bronchitis and acute lower respiratory tract infections. Our pilot study quantified pollution known to cause adverse health-effects, including: PM2.5, black carbon, PM mass (≤40 µm), bioaerosols, and microbiological species from kitchens of primary cooks in Thanjavur, India. Methods: We assessed 24-hour PM2.5 concentrations via real-time (Pats+) and gravimetric (UPAS) measurements and PM-loaded filter light absorbance using a photo-optical method in 34 biomass (n=27) and LPG (n=7) kitchens, and XRF imaging of filters. We vacuumed settled (over 7 days) particles (≤40 µm) from kitchens into a filter cartridge (DUSTREAM®) from a rug placed in the cooking area and quantified microbial content from the settled particles. Next, we assessed within-kitchen endotoxin (EU) levels using electrostatic dust collectors (EDC) deployed over 24-hrs and 7 days. Results: Peak PM2.5 concentrations correlated with cooking, and 24-hour arithmetic means ranged 18-732 µg/m3. Pats+ and UPAS inter-sampler correlation was high ($r^2=0.74$). Primary fuel-type predicted daily PM2.5 exposures – with biomass exposures higher than LPG – using both measurement methods ($p<0.05$). Filter light absorption coefficients (Fabs) were significantly higher in biomass homes than LPG ($p<0.01$). A trend was observed toward higher rug deposition in biomass homes ($p=0.18$). 24-hour EDC deposition concentrations ranged from 5,779-2,280,407 EU/m2, with a trend of higher concentrations ($p=0.10$) in biomass kitchens, while 7-day endotoxin concentrations were significantly higher in biomass kitchens ($p<0.01$). Conclusions: Field UPAS and Pats+ 24-hour PM2.5 concentrations were well correlated. 24-hour endotoxin concentrations were exceedingly high, and fuel-type was predictive of concentrations over 7-days. Biomass (wood) cook-stove kitchen concentrations were greater than those of LPG cook-stoves when measuring multiple aerosol types.

Keywords: Biomass, LPG, Particulate matter, Tamil Nadu, India

TU-PO-50
CFD simulation of bacteria distribution in an operating room with mixing ventilation at St. Olavs hospital

Zhang Y., Cao G., Pedersen C., Sternstad L., Drangsholt F., Feng G., Skogås G.

Background: Transmission of airborne bacteria is one of these main factors causing surgical site infection (SSI), which is harmful to patient's health and even life, and would cause great economic burden. Many earlier studies have provided evidence of relationships between ventilation of operating room and incidence of SSI. Method: This paper focuses on the airborne transmission of bacteria under mixing ventilation in hospital operating room by using CFD simulation, which is carried out to calculate the bacteria trajectories using the large eddy simulation (LES) coupling with Lagrangian method. Under operating conditions, one patient, two surgeons and three nurses were simulated by relatively realistic three-dimensional thermal manikins. In the simulation, the bacteria are produced from the skin and nose breathing. Field measurements were conducted at St. Olavs hospital to validate the simulation. The influence of other parameters on the bacteria distribution in the wound zone are analyzed, including ventilation rate, supply air velocity and supply vent area. Results: The numerical simulation results of wound zone are analyzed by using velocity vector diagram, temperature contour diagram and bacteria distribution. The results indicate that the interaction of thermal plumes from patient and surgeon, the use of surgical light and the downward airflow may dominate the operating microenvironment and influence the bacteria distribution. Conclusions: The bacteria distribution in the wound zone may be influenced by both the surgical facility and the presence of medical staff. Simply increasing supply air velocity and supply airflow area may reduce bacteria deposition on wound surface, but its effect may be less evident than increasing clean air supply directly to the operating microenvironment.

Keywords: air, particles, ventilation

TU-PO-51 MOVED TO TU-PL-D2-49

TU-PO-52

Development of a prediction model for indoor PM2.5 concentrations – the case in Tainan households

Su, H-J.[1], Jung, C-C.[2], Lin, W-Y.[1], Hsu, N-Y.[1], Chang C-T.[1], Wu C-D.[3]; [1]Department of Environmental and Occupational Health, National Cheng Kung University, Tainan City, Taiwan, [2]Research Center for Environmental Trace Toxic Substances, National Cheng Kung University, Tainan City, Taiwan, [3]Department of Geomatics, National Cheng-Kung University, Tainan City, Taiwan

Long-term PM2.5 exposures may contribute to the elevated incidence of multiple organs' damage. However, people spend approximately 90% of their time staying indoors, over- or under-estimates of health risks might be caused due to the difference of PM2.5 level between indoors and outdoors. Therefore, it is necessary to clarify the relationship between indoor and outdoor PM2.5 concentrations; especially, this study using a preliminary database to clarify the relationship between indoor and outdoor PM2.5 concentrations. From August 2008 to September 2009, study households contained 93 preschool children's bedroom were selected from Tainan. Research material involved indoor human actives, building characteristic, meteorological data, land use variables, and indoor/outdoor PM2.5 concentrations. The ambient PM2.5 concentrations were collected from the EPA's air quality monitoring stations and interpolated by Kriging method to be outdoor PM2.5 concentrations. Finally, the Multiple Linear Regression model was used to establish a predictive model for indoor PM2.5 concentrations. And model validation was conducted by using cross-validation and external validation to ensure the accuracy of the model. After establishing the predicted model, we can find that outdoor PM2.5 concentrations can explain most of the indoor PM2.5 concentrations. In addition, the remaining predictive power is regarding the infiltration rate. So, if any research wants to establish a model to find the correlation between indoor and outdoor environment in the future, these variables cannot be ignored.

Keywords: air, built/indoor environment, exposure models, particles

TU-PO-53

Development of predictive models for PM2.5 concentration in office

Wang W.[1], Jung C.[2], Su H.[1]; [1]Department of Environmental and Occupational Health, College of Medicine, National Cheng Kung University, Tainan, Taiwan, [2]Research Center of Environmental Trace Toxic Substances, National Cheng Kung University, Tainan, Taiwan

Background: Studies have shown that fine particulate matter (PM2.5) exposure increases both the mortality and morbidity of cardiovascular or respiratory diseases. People, meanwhile, often spend more than 85% of their time indoors, and for most adults, office setting becomes the predominant type of indoor environment where they reside. This study aimed to develop a predictive model for PM2.5 concentration in offices to better facilitate the needs for exposure assessment and epidemiological studies. Method: We collected indoor PM2.5 concentration from 39 offices from 2001 to 2010 in Taiwan and used a Geographic Information System Kriging model to estimate the outdoor concentration around these offices. We also acquired the building characteristics, indoor pollution sources (such as printer usage) by questionnaire, and outdoor pollution sources from land uses data of different buffer zone analysis through GIS. Finally, a stepwise regression model was used to establish the predictive model for PM2.5 concentrations in office. Results: Results showed that kriging maximum outdoor PM2.5 concentration and restaurant within 500 meters of radius can explain the most of 8-hour average PM2.5 concentration (R2 = 0.66). We also establish a predictive model for 24-hour average PM2.5 concentration and found that medium outdoor PM2.5 concentration, numbers of express delivery services industry within 3000 meters of radius, and the curtains with metal were major effect factors (R2 = 0.88). Conclusions: We suggest that it is necessary to collect outdoor PM2.5 concentrations, sources of outdoor pollution and indoor decoration, in order to establish accurate 24-hour average PM2.5 concentration in office. However, more information is warranted for increasing the predictive power for 8-hour average PM2.5 concentration in office, such as the frequency of occupants' activities indoors.

Keywords: air, behavior, built/indoor environment

TU-PO-54
Investigating the influence of human activities on exposure to semi-volatile organic compounds through multimedia modeling

Kvasnicka J., Diamond M.[1]; [1]University of Toronto, Toronto, Ontario, Canada

Multimedia models have proven to be useful predictive tools for assessing the fate of neutral organic compounds released in indoor environments and associated human exposures. The characterization of human occupants in these models, however, has been limited to treating them as passive individuals with static and continuous behaviors. Using an indoor multimedia model, we investigated how typical human activities, such as household cleaning and use of furniture, influence exposure to semi-volatile organic compounds under static conditions. We then conducted a dynamic simulation by incorporating a time-activity schedule for the human occupant. The results of these simulations present a tentative picture of the differences in exposure estimation that may arise from different modeling assumptions pertaining to the influence of human occupants on the indoor system. Further research is underway, aimed at better accounting for temporal variation in exposure-relevant human behaviors through dynamic, multimedia modeling.

Keywords: SVOCs, activity patterns, behavior, built/indoor environment, exposure models

TU-PO-55 MOVED TO MO-PO-94
Recovery of viable Staphylococcus aureus bioaerosols

Chang C-W.[1], Lin M-H.[1], Horng Y-J.[1]; [1]National Taiwan University, Taipei, Taiwan

Staphylococcus aureus aerosols are present in facilities (e.g., animal barns and hospitals) and may threaten human health. Air monitoring by appropriate sampling techniques is required for exposure assessment. This study thus aims to assess the recovery efficiency (R) of liquid-based samplers for viable S. aureus based on cellular membrane integrity. By using propidium monoazide and qPCR, the R of six sampling methods, i.e. two widely-used samplers (all-glass impinger-30 (AGI-30) and BioSampler) filled with three collection liquids (deionized water (DW), Tween mixture (TM) and phosphate-buffered saline (PBS)), were determined at five sampling times (3, 6, 15, 30 and 60 min). Effects of liquid refilling every 15 min during 60-min sampling were also evaluated. Irrespective of sampler type, PBS presented the lowest cell recovery, statistically different from DW ($P<0.05$). Liquid replenishment improved cell recovery in DW by AGI-30 (P0.05). In contrast, samplings for 15 min or greater resulted in decreased recoveries in PBS or TM by both samplers compared to 3-min sampling. Overall, BioSampler recovered S. aureus greater than AGI-30 ($P<0.001$). In particular, BioSampler with DW presented great R independent of effects of sampling time and liquid refilling, supporting it as an appropriate sampling method. This study elucidates the factors affecting the collection of S. aureus aerosols, which helps professionals possess greater certainty of successful capture of S. aureus and collect reliable data.

Keywords: sampling methods

TU-PO-56

An Evaluation of Radon Exhalation Rate from Building Materials

Lee H.W. [1], Nam Goung S.J. [1], Lim H.B.[2], Park S.H. [1], Hong H.J. [1], Yoon D.K.[1], Choi G.Y.[1], Lee C.M. [1,2]; [1]Institute of Risk Assessment, Seokyeong University, Seoul, South Korea, [2]Department of Chemical & Biological Engineering, Seokyeong University, Seoul, South Korea

Recently, the concern on indoor radon has increased since building materials were reported to cause indoor radon. Building materials refer to all products for assembling constructions permanently; they can affect the exposure on gamma ray and indoor radon concentrations as having radionuclide(Trevisi et al., 2018). However, there are no regulations or standards of process tests able to evaluate and constrict radon exhalation rate of building materials in Korea. Therefore, this study aimed to suggest the protocol of assessment for radon exhalation, as based on ISO 11665 and standards of process tests on indoor air quality. The assessments on radon exhalation rate of building materials were performed using the presented methods and the results of this study will be able to be used for data base of policy. This study suggests two test protocols, as follows. The one is "Closed chamber test", the samples(16 cm × 16 cm × h) were put in purified and sealed chamber connected with real-time measurement instrument(RAD7). The other one is "Circulated chamber test", the chamber's condition is same with "Closed chamber test" but there is difference that nitrogen is injected continuously to the chamber as a flow rate of 0.167 L/min. Acknowledgements This subject is supported by the Environmental Health Action Program of the Ministry of Environment (No. 2015001350004).

Keywords: Building materials; Radon; Radon exhalation; Radioactivity

TU-PO-57

The Indoor Chemical Human Emissions and Reactivity Project (ICHEAR): Methods

Wargocki P.[1], Bekö G.[1], Ernle L.[2], Langer S.[3], Li M.[2], Licina D.[4], Morrison G.[5], Wang N.[2], Weschler CJ.[6], Williams J.[2], Yang S.[4], Zannoni N.[2]; [1]Technical University of Denmark, Lyngby, Denmark, [2]MPIC, Mainz, Germany, [3]IVL, Gothenburg, Sweden, [4]École Polytechnique Fédérale de Lausanne, Lausanne, Switzerland, [5]University of North Carolina, Chapel Hill, NC, USA, [6]Rutgers University, NJ, USA

Here we present the new project "Indoor Chemical Human Emissions and Reactivity" (ICHEAR), which is focused on examining the role of human emissions on indoor air chemistry. The specific goal of this project is to investigate the impact of exhaled and dermally emitted human bioeffluents on the chemical compounds present in indoor air, their chemical transformations and total OH reactivity (overall loss rate of OH radical) under different conditions comprising a variety of factors (temperature, relative humidity, ozone, clothing level, age of human participants). The measurements were conducted in stainless steel climate chambers at the Technical University of Denmark (DTU) using state-of-the-art instruments from the Max Planck Institute for Chemistry (MPIC), the Swedish Environmental Research Institute (IVL), and École Polytechnique Fédérale de Lausanne (EPFL). During April-May 2019, groups of four persons (two males and two females) occupied the chamber daily and their emissions were quantified. This poster describes the chambers, instrumentation and the overall experimental approach used in ICHEAR.

Keywords: indoor air, human emissions, VOC, OH reactivity, secondary organic aerosols

TU-PO-58

The Indoor Chemical Human Emissions and Reactivity Project (ICHEAR): First Results

Bekö G.[1], Ernle L.[2], Langer S.[3], Li M.[2], Licina D.[4], Morrison G.[5], Wang N.[2], Wargocki P.[6], Weschler C.J.[7], Williams J.[2], Yang S.[4], Zannoni N.[2]; [1]Technical University of Denmark, Lyngby, Denmark, [2]MPIC, Mainz, Germany, [3]IVL, Gothenburg, Sweden, [4]École Polytechnique Fédérale de Lausanne, Lausanne, Switzerland, [5]University of North Carolina, Chapel Hill, NC, USA, [6]DTU, Lyngby, Denmark, [7]Rutgers University, NJ, USA

Here we present first results from the project "Indoor Chemical Human Emissions and Reactivity" (ICHEAR). The aim of this project is to chemically characterize how human emissions, the single common factor in all dwellings, impact indoor air chemistry. The measurements were conducted in stainless steel climate chambers at the Technical University of Denmark (DTU) using state-of-the-art instruments from the Max Planck Institute for Chemistry (MPIC), the Swedish Environmental Research Institute (IVL), and École Polytechnique Fédérale de Lausanne (EPFL). During April-May 2019, groups of four persons (two males and two females) wearing pants and long-sleeve shirts or shorts and t-shirts occupied a 22m3 stainless steel climate chamber daily (air exchange rate 3 h-1). The emissions and their chemical transformations were quantified under various environmental conditions with and without ozone present in the chamber air. This poster shows first results for selected volatile organic compounds, OH reactivity and secondary organic aerosol concentrations. Variations were seen as a function of ozone concentration, fraction of exposed skin, temperature and humidity.

Keywords: indoor air, human emissions, VOC, OH reactivity, secondary organic aerosols

TU-PO-59
Overheating Risk Assessment: a Real-life Case of a Public Library in Auckland

Lesley, M.; Victoria University Wellington, Wellington, New Zealand

Very few data and scientific studies on measurements of public library environments, especially on thermal conditions in relation to the risk of overheating are available. Also, the indoor environments in public libraries are rarely studied in comparison to other building types such as offices. This raises issues on how to further improve indoor thermal condition and productivity. This paper examines the thermal condition of occupants and the risk of overheating in an eco-friendly public library in Auckland region of New Zealand. The study utilises long term environmental monitoring of the building performance during the summertime and occupants' survey of staff who have occupied the buildings for more than two years. The results obtained with the overheating assessment demonstrate that the library has low risk of overheating. These results were aligned with the survey responses since all the responses coincide that the environment is slightly warm which is acceptable. Furthermore, results show that a considerable amount of hours would be below the lower temperature and humidity ratio limit although none of the occupants claimed the environment as slightly cold. The study stressed the need to assess the thermal comfort responses of people who are not adapted to the building thermal condition. In this regard, commonly used methods might not be the most appropriate tool since they are based on the adaptive assumption that occupants will adapt their behavior according to previous experiences. Therefore, in-depth interviews might shed light of how people who are not adapted to the building finds the thermal sensation of the building.

Keywords: thermal environment

TU-PO-60
Field investigation of thermal comfort level of patients and surgical staff in operating rooms at St Olavs hospital

Kuivõgi H.[1], Cao G.[2], Bogdan A.[3], Thalfeldt M.[1], Pedersen C.[2], Zhang Y.[4], Fan M.[6], Stenstad L-I.[5], Skogås, J.G.[5]; [1]Department of Civil Engineering and Architecture, Tallinn University of Technology, Tallinn, Estonia, [2]Department of Energy and Process Engineering, Norwegian University of Science and Technology, Trondheim, Norway, [3]Department of Air-Conditioning and Heating, Warsaw University of Technology, Warsaw, Poland, [4]Urban Construction and Environmental Engineering, Chongqing University, Chongqing, China, [5]St. Olavs Hospital Operating Room of the Future, Trondheim, Norway, [6]School of Environmental Science and Engineering, Tianjin University, Tianjin, China

In healthcare facilities and hospital environments, it is essential to be able to predict thermal comfort as a parameter of indoor air quality. Unstable thermal conditions are critical in the operating room (OR) environment and may reduce the work efficiency and increase the possibility of surgical errors. At the same time, the main purpose of ventilation systems in operating rooms is to enable clean and comfort indoor environment. Likewise, the metabolic rates of occupants have a substantial impact to thermal comfort and it is completely different for each occupant in operating room. Same is for surgical clothing parameters (as vapor and thermal resistance) – clothing, made by qualified material, should protect wound from infection. Although, there is an increasing number of studies covering the area of thermal comfort, there have been few literature reports written on the comfort of healthcare staff and patients in hospitals and in the OR environment. Therefore, the purpose of this study is to estimate the thermal comfort level for every occupant and analyze the thermal environment in the operating room. The target is to base on PMV-PPD level and deduce the thermal sensation from environmental conditions (temperature, humidity and airflow), and from personal conditions (their metabolic rate and clothing insulation). Research methods includes the thermal environment measurements in hospital operating room after mock surgeries, a survey among surgical staff and patients, to get information about real sensation, and as well, the estimation of the activity level and clothing vapor and thermal resistance of every occupants in operating room. The expected result to be obtained, is the PMV-PPD theoretical and real level of four main occupant group in operating room. As a conclusion, the purpose is to provide suggestions and guidance of indoor environment control in hospitals improving thermal comfort level of patients and surgical staff.

Keywords: built/indoor environment, thermal environment, ventilation, workplace

TU-PO-61
Quantifying the Contribution of Temperature Anomaly to Stroke Risk in China

Guan T.; Peking University, Peking, China

Stroke, an important public health concern, has been linked to outdoor temperatures. However, the association between stroke and temperature anomaly (TA), a key measure of climate change, has not been examined thoroughly on a nationwide scale in China. A total of 11,144 stroke (both ischemic and hemorrhagic) and transient ischemic attack (TIA) patients from 339 communities were included in this case-crossover study between 2000 and 2016. Temperature estimates were gathered from an assimilation of in situ observations, satellite measurements, and weather research forecast model simulations. Occurrence of stroke or TIA was associated with TAs using a lag-distributed nonlinear model. Variations in this association were further explored by strata of subpopulations and by geographic region. The attributable fraction was derived to quantify TA-associated risk burden of stroke. Ischemic and hemorrhagic stroke were strongly associated with both cold and hot temperatures (defined as negative and positive TAs, respectively) at different lags, while TIA was weakly associated with cold weather with no lag. With a lag of up to 2 days, the risk for stroke increased by 20% (95% confidence interval [CI] 3–40%) for extreme cold and 19% (CI 3–37%) for extreme hot. Recurrent stroke was more strongly associated with TA than an initial stroke. In China, TAs may contribute to 3.42% of strokes, and more strokes were attributable to hot (2.05%) than cold (1.38%). During 2000–2016, stroke risk attributable to TA was estimated to increase by 0.02% per year. Our findings enrich the epidemiological evidence on the association between stroke and TA, which support China's policy for stroke prevention and climate change mitigation.

Keywords: environmental health

TU-PO-62
Testing Options for Mold: Each Producing a Different Result

Grimes C.; Hayward Healthy Home Institute, Monterey, California, USA

The use of the word "mold" by the public, practitioners, and some researchers has increasingly slipped from its scientific properties and identifying taxonomy to the "street" use of "mold problem," including perceived health claims. The influence of the shift derives from the conflation of the multiplicity of possible results attainable by sampling method, sample location, and choice of analytical method. This table compiles the 16 most common combinations of testing options along with the unit of measure, which component of the mold organism is detectable, and the bias of sample location. The combinations can be used as a check list to compare sets of test results for consistency and initially for the selection of the most appropriate method for a specific sampling need.

Keywords:

TU-PO-63

Transformation of occupational hygiene in to exposure science; Adjusting training and learning programs to meet 21st century needs

Brouwer D.; University of the Witwatersrand, Johannesburg, South Africa

Despite the similarity between the definitions of exposure science and occupational hygiene (OH), apparently the latter field of expertise operates somewhat isolated from the first. Since in occupational settings mostly the exposure emanates from near-field sources, it is expected that OH would focus on understanding the underlying emission, transmission and immission processes to control exposures. Unfortunately, triggered by South African regulations, OH practitioners focus more on exposure and compliance measurements rather than exposure control. Moreover, risk assessment is in most cases to limited to a simple comparison of the exposure level with the occupational exposure limit. Consequently, Teaching & Learning programmes for OH should play an important role to provide OHs with a more holistic view with regard to exposure and risks , e.g. the exposome perspective. Recently, the School of Public Health of the University of the Witwatersrand, Johannesburg launched its MSc Exposure Science programme targeted on OHs and Environmental Health Practitioners (EHPs). It is 2.5 years part-time course-research programme aiming to prepare OHs and EHPs meeting 21st-century demands with regard to risk assessment and –management. The central part of the course part of the programme consists of knowledge, understanding and applying the principles of exposure science. Students are taught in source-receptor approaches to understand occupational, consumer and residential exposures and associated health-risks. In addition to core courses such as exposure and risk assessment methods, stressor-induced heath-outcomes, supportive courses are amongst others, computational exposure assessment, transmission and transformation of stressors and selection and implementation of exposure controls. To emphasize and demonstrate the need for understanding the processes resulting in exposure, the research component will be focused on the development of tailor-made effective and efficient exposure control interventions.

Keywords: aggregate exposure, exposure factors, industrial hygiene, workplace

TU-PO-64

A single-arm trial examining the feasibility, acceptability and distribution of LPG cookstoves and gas cylinders to pregnant women living in rural Bangladesh, in preparation for The Poriborton: Change Trial

Raynes-Greenow, C.[1], Islam S.[2], Nisha M.[1], Khan J.[2], Tasnim F.[2], Thornburg J.[3], Billah M.[2], Alam A.[1]; [1]The University of Sydney, Sydney, Australia, [2]International Centre for Diarrhoeal Disease Research, Daka, Bangladesh, [3]RTI International, NC, USA

Background/Aim: The burden of household air pollution on perinatal mortality has not been accurately defined. Our overall aim was to develop a protocol for a cluster randomised controlled trial (cRCT) to assess the impact of liquid petroleum gas (LPG) cooking compared to usual cooking on perinatal mortality in pregnant women in rural Bangladesh (ACTRN12618001214224). For this study we aimed to assess the feasibility of the planned trial and the barriers/facilitators of distributing LPG to rural households. Methods: We used a mixed method iterative design. We included pregnant women, their families, and local LPG stakeholders. We distributed LPG to households for 3 months and assessed process issues, acceptability and cooking/food behaviours. We interviewed

LPG stakeholders, and conducted focus groups and interviews with users. The study had ethical approval and was conducted in Basail, north of Dhaka. Results: 50 pregnant women, and their family members trialled LPG. Initial distribution of LPG was hampered by process issues (fees, storage, timing) most of these were due to the lack of an established supply chain. Initial safety concerns were overcome with safety measures and training, LPG cooking was very acceptable, and all users reported a preference for continued use, fuel-sparing was heavily practised. Continuous LPG supply was hampered by the ad hoc supply chain. LPG stakeholders reported that LPG demand differed by season. Conclusions: Our iterative study meant that all issues were manageable. Implications for a larger trial include: ensuring safety, information to support continuous LPG use to ensure efficacy of the intervention is critical, a well-developed supply chain is essential for continuous supply, and changes in LPG demand due to season will need to be monitored to ensure timely cylinder replenishment. This was a very beneficial study for our protocol development, and overall, distributing LPG to pregnant women in rural Bangladesh was feasible.

Keywords: air, behavior, built/indoor environment, environmental health, epidemiology

TU-PO-65
Evaluation on the Suitability of Different Ventilation Strategies for Operating Rooms

Fan M.[1], Cao G.[2], Pedersen C.[2], Stenstad L.[2], Lv S.[1], Skogås J.G.[2]; [1]Tianjin University, Tianjin, China, [2]Norwegian University of Science and Technology, Trondheim, Norway

The performance of ventilation system in operating room is closely related to the quality of operations, since it not only controls the concentration of bacteria and particles around the wound area, but also dedicates to create an appropriate indoor environment for patients and doctors. However, high energy consumption and costs caused by the large ventilation rate are also problems that trouble the users and energy-saving department. This study aims to develop a comprehensive evaluation framework on the suitability of different ventilation strategies for existing operating rooms. Evaluation indices are selected from the aspects of ventilation effectiveness, energy consumption, lifecycle costs and users' satisfaction. Analytic hierarchy process (AHP) and fuzzy comprehensive evaluation (FCE) methods are adopted in this framework. To verify the applicability of this evaluation framework, operating rooms with laminar air flow (LAF) and mixing ventilation (MV) system in St. Olavs Hospital, Norway are used as case studies. Measurements including bacteria, particle concentration, air velocity, temperature as well as humidity etc. are conducted through mock surgeries. Also, questionnaires about thermal and acoustic comfort are distributed to surgical team staffs and patients. Meanwhile, it is expected that energy consumption benchmark of the ventilation system in the operating room in St.Olavs Hospital will be established through the analysis of actual working statistics. The significance of this study is that it may help to provide guidance for the adjustment of ventilation system in operating rooms and the selection of reasonable ventilation system for future designers.

Keywords: ventilation

TU-PO-66
Systematic Analysis of Dermal Exposure to Hazardous Chemical Agents at the Workplace (SysDEA): Study Design

Schlueter U.[1], Rother D.[1], Roitzsch M.[1], Poppek U.[1], Spaan S.[2], Franken R.[2], Gerritsen-Ebben R.[2], Kasiotis K.[3], Tsakirakis A[3], Chartzala I.[3]; [1]BAuA, Dortmund, Germany, [2]TNO, Zeist, The Netherlands, [3]BPI, Anthens, Greece

Several methods to measure dermal exposure exist, but there is no common standard. The goal of the SysDEA project was to generate scientific knowledge to improve and standardize measurement methods for dermal exposures to chemicals at the workplace. The design of an experimental setup to focus on determination of differences between sampling methods and pinpointing influencing factors is presented here. In this project, a range of experiments – considered representative for real workplace exposure situations – were carried out within five task groups: A: Transfer (pouring of low and high viscosity liquids, dumping of a dusty powder) B: Spreading (rolling low and high viscosity liquids) C: Spraying low and high viscosity liquids D: Handling of objects immersed in low and high viscosity liquids E: Handling of objects contaminated with a dusty powder In order to identify differences between the sampling methods, the experiments were performed following detailed standard operating procedures in a laboratory environment to reduce the variability in the way activities were carried out. In addition, the test room was cleaned before each experiment to prevent cross-contamination and kept at steady conditions (e.g. temperature, ventilation). Each experiment was repeated four times by four volunteers. Furthermore, this was done for three different combinations of measurement methods: • whole body dosimetry (coverall) / gloves / headband •patch sampling / hand washing / wiping of forehead • fluorescence method (photos were taken of the volunteers before and after each experiment under UV-light) This way, a total of 320 experiments were performed in the course of the SysDEA project. A comprehensive database was produced that will contribute to advance and standardize monitoring of dermal exposure to chemicals at workplaces.

Keywords: exposure factors, occupational, sampling methods, workplace

TU-PO-67

Systematic Analysis of Dermal Exposure to Hazardous Chemical Agents at the Workplace (SysDEA): Detailed Results

Meyer J.[1], Roitzsch M.[1], Poppek U.[1], Rother D.[1], Spaan S.[2], Franken R.[2], Kasiotis K.[3], Tsakirakis A.[3], Chartzala I.[3], Gerritsen-Ebben R.[2]; [1]Federal Institute for Occupational Safety and Health (BAuA), Dortmund, Germany, [2]Netherlands Organisation for Applied Scientific Research (TNO), Zeist, The Netherlands, [3]Benaki Phytopathological Institute (BPI), Athens, Greece

For the SysDEA research project, a series of standardized laboratory experiments were carried out in order to assess the effects of different sampling methods on measured dermal exposure levels. The sampling methods compared were whole body dosimetry (coveralls) vs. patch sampling, gloves vs. hand washing, and headbands vs. wipes of the forehead. The experiments were carried out with different test products: a low and a high viscous liquid and a dusty powder. In addition, a fluorescence method was tested by taking photos of the samplers worn by the volunteers before and after each experiment under UV-light. The design of the standardized experiments was inspired by activities commonly performed at workplaces and included transfer (pouring of liquids, dumping of powder), spreading of liquids by rolling, spraying of liquids, handling of objects immersed in liquids, and handling of objects contaminated with powder. This contribution aims at giving a more detailed insight into the results of the SysDEA project by means of presenting observed statistical trends with regard to: 1.) sampling technique, 2.) exposure patterns, 3.) physical state of the test substance applied, 4.) influences of exposure levels, 5.) within and between user variability. In addition, the SysDEA results will be compared with advantages and disadvantages of the different sampling methods as published in the literature.

Keywords: exposure factors, occupational, sampling methods, statistical methods, workplace

TU-PO-68

Developing a Specification for Representing Exposure Health Semantics

Gouripeddi R.[1], Habre R.[2], Prisms D.M.W.G.[3]; [1]University of Utah, Salt Lake City, UT, USA, [2]University of Southern California, Los Angeles, CA, USA, [3]NIH/NIBIB PRISMS Consortium

Representing exposure health studies and the exposome often require the use of semantic standards in order to provide a shared representation, support performance of secondary analysis and achieve reproducibility of the data and results. In order to specify appropriate semantic vocabularies and ontologies for use in research studies, and discover existing gaps needing further development of semantics, the Pediatric Research Using Integrated

Sensor Monitoring Systems (PRISMS) Data Model Working Group is following a multi-step process. First, we identified available semantic resources and literature for exposomic research. Examples of these include (1) Children's Health Exposure Analysis Resource ontology, (2) Chemical Entities of Biological Interest Ontology, (3) Environment Ontology, (4) Exposure ontology, and (5) Unified Medical Language System Metathesaurus and its component vocabularies. We are evaluating these identified resource by mapping them to data being collected within the PRISMS informatics platforms and stored in the program's data coordinating center. These include data being collected for two pilot study cohorts based in the state of Utah and in the city of Los Angeles respectively, and includes exposure data collected from existing commercial air quality sensor devices at these two study locations, and six novel air quality sensors being developed as a part of the PRISMS program. We are evaluating the identified semantic resources for their representativeness and completeness of all domains present within these pilot exposure health studies. We supplemented initially identified resources as required in order to develop a comprehensive recommendation of semantics for each domain. A draft version of the specification will be released for public comment. We envision, when completely developed, the specification will help exposure health researchers to select appropriate semantic standards for representing their data, and it will be used within informatics platforms for harmonizing results arising from multiple sensors deployed in various studies.

Keywords: environmental health, epidemiology, exposure models, sensor technology

TU-PO-69

Exposure to Environmental Chemicals in Canada: Recent data from the Canadian Health Measures Survey

Yome J.[1], St-Amand A.[1]; [1]Health Canada, Ottawa, Canada

Canada monitors exposure to environmental chemicals as part of the Canadian Health Measures Survey (CHMS), a national direct measures health survey led by Statistics Canada, in partnership with Health Canada and the Public Health Agency of Canada. This national biomonitoring program provides levels of a range of environmental chemicals measured in blood, urine, or hair. This nationally-representative data provide a base to compare with sub-populations and with other countries. It is also the only baseline data for tracking trends in environmental chemical levels over time, and to explore relationships between exposure to environmental chemicals and other health measures (e.g., blood pressure, nutrition). The CHMS is a cross sectional survey carried out in ongoing two-year cycles. Every cycle, blood and urine are collected from around 5,000 participants aged 3-79 years at 16 sites across the country. Along with biomonitoring data, the CHMS also collects information on the general health and lifestyles of Canadians to provide information on chronic and infectious disease, physical fitness, nutrition, and other factors that influence health. Five cycles of data collection have been completed over the past decade, measuring over 250 chemicals (metals, persistent organic pollutants, environmental phenols, acrylamide, pesticides, phthalates) in over 30,000 Canadians at 81 sites across the country. For many chemicals, this data provides the first national level baseline of exposure. For others, repeated measures can help track exposure trends over time. Tools have also been developed to interpret and communicate biomonitoring data in a public health context. Biomonitoring equivalents (BEs) are risk-based tools used to derive hazard quotients to identify priority chemicals for further follow up. Arsenic is a case study that demonstrates how biomonitoring data helped prioritize chemicals for further analysis. Human biomonitoring data are used to quantify human exposure and to provide information for setting priorities to protect public health.

Keywords: biomarkers, biomonitoring, chemical prioritization, environmental health, metals, source control

TU-PO-70

Influence of living in the same home on biomonitoring levels of consumer product chemicals in the Detox Me Action Kit crowdsourced study

Dodson R.[1], Boronow K. [1], Cedeño Laurent J.[2], Spengler J. [2], Brody J.[1],Rudel R.; [1]Silent Spring Institute, Newton, MA USA, [2]Harvard T.H. Chan School of Public Health, Boston, MA USA

Detox Me Action Kit (DMAK) is a biomonitoring study of volunteers in the U.S. and uses an innovative crowd-sourced recruitment strategy. Participants were recruited largely online via email and social media, contributed the cost of sample collection and laboratory analysis for environmental phenols, and were given the option to receive an interactive web-report of their personal biomonitoring levels. Of the 725 participants sampled between January 2017 and October 2018, 195 lived in the same home with one or more other DMAK participants (n = 85 pairs). In addition to homes with similar-aged adults, some homes were multigenerational (i.e., elderly adults, adults, and/or children). We analyzed data from participants from the same home to investigate the influence of shared products and behaviors or home-based exposures on biomonitoring levels. We estimated the Kendall's rank correlation, which accounts for multiply left-censored data, between pair-members. Concentrations between pair-members were significantly positively correlated for all ten chemicals. Concentrations of 2,5-dichlorophenol, a metabolite of the deodorizer and pesticide 1,4-dichlorobenzene, was the most strongly correlated between pair-members (tau = 0.41), followed by benzophenone-3, a UV filter (tau = 0.33), and bisphenol A (tau = 0.20). Between-pair correlations suggest shared exposure sources, either through similar behaviors and product use or indirect exposures to contaminated air or dust within the home. Several self-reported behaviors were similar between pair-members, including drinking from a plastic water bottle and eating canned food in the last 24 hours. The majority of participants in the study reported avoiding at least one of four chemical types. Avoidance behavior was significantly concordant within the pairs (kappas = 0.28 – 0.43), meaning that pair-members had similar attitudes about avoiding chemicals. Adults may also compel environmental behaviors for dependents through household purchasing. Results suggest that intervention strategies to reduce exposure to environmental phenols may have impacts beyond the individual.

Keywords: biomonitoring, consumer product chemicals, phenols, shared environment

TU-PO-71 MOVED TO MO-SY-C2-90

WEDNESDAY SESSIONS

Wednesday, August 21st

WE-PL-A1-EHR: Health Impacts from Heat and Chemicals

WE-PL-A1-01

Occupational Heat Stress induced health impacts – A cross-sectional study from Southern India

Venugopal V.[1], Latha P.K.[1], Rekha S.[1], Krishnamoorthy M.[1], Nandhini K.[1], Johnson P.[2]; [1]Public Health, Sri Ramachandra Institute of Higher Education and Research, Chennai, India, [2]Department of Physiology, Sri Ramachandra Institute of Higher Education and Research

Background: Changing climate and rising temperatures will potentially affect millions of workers due to risks of heat stress in tropical settings. Chronic high-heat exposures impose physiological strain on the workers and the impacts are varied, influenced by variety of factors. Methodology: We used a cross-sectional study design to profile the heat stress exposures of ~1900 workers from 7-industrial sectors using a QuesTemp WBGT monitor and objectively quantified heat strain indicators viz., Core Body Temperature, Sweat Rate and Urine Specific Gravity. We evaluated the health and social impacts by a structured interview. Findings: Heat exposures exceeded the threshold limit value for safe manual work for moderate/heavy work (Avg. WBGT of 30.1°C±2.6°C) in almost all workplaces. The workplace WBGT was positively associated with rise in CBT >1°C (13%) and elevated USG >1.020 (9%) in workers. Widespread concerns about heat-related health issues were reported by the workers (86%) with excessive sweating, fatigue, tiredness and fainting being common issues. Heat-exposed workers had higher odds of adverse health outcomes compared to workers with exposures to WBGTs less than >27.5°C (Adj.OR=2.43, 95% CI 1.88-3.13, p=<0.0001). Prevalence of kidney stones among steel workers (9%) and adverse renal health in salt-pan workers (OR:2.8, 95% CI 1.02-7.73, p=<0.004) was significantly associated with exposure to higher WBGT. Years of exposure, workload and type of sector had major influencing role in disease development. Conclusion: The findings clearly identify that high heat exposures causes adverse health outcomes in the exposed worker population and protection via cooling intervention from heat stress is mandatory as climate change proceeds. Further in-depth assessments with a seasonal approach are warranted to develop strategies for interventions and comprehensive protective labor policies to avert adverse occupational health and productivity consequences for millions of workers globally.

Keywords: thermal environment

WE-PL-A1-02

Association of short-term exposure to air pollution particulate matter with atrial fibrillation in high risk patients

Gallo E.[1], Folino F.[2], Bottigliengo D.[1], Lanera C.[1], Gregori D.[1], Iliceto S.[2]; [1]Unit of Biostatistics, Epidemiology and Public Health, Department of Cardiac, Thoracic, Vascular Sciences and Public Health, University of Padova, Padove, Italy, [2]Department of Cardiac, Thoracic, Vascular Sciences and Public Health, University of Padova, Padova, Italy

Background: Several epidemiological studies have found an association between short-term exposure to fine particulate matter of less than 2.5 µm and 10 µm in aerodynamic diameter (PM2.5 and PM10) and cardiovascular mortality, cardiovascular diseases and ventricular fibrillation incidence. Effects of air pollution on atrial fibrillation (AF) beyond the first few hours of exposure are still controversial. Methods: Patients with implantable cardioverter-defibrillators (ICDs), cardiac resynchronization therapy defibrillators (ICD-CRT) or pacemakers were prospectively enrolled in this multicenter study. Median daily levels of PM2.5, PM10, temperature, pressure and humidity were collected from monitoring stations within 20 km from the patient's residence. A bias-reduces logistic regression model was implemented to evaluate the association between AF and daily exposure to PM2.5 and PM10. Results: Of one-hundred-forty-five patients enrolled, thirteen of them developed AF. Exposure levels to PM2.5 and PM10 were above the World Health Organization (WHO) thresholds of 25 µg/m3 and 50 µg/m3 respectively in 26% and 18% of the follow-up days. An association was found between daily levels of both PM2.5 and PM10 and AF (95% confidence interval (CI) respectively 1.34 – 2.40 and 1.44 – 4.28). Conclusion: Daily exposure to moderate PM2.5 and PM10 levels seems to be associated with AF.

Keywords: air, health, other/general, risk assessment

WE-PL-A1-03
Air pollution is associated with Pediatric Emergency Room admissions for bronchiolitis

Gallo E.[1], Bottigliengo D.[1], Bressan S.[2], Geremia S.[3], Lanera C.[1], Zare Jeddi M.[1], Gregori D.[1]; [1]Unit of Biostatistics, Epidemiology and Public Health, Department of Cardiac, Thoracic, Vascular Sciences and Public Health, University of Padova, Padove, Italy, [2]Paediatric Emergency Unit, Department of Women's and Children's Health, University Hospital of Padova, Padova, Italy, [3]University of Bologna, Bologna, Italy,

Background: in the last decade air pollution has been linked to an increased risk of several respiratory diseases in children, in particular respiratory tract infections. The present study aimed to evaluate the association between Pediatric Emergency Room (PER) admissions for bronchiolitis and air pollution. Methods: PER admissions due to bronchiolitis in children aged less than 1 year were retrospectively collected from 2007 to 2018 in Padua, Italy. Daily environmental data were obtained from monitoring stations located within 30 km from the children residence. A conditional logistic regression based on a time-stratified case-crossover design was performed to evaluate the association between PER admissions and exposure to nitrogen dioxide (NO2), particulate matter of less than 2.5 µm (PM2.5) and less than 10 µm (PM10) in aerodynamic diameter. Models were adjusted for atmospheric pressure, temperature, humidity and national holidays. Odds Ratio for lags from 0 to 15 and the average of the 7 days before the access (lag07) were obtained. A secondary analysis was focused only on children that were hospitalized. Results: 2333 children presented to the PER for bronchiolitis. An association was found on lag 0, lag 1, lag 3 and lag07 with both PM2.5 and PM10. Considering only children with a more severe outcome that led to hospitalization, the association was statistically significant until lag 5 and on lag07. Major effects of NO2 on PER admission and hospitalization were found from lag 5 to lag 13. Conclusion: PM and NO2 are associated with PER admissions for bronchiolitis. Exposure of infants to air pollution could damage the respiratory tract mucosa facilitating viral infections and exacerbating symptoms.

Keywords: air, children, respiratory health, risk assessment

WE-PL-A1-04
Potential reductions in premature mortality attributable to PM2.5 by reducing indoor pollution: a model analysis for Beijing-Tianjin-Hebei of China

Ji W.; Civil and Resource Engineering School, China

China has one of the highest PM2.5 (particulate matter with an aerodynamic diameter smaller than 2.5 µm) pollution levels in the world. It might still be long before air quality reaches the National Class II standard 35 µg/m3. We combined outdoor PM2.5 migration model and the Global Burden of Disease 2016 methodology to estimate potential reductions in premature mortality attributable to PM2.5 by reducing indoor PM2.5 to National Class I standard 15 µg/m3, and compared with reducing outdoor PM2.5 to Government 2020 Interim target of 64 µg/m3 or National Class II standard 35 µg/m3. Causes of deaths included ischemic heart disease (IHD), stroke, chronic obstructive pulmonary disease (COPD), and lung cancer (LC) for adults; and acute lower respiratory illness (ALRI) for children. A total of 74,000 (95% uncertainty interval: 43,000-111,000) premature deaths were attributable to PM2.5 exposure in 2013. Thirty percent, or 22,000 (17,000-32,000) deaths, would have been averted if indoor PM2.5 had reached the National Class I standard. The benefit is greater than that from reaching the Government 2020 Interim target for outdoor PM2.5 [22%, or 16,000 (12,000-23,000), deaths], although still smaller than that from reaching the National Class II standard [42%, or 31,000 (24,000-45,000), deaths]. A larger relative reduction in deaths was seen in regions with higher PM2.5 levels, e.g., in the southern part of the region, and among children and older people. Reaching the National Class I level of indoor PM2.5 at current outdoor pollution levels could bring considerable health benefits, which are comparable to those from reaching the Government 2020 Interim target for outdoor PM2.5. A strategy that involves both indoor and outdoor cleaning is needed to balance the immediate reduction in PM2.5 exposure and the eventual clean-up of air pollution.

Keywords: aggregate exposure, air cleaning, built/indoor environment, particles, risk assessment

WE-PL-A1-05
Health risk of indoor air pollutants in primary school buildings in Central Europe

Szabados M.[1], Kotlík B.[2], Kazmarová H.[2], Kozajda A.[3], Jutraz A.[4], Kukec A.[4], Otorepec P.[4], Dongiovanni A.[5], Di Maggio A.[5], Szigeti T.[1]; [1]National Public Health Center, Budapest, Hungary, [2]National Institute of Public Health, Prague, Czech Republic, [3]2 Nofer Institute of Occupational Medicine, Lodz, Poland, [4]Slovenian National Institute of Public Health, Ljubljana, Slovenia, [5]LINKS Foundation, Turin, Italy

Children are exposed to wide variety of hazardous chemicals when they spend a certain amount of their time in primary schools. In order to develop action plans to improve indoor air quality (IAQ) in schools, it is necessary to investigate the concentration of chemical and biological air pollutants and physical parameters as well as to assess the health risk related to them. IAQ was investigated in 64 primary school buildings in 5 Central European countries during the heating season of 2017/2018 in the frame of the InAirQ project. Risk assessment was conducted to assess both carcinogenic and non-carcinogenic risks attributable to air pollution for the time period when children present at school. The excess lifetime cancer risk (ELCR) was estimated by the product of the exposure concentration and the inhalation unit risk. Hazardous quotient (HQ – ratio of exposure concentration to an inhalation reference concentration) was calculated to evaluate the non-carcinogenic risk for each compound. The health risk associated with the co-exposure to multiple air pollutants was assessed by the hazard index (HI - sum of the HQ values). Furthermore, the maximum cumulative ratio (MCR) approach was applied to determine whether one compound is responsible for the health risk or more. The sum of ELCR values was higher than 10-6 in the case of several primary schools which means that a certain portion of the Central European pupils are exposed to unacceptable risk. Neither the HQ nor the HI values exceeded one which means that the non-carcinogenic risk associated with the investigated air pollutants might not cause adverse health effects during the time when the schoolchildren are in the school building. The MRC values were mostly between 2 or 3 which means that, in general, the only a couple of compounds having HQ values are relevant for the human health.

Keywords: built/indoor environment, environmental health, health, other/general, risk assessment

WE-PL-A1-06
Impact of seasonal trends and external windows structures on both indoor and outdoor PM2.5 mass concentrations at several residential buildings in Beijing, China

Chen Z.[1], Cao G.[2], Chen F.[1], Hui Zhao[1], Gao P.[1]; [1]Beijing University of Technology, Beijing, China, [2]Institute of Building Environment and Energy, China Academy of Building Research, Beijing, China

This paper investigated the variation characteristics of ambient atmospheric PM2.5 and its influence on indoor environment, with considering several monitored residential buildings in Beijing, China. Conditions were considered for the room by air infiltration, with all the external windows closed and none indoor particle sources. Real-time monitoring of indoor and outdoor PM2.5 mass concentrations was respectively made for one year from March 2016 to February 2017. Meteorological parameters were monitored simultaneously to be correlated with

PM2.5 mass concentrations and indoor-outdoor concentration ratios. The results indicated: 1) Size of more than 90% of atmospheric particles were under 1μm. 2) The monthly-average concentrations of outdoor and indoor PM2.5 tended to be highest in winter, followed by autumn, spring and summer. Additionally, the average I/O ratios in summer were lower compared to other three seasons. 3) PM2.5 mass concentrations were closely correlated with wind speed and relative humidity. However, outdoor temperature played a negligible role in PM2.5 mass concentrations. 4) Better window airtightness will help to remain good indoor air quality by preventing more outdoor PM2.5 going indoors. The prevention ability of windows on outdoor PM2.5 is dependent on the dust diameter.

Keywords: air, built/indoor environment, environmental health, exposure factors, particles, sampling methods

WE-SY-B1: Exposure Science Studies from Asian Perspectives – Environmental and Study Diversities Among Asian Countries

WE-SY-B1-07

Exposure to organophosphate esters in Japan: associations among their concentrations in house dust, urinary metabolite levels, and allergies

Araki A.[1], Ait Bamai Y.[1], Kishi R.[1]; [1]Hokkaido University Center for Environmental and Health Sciences, Sapporo, Hokkaido/Japan

Organophosphate esters are a class of chemicals predominantly used as additives in flame retardants and plasticizers. Phosphorus-based flame retardants exhibit fastest market gains of any major product type. According to the HIS Consulting report, Asia consumed the largest volume of flame retardants in 2016 with a 50% share. In Japan, the use of polybrominated diphenyl ethers diminished in the early 1990s for voluntary controls, and phosphate flame retardants (PFRs) are now the most frequently used organic flame retardants. Consequently, people are likely to be exposed to PFRs ubiquitously, as many PFRs are found in indoor environments such as indoor air and house dust, as well as urine. We conducted 2 studies in newly built detached houses in 6 regions of Japan and a school children's home in Sapporo. House dust and children's urine were collected to measure PFRs and their metabolites. Levels of tris(2-butoxyethyl)phosphate (TBOEP) are extremely high in Japan, whereas other PFR levels are comparable with those in other countries. Levels of TBOEP and tris(1-chloro-2-propyl)phosphate (TCIPP) in dust and their metabolite in urine, respectively, showed significant correlations. A potential TBOEP exposure source is wooden flooring material. We found significant associations among increasing levels of tris(1,3-dichloro-2-propyl)phosphate (TDCIPP) and TCIPP in house dust and increasing risk of atopic dermatitis, as well as increasing levels of tri(n-butyl)phosphate and asthma among participants of the 6-region study. Among elementary school children in Sapporo, we found an association between TDCIPP in dust and children's eczema, as well as TCIPP and TBOEP metabolites in children's urine and any allergic symptoms. One major limitation is that these studies are cross-sectional, so we cannot confirm the causality of these associations. The Asia-Pacific region dominates the global phosphorus-based flame retardant market. Considering these points, more studies are needed to determine exposure levels and their health effects.

Keywords: biomonitoring, environmental health, epidemiology, flame retardants, health, symptoms

WE-SY-B1-08

Can we estimate personal PM exposure using GPS?

Lee K.[1], Hwang Y.[1], Park J.-Y.[1]; [1]Seoul National University, Seoul, Korea

[Background/Aim] Indirect estimation of personal exposure could be improved with new technology such as global positioning system (GPS). The main purpose of this study was to estimate personal PM10 exposure using personal GPS information and PM10 data from the fixed site monitoring stations.

Keywords: exposure models

WE-SY-B1-09

Air Pollution and Health Impact in Thailand

Tantrakarnapa K.[1], Mueller W.[2], Steinle S.[2], Loh M.[2], Sotiris V.[2], Precha N.[1], Kliengchuay W.[1], Sahanavin N.[3], Sillaparassamee R.[4], Nakhapakorn K.[5], Cherrie J.[6]; [1]Department of Social and Environmental Medicine, Faculty of Tropical Medicine, Mahidol University, Bangkok, Thailand, Bangkok, Thailand, [2]Research Division, Institute of Occupational Medicine, Edinburgh, Scotland, UK, [3]Srinakharintarawiroj University, Bangkok, Thailand, [4]Valaya Alongkorn Rajabhat University, Pathumthani, Thailand, [5]Environment and Natural Resources, Mahidol University, Bangkok, Thailand, [6]Research Division, Institute of Occupational Medicine, Institute of Biological Chemistry, Biophysics and Bioengineering, Heriot Watt University, Edinburgh, Scotland, UK

Thailand is facing to air pollution from traffic, biomass burning and other sources that constitute a major threat to human health. However, the adverse health impacts and underlying mechanisms of different air pollutants are not clearly understood. Particulate matter has been observed in high concentration in many places, including Bangkok, the capital of Thailand. Our study used the secondary data of particulate matter from automatic monitoring stations located in Bangkok operated by the Pollution Control Department and health related data from Ministry of Public Health during the same period. The particulate matter data are available for PM2.5 and PM10, although PM2.5 monitoring only commenced in 2011. We found that average PM2.5 concentrations in Bangkok have decreased from 41.5 to 18.5 μg/m3 between 2011 and 2017. However, during a haze episode in Bangkok the daily concentration of PM2.5 and/or PM10 were over the ambient air quality standard; the average ratio of PM2.5/PM10 was 0.6. In addition, human exposure was measured among groups of people in Bangkok by using MicroPem monitors with associated the time activity data. The results indicated that PM2.5 exposure in residential and working spaces were less than outdoor exposure levels. These data will be used in a Health Impact Assessment in different regions of Thailand with various proposed future scenarios involving mitigation measures.

Keywords: health, other/general

WE-SY-B1-10

Levels and temporal variations of urinary lead, cadmium, cobalt, and copper exposure in the general population of Taiwan

Huang P.C.[1], Liao K-W.[1], Pan W-H.[1], Wang S-L.[1]; [1]National Institute of Environmental Health Sciences/National Health Research Institutes, Miaoli, Taiwan

Toxic metal contamination in food products and the environment is a public health concern. Therefore, understanding human exposure to cadmium (Cd), lead (Pb), cobalt (Co), and copper (Cu) levels in the general population of Taiwan is necessary and urgent. We aimed to establish the human biomonitoring data of urine toxic metals, exposure profile changes, and factors associated with metal levels in the general population of Taiwan. We randomly selected 1601 participants older than 7 years of age (36.9 ± 18.7 years (7–84 years)) from the Nutrition and Health Survey in Taiwan (NAHSIT) conducted during 1993–1996 (93–96) and 2005–2008 (05–08) periods and measured the levels of four metals in the participants' urine samples using inductively coupled plasma-mass spectrometry. The median (range) levels of urinary Cd, Pb, Co, and Cu in participants from the NAHSIT 93–96 (N = 821)/05–08 (N = 780) were 0.60 (ND–13.90)/0.72 (ND–7.44), 2.28 (ND–63.60)/1.09 (0.04–48.88), 0.91 (0.08–17.30)/1.05 (0.05–22.43), and 16.87 (2.62–158.28)/13.66 (1.67–189.70) μg/L, respectively. We found that the urinary median levels of Pb and Cu in our participants were significantly lower in the NAHSIT 05–08 (Pb 1.09 μg/L, Cu 13.66 μg/L) than in the NAHSIT 93– 96 (Pb 2.28 μg/L, Cu 16.87 μg/L; $P < 0.01$), whereas those of Cd and Co were significantly higher in the NAHSIT 05–08 (Cd 0.72 μg/L, Co 1.05 μg/L; $P < 0.01$). Youths had higher exposure levels of Pb, Co, and Cu than adults. Participants with alcohol consumption, betel quid chewing, or cigarette smoking had significantly higher median levels of urinary Pb or Cu ($P < 0.01$) than those without. Principal components and cluster analysis revealed that sex had different exposure profiles of metals. We concluded that levels of urinary Cd, Pb, Co, and Cu exposure in the general Taiwanese varied by age, sex, and lifestyles.

Keywords: biomonitoring, environmental health, exposure factors, metals

WE-SY-B1-11
Evaluation of nanomaterial exposure and estimation at the respiratory track deposition during the use of spray products

Yoon C.S.; Graduate School of Public Health, Seoul National University, Seoul, South Korea

Nanotechnology-based consumer products have been used wide over the past decade. We evaluated the airborne nanomaterials during the use of spray consumer products and estimated the deposited dose at the lung. Five propellant and three pump types of spray products were tested to evaluate the airborne nanoparticles across time and space in a well-controlled cleanroom. Direct-reading instruments (Scanning Mobility Particle Sizer & Optical Particle Counter) with a thermodesorber unit were used to measure the airborne particles. Also. filter sampling was conducted to examine physicochemical characteristics of the particles. Sampling was conducted simultaneously by distance (1, 2, 3 m) from the spraying source. The inhaled doses at the breathing zone and the deposited dose in each part of the respiratory tract were estimated using the experimental data and mathematical models. High concentration of nanoparticles released from the propellant sprays persisted long in the air and dispersed over a far distance due to their small size. But, the pump spray products produced larger droplets that settled down quickly from the air, so the concentration was similar to background level. The estimates of inhalation exposure also suggested that exposure to nanoparticles was much higher with propellant spray products than pump spray products. We found that much higher number of nanoparticles could be deposited in alveolar region during propellant types use than during pump types use. We found that the spray nozzle type was a crucial factor determining inhalation exposure. Propellant sprays released a larger quantity of nanoparticles which dispersed over a greater distance and persisted for a longer time in the air due to their small size. Also, higher number concentration of them could go and deposited at the alveolar region. The concentration and deposition of nanomaterial at the respiratory track from the propellant spray products were much higher than the pump types.

Keywords: consumer and personal care products, exposure models, nanoparticles, particles, respiratory health

WE-SY-C1-12

Microbial and non-microbial volatile organic compounds from indoor materials subjected to dust and moisture

Marciniak K., Haines S., Adams R., Dannemiller K., Goldstein A., Heal M., Misztal P.

Humans spend the majority of their time in an indoor environment where they are exposed to microbiological and chemical processes that influence the air quality. Volatile organic compounds (VOCs) are emitted from various indoor sources and derive from chemical processes and microbial metabolism. Microbial VOC (mVOC) are produced by bacteria and fungi and their production is strongly correlated with water activity and the surface that they grow on such as carpet, wood or drywall. The goal of the Chemistry of Homes – Environmental Microbes and Moisture (CHEMM) project funded by the Alfred P Sloan Foundation is to understand the roles that microorganisms play in shaping the indoor environment, with a specific focus on the role of humidity and how it affects chemical activity of the microbiome in terms of changes in the magnitude and composition of mVOC emissions. Here, we show recent results from chemistry and microbiology experiments. We focus on the total and speciated time-resolved emission rates of VOCs from carpets and other substrates inoculated with natural and sterile indoor dust. We show the transition in chemical composition and VOC abundance dynamics as a function of air humidity. We track the enhancements and depletions of volatile constituents over time and discriminate between microbial and non-microbial VOCs to ultimately understand unknown mechanisms behind the uptake and production of VOCs and to quantify the role of air humidity on shaping microbial life on these microhabitats.

Keywords: VOCs

WE-SY-C1-13

Indoor measurements and chemistry of chlorine bleach-related compounds

Mattila J.[1], Farmer D.[1], Lakey P.[2], Shiraiwa M.[2], Wang C.[3], Abbatt J.[3], Ampollini L.[4], DeCarlo P.[4], Zhou S.[5], Kahan T.[5]; [1]Colorado State University, Fort Collins, CO, USA, [2]UC Irvine, Irvine, CA, USA, [3]University of Toronto, Toronto, Canada, [4]Drexel University, Philadelphia, PA, USA, [5]Syracuse University, Syracuse, NY, USA

We measured indoor and outdoor gas-phase compounds with time-of-flight chemical ionization mass spectrometry (TOF-CIMS) during the House Observations of Microbial and Environmental Chemistry (HOMEChem) field campaign at the University of Texas Austin in the USA in June 2018. One set of experiments included sequential bleach mopping, while another set involved bleach mopping after large cooking events. Indoor gas-phase mixing ratios of hypochlorous acid (HOCl), chlorine gas (Cl2) and nitryl chloride (ClNO2) increase to part-per-billion by volume (ppbv) levels during these periods, substantially higher than outdoor concentrations. These compounds and their associated chemistry influences organic chemistry and radical oxidation in indoor air. The photolysis of these compounds can produce hydroxyl (OH) and chlorine atom (Cl) radicals indoors, thereby impacting the oxidative capacity of air in the built environment. We observe stark decreases in indoor isoprene, α-pinene, and limonene mixing ratios during periods of bleach mopping, consistent with this oxidation chemistry. Indoor multi-phase reactions of ammonia (NH3) and nitrous acid (HONO) with bleach-related compounds produce chloramines (NH2Cl, NHCl2, NCl3) and ClNO2, respectively. We observe elevated indoor concentrations of potentially toxic compounds from bleach mopping, including isocyanates (HNCO, CH3NCO, C2H5NCO; toxic and irritating to a variety of organs), nitrogen trichloride (NCl3; respiratory irritant), and chloroform (CHCl3; carcinogenic). Future directions include chemical modeling to better understand the oxidant budget and multi-phase chemistry during bleach cleaning.

Keywords: air, analytical methods, built/indoor environment, health, other/general

WE-SY-C1-14

Are Indoor Surfaces Dropping Acid or Dropping the Base? Insights into Water Films and pH for Model and Authentic Indoor Samples

Ault A.[1], Cooke M.[1]; [1]University of Michigan, Department of Chemistry, Ann Arbor, MI, USA

Indoor surfaces are complex environments that play an important role in driving gas and particle phase chemistry. These surfaces can have heterogeneous uptake, condensed phase reactions, and emitted reaction products altering the composition of indoor environments. Despite their important role, indoor surfaces remain a mystery due to their complex and difficult to probe nature. A fundamental question that is not well understood is the acidity or basicity of indoor environments and the impact that may have on indoor gas and particle chemistry. This presentation will discuss initial results aimed at probing the properties, composition, and pH of water films on common indoor materials and model systems through systematic studies with Raman microspectroscopy, atomic force microscopy with photothermal infrared spectroscopy, and other characterization methods. These initial results offer an early glimpse into these complex systems and their role in indoor atmospheric chemistry.

Keywords: air, analytical methods, sampling methods, surfaces

WE-SY-C1-15

Antibacterial Toothbrushes – What Effects do Nano-Enabled Products Have on Microbial Communities?

Johnson, C., Hartmann, E.

Over the past two decades, silver and zinc nanoparticles have been integrated into various consumer products for the purpose of killing bacteria associated with these products. While metal nanoparticles have antibacterial properties, their effects on the complex microbial communities found in indoor environments, as well as the downstream human exposures and health outcomes, are unknown. In this study, we use a multifaceted approach to examine a nanoparticle-enabled product that also serves as a conduit for human exposure to bacteria: toothbrushes. We combine molecular biology, microscopy, and chemical techniques as a means to better understand the interactions between typical toothbrush-associated bacteria and the nanoparticles embedded in toothbrush bristles. More specifically, we determine characteristics relevant to the antibacterial properties of the nano-enabled toothbrushes including: nanoparticle concentration, size, release, and solubility, as well as toothbrush bacterial adsorption capacity. We then evaluate the effects that these toothbrush characteristics have on the toothbrush microbiome. Particularly, we study bacterial community dynamics, bacterial pathogen propagation, and mobile genetic element dissemination, all of which relate bacterial infection risks associated with using nanoparticle-enabled toothbrushes. The results of our study help to better inform us of the potential health effects associated with the inclusion of nanoparticles in consumer products.

Keywords: consumer and personal care products, metals, microbial agents, nanoparticles, particles

WE-SY-C1-16

Confined built environments reveal surprising dynamics of their microbiome and resistome

Mahnert A.[1], Verseux C.[2], Koskinen K.[1], Wink L.[1], Brunner D.[1], Gößler T.[1], Schwarz A.[1], Kumpitsch C.[1], Blohs M.[1], Moissl-Eichinger C.[1]; [1]Interactive Microbiome Research, Department of Internal Medicine, Medical University of Graz, Center for Medical Research (ZMF), Graz, Austria, [2]Center of Applied Space Technology and Microgravity (ZARM), University of Bremen, Bremen, Germany

During space simulation experiments the crew's safety and health are challenged by an extreme confinement from the surrounding hostile environment with specific implications and exposures for the human microbiome. To better understand microbial dynamics in isolated and confined built environments (ICE), front torso skin of 6 crewmembers and frequently interacted surfaces of the furniture in an only 11-meter-in-diameter dome located on the barren slopes of the Mauna-Loa volcano in Hawaii were sampled for a whole year (HI-SEAS IV). The crew was isolated and confined, and hygiene practices were restricted. 181 microbiome samples were analyzed by 16S rRNA gene profiling and qPCR as well as the antimicrobial resistance profile (covering detergents and multidrug resistances e.g. qacE-delta1, int1, bla and tetM). In contrast to our previous study (Schwendner et al. 2017) the microbial diversity increased slightly in the built environment, and even stronger and significantly ($P = 8.7 \times 10E\text{-}19$) on the skin of the crew members. Metadata predictions showed high model accuracy (97%) for different sampling categories and the day of sampling ($R = 0.8$, $P = 2.3 \times 10E\text{-}8$). Differential abundance analysis indicated a discriminative pattern for microbial communities at the beginning (Gardnerella) and end (Kocuria, Dermacoccaceae, Brevundimonas) of the isolation period. Important microbial and metabolic features to predict sample states were identified by a supervised learning regressor. Hence, detected microbes faced their exposure to this ICE by many detoxification mechanisms against stress inducing factors from the environment or its maintenance by the crew (e.g. mycothiol and ectoine biosynthesis). Overall the interaction of the skin microbiome and the microbiome of the built environment seems to be highly dynamic especially in this confined habitat. Our attempts to understand the dynamics of the microbiome and resistome could present a first step to manage such built environments more expedient in the future.

Keywords: biomonitoring, built/indoor environment, longitudinal metrics, statistical methods

WE-SY-C1-17
Chemical properties and transformations of complex organic mixtures deposited on indoor surfaces

O'Brien R.[1], Przelomski H.[1]; [1]College of William and Mary, Williamsburg, VA, USA

Indoor areas have larger surface area to volume ratios and slower air exchange rates compared to outdoors, thus increasing the importance of surfaces in indoor chemistry. Semi-volatile and lower volatility organic material can deposit on indoor surfaces, building up a film that serves as both a source and a sink for organic molecules in indoor air. The chemical composition of this material and its physical properties, including viscosity, will play a large role in the interactions between the gas-phase and the surface films. Here, the chemical composition of the lower volatility fraction of material deposited on glass surfaces in a kitchen is characterized using both ultra-high resolution mass spectrometry (UHR-MS) and offline-Aerosol Mass Spectrometry (AMS). The average molecular weight of these compounds is around 900 amu with an average carbon number of around 50. A comparison with online-AMS shows good overlap between the composition of surface film and aerosol particles generated by food cooking over the same time period. This organic mixture is also sensitive to aging processes including photolysis and ozonolysis. Here, a decrease in the average molecular weight observed in both cases, consistent with fragmentation and volatilization of a fraction of the material. This work demonstrates the advantages of UHR-MS and offline AMS to characterize these films from indoor spaces. A better understanding of the composition and transformations of these films will provide insights into indoor chemical transport and exposure with impacts on human health in indoor spaces.

Keywords: SVOCs, analytical methods, built/indoor environment, particles

WE-PL-D1-18

The hexadecane/air distribution coefficient - a key parameter for predicting the gas/particle distribution of SVOCs

Salthammer T., Goss K.-U.[2]; [1]Fraunhofer WKI, Braunschweig, Germany, [2]Helmholtz Centre for Environmental Research - UFZ, Leipzig, Germany

We used poly-parameter Linear Free Energy Relationships (pp-LFER) to describe the partitioning behavior of 14 non-ionizable SVOCs with high relevance for the indoor environment. The pp-LFER concept is based on Abraham descriptors and considers interactions between molecule and particle by separate parameters. Van der Waals interactions can be approximated by the logarithm of the hexadecane/air partitioning coefficient L, which is the single most important descriptor for systems where one of the phases is a gas. L is experimentally accessible because the gas chromatographic retention time of an organic substance is proportional to the stationary phase/gas partition coefficient in an isothermal gas chromatographic system with a suitable, completely non-polar column. For the 14 compounds of interest we saw the necessity to determine the L-values experimentally although QSPR estimations are available, because the accuracy of L is crucial for the overall result. Moreover, the expected L-values of our target compounds lie at the high end or even beyond the existing calibration data set of the QSPR method. In the gas chromatographic experiments the known L-values of the calibration compounds DEP, DnBP and DEHP at 298 K were directly correlated with the logarithm of the retention time at 250 °C and 280 °C, respectively. For reasons of quality control, the experiments were performed at two different temperatures, because the calculated L-values should be independent of the absolute retention times. The L-values could be determined with high accuracy and ranged between 6.75 (DEP) and 18.87 (TOTM). The experimental data were compared with calculated L-values from QSPR and with calculated octanol/air partitioning coefficients. They were found to be very useful for describing of the gas/particle partitioning of SVOCs and for describing the sorption of SVOCs into a pure organic phase.

Keywords: SVOCs, air, flame retardants, particles, phthalates

WE-PL-D1-19

The Use of Probability of Addition to Improve Dietary Intake Estimates of Flavour Substances

Vilone G.[1], Slickers N.[2], Smith B.[3], Hickey C.[2]; [1]Creme Global, Dublin, Ireland, [2]Firmenich Inc., Plainsboro, NJ, USA, [3]Innovations in Food and Chemical Safety Programme Agency for Science, Technology and Research, Republic of Singapore

A common source of overestimation of dietary exposure is the assumption that a flavouring substance (FS) is always present in a given food category. Currently, intake estimates are calculated based on the FS concentrations in a limited set of food categories and the amount of food in that category consumed daily. It is assumed that all foods in a specific food category contain the FS at the same concentration. To refine this assumption in exposure models, the probability of addition (POA), i.e. the percentage of foods in a given food category that contains the FS, can be utilised. The goal of this study was to develop a methodology to estimate the POA for candidate FSs and to measure their impact in a dietary exposure assessment. This analysis was based on a set of data for benzaldehyde and bornyl acetate, used in flavour formulae that would be incorporated in several food categories. Firmenich provided data on the concentration of these two FS in sold flavourings; the typical concentration of flavourings in a specific food category; the volume of flavourings containing these two FSs sold for specific food categories; and the total volume of all flavourings sold for a specific food category. The POA for each food category was estimated by calculating the fraction of sold flavourings containing the FS over the total volume of flavours sold. The estimated POAs and the concentration data were used as inputs in Creme

Global's Expert Models® (EM) package. The model was run on the dietary part of NHANES 2013-14 to assess the exposure to the two FS in the entire US population. Resulting exposure estimates for both FSs with POA are significantly lower than those without. This analysis made clear that the use of POA's provided a more refined estimation of exposure from flavour consumption.

Keywords: aggregate exposure, exposure factors, exposure models, food

WE-PL-D1-20

Quantifying how Mobility Impact Air Pollution Exposure Estimation Using a Large Cell Phone Location Dataset

Ivey C.[1], Yu X.[2], Henneman L.[3], Yu H.[2]; [1]University of California, Riverside, CA, USA, [2]University of Central Florida, Orlando, FL, USA, [3]Harvard University, Boston, MA, USA

The spatiotemporal movement of human individuals substantially impacts air pollution exposure estimation. However, mobility is usually neglected in exposure estimation due to lack of data, though exposure misclassification errors are likely. How mobility impacts the results of exposure estimation at population and individual levels remains under-investigated. In this study, we applied a large cell phone location dataset containing over 35 million location records collected from 310,989 subjects to investigate how different levels of mobility impact each subject's estimated exposures for five chosen ambient pollutants. Additionally, we also applied and compared the results of exposure estimates based on concentration fields developed using two different methods, including one based on CMAQ model outputs and another based on the inverse distance weighting (IDW) method. Our results showed that, including detailed mobility information does not considerably impact the estimated exposures averaged across the entire population, although the impact of mobility at the individual level can be substantial. We found that the errors in exposure estimates and exposure misclassifications generally increase with increased mobility when the home-based method was applied. Neglecting mobility will also likely underestimate exposures to traffic-related pollutants particularly during afternoon rush-hour, but overestimate exposures to ozone during mid-afternoon. We also found that pollutant concentration fields generated using the IDW method are smooth and are not suitable for exposure estimation when detailed mobility data were considered. Our findings highlighted the tremendous potentials of using cell phone location data in air pollution exposure estimation for a large population, and our results have significant implications for future air pollution health studies.

Keywords: activity patterns, air, exposure models

WE-PL-D1-21

Exposure assessment of urban greenness using satellite-derived normalized difference vegetation index (NDVI): the impact of spatial resolution on exposure misclassification

Jimenez R.[1], Lane K.[1], Fabian P.[1]; [1]Boston University School of Public Health, Boston, MA, USA

Epidemiological studies analyzing the association between urban greenspace and health are increasingly using satellite-derived normalized difference vegetation index (NDVI) to assess greenness exposure. In this study we quantified the effect of using NDVI data at different spatial resolutions and aggregation areas to assess its potential for greenness exposure misclassification. We estimated NDVI from satellite data at different spatial resolutions across different levels of geographic aggregation (census block, block group, and tract) for the 2016 summer season in the metro Boston area. NDVI was calculated using remote sensing data from MODIS (250m x 250m), Landsat 8 OLI/TIRS (30m x 30m) and Sentinel 2 (10m x 10m). We calculated mean NDVI at each census geography class, and analyzed their distributions as continuous and categorical variables (quantile). We explored patterns and differences in continuous variables using descriptive statistics, t-tests and Pearson correlation analysis, and compared NDVI categories to identify discrepancies in exposure classification. The largest differences in NDVI range and distribution were observed at the block level, followed by block group and census tract. Mean estimates of NDVI from MODIS were consistently higher than estimates from Landsat 8 and Sentinel 2 across census geographies in the study area. NDVI quantiles varied across spatial resolution and census geography, yielding substantial differences in categorical exposure classification. The level of agreement between NDVI exposure categories for the three satellite products and census geographies ranged from 48% (quintiles, census blocks) to 89% (tertiles, census tracts). Greenness exposure estimates from satellite-derived NDVI are not invariant to spatial resolution of remote sensing data. Researchers studying the relationship between greenness and health should consider the appropriate NDVI spatial resolution in relation to the size of aggregation area in order to reduce the risk of exposure misclassification in epidemiologic studies.

Keywords: environmental health, geospatial analysis/GIS, other (specify)

WE-PL-D1-22

Impact of air quality model resolution on health risk estimates associated with particulate matter exposure

Parvez A.[1], Wagstrom, K.[1]; [1]University of Connecticut, CN, USA

This study applies a hybrid-modeling framework (HYCAMR) that combines a regional chemical transport model, CAMx, and a road dispersion model, R-LINE, to estimate hourly combined pollutant concentrations at 40mx40m resolution (Parvez and Wagstrom 2019) and compare the estimated health impacts with those using just a regional model. This study investigates the sensitivity of the health impacts estimates from PM2.5 exposure to air quality modeling resolution in three locations in Connecticut: Hartford, New Haven, and Windham. This study uses concentration estimates from both the CAMx regional model and the HYCAMR hybrid model to calculate the change in estimated health impacts from PM2.5 exposure in the year 2011 using BenMAP-CE. This provides an estimate of the potential under-prediction of health impacts resulting from not accounting for the sharp concentration gradients in near road environments in urban areas where a significant fraction of the population lives in close proximity to roadways. This fine-scale modeling technique captures the elevated concentrations near the roadways which leads to increased estimates of mortality and morbidity. There is also an observed increase in the likelihood of emergency department visits, deaths from cardiovascular disease, and deaths from respiratory disease in the urban core where population density is highest. This analysis also compares the how the health impacts estimates change for different demographic groups. The only location that showed substantial differences in the changes between demographic groups was Willimantic, CT. These results indicate that using regional air pollutants concentrations may lead to an under-prediction of the health impacts from air pollution exposure.

Keywords: air, environmental health, environmental justice, exposure models, particles

WE-PL-D1-23

Relationship between exposure to humidifier disinfectants and lung disease by analyzing the errors of exposure classification

Ryu H.[1], Park J., Heo J., Kim E., Choe Y, Choi Y.[1], Yang W.[1]; [1]Catholic University of Daegu, Gyengsan, South Korea

Lung disease caused by chemical substances (PHMG, PGH, CMIT/MIT) used in humidifier disinfectants (HDs) occurred in 2011, Korea. The Ministry of Environment of Korea has started to accept damage request for victims, and has investigated the causal relationship between HDs and lung injury using environment exposure assessment and medical records. For the HDs victims, environmental exposure assessment and clinical diagnosis were conducted to determine the levels of damage by HDs. However the results of environmental exposure and medical assessment are different from each other. It is needed to supplement for this method. The affecting factors of exposure levels were analyzed by characterizing exposure and demographics. Exposure concentration and cumulative time, exposure level were classified and compared by percentage of clinical diagnosis rating. The high exposure and low clinical diagnosis rating group, and low exposure and high clinical diagnosis rating group were analyzed to consider the limitation of past exposure assessment such as recall bias. The damages by HDs exposure were concentrated on these susceptible populations such as under 10 years, pregnant women with many deaths and showed low exposure and high clinical diagnosis rating. On the other hand, the groups classified by high exposure and low clinical diagnosis rating showed different characteristics such as never smokers, and a large number of elderly people. In cases of victim was child at the time of the injury or death, the environmental exposure assessment and causal relationship investigation between exposure and lung injury were getting harder over time by recall-bias. However, the dose-response relationship between classified exposure levels and clinical diagnosis rating showed positive correlation if the exposure assessment errors were analyzed and controlled. The relationship between dose(exposure level) and response (diagnosis rating) was significant (adjusted $R2= 0.58$)

Keywords: epidemiology, exposure factors, exposure models, respiratory health, risk assessment

WE-PL-E1-SEM: Fine Particles from Burning and Exhaust

WE-PL-E1-24
Chemical characteristics of primary and aged fine particulate matter from biomass burning smoke

Ho K.K-F.[1], Li J.[2], Wang Q.[2], Cao J.[2]; [1]Shenzhen Research Institute of The Chinese University of Hong Kong, Shenzhen, China, [2]Key Lab of Aerosol Chemistry & Physics, Institute of Earth Environment, Chinese Academy of Sciences, Xi'an, China

The aim of this proposal is to investigate the chemical characteristics of the freshly emitted and photo-chemically aged PM2.5 from major crop residues burning. Biomass burning has been recognized as an important source of PM pollutants in local, regional and global atmospheric environments and potentially contributes to respiratory diseases through inhalation exposure. Reliable scientific data on the levels of the freshly and aged biomass burning smoke and their chemical compositions are essential in the efforts to evaluate the potential effects on human health. In this study, fresh and aged PM2.5 biomass burning samples from three crop residues (including rice wheat, corn straws) were collected. Combustions of biomass were conducted in the combustion chamber (~10 m3). Known amounts (0.1–1.0 kg) of biomass were ignited by electric heating coils treated with a small amount of ethanol (~5 ml). The biomass were allowed to burn to completion, filling the combustion chamber with smoke from the entire burn event (flaming phase). Fresh emissions were sampled from the sample outlet at the top of combustion chamber and diluted by a factor of ~50-100. The photo-chemically aging was performed in parallel in Potential Aerosol Mass (PAM) reactor. Chemical measurements (organic carbon (OC), elemental carbon (EC), and organic species) were conducted on filters collected. For fresh emission samples, organic carbon (OC) and elemental carbon (EC) were the most abundant components and contributed in a range of 41.7-54.9% in composition. The details chemical profiles of freshly emitted and photo-chemically aged PM2.5 from major crop residues burning will be reported and compared. Currently limited amount of photo-chemically aged emission data is available in PRD region. This research can contribute to better understanding on how photo-chemically aged sources contribute to atmospheric pollution. It also serves as a scientific baseline for formulating effective control strategies regarding biomass burning pollution.

Keywords: VOCs, air, environmental health, particles, sampling methods

WE-PL-E1-25

The effect of efficiency of local exhaust ventilation on reducing the concentration of airborne nanoparticles emitted during welding

Jankowski T.; Central Institute for Labour Protection - National Research Institute, Czerniakowska, Warsaw, Poland

Welding, cutting and related procedures, such as thermal spraying or soldering, emit fumes, gases and particles. The particles emitted into the respiratory tract and further into the pulmonary alveoli, and - depending on the chemical composition - may pose a threat to health and even act carcinogenically. Emitted substances also contain a significant amount of nanoparticles that penetrate inside the cells and can cause diseases of unknown nature. Due to the protection of work, it is necessary to implement measures to capture pollution and purify the air. The capture of nanoparticles emissions at source (by LEV – local exhaust ventilation) is the best way to protect. Measurements of particles concentration were made using a OPS 3330 (TSI Inc.). Face velocity tests were carried on using Testo 435-4 Transducers (Testo AG). The directed air flow from the emission source of nanoparticles to the suction was strongly modified by the interaction of exhaust and supply ventilation, which resulted in a greater variation of air velocity in a wider range of measurement planes. It was found that shaping the air movement as part of the supply and exhaust ventilation can capture welding fumes with efficiency by more than 20%. This paper has been based on the results of a research task carried out within the scope of the fourth stage of the National Programme "Improvement of safety and working conditions" partly supported in 2017–2019 — within the scope of research and development — by the Ministry of Science and Higher Education/National Centre for Research and Development. The CIOP-PIB is the Programme's main co-ordinator.

Keywords: workplace

WE-PL-E1-26

IAQ and Emission s - Impacts of Wood Stove Replacements in Alaska

Francisco P.[1], Merrin Z.[1]; [1]University of Illinois at Urbana-Champaign, Champaign, IL, USA

Wood fueled space heating is common in parts of the world where traditional heating fuel prices and availability make it a logical choice. Although wood may be the preferred choice from these perspectives, it is not without drawbacks. Burning wood results in incomplete combustion generating, amongst other things, carbon monoxide (CO) and particulate matter. These undesirable combustion byproducts can negatively influence the health of people directly exposed to them, and affect the environment as they contribute to overall atmospheric pollution. In remote Alaskan communities, wood is often the primary or exclusive heating fuel. A study to investigate the emission and indoor air quality (IAQ) implications of heating with wood stoves, and the potential for improvement by upgrading to modern designs certified by the U.S. Environmental Protection Agency (EPA) was conducted in the predominantly Native Alaskan village of Kake in southeast Alaska, approximately 100 miles south of Juneau. Measured stack emissions included carbon dioxide (CO2), CO, total particulate matter (PM), fine particulate matter (PM2.5), elemental/organic carbon ratio (EC/OC), and moisture. IAQ metrics included indoor measurements of PM2.5, CO, and moisture. A modified kitchen performance test (KPT) was conducted at each home, by determining the weight and moisture content of the wood used for heating at daily intervals, to calculate the fuel consumption rate. Blower door tests were conducted to determine the leakage rate of each home. In total testing occurred with different 56 stoves at 30 sites. Twenty-five of the original sites experienced testing of both the original and the replacement stove; the remainder withdrew from the study for various reasons following preliminary testing. Results for IAQ metrics and stack emissions as well as relationships to building tightness and other characteristics will be discussed.

Keywords: built/indoor environment, other (specify), particles

WE-PL-E1-27
Methane Emissions from Residential Natural Gas Appliances

Francisco P.[1], Merrin Z.[1]; [1]University of Illinois at Urbana-Champaign, Champaign, IL, USA

Methane, the primary component of natural gas, is a potent greenhouse gas. Natural gas is a common fuel for residential appliances because of low cost, high energy density, and relatively clean combustion. Natural gas appliance exhaust inevitably contains some unburned methane due to incomplete combustion. A field campaign measuring methane concentrations in exhaust from residential natural gas appliances was conducted to determine the magnitude of their emissions. Natural gas space heating, water heating, and cooking appliances were measured in 100 homes, concentrating in Boston Massachusetts and Indianapolis Indiana USA (areas that previous research has indicated relatively high and low levels of ambient methane, respectively), and homes with tankless hot water heaters (appliances suspected of emitting high methane concentrations). Appliance exhaust typically exhibits a brief methane concentration spike during ignition and extinguishment, and relatively low concentrations during steady-state operation. Exceptions to this pattern include ovens, suboptimal stove burners, and tankless water heaters, which either have a different operating pattern, or non-trivial steady-state concentrations. Findings were combined with appliance usage and prevalence assumptions to estimate total emissions. Annually, ~30 [97.5% CI: 19-160] Gg of methane emissions can be attributed to U.S. residential natural gas appliances, corresponding to ~830 [530-4500] Gg carbon dioxide equivalent (CO2e100). This accounts for ~0.1% [0.08-0.7%] of US anthropogenic methane emissions (which account for ~10% of total US greenhouse gas emissions), and corresponds to an emission factor of 0.38 g/kg of natural gas consumed (0.038% [97.5% CI: 0.024%-0.21%]).

Keywords: built/indoor environment, climate change, other (specify)

WE-PL-E1-28
Pregnant Women's Exposure to Household Air Pollution During a LPG Cookstove Intervention in Rural Bangladesh: The Value of Real-Time Data

Thornburg J.[1], Raynes-Greenow C.[2], Abbott M.[1], Cho S-H.[1], Islam S..[3], Tasnim F.[3], Billah M.[3], Neeloy A.[2]; [1]RTI International, Research Triangle Park, NC, USA, [2]University of Sydney, Sydney, Austrailia, [3]iccdr,b, Dahka, Bangladesh

Background/Aim: We developed a protocol for a cluster randomised controlled trial to assess the impact of liquid petroleum gas (LPG) cooking compared to usual cooking on perinatal mortality in pregnant women in rural Bangladesh (ACTRN12618001214224). One aspect of protocol development was to assess the value of personal exposure monitoring to measure reductions in HAP eposure before and after the intervention. Methods: We measured PM2.5 exposure with the MicroPEM. Gravimetric and optical density analyses of the filter provided total mass and black carbon mass concentrations, respectively. Analysis of MicroPEM nephelometer data provided temporal insights into the participants' exposure. We enrolled 30 pregnant women for 24 hours of exposure assessment when following traditional cooking practices (baseline) and repeated after receiving their LPG cookstove (intervention). Results: We collected valid baseline-intervention MicroPEM data from 22 of the 30 women. Participants wore the MicroPEM 77% and 69% of the time during baseline and intervention. Mean gravimetric PM2.5 concentrations during baseline and intervention were similar: 81.3 ug/m3 vs. 75.3 ug/m3. BC concentrations were also similar: 56.4 ug/m3 vs. 68.7 ug/m3. A temporal pattern in PM2.5 concentrations indicated PM2.5 exposure was lower when cooking with the LPG stove. The mean MicroPEM nephelometer concentration during cooking periods with LPG was 84.6 ug/m3 vs.136.5 ug/m3 with traditional stoves ($p<0.0001$). Conclusions: External factors to the LPG cookstove intervention influenced the time-integrated PM2.5 concentrations. Baseline measurements occurred in May-June, monsoon season in Bangladesh. The LPG intervention occurred in November, a dry month with widespread agricultural activities. The consistent rain during the monsoon lowers ambient PM2.5, whereas crop harvest increased ambient PM2.5. These factors elevated the background PM2.5 exposures during the intervention that are not associated with the LPG cookstove. In complex air pollution environments, real-time PM2.5 exposure data will help evaluate the effectiveness of a clean-cooking intervention.

Keywords: built/indoor environment, epidemiology, exposure factors, particles, risk assessment

WE-PL-E1-29

Low-cost Sensor Array Devices as a Method for Reliable Assessment of Exposure to Traffic-related Air Pollution

Mykhaylova N.; WeavAir, Toronto, Canada

The exposure to air pollutant mixtures is a well-known risk factor for inducing and increasing the severity of diseases. Combining low-cost sensors in arrays holds great potential for real-time monitoring of pollutant exposure because of versatility and aptitude for tracking multi-pollutant exposure. While many low-cost air pollution monitoring devices have been proposed, several underexplored opportunities remain, including sensor-derived pollution indices, source analysis and exposure assessment. A thorough investigation of different low-cost commercial gas and particulate matter sensors from 5 manufacturers has been conducted and best-performing sensors were identified. A device for monitoring air quality has been developed and tested. Each device consists of an array of commercially available metal oxide semiconductor for monitoring NOx and O3, CO, CO2 and optical sensors for monitoring PM2.5. Level of pollutant exposure has been characterized at different locations in Toronto over 3 different campaigns between 2013 and 2016. These deployments allowed long-term sensor performance to be evaluated under different meteorological conditions as well as different ranges of pollutant concentrations. Analysis of a large range of gas sensors revealed several key challenges, including high intra-sensor variability, interference from temperature and nonlinearity. Air quality health index estimation from sensor readings was successfully demonstrated. Three aspects of device reproducibility were evaluated: drift over time, impact of interferences and impact of site-specific mixtures. Three categories of approaches for improving sensor accuracy and reproducibility were tested: nonlinear calibration models, variable transformations and training data selection. Model reproducibility, ability to adjust for multiple combinations of interferences and ability to resolve sites was improved when devices were calibrated at multiple sites. Analysis showed that both short-term and long-term temporal patterns could be resolved and compared at different sites. Background subtraction helped further emphasize the differences and rank sites in terms traffic-related pollution.

Keywords: VOCs, air, analytical methods, behavior, built/indoor environment, environmental health, exposure factors, sensor technology, ventilation

WE-PL-A2-EHR: Metals: Exposure and Risk

WE-PL-A2-30

The Safe Urban Harvests Study: Assessing exposure to six toxic metals among urban agriculture participants in Baltimore, MD

Lupolt S. [1], Santo R. [2], Kim B. [2], Green C. [3], Codling E. [3], Chen R. [1], Rule A. [1], Nachman K. [4]; [1]Department of Environmental Health & Engineering, Johns Hopkins Bloomberg School of Public Health, Baltimore, MD, USA, [2]Center for a Livable Future, Johns Hopkins Bloomberg School of Public Health, Baltimore, MD, USA, [3]Agricultural Research Service, US Department of Agriculture, Beltsville, MD, USA [4]Department of Environmental Health & Engineering, Center for a Livable Future, Risk Sciences and Public Policy Institute, Johns Hopkins Bloomberg School of Public Health, Baltimore, MD, USA

Background: Urban agriculture (UA) is increasing in popularity due to its numerous sociocultural, health, environmental, and economic benefits, but few groups have evaluated exposure to soil contaminants among urban agriculture participants (UAPs). The community-driven Safe Urban Harvests Study aims to characterize potential exposures to six toxic metals (arsenic, barium, cadmium, chromium, lead, and nickel) in soil, irrigation water, and urban grown produce among UAPs in Baltimore, MD. Methods: We recruited, surveyed, and collected samples of water and soil from 104 (70% of eligible) farms and community gardens growing food during the 2017 season. At 72 sites, we collected 1-7 of the thirteen most commonly grown produce items (e.g., kale, carrots, sweet potatoes, tomatoes, peppers, and cucumbers) and purchased those same items from grocery stores and farmers markets in Baltimore. We measured total metal concentrations in water using ICP-MS and in soil using ICP-Optima. Produce analysis is in progress at the time of abstract submission. Results: Five (<5%) water samples (n=114) exceeded US EPA's drinking water standard for lead (15ppb). We compared concentrations of all six metals in soil to several health-based guidelines. Thirty eight (6.2%) of soil samples (n= 616) exceeded the New York Soil Cleanup Objective, (NY SCO) for lead (400ppm) and 238 (45%) of samples exceeded the NY SCO for chromium (III) (36ppm). Measured produce concentrations will be compared to concentrations reported in US FDA's Total Diet Study. Discussion: These results are the first to evaluate metal contamination among UA sites in Baltimore to address safety perceptions and will inform specific recommendations for reducing exposure. Our long-term goal is to describe and quantify UA activities that may increase soil exposure, in order to evaluate the adequacy of existing guidelines for soil contaminants to protect UAPs who have frequent, direct contact with urban soil and consume urban-grown produce.

Keywords: community, exposure factors, food, metals, risk assessment, soil

WE-PL-A2-31

Systematic Review for Exposure Assessments under the United States' Amended Toxic Substances Control Act

Orentas N., Wong., Fehrenbacher.; [1]University of Tennessee, Knoxville, TN, USA

On June 22, 2016, the Frank R. Lautenberg Chemical Safety for the 21st Century Act was signed into United States law thereby amending the Toxic Substance Control Act (TSCA), the Nation's primary chemicals management law. TSCA requires that chemicals currently in commerce be evaluated in terms of hazard and exposure using best-available scientific information. EPA is implementing a systematic review process to identify the available relevant information and data to be used to support risk evaluations, thereby providing a compilation of fit-for-purpose data. Risk evaluations are under development for the first 10 chemicals. Currently, as required under TSCA, EPA is reviewing chemicals based on criteria EPA established during rule-making, to initiate the next 20 chemicals as high-priority substances as well as to propose 20 chemicals as low-priority substances. The high-priority substances would be moved to risk evaluation whereas the low-priority substances would be designated as not an unreasonable risk to human health and the environment. TSCA requires that the risk evaluation process last no longer than three years (with a possible six-month extension) and includes, for the conditions of use of the identified chemical, development of a problem formulation outlining the scope of the evaluation, a hazard assessment, exposure assessment, and risk characterization with statutory timeframes for the scope and risk evaluation. Given the timeframes mandated by TSCA, EPA has been working to develop strategies to more efficiently assess exposures to environmental receptors, consumers, the general population and susceptible subpopulations. Exposure assessments will rely on the weight of the scientific evidence. This presentation will discuss EPA's implementation of the systematic review process in the context of the TSCA existing chemical risk evaluations. EPA is implementing the systematic review methodology using machine-learning techniques and tools to transparently identify, collect, screen, review, evaluate, and integrate exposure-related studies and data for risk evaluations.

Keywords:

WE-PL-A2-32

Metal concentrations of e-cigarette liquid and aerosol samples across the different generation devices

Aherrera A.[1], Aravindakshan A.[1], Olmedo P.[2], Chen R.[1], Jarmul S.[1], Navas-Acien A.[3], Rule A.[1]; [1]John Hopkins University, Baltimore, MD, USA, [2]The University of Granada, Granada, Spain, [3]Columbia University Irving Medical Center, New York, NY, USA

Background: Toxic metal concentrations have been detected in the aerosol of electronic cigarettes (e-cigarettes); however, metal concentrations by device type (generation) have not been described. We investigated metal transfer from e-liquid to aerosol, and compared among different device types. Methods: We sampled 138 e-cigarette devices: 20 first-generation (cig-a-likes), 109 2nd or 3rd generation devices (MODs), and nine 4th generation devices (PODs), including Juul. Samples from the refilling dispenser/cartridge, aerosol, and remaining e-liquid in the tank (MODs only) were obtained. Metals were measured using ICP-MS and reported as mass fractions (μg/kg) in liquids. Results: The median (IQR) concentrations (μg/kg) measured in the aerosol of MOD devices were 62.3 (8.00, 309) for Ni, 13.1 (3.74, 41.4) for Cr, 9.99 (2.70, 34.7) for Pb, 2.81 (0.90, 9.50) for Mn; in aerosol of POD devices we found 42.0 (27.0, 421) for Ni, 10.3 (8.40, 13.2) for Cr, 4.40 (2.75, 217) for Pb, 2.30 (1.45, 3.00) for Mn. Cig-a-likes and PODs had lower mean concentrations (80% and 50%, respectively) of Ni, Cr, Pb, Mn in aerosol compared to the mean concentrations in the dispenser (p-values <0.005). MODs, however, had 90% higher mean concentrations in the aerosol for all metals compared to the dispenser, and 375% higher in the tank compared to the aerosol (p-values 300% higher in MODs and PODs than cig-a-likes. These results demonstrate that all e-cigarettes are a potential source of exposure to toxic metals.

Keywords: exposure factors, metals, other (specify)

WE-PL-A2-33

Quantitative assessment of lung deposition exposure to heavy metals from e-cigarettes– a simulation study

Shao Y.[1], Chen R.[1], Koehler K.[1], Rule A.[1]; [1]Johns Hopkins Bloomberg School of Public Health, Baltimore, MD, USA,

Background To better understand exposure to inhalation hazards from e-cigarette aerosols, it is important to characterize particle size-selective deposition in the airways A foam sampler has been developed to capture the fraction of aerosols that deposits in the human respiratory system as a function of particle size. The objective of this study was to simulate typical e-cigarette vaping, and to quantify the fraction of heavy metals deposited in the human respiratory tract. Methods We evaluated aerosols generated by JUUL devices (mint flavor) using foam lung deposition samplers (LDS). An automated e-cigarette aerosol sampling system was designed for this study. The system generates a 50 mL puff at 0.6 LPM for 5 seconds. This puff is stored in a 0.6L Tedlar bag for less than 1 second, then pushed through an LDS at 4 LPM. The aerosol not collected in the LDS was collected with a 25mm Teflon filter to represent the exhaled fraction. JUUL samples from 0, 20, 40 and 60 puff cycles were collected, each in triplicate. Total metal concentrations in the LDS and filters were analyzed using ICP-MS. Results The median (range) collected in the LDS was 2645.46(1818.04, 5546.19) ng/puff for Al; 1.42(0.77, 2.06) ng/puff for Cr; 0.49 (0.00, 4.80) ng/puff for Mn, 1.48 (0.60, 5.03) ng/puff for Ni; and 0.31 (0.29, 13.66) ng/puff for Pb. For the filters, all metals except Ni were < LOD, and the median (range) of Ni in the filter was 0.12 (0.05,0.81) ng/puff. Conclusion E-cigarettes are a source of aerosolized metals. We found that nearly all Al, Cr, Mn, Ni and Pb were collected in the lung deposition sampler, indicating that metals reach and nearly 100% deposit in the human respiratory tract during e-cigarette use. Additional studies are needed to evaluate the deposited mass for other popular e-cigarette devices.

Keywords: exposure models, metals, respiratory health, risk assessment

WE-PL-B2-34

Perturbations of the Arginine Metabolome Following Exposures to Traffic-Related Air Pollution in a Panel of Commuters with and without Asthma

Liang D., Ladva C.N.[1], Golan R.[2], Yu T.[3], Walker D.I.[4], Sarnat S.E.[3], Greenwald R.[3], Uppal K.[3], Tran VT.[3], Jones D.P.[3], Russell A.G.[5], Sarnat J.A.[3]; [1]Centers for Disease Control and Prevention, Atlanta, GA, USA, [2]Ben Gurion University of the Negev, Beershaba, Israel, [3]Emory University, Atlanta, GA, USA, [4]Icahn School of Medicine at Mount Sinai, New York, NY, USA, [5]Georgia Institute of Technology, Atlanta, GA, USA

Background. Mechanisms underlying the effects of traffic-related air pollution on asthma etiology remain largely unknown, despite the abundance of observational and controlled studies reporting associations between traffic sources and asthma exacerbation and hospitalizations. Objectives. We conducted the Atlanta Commuters Exposure (ACE-2) study: a crossover panel of commuters with and without asthma, to identify molecular pathways perturbed following traffic pollution exposures. Methods. We measured 27 air pollutants and conducted high-resolution metabolomics profiling on blood samples from 45 commuters before and after each exposure session. We evaluated metabolite and metabolic pathway perturbations using an untargeted metabolome-wide association study framework with pathway analyses and chemical annotation. Results. Most of the measured pollutants were elevated in highway commutes ($p < 0.05$). From both negative and positive ionization modes, 17,586 and 9,087 metabolic features were extracted from plasma, respectively. 494 and 220 unique features were associated with at least 3 of the 27 exposures, respectively ($p<0.05$), after controlling confounders and false discovery rates. Pathway analysis indicated alteration of several inflammatory and oxidative stress related metabolic pathways, including leukotriene, vitamin E, cytochrome P450, and tryptophan metabolism. We identified and annotated 45 unique metabolites enriched in these pathways, including arginine, histidine, and methionine. Most of these metabolites were not only associated with multiple pollutants, but also differentially expressed between participants with and without asthma. The analysis indicated that these metabolites collectively participated in an interrelated molecular network centering on arginine metabolism, underlying a potential etiology of traffic-related pollutants toxicity in individuals with asthma. Conclusions. We detected numerous significant metabolic perturbations associated with in-vehicle exposures during commuting and validated metabolites that were closely linked to several inflammatory and redox pathways, elucidating the potential molecular mechanisms of traffic-related air pollution toxicity. These results support future studies of metabolic markers of traffic exposures, their corresponding molecular mechanisms, and asthma etiology.

Keywords: air, biomarkers, environmental health, health, other/general, other (specify)

WE-PL-B2-35

Childhood Exposure to Traffic-Related Air Pollution and the Lower Respiratory Tract Microbiome

Uebel-Niemeier C.[1], Ryan P.[2], Meller J.[1], Ollberding N.[2], Adhikari A.[3], Reponen T.[1]; [1]Department of Environmental Health, University of Cincinnati, Cincinnati, OH, USA, [2]Cincinnati Children's Hospital Medical Center, Cincinnati, OH, USA, [3]Jiann-Ping Hsu College of Public Health, Georgia Southern University, Statesboro, GA, USA

Exposure to traffic-related air pollution (TRAP) is associated with incident and exacerbation of existing asthma. Urban PM has been shown to increase the adhesion of bacteria to human airway epithelial cells and associations of bacterial colonization with the initiating events of early asthma were reported previously. Therefore, we hypothesized that chronic exposure to TRAP increases the adhesion of microorganisms to the respiratory tract resulting in an altered microbiome. Forty children were recruited through the Cincinnati Childhood Allergy and Air Pollution Study, a longitudinal cohort followed from birth through early adolescence. Saliva and induced sputum were collected at approximately age 14 years from each participant. Average exposure to TRAP was characterized from birth through time of saliva and sputum collection using a previously validated land-use regression model and high or low exposure was defined as above or below the median, respectively. Sequencing of the 16S rRNA gene was performed on the DNA extracted from the sputum and saliva samples. Relative abundance of taxa and diversity indices were compared using the R Statistical Analysis Software package, phyloseq. Results demonstrated distinct communities between saliva and sputum sample types. There was little difference in the relative abundance of phyla in sputum between the two TRAP exposure groups. However, the number of observed amplicon sequence variants (ASVs) in the high exposure group was significantly greater than the low exposure group (Wilcoxon sum rank test; p=0.03). Shannon's diversity index also showed a trend of greater diversity in the high exposure group (Wilcoxon sum rank test; p=0.08). We observed no statistically significant difference in richness or Shannon diversity between asthmatics and non-asthmatics. Thus, asthma status does not appear to confound the observed differences. These findings indicate that exposure to TRAP in early childhood and adolescence is associated with greater diversity in the lower respiratory tract.

Keywords: built/indoor environment, children, environmental health, epidemiology, industrial hygiene, microbial agents

WE-PL-B2-36

Effects of short-term exposure to traffic related air pollution on blood pressure

Hudda N.[1], Eliasziw M.[1], Reisner E.[2], Zamore W.[2], Durant J.[1], Brugge D.[3]; [1]Tufts University, Medford, MA, USA, [2]STEP, Somerville, MA, USA, [3]Univ. of Connecticut, Farmington, CT, USA

Living close to major roadways is associated with elevated blood pressure (BP) amongst other adverse health effects. There is growing evidence that ultrafine particles (UFP, <100 nm in diameter), which are elevated near major roads and highways, might contribute to such risks. We assessed the efficacy of high efficiency particulate arrestance (HEPA) filtration and building envelope adjustment at reducing exposure to and health effects of air pollution next to major highways. We used a randomized three-period crossover trial design to assign 77 participants to three two-hour TRAP exposure sessions conducted in buildings adjacent to highways. High, medium and low exposures were attained by varying the degree of ventilation and amount of HEPA filtration in the room. Indoor particle number and black carbon (BC) concentrations (i.e., markers of traffic-related air pollution) were monitored continuously. During each session participants sat quietly and wore noise-cancelling headphones, while their BP was monitored every ten minutes using an ambulatory BP monitor. We monitored pulse and oxygen saturation continuously. Median particle number and BC concentrations for low, medium and high exposure sessions were 2,300, 9,400, and 28,800 particles/cm3 and 120, 380 and 680 ng/m3, respectively. They were relatively steady during sessions with a coefficient of variation of about ±20%. We used a linear mixed model, consisting of a random intercept, a Toeplitz covariance structure, and effects for exposure, period, and sequence to analyze mean changes in systolic BP. Despite systolic BP rising linearly over time during the exposure sessions ($p < 0.001$), enhanced HEPA filtration was significantly effective at minimizing changes in BP: the overall mean change in systolic BP over the course of the sessions was 0.6 mmHg during low (maximum HEPA filtration), 1.2 mmHg during medium (moderate HEPA filtration), and 2.5 mmHg during high exposure (no HEPA filtration) sessions (linear trend p-value = 0.019).

Keywords: air, built/indoor environment, environmental health, particles

WE-PL-B2-37

Assessment of exposure of fuel station attendants to benzene, toluene and xylenes by analysis of end-exhaled air

Scheepers P.; Radboud University Medical Center, Nijmegen, Netherlands

Exposure to benzene, toluene and p-, m-, o-xylene (BTX) was studied in 29 fuel station attendants and 16 office workers in Sri Lanka with the aim to assess the exposure level and identify potential exposure mitigating measures. Pre- and post-shift samples of end-exhaled air were collected and analyzed for BTX on a thermal desorption gas chromatography mass spectrometry system (TD-GC-MS). In addition, exposure was measured by personal air sampling using both diffusive and active sampling and analyzed by gas chromatography flame ionisation detection (GC-FID). Pre-shift BTX concentrations in end-exhaled air were higher in smokers compared to non-smokers ($p<0.01$) but storage of fossil fuels or motor vehicles in homes and staying overnight at the workplace did not contribute significantly. The median change from pre- to post-shift concentrations of benzene, toluene, m/p-xylene in end-exhaled air was 7.4, 11.4, 0.79 and 0.71 ng/L in fuel station attendants and 1.5, 0.68, -0.06 and 0.59 ng/L in office workers, respectively. These concentrations were higher after self-reported dermal contact related to spills of gasoline ($p<0.05$). Median (range) breathing zone air concentrations were 608.8 (65.1 – 1,957.7) µg/m3 for benzene and 1,372 (267.5 – 3,548) µg/m3 for toluene. Xylenes were not detected. Diffusive sampling overestimated results compared to active sampling by 26% for benzene and 10% for toluene. Benzene in exhaled air was moderately associated with benzene in the breathing zone ($r=0.422$; $p<0.001$). After adjustment for long work hours, 28% of the measured exposures exceeded the ACGIH threshold limit value (TLV) for an 8-h time-weighted average of 1.6 mg/m3 for benzene exposure. In conclusion, working as a fuel station attendant leads to inhalation exposure and occasional skin exposure to BTX. Smoking was identified as the most important co-exposure. Use of protective equipment and avoiding long work are expected to reduce exposure for benzene below the current TLV.

Keywords: biomonitoring

WE-SY-C2: Integration of Biology and Chemistry for Indoor Environmental Quality (continued)

WE-SY-C2-38

House Observations of Microbial and Environmental Chemistry (HOMEChem): Insights into Particulate Matter Concentrations and Exposure

Vance M.E.[1], Patel S.[1], Sankhyan S.[1], Tian Y.[2], Farmer D.K.[3], Goldstein A. H..[2], Vanhanen J.[4]; [1]University of Colorado Boulder, CO, USA, [2]University of California, Berkeley, CA, USA, [3]Colorado State University, Fort Collins, CO, USA, [1]Airmodus Oy, Helsinki, Finland

Although people spend most of their time in indoor environments in the developed world and exposure to particulate matter (PM) is known to cause an array of deleterious health effects, detailed investigations into PM emissions resulting from everyday activities in the home—such as cooking and cleaning—are still needed. The HOMEChem (House Observations of Microbial and Environmental Chemistry) study was a month-long collaborative investigation into the chemistry of indoor environments that took place in a manufactured test house. By incorporating state-of-the-art measurements by over 20 research groups from the fields of chemistry, microbiology, and engineering, this experiment engaged far more institutions and disciplines than ever applied to the study of the air quality of a home. A range of instruments characterized both gaseous and particulate species during simulated everyday activities such as cooking, cleaning, human occupancy and the use of personal care products, and opening doors and windows. This work includes results from time-resolved observations of PM, black carbon (BC) and brown carbon (BrC) during the HOMEChem study. While indoor concentrations of BC and BrC were 2-3 times lower than outdoors during moments of no activities, these raised to almost 30 times higher than outdoors during a cooking-intensive activity: a simulated Thanksgiving meal preparation. During these events, PM0.1 and PM2.5 exceeded 100 μg m-3 and 300 μg m-3, respectively. The estimated PM mass deposited in the respiratory system of an adult residing in the test house during a routine day (breakfast, lunch, and dinner followed by cleaning activities) and a simulated Thanksgiving meal preparation were almost five (70.5 μg) and ten times higher (113.7 μg), respectively, than in a day with no activities (13.7 μg). These results are expected to bring insights into the people's exposure to PM from cooking activities and to inform doses for toxicity studies.

Keywords: air, built/indoor environment, cumulative exposure, exposure factors, particles

WE-SY-C2-39

Real-time air monitoring of occupational exposures to indoor pollutants among U.S. hairdressers

Kavi L.[1], Shao Y.[2], Boyle M.[1], Pool W.[3], Thomas S.B.[4], Wilson S.M.[1], Rule A.M.[2], Quiros-Alcala, L.[2]; [1]Maryland Institute of Applied Environmental Health, University of Maryland School of Public Health, College Park, MD, USA, [2]Department of Environmental Health and Engineering, Johns Hopkins Bloomberg School of Public Health, Baltimore, MD, USA, [3]Centro de Apoyo Familiar, Riverdale, MD, USA, [4]Maryland Center for Health Equity, University of Maryland School of Public Health, College Park, MD, USA

Background: Hairdressers are exposed to many chemicals found in or emitted from hair products, including volatile organic compounds (VOCs) and particulate matter (PM), which have been associated with adverse health effects. Still, studies on occupational exposures among U.S. hairdressers, including workers servicing African American and Latino clients, are scarce. Exposures among hairdressers servicing African American and Latino clients are of growing concern, as the formulated products used to service their clientele are suspected of containing more toxic ingredients than products used on Caucasian clients. Objective: To characterize indoor air concentrations of select contaminants in six hair salons in Maryland primarily serving African American and Latino clients. Methods: Using a combination of low cost and state of the art real-time instruments, we conducted air monitoring in each of the salons over three consecutive days to quantify concentrations of formaldehyde, total VOCs (TVOCs), CO, CO2, PM1, PM2.5 and PM10 during the entire work shift. We calculated 8-hour time weighted average (TWA) concentrations and compared them to occupational exposure guidelines available. Results: The daily median (p25, p50) 8-hr TWA for formaldehyde was 0.032 ppm (0.01, 0.05 ppm). Two of the salons monitored exceeded the NIOSH recommended exposure level (REL) for formaldehyde. TVOC concentrations ranged from 0.006-76.8 ppm. Daily CO median concentrations in each salon were generally low (<8.00 ppm). Daily CO2 median concentrations in each salon ranged from 656-1,280 ppm and were below the OSHA permissible exposure limit. Daily PM1, PM2.5, and PM10 median concentrations observed over the 3-day sampling period in each salon ranged from 5-1,010 ug-m-3, 6-1,040 ug-m-3 and 9-1,050 ug-m-3, respectively. Conclusions: Findings suggest that formaldehyde and PM concentrations could be of concern in hair salons primarily serving African American and Latino clients. Further studies are needed to characterize concentrations of select indoor air pollutants in these settings.

Keywords: VOCs, built/indoor environment, industrial hygiene, particles, workplace

WE-SY-C2-40

Evaluation of air sampling methods to assess pathogen contamination in agricultural animal environments

Rule A.[1], Rhodes S.[2], Sexton K.[2], Wiesner C.[2], Davis M.[1], Heaney C.[2], Stewart J.; [1]John Hopkins University, Baltimore, MD, USA, [2]The University of North Carolina at Chapel Hill, Chapel Hill, NC, USA

Background: Antibiotic-resistant bacteria,i.e. Staphylococcus aureus (SA), and other zoonotic pathogens have been detected within Concentrated Animal Feeding Operation (CAFO) environments (e.g. barn surfaces), the ambient environment near CAFOs (e.g. air, soil, water), CAFO workers, and in community residents without direct livestock contact. Additionally, transmission between animals and humans has been documented. For this reason, it is important to measure contamination of animals, people, and the environment simultaneously to assess transmission dynamics of pathogens, and to develop public health interventions. Methods: In a pilot study, we deployed personal and area air samplers inside one CAFO and in the surrounding environment, using three well-established biological air sampling methodologies: (1) Inhalable air samplers (Button®, SKC Inc.) with 25mm gelatin filters (Sartorius, Germany), (2) Sterile all-glass impingers (BioSampler®, SKC) with 20 mL sterile PBS, and (3) single stage impactor (N6, Thermo Scientific, Inc) with selective media plates or MD8 (Sartorius) with gelatin filters. The same samplers were deployed in and around four antibiotic-free hog farms (AFHO). Results: Personal button samples collected inside the CAFO were positive for viable SA (2/2, 100%) whereas none from AFHO (0/6). Anderson impactors had the highest proportion of samples positive for viable SA (5/6, 83%) from CAFO samples and 0/9 from AFHO. Two of four (50%) of impinger, and 1/4 (25%) area inhalable samples were positive for viable SA in CAFOs, none for AFHOs. All samples collected on the CAFO and AFHO were negative for SA genes. Conclusion: The choice of sampler is determined by the specific environment to be measured. Personal samplers are best to capture the individual exposures (workers or community members). Impingers are preferred for molecular assays, and impactor samplers are preferred when sampling low-concentration environments.

Keywords: air, microbial agents, occupational, sampling methods

WE-SY-C2-41

Source and composition of indoor particles: transport from outside, emissions from inside, and residual smoke in the indoor environment

DeCarlo P.; Drexel University, Philadelphia, PA, USA

We spend roughly 22 hours indoors per day, as such our exposure to outdoor air pollutants predominantly occurs in the indoor environment. This presentation will focus on high-time resolution indoor and outdoor submicron (PM1) particle composition measurements made with an aerosol mass spectrometer. Measurements were made in both a classroom at Drexel University in Philadelphia, PA, USA, and at a model home as part of the HOMEChem project. The data analysis provides a detailed understanding of outdoor particle penetration into the indoor environment as a function of temperature and relative humidity gradients. Additionally, by comparing the speciated chemical composition of both indoor and outdoor particulates, we are able to identify the chemical signatures of indoor orginated components including revolatilization of residual cigarette smoke (termed thirdhand smoke), and contributions from human occupants to the indoor aerosol. Direct emissions from cooking and cleaning were observed and characterized during the HOMEChem experiement, and will also be discussed. Observations indicate that in the absence of strong indoor sources, outdoor penetration remains the main source of submicron aerosol in the indoor environment. However, indoor sources can dominate the overall mass concentration when present, and an understanding of the chemical differences between outdoor and indoor aerosol is important in assessing the potential health impacts.

Keywords: analytical methods, particles, surfaces, ventilation

WE-PL-D2-42

Development and evaluation of a pregnancy Physiologically Based Toxicokinetic (pPBTK) model for high-throughput application

Codaccioni M.[1], Brochot C.[1]; [1]French national institute for industrial environment and risks (Ineris), Verneui-en-Halatte, France,

In utero exposures to chemicals have been associated with adverse effects that can be observed over the whole lifetime. These exposures are commonly assessed with cord sampling at delivery or maternal sampling over the pregnancy. Pregnancy physiologically-based toxicokinetic (pPBTK) models are adequate tools for estimating the internal exposure of the fetus throughout pregnancy. We reviewed the placental transfers of chemicals described in published pPBTK models and compared the impact of the model structure on the fetal dosimetry. We also proposed a placental transfer structure accounting for both histiotrophic (trophoblastic plugs block the spiral arteries in the 1st trimester) and hemotrophic (spiral arteries inflow the intervillous space with maternal blood in the 2nd and 3rd trimesters) nutrition phases. It comprises two diffusional constants separating maternal placenta, intervillous space and fetal placenta. Under a single dose exposure, four patterns of fetal dosimetry were observed and related to assumptions about the number of diffusion constants between maternal and fetal blood. The placental transfer model was integrated in a pPBTK model that is an extension of the generic PBTK model included in the httk R package. Two types of parameterization were tested for the placental diffusional constant between intervillous space and fetal placental compartment, either with a semi-empirical equation or with a passive diffusion clearance scaling method based on in vitro apparent permeabilities. The model predictions were evaluated with published data. It shows a good predictive capacity as predictions of maternal and cord serum concentrations felt into the 10-fold range of observed concentrations. The type of placental diffusion constant parameterization does not influence the classification of fetal to maternal exposure ratios. Our pPBTK model coupled with in silico tools or in vitro data for parametrization can be used to assess the in utero exposures for a wide range of chemicals throughout pregnancy.

Keywords: biomarkers, chemical prioritization, prenatal

WE-PL-D2-43

An Integrated and Spatialized Approach to Identify Potentially Overexposed Population: A Case Study of Chlorpyrifos Contamination

Regrain C.[1], Zeman F.[1], Chardon K.[2], Haraux E.[2], Guedda M.[3], Caudeville J.[1]; [1]Institut National de l'Environnement Industriel et des Risques (INERIS), Verneuil-en-Halatte, France, [2]PériTox - UMR_I 01, CURS, Université de Picardie Jules Verne, Chemin du Thil, Amiens, France, [3]Laboratoire Amiénois de Mathématiques Fondamentales et Appliquées (LAMFA) - CNRS, UMR 7352, Université de Picardie Jules Verne

Exposure to pesticides is characterized by the multiplicity of exposure routes (food, water, soil, air) related to their presence in all environmental media. First, for a fine characterization of the environmental exposure, it is necessary to gather within the same analysis system a dataset integrating the population behavior and the local contamination of the environmental media at fine resolutions and on large territories. This integration requires a sufficient toolchain to produce or process these data in a rigorous manner, characterize the entire continuum from the source of contamination, transport, and transfer of contaminants in environmental media, and estimate external and internal exposure doses for the general population. The aim of this work is to establish an integrated methodology for the mapping of exposure indicators at fine spatial and temporal resolutions (1 km^2 and week) directly comparable with individual biomonitoring measurements from French cohorts. To illustrate the approach, we consider the levels of chlorpyrifos among pregnant women in the Picardy region (France). Results provide insights into the identification of areas of potential overexposure and analysis of environmental determinants, elements useful for public health decision to reduce exposure and thus health and environmental impacts.

However, the exposure characterization involves uncertainties regarding modeling assumptions and data measurements. Moreover, the efficiency of exposure modeling can be limited by calculation times. To optimize numerical resolution and reduce uncertainties related to knowledge limits and data representativeness, modeling assumptions and tools need to be reconsidered to improve the methodology and allow future developments.

Keywords: biomarkers, exposure models, geospatial analysis/GIS, pesticides, prenatal

WE-PL-D2-44

Per- and Polyfluoroalkyl Substances (PFAS) Exposure Assessments in Communities with Drinking Water Exposures

Breysse P.[1], Johnson J.M.[2], Rogers R.D.[2], Goodwin B.P.[2], Karwowski M.P.[3], Kowalski P.[2], Roobol H.E.[1], Choudhary E.[3], Mortensen M.[3], Chang A.[3], Olson D.R.[3], Gaines C.R.[1], Reh C.M.[2]; [1]National Center for Environmental Health/ Agency for Toxic Substances and Disease Registry, Atlanta, GA, USA, [2]Agency for Toxic Substances and Disease Registry, Atlanta, GA, USA, [3]National Center for Environmental Health, Atlanta, GA, USA

Per- and polyfluoroalkyl substances (PFAS) are a large class of approximately 5000 man-made chemicals that have been used in industry and consumer products worldwide since the 1950s. Because of their widespread use and their persistence in the environment, certain PFAS are found in the blood of people and animals all over the world. PFAS have been detected in numerous public and private drinking water supplies throughout the United States. The Centers for Disease Control and Prevention (CDC) and Agency for Toxic Substances and Disease Registry (ATSDR) recognized a need for tools that health departments could use to conduct PFAS biomonitoring activities in communities exposed to PFAS through drinking water and created the PFAS Exposure Assessment Technical Tools (PEATT). The PEATT includes a protocol for statistically based representative sampling, risk communication materials, questionnaires, and EPA's drinking water sampling protocol. In 2018, CDC/ATSDR, in partnership with the Association of State and Territorial Health Officials (ASTHO), funded two grants to pilot the exposure assessment protocol – one to the Pennsylvania Department of Health (PADOH) and one to the New York State Department of Health (NYSDOH). The PEATT evaluation aimed to identify barriers and facilities, recommendations, and outcomes of utilizing the PEATT. The two grantees conducted the pilots and provided feedback through surveys, brief interviews, regional communications, and a comprehensive interview. During this presentation I will present the exposure assessment protocol, data on PFAS serum levels from the two communities where the pilot exposure assessments took place, and the lessons learned from the pilots. CDC/ATSDR has used the feedback provided by the grantees to refine the PEATT and provide the best support to states, territories and affected communities who are dealing with this important health issue. The feedback will also inform CDC/ATSDR's upcoming additional exposure assessments that will take place in 2019 through 2020.

Keywords: biomonitoring, community, environmental health, water

WE-PL-D2-45

Evaluation of Effectiveness of an Intervention for Communities Exposed to PFNA-contaminated Drinking Water in New Jersey

Yu C.H.[1], Weisel C.[2], Alimokhtari S.[2], Riker D.[1], Georgopoulos P.[2], Fan Z (T).[1]; [1]New Jersey Department of Health, Ewing, NJ, USA, [2]Environmental & Occupational Health Sciences Institute, Piscataway, NJ, USA

Elevated perfluorononanoic acid (PFNA), one of the per-/poly-fluoroalkyl substances (PFAS), was detected in public water systems (up to 96 ng/L) and private wells (up to 1,500 ng/L) in Paulsboro/West Deptford communities in New Jersey (NJ). Since 2014, the water supplies have been filtered with activated charcoal. To assess the effectiveness of this water intervention on reducing exposure to PFNA, a community exposure study was initiated to monitor the change of PFNA in serum of residents after interventions. A convenience sample of

120 adult residents in the communities have been recruited and PFAS were measured in serum, household tap water and household dust. A questionnaire was administered to collect information on demographics and potential PFNA sources. Three visits were planned to take place approximately a year apart. The first visits have been completed with recruiting 120 subjects (105 homes), and the second visits are in progress. The first-year PFNA results indicated 85 subjects were elevated than the 95% of PFNA in general U.S. population (1.90 ng/mL; NHANES in 2015-2016), and consecutive second-year paired PFNA results showed changes between 1% (Q1) and -21% (Q3) from Year 1 to 2 among the tested 79 subjects. Computational simulations, conducted for all the subjects on integrated exposure data and physiologically-based pharmacokinetic (PBPK) modeling approach, also demonstrated estimated PFNA serum levels were consistent with determined blood test data. All PFNA water levels were lower than the NJ drinking water standard (13 ng/L), indicating that the water intervention has effectively reduced PFNA exposure through this route. PFNA dust levels were not significantly different with levels found in 20 control homes, indicating house dust was not significant contributing sources. This study suggests serum levels of PFNA will be declining over the next few years, if the drinking water was the main exposure source for the communities.

Keywords: biomonitoring, emergency response, environmental health, source control

WE-PL-E2-SEM: Effects of Climate Change on Human Health and Exposure

WE-PL-E2-46

Impact of urban heat island on indoor heat exposure in London coupling crowdsourcing data and building simulation

Luo Z.[1], Xie X.[1], Benjamin K.[1], Grimmond S.[1]; [1]University of Reading, Reading, United Kingdom

Traditional epidemiology study links heat-related mortality and mobility with outdoor temperature, but overlooks the fact that the most exposure occurs indoors as people spend 90% of their time indoors. The heat exposure indoors could further exacerbated by urban heat island and location of the building. During the heatwave, the building in the city centre will experience higher heat exposure compared with the rural one given the same building design. Traditional UHI investigation methods are often incapable of providing the high spatial density of observations required to distinguish small-scale temperature differences within the UHI. Here we report our study, i.e., an enhanced/improved indoor heat exposure estimation in London by coupling crowdsourcing weather data and building energy simulation using EnergyPlus. We collected weather data from 1300 Netatmo person weather stations to understand the spatial-temporal urban heat island in London during a heatwave event in July 2018. The highly spatially resolved weather data are used as inputs to drive building energy simulation using a building archetype in London and therefore predict the indoor exposure to heat during heatwave in London. Our simulation results demonstrate a high heterogeneity of indoor heat exposure risk in Greater London due to urban heat island.

Keywords: built/indoor environment, geospatial analysis/GIS, thermal environment

WE-PL-E2-47

Application of Distributed Low-cost Air Quality Sensors to Investigate Heatwave Effect on Indoor Air Quality

Mainelis G.[1], He R.[1], Tsoulou I. [1], Thirumurugesan S.[1], Morgan B.[1], Gonzalezmena S.[1], Plotnik D.[1], Senick J.[1], Andrews C.[1]; [1]Rutgers University, New Brunswick, NJ, USA

The proliferation of low-cost air quality sensors offers new opportunities to deploy such sensors en masse and investigate factors affecting the presence of and exposures to air pollutants, including in indoor environments. In this case, we focused on the effect of heat waves on indoor air quality. The changing climate in the Northeastern US is predicted to lead to a higher number of days when the air temperatures exceed 33C (90F) for several consecutive days. The effect of such conditions on the indoor air quality has not been studied widely. Thus, we investigated the effect of heat waves on indoor air quality, specifically in vulnerable populations. We recruited 24 seniors residing in three different low-income multi-resident buildings in Elizabeth, NJ, USA to participate in a study. We deployed AirVisual air quality sensors (IQAir North America) in their apartments to monitor indoor and outdoor air levels of PM2.5, air temperature, relative humidity, and CO2. Another set of sensors monitored air conditioner (AC) use, window operation, and ozone. We found that air pollution levels indoors were affected not only by outdoor pollutant levels and outdoor temperatures but also by building conditions and individual behaviors of seniors, including smoking, use of windows, and their financial ability to use AC. For example, in non-smoking apartments, average PM2.5 levels were 15 μg/m3, while in smoking apartments, those levels approached 100 μg/m3. Our experience demonstrated the utility and usability of consumer-grade air quality sensors to study factors affecting indoor air pollution. At the same time, data suggest that long-term adaptation and resilience to climate change could benefit from an integrated resident-to-building-to-neighborhood evaluation. This research was funded by NSF grant AGS-1645786.

Keywords: aging/elderly, air, built/indoor environment, environmental health, exposure factors, particles, sensor technology

WE-PL-E2-48

Assessing the Potential Impact of Climate Change on Mortality for the Greater Boston Area

Lee W.C.[1], Mukherjee R.[2], Zanobetti A.[2], Koutrakis P.[2]; [1]National Taiwan University, Taipei, Taiwan, [2]Harvard T.H. Chan School of Public Health, Boston, MA, USA

Rising temperatures associated with climate change are expected to impact human adaptive responses, including changes to home cooling and ventilation patterns. These changes in turn will alter the contributions of outdoor fine particulate matter (PM2.5) sources to indoor air quality, and subsequently lead to differential effects of PM2.5 exposures on human health. Therefore, the objective of this study was to examine the impact of climate change-related differences in exposures to indoor particle of outdoor origin on PM2.5-related mortality. We gathered all-cause non-accidental daily mortality data in the greater Boston area for the years 2000-2009 as the health outcome in the epidemiologic modeling. We used a negative binomial model to assess the association between the same day indoor PM2.5 concentration and total mortality, adjusted for long-term trend and ambient temperature through smoothing splines, as well as seasons using indicator variables for the past period. For the future climate change scenario, we used the health model to predict the daily total mortality under the climate of for 2056-2065. A strong, positive relationship was found between indoor PM2.5 concentration and total mortality, whereas a U-shaped relationship was seen for ambient temperature with higher mortality for colder months, followed by high temperature range. Under the future climate, there was an increase in indoor PM2.5 concentrations in warmer seasons, ranging from late spring (May) to mid-autumn (October). Overall, a similar trend was found for monthly averaged total mortality, with up to 3.6% and 3.7% increase in May and September, respectively. On the contrary, increase in temperature showed an overall protective effect (1.6 - 4.4% decrease) on mortality for colder months (January to April). Our findings suggest different adaptation strategies should be considered for colder and warmer months, respectively, to protect population health under the future climate change scenario.

Keywords: climate change

WE-PL-E2-49
The impact of regional weather on residential energy consumption and indoor air quality changes due to energy retrofits

Connolly C.[1], Milando C.[1], Dols W.[2], Levy J.[1], Fabian M.[1]; [1]Boston University, Boston, MA, USA, [2]NIST, Gaithersburg, MD, USA

Climate action plans in cities in the United States include updated requirements for building retrofits that decrease energy demand but without considering the subsequent changes on indoor air quality (IAQ) for residents. While previous studies have shown that energy retrofits paired with proper ventilation and filtration can improve health outcomes for residents, little is known of how the energy/IAQ tradeoffs change in different climate regions. In this study, we used a coupled energy-IAQ model to quantify these tradeoffs in multiple regions of the US. We modeled the effects of regional climate differences on residential energy consumption and indoor air quality (e.g. ambient and indoor sourced PM2.5) using a coupled model integrating an energy simulation program (EnergyPlus) and a multi-zone indoor air quality and ventilation program (CONTAM). The co-simulated framework incorporates multiple factors that affect indoor air quality and energy use such as indoor human behavior, weather, ambient air pollution, and building characteristics. Energy and IAQ were estimated for a residential building located in northeast, southern, northwest, and southwest regions. We found that regions with warmer summers (i.e. Florida, Arizona) had better annual indoor air quality from increased use of the HVAC cooling system in the summer, but at the expense of requiring significantly more energy to cool the residence. Regions with colder winters (i.e. Boston) had worse average yearly IAQ and higher energy consumption due to heating demands in the winter. Indoor sources of PM2.5 had a greater impact on indoor air pollution compared to outdoor sourced ambient pollution, even with increased exhaust ventilation. This co-simulation provides a novel way to account for the multi-faceted reasons for human exposure to pollutants in their residences. With increasing energy demands due to climate change, we can use this structure to model these energy/IAQ tradeoffs.

Keywords: air, built/indoor environment, climate change, ventilation

WE-PL-A3-EHR: Inhalation Exposure

WE-PL-A3-50
PM2.5 exposures in different routes for Bike and Scooter Commuters in Taipei, Taiwan

Wu T.G.[1], Chang J.C.[2], Wu C.F.[2], Chan C.C.[3]; [1]Institute of Environmental Health, College of Public Health, National Taiwan University, Taipei, Taiwan, [2]Institute of Occupational Medicine and Industrial Hygiene, College of Public Health, National Taiwan University, Taipei, Taiwan, [3]Innovation and Policy Center for Population Health and Sustainable Environment, College of Public Health, National Taiwan University, Taipei, Taiwan

In urban cities, fine particle (PM2.5) is a major pollutant emitted from the vehicle engine combustion that causes adverse effects on human beings. To reduce vehicular emissions, promoting bike commuting is considered a healthy way to reach the goal. Furthermore, the survey from Taiwan EPA showed that there are about 36% of people who own the scooters in Taipei in 2017. In this study, we aimed to evaluate the PM2.5 exposure of the bike and scooter commuters in Taipei. Two types of mobile platform, bicycle and electric scooter based, were used to monitor the PM2.5 along three paired routes in Taipei, Taiwan. A total of 82 runs with approximate 68 riding hours were conducted in morning rush-hours and afternoon non-rush hours in different paired routes. The mean PM2.5 measurements for bike and scooter platforms ranged between 16.01 and 25.24 μg/m3 and 16.52 and 29.26 μg/m3, respectively. By applying the generalized linear models, it was found that riding in routes in small lane had lower PM2.5 exposures, as compared to riding along major roads. After controlling other effects in the models, the level of exposure reduced ranged between 1.36 to 1.48 μg/m3. Moreover, riding scooter could have higher PM2.5 concentrations than riding bike in the routes that the level of exposure increased ranged between 1.15 to 1.16 μg/m3. The study results suggest that commuters could choose the small lanes to avoid riding in the routes frequently with traffic congestions in the rush hours.

Keywords: air, exposure factors, other (specify), particles, sampling methods

WE-PL-A3-51

Development of hazard and emission/exposure assessment methods for cellulose nanofibers

Kajihara H.[1], Fujita K.[1], Ogura I.[1], Oguri T.[11]; [1]National Institute of Advanced Industrial Science and Technology

Cellulose nanofibers (CNFs) are new materials that are expected to be useful for various applications. To accelerate the practical use of CNFs in society, the development of safety assessment methods for CNFs, such as hazard and emission/exposure assessment methods, is required. We are developing methods for the analysis, hazard assessment, and emission/exposure assessment of CNFs with the cooperation of the CNF-manufacturing companies. Three types of CNFs, namely, TEMPO-oxidized CNFs, phosphoric acid-esterified CNFs, and mechanically fibrillated CNFs are selected as the main target materials in this study. In addition, ligno CNF reinforced composites are also subject to emission/exposure assessment of applied products. We examined a method to identify and quantify CNFs as decomposition products, as a method to detect and quantify trace amounts of CNF in living bodies and environment. We confirmed the state of rat (asphyxiation, behavioral abnormalities, etc.) intratracheally instilled with CNF samples and decided the maximum dose of CNFs for establishment of appropriate intratracheal instillation method. We investigated CNFs sample conditions for skin permeability test using high performance synthetic membrane and 3D human skin model and examined the experimental procedure of fluorescent labeling technique for development of CNF detection method in 3D human skin model. We developed an emission simulation test system for transferring CNF dry powder. Data on release properties of CNFs were obtained by the simulation test. We extracted representative product cases and their exposure scenarios by interview and literature survey on CNFs applied products. The expected exposure amount when CNFs were used as a cosmetic spray was estimated using an exposure model. The result of biodegradability tests (OECD TG 301C method) showed that CNFs were readily degradable, and the biodegradability of CNFs was higher than that of non-nano cellulose powder. This study is supported by the New Energy and Industrial Technology Development Organization (NEDO), Japan.

Keywords: consumer and personal care products, nanoparticles, nanotechnology, occupational, workplace

WE-PL-A3-52

Online (continuous) measurement of indoor/outdoor reactive oxygen species (ROS) in a US home

Eftekhari A. [1], *Morrison G.* [1]; [1]*University of North Carolina at Chapel Hill, Chapel Hill, NC, USA*

Excessive reactive oxygen species (ROS) can disturb pro-oxidant/antioxidant balance and lead to oxidative stress causing different chronic obstructive pulmonary, cardiovascular, and metabolic diseases. ROS is thought to be associated with particulate matter toxicity in ambi-ent (outdoor) air. There are few measurements of ROS in indoor air, despite infiltration of outdoor ROS and generation indoors by oxidative chemistry. In this research, we designed, tested and field-deployed an online (continuous) instrument based on particle collection after condensational growth using low temperature water vapor. Continuous measurement of in-door and outdoor paricle phase (denuded) and gas+particle phase ROS (undenuded or "to-tal") was conducted in an unoccupied residential house in St. Louis, MO during August 2018. The system was operated under conditions of open and closed windows. The fluorescent in-tensity of ROS-2',7'- dichlorodihydrofluorescin diacetate (DCFH) in the presence of Horse-radish peroxidase (HRP) was measured using a fluorescence detector and calibrated by hy-drogen peroxide solutions. Results of laboratory tests showed a particle collection efficiency of 90±5% using ammonium sulfate seeds. During field sampling, we measured total phase indoor ROS and total phase outdoor ROS in the range of 0.9-4.1 and 1.5-5.1 nmol H2O2 equivalents m-3 air respectively. On separate measurement days, particle phase ROS of in-door and outdoor ranged 2.6-3.6 and 2.6-3.3 nmol H2O2 equivalents m-3 air; respectively. Both indoor and outdoor total peaked late in the afternoon. Indoor particulate phase ROS was 90-120% of outdoor ROS, and fairly insensitive to window opening or air exchange rate. This suggests most ROS resides in particles that are not readily removed by deposition in the building shell or on indoor surfaces.

Keywords: air, built/indoor environment, other (specify), particles

WE-PL-A3-53

Human CO2 generation rate: Effect of room temperature and elevated background CO2

Bivolarova M. [1], *Melikov A.* [1], *Izydorczyk P.* [1], *Markov D.* [2]; [1]*International Centre for Indoor Environment and Energy, Department of Civil Enginnering, Technical University of Denmark, Kgs. Lyngby, Denmark,* [2]*Department of Hydroaerodynamics and hydraulic machines, Technical University of Sofia, Sofia, Bulgaria*

Indoor carbon dioxide (CO2) concentration is used in the current standards to characterize indoor air quality. In addition, a widely used method to calculate supply outdoor airflow rate is based on the human CO2 generation rate. Insufficient studies as well as uncertainty associated with the common approach to estimate the CO2 production have triggered the current research. The current study investigated the impact of the room temperature on the metabolic CO2 production and the resulting ventilation design. In addition, the influence of elevated CO2 concentration in the inhaled air was examined. For this purpose, a human subject experiment was conducted in two identical climate chambers. Thirty subjects, both female and male, were selected. Participants were exposed to three different temperatures: 18°C, 22°C and 26°C. At 26°C, they were exposed to low (800 ppm) and to high (1700 ppm) concentration of CO2. There was only one person at a time in each chamber. 10 L/s of outdoor air was supplied to each chamber. A mixing fan was used to ensure complete mixing inside each chamber. Measurements of the CO2 concentration generated by the test subjects were performed in the exhaust air of each chamber. Based on the obtained results, the CO2 generation rate for each participant at the steady-state condition inside the chamber was calculated. There was no statistically significant difference of the CO2 generation rates under different temperatures. Hence, CO2 generation rate is not temperature-related feature at least after 60 minute of exposure. Contrary to that, it was observed that the high CO2 concentration in the inhaled air decreased the amount of CO_2 released to the environment. The comparison of the CO_2 generation rate obtained from the current experiment with the one predicted based on the equation in the standard showed that the experimental values were about 23% lower.

Keywords: other (specify)

WE-PL-A3-54
Indoor Air Quality in First Nations Homes in the Sioux Lookout Zone, Canada

Mallach G.[1], Kovesi T.[2], Miller J.D.[3], Mamakwa S.[4], McKay M.[5], Kirlew M.[6], Kelly L.[7]; [1]Health Canada, Ottawa, Canada, [2]University of Ottawa, Ottawa, Ontario, Canada, [3]Department of Chemistry, Carleton University, Ottawa, Ontario, Canada, [4]Sioux Lookout, Ontario, Canada, [5]Nishnawbe Aski Nation Council, Thunder Bay, Ontario, Canada, [6]Sioux Lookout First Nations Health Authority, Sioux Lookout, Ontario, Canada, [7]Sioux Lookout Meno Ya Win Health Centre, Sioux Lookout, Ontario, Canada

Background: Few studies to date have examined indoor and outdoor air quality in First Nations communities in Canada, and there remains a need to characterize air pollution levels in these areas. Rates of asthma, bronchiolitis and pneumonia are elevated in children living on-reserve in the Sioux Lookout Zone (SLZ) and indoor environmental quality (IEQ) is likely to be a contributing factor. In partnership with local First Nations communities, this study characterized IEQ in a sample of 100 homes of young First Nations children living in reserve housing in 4 remote communities in the SLZ, and collected respiratory health information for residents aged 3 and under. Methods: Air quality measurements were conducted for each home for five days. Measurements included particulate matter, VOCs, aldehydes, settled dust allergens, CO_2, temperature and relative humidity. A housing inspection provided data on ventilation systems, mould growth, and building characteristics. Questionnaires and medical chart reviews documented the frequency, duration and level of health services utilization by children in participating households. Results: Two communities participated during each of winter 2017/18 and 2018/19, when homes are most tightly sealed. Preliminary results suggest that, on average, particulate matter and CO_2 concentrations in on-reserve First Nations houses exceed those measured in other Canadian homes, likely owing to factors including inadequate ventilation, low uptake of heat recovery ventilators, high occupancy, and the reliance on wood stoves for heating. Relevance: The results will be used to expand local First Nations capacity and expertise related to housing, and to identify simple home improvements and other building interventions that could improve the respiratory health of this vulnerable population.

Keywords: air, built/indoor environment, children, particles, respiratory health, ventilation

WE-PL-A3-55
Temporal change of formaldehyde concentration in outdoor environments

Mizukoshi A.[1], Nagano A.[2], Azuma K.[1], Okumura J.[1]; [1]Department of Environmental Medicine and Behavioral Science, Kindai University Faculty of Medicine, Osakasayama, Japan, [2]Research Group of Environmental Measurement, Osaka, Japan

Formaldehyde is one of the typical pollutants in indoor air whose possible sources are artificial materials, furniture, etc. Moreover, formaldehyde is generated secondarily from the reaction between ozone and volatile organic compounds in both indoor and outdoor environments. In outdoor environments, the generation of formaldehyde may vary temporally depending on fluctuation of the ultraviolet radiation intensity and the photochemical oxidant concentration. Considering the health effects of formaldehyde occurred by acute exposure for 30 min, information on temporal concentration change is necessary. In this study, we measured formaldehyde concentration continuously in outdoor environment using Formaldemeter Htv-m in summer (June) and winter (November to January). In addition, we referred corresponding data of the solar radiation intensity, and the concentration of photochemical oxidant, non-methane hydrocarbon (NMHC), and particulate matter 2.5 (PM2.5) form nearest ambient monitoring stations. As results, formaldehyde concentration fluctuated during one day, indicating the concentration increased from 8:00-

10:00, was highest in afternoon, and decreased to middle of the night. These tendencies synchronized to the change of solar radiation intensity and oxidant concentrations, though NMHC, PM2.5 and nitrogen dioxide indicated different tendencies. Timing of formaldehyde concentration increase was delayed compared to those of the solar radiation intensity and photochemical oxidant concentrations, suggesting the generation of formaldehyde was attributed by photochemical oxidant including ozone.

Keywords: VOCs

WE-SY-B3: Modeling Exposure by Spraying and Foaming Activities

WE-SY-B3-56
Basic relevant parameters in modelling spraying activities

Koch W.[1], Hahn S.[1], Schwarz K.[1]; [1]Fraunhofer ITEM, Hannover, Germany

Spray- and foam application of substances in liquid formulations is a widely used industrial and professional process to cover surfaces with specific chemical compounds. The formulations may contain solvents and/or propellants and non-volatile chemical entities. Modelling the exposure during spraying activities via deterministic models requires information on the spraying process as well as on the underlying application and behavior of the applicator. Several Tools are available to estimate exposure such as SprayExpo or ConsExpo. These tools need specific input parameters to estimate the inhalable concentration. The first part of the presentation covers our approach used for the determination of the parameters related to the process. The most important quantity is the source strength of inhalable mass characterized by compound specific release fraction defined as the airborne mass of the compound normalized to the entire sprayed mass of this compound. For aerosols (non-volatile compounds) different particle size fractions are of interest as well as the dependence of the release fractions on the process parameters of the spraying. Among them the most important one is the droplet size distribution of the spray. For solvents their vapor pressure and the geometry of the spray cone are parameters controlling the release fractions and the partitioning between aerosol and vapor phase. Data for aerosols originating from spraying and foaming as well as for the volatiles generated in spray processes are provided. In the second part, the aspect of practical use of this information is discussed. This covers the grouping of release fractions in classes characterized by easily accessible spray and foam parameters. The influence of the uncertainty when using this grouping scheme in deterministic exposure models is finally addressed.

Keywords: consumer and personal care products, exposure models, respiratory health, workplace

WE-SY-B3-57
Modelling approaches and desired improvements for workplace exposure assessment of spray processes

Meyer J.[1], Roitzsch M.[1]; [1]Federal Institute for Occupational Safety and Health (BAuA), Dortmund, Germany

At workplaces, chemicals are often applied by spraying, which often bears a high exposure potential. Therefore, spray applications are highly relevant for workplace risk assessments and regulatory assessments under legal frameworks such as the European REACH or biocides regulations. Assessment of these applications is commonly performed employing models based on workplace measurements, or on mechanistic simulations. For regulatory assessments, this provides some important challenges. Focussing on mechanistic approaches, this contribution will highlight two major challenges: 1. Professional users may be exposed to aerosols, but depending on the volatility of the substances, also the exposure to vapours may be relevant. However, most models used for assessment of occupational spray processes (e.g. SprayExpo or ConsExpo) only consider the exposure to aerosols. 2. Regulatory assessments should cover an application as far as possible with its full range. However, the models are usually designed to simulate a concrete situation based on a discrete set of parameters. Furthermore the definition of a set of parameters which can be considered representative for a broad spectrum of scenarios is extremely difficult, if not impossible. We will outline how these challenges are currently addressed under the European regulatory frameworks and provide suggestions for model improvements that we consider suitable to tackle these issues. By means of the occupational exposure model SprayExpo, an overview of the input parameters that seem to be most challenging for users of mechanistic spray models will be given. Approaches that may increase user acceptability will be discussed. Overall, the authors suggest a reduction to the most relevant exposure determining parameters and consider the replacement of distinct input parameter values in favour of an input parameter spectrum. In addition, the pre-determination of sector specific parameters, e.g., for description of common spray equipment, may considerably improve model usability.

Keywords: activity patterns, exposure models, occupational, workplace

WE-SY-B3-58

Experiences and Challenges of modelling consumer exposure by sprays

Jung, C.; German Federal Instittue for Risk Assessment, Berlin, Germany

Consumer exposure for sprays, e.g. for cleaning products, may be assessed using mechanistic models which predict exposure based on various input parameters, especially the particle size distribution of the sprayed product. There exist several models, e.g. the spray models within ConsExpo or the SprayExpo tool. The presentation reflects experiences with using these tools in various regulatory areas, in particular within the framework of the REACH regulation. The specific challenges in describing the input parameter characteristics for consumer spray products are presented. The applicability in the domain of consumer exposure estimation is discussed. For example the included spray nozzles within SprayExpo are not necessarily applicable for consumer spray products. Overall, ideas and wishes for further model development are discussed. Also, areas of spraying for which no readily accessible mechanistic models exist, e.g. spraying outdoors, are highlighted.

Keywords: consumer and personal care products, exposure models

WE-SY-C3: Estimating and Comparing the Burden of Disease of Indoor Air Pollutants in European Countries, USA, and China

WE-SY-C3-59

Burden of disease due to indoor exposures in Europe

Hänninen O.; National Institute for Health and Welfare, Kuopio, Finland

Poor air quality is a leading contributor to the global burden of disease and mortality worldwide. People spend most of their time in built indoor environments, highlighting the special role of indoor air quality. While the importance of environmental pollution in general is well-linked to the global burden of disease, the quantitative understanding of the role of indoor exposure is lagging behind. Known contributing indoor exposures include particles and ozone from outdoor and indoor sources, and indoor sources such as tobacco smoking as well as building-related emissions from construction materials (e.g. volatile organic compounds), toxic and harmful emissions due to dampness and mould, radon from soil and carbon monoxide from combustion devices. The overall aim of this presentation is to give an overview of the European estimates from EnVIE, IAIAQ, HEALTHVENT and EBODE projects, supplemented by exposure/infiltration results for PM2.5. The Global Burden of Disease community, with World Health Organization (WHO) and Institute of Health Metrics and Evaluation (IHME) in the lead, has moved towards non-discounted, non-age-weighted estimates and shifted from incidence to prevalence based estimation. These changes do have a substantial impact on the numerical values. Nevertheless, the overall burden remains dominated by fine particulate matter (PM2.5) from outdoor air sources, followed by indoor generated particles, radon, bioaerosols (as allergens), and second hand smoke. Wider application of burden of disease methods would allow for prioritization of preventive measures and optimized uses of resources. There remains a large potential for improvement in accounting for morbidity and better understanding of effects on wellbeing, happiness and productivity.

Keywords: air, built/indoor environment, chemical prioritization, dampness/mold, health, chronic diseases, health, other/general, particles, respiratory health, risk assessment, secondhand smoke

WE-SY-C3-60
Indoor air pollution burden of disease in France

Mandin C.; Scientific and Technical Center for Building (CSTB), Paris, France

An evaluation of the socio-economic costs of indoor air pollution can facilitate development of the appropriate public health policies. For the first time in France, such an evaluation was conducted for six selected pollutants: benzene, trichloroethylene, radon, carbon monoxide (CO), particles (PM2.5), and environmental tobacco smoke (ETS). The health impacts of indoor exposure (mortality and morbidity) were already available for radon, CO and ETS. For the other target pollutants, the health impacts of indoor exposure were calculated. For these calculations, two approaches were followed depending on the available data: the first followed the principles of quantitative health risk assessment based on the US National Research Council method, and the second was based on concepts and methods related to health impact assessment. For both approaches, toxicological data and indoor concentrations related to each target pollutant were used. External costs resulting from mortality, morbidity (life quality loss) and production losses attributable to these health impacts were assessed. In addition, the monetary costs for the public were determined. Lastly, a sensitivity analysis was performed. Indoor pollution associated with the selected pollutants was estimated to be associated with more than 28,000 new cases of disease and around 20,000 deaths annually in France. Particulate matter (PM) contributed the most to the total burden of disease (75%), followed by radon. The economic costs attributable to indoor exposure to these six air pollutants was approximately €20 billion per year in France. Despite its limitations, the results of this study are consistent with the orders of magnitude reported in previous national or European evaluations.

Keywords: VOCs, air, built/indoor environment, chemical prioritization, environmental health, particles, risk assessment

WE-SY-C3-61
The Burden of Disease for Indoor Air Pollutants in Urban China

Liu W.[1], Zhang Y.[1]; [1]School of Environment, Tsinghua University, Beijing, China

In October 2017, we launched a study on Disease Burden of Indoor Air Pollutants (DB-IAPs) in urban China. The main objectives of this study are to: (1) Clarify the detailed lists of indoor air pollutants and their concentration distributions in different kinds of buildings in urban China and in different climate regions of China, especially in the most densely-populated regions (the Jing-Jin-Ji capital region, the Yangtze River Delta region, and the Pearl River Delta region); (2) Characterize the incidence of major non-communicable diseases in the Chinese population and quantify the burden of these diseases in urban China, according to year, sex, age, and region; (3) Systematically review and meta-analyse the associations between indoor air pollutants and the selected diseases; (4) Calculate DB-IAPs and provide a rank list; (5) Provide reasonable thresholds of concentrations for various IAQ related standards according to the DB-IAP. Our methods will include systematic literature review and meta-analysis, field measurement, and updated methods from the Global Burden of Diseases (GBD) study and World Health Organization (WHO) for the calculation of DB-IAPs. This study is on-going. The preliminary results from the systematic literature review and meta-analysis will be introduced in the symposium. This study will provide scientific data for updating the available indoor air quality (IAQ) related standards "Indoor Air Quality GB18883-2002)" and "Assessment Standard for Healthy Building" and for future IAQ related standards.

Keywords: built/indoor environment, environmental health, exposure models, health, chronic diseases, respiratory health

WE-SY-C3-62
Revision of China's national indoor air quality standards

Xu D.[1], Xu C.[1], Zhou J.[1]; [1]National Institute of Environmental Health, Chinese Center for Disease Control and Prevention, Beijing, China,

China's national indoor air quality standard GB/ T18883 was issued in November 2002 and was implemented on March 1, 2003. This standard specifies limits and ranges for a variety of factors. Physical metrics: temperature, relative humidity, air flow rate and fresh air volume; chemicals: sulfur dioxide, nitrogen dioxide, carbon dioxide, ozone, ammonia, formaldehyde, carbon monoxide, benzene, toluene, xylene, TVOC, benzo[a]pyrene, respirable particulate matter (PM10); biological indicator: total number of bacteria colonies; radioactive indicator: Rn222. The standard mainly applies to residential and office building indoor environments. This standard has been used for more than 15 years. With the rapid development of China's economy and society, indoor and outdoor environmental air pollution and people's lifestyles have changed greatly, and many new indoor air quality problems have emerged. Both the indicators and their limits need to be revised based on the latest evidence from air pollution health impact studies. Accordingly, the National Health Commission of the People's Republic of China initiated revision of the standard in November 2018. To conduct this revision, we must first assess trends of indoor air quality and pollution sources in China. Do the existing standards address current issues of indoor air pollution in China? The WHO guidelines include standards for indoor benzene, carbon monoxide, formaldehyde, naphthalene, nitrogen dioxide, polycyclic aromatic hydrocarbons, especially benzo[a]pyrene, radon, trichloroethylene, tetrachloroethylene, dampness and mold. Is this revision consistent with WHO? Indoor air pollutants can be ranked according to their toxicity, concentration and disease burden. We should determine if new air pollutants need to be added and whether the original limits of air pollutants need to be modified based on the latest evidence from health impact studies.

Keywords: built/indoor environment, environmental health, environmental regulation, exposure factors, health, other/general

WE-SY-C3-63
Modern Life Makes Children Sick

Sun Y.[1], Hou J.[1], Weschler L.[2], Sundell J.[3]; [1]Tianjin University, Tianjin, China, [1]Independent Researcher, Colts Neck, NJ, USA, [1]Tsinghua University, Beijing, China

In the past 60 years, the prevalences of asthma and allergy among children has increased all around the world. Neither genetic nor outdoor environmental factors can explain this increase. We performed a cross-sectional study of 7366 children in Tianjin, China, on associations of home environment and life styles with asthma and allergy. The prevalences of diagnosed asthma, rhinitis and eczema among 0-8 year-old children in the Tianjin area were 4%, 9% and 39%. For asthma and allergy, the population attributable fraction (PAF) due to modern floors and wall coverings (i.e. laminated wooden floors and painted walls compared to tile floors and lime-coated walls) was 23%. Window condensation in winter and air conditioner use in summer, which are proxies of less ventilation, accounted for 4%-18% of rhinitis and eczema. Caesarean delivery accounted for 8%-12% of rhinitis and eczema symptoms. We developed a modern life index from appropriate home characteristics and lifestyle and food consumption habits and found it to have a clear dose-response relationship with asthma and allergy in Tianjin children. It indicated that a "modern" home environment together with a modern lifestyle was associated with increased prevalence of asthma and allergies among children.

Keywords: built/indoor environment, children, environmental health, epidemiology, ventilation

WE-PL-D3-EAM: Air Pollution Modeling

WE-PL-D3-64

Incorporation of air pollution dispersion principals into regression models - how far can we take it?

Chen S.[1], Yuval.[1], Broday D.[1]; [1]Technion - Israel Institute of Technology, Haifa, Israel

Regression modelling is the most commonly used method for personalised air pollution exposure assessment. These models often incorporate a large number of land-use variables for which the causal relationship to air pollution is not clear. Moreover, it has been challenging to apply them to exposure assessment at a high temporal resolution. Formulating a regression model which takes into account well-established dispersion principals may improve previous capabilities. In this work, we designed a non-linear regression model which is based on traditional Gaussian dispersion. This model has 9 regression parameters which were optimised to provide the best fit between modelled and observed nitrogen oxides concentrations at the 39 regulatory air quality monitoring locations in our study area. Using a single representative wind direction for the entire study area, the new model performed better than Inverse-Distance-Weighing (IDW) interpolation and a simpler regression model (TI-ODM). In particular, the cross-validated Mean Spatial Pearson Correlation (MSPC) was 0.23 over a dataset of 3500 unique half-hourly time-points, versus 0.19 for TI-ODM and 0.14 of IDW. The MSPC is particularly interesting when considering the epidemiological applications of exposure models. It was striking that all models performed worse in predicting total nitrogen oxides than nitrogen dioxide, as shown in previous studies over the same area. A heterogeneous wind field was incorporated through a puff-like regression model. This model was applied both dynamically (i.e. having memory terms that propagate from each half-hourly time-point to the next) and statically (each time-point modelled separately as a "clean slate"). The static version performed much better (Nash-Sutcliffe efficiency=0.3 in the static version vs. -537.9 in the dynamic version). However, the static version could be run only in cross-validation mode due to computational limitations. In future work, this model must be qualitatively validated. If successful, it can be applied to epidemiological studies.

Keywords: air, exposure models

WE-PL-D3-65

Predicting indoor PM2.5 levels in Finnish buildings

della Vecchia S.[1], Clinchard S.[1], Aller R.[1], Haverinen-Shaughnessy U.[1], Espinosa Leal, L.[1]; [1]720°, Helsinki, Finland

Fine particulate matter (PM2.5) is considered one of the most harmful air pollutants. While a large proportion of the particles is originating from outdoor sources, people are mostly exposed while indoors. Predicting future trends of PM2.5 concentrations could help buildings owners and operators developing better control strategies, and minimizing delays in responding to potential indoor air quality (IAQ) issues. Machine Learning and Deep Learning methods, in particular Long-Short Term Memory Neural Networks (LSTM), have shown good results in predicting sequential data. In this study, PM2.5 data from 260 sensors in 119 Finnish buildings were collected during the period 2014/09 - 2019/01. The mean PM2.5 concentration observed was 1.01 μg/m3 (SD2.41 μg/m3). Different methods were compared to predict from one hour up to 8 hours lead times. Three methods were tested for short term predictions (+1 hr): Autoregression, Random forest for regression, and LSTM, while the latter two were tested for long-term predictions (+8 hr). For short term prediction, all methods used a univariate time series (historical hourly indoor PM2.5 average). The best prediction was obtained using LSTM with only one lag variable (mean absolute error 0.21 μg/m3, mean squared error 0.85 μg/m3). For long term prediction, both methods were first tested using a multivariate time series (historical indoor and outdoor PM2.5). An additional time series containing the outdoor PM forecasts for the next eight hours was then added to the models, which significantly improved the model accuracy. The lowest Mean Absolute Error (0.49 μg/m3) and Mean Squared Error (1.83 μg/m3) were obtained using LSTM with eight lag variables and eight forecasts. In conclusion, long-term predictions are more challenging, but the predictions can be improved by multivariate methods.

Keywords: analytical methods, environmental health, particles, workplace

WE-PL-D3-66

Land Use Regression models for Ultrafine Particles: development and transferability within a mega-city

Yang Z.C. [1], Freni-Sterrantino A. [1], Fuller G. [1], Gulliver J. [2]; [1]MRC-PHE Centre for Environment and Health, Imperial College London, London, United Kingdom; [2]Centre for Environmental Health and Sustainability, University of Leicester, Leicester, United Kingdom

Background: A lack of routine monitoring for ultrafine particle (UFP) means that bespoke monitoring (e.g. mobile measurements on vehicles or repeated short-term monitoring over a spatially distributed network of sites) has to be undertaken to develop land use regression (LUR) models. This is especially challenging in mega-cities due to the large spatial area. Methods: As an alternative to monitoring across a mega-city, we developed single- and two-district models in London, and then tested the transferability (i.e. generalisability) of models between districts. In each district, repeated 30-min monitoring was undertaken at 40 sites in three different seasons (2016-2018) to estimate annual mean UFP (particles cm-3). We constructed ten single- and two-district LUR models by allocating monitoring sites to one of ten groups and then used nine groups of sites to develop each model, iteratively, with the remaining group held-out for model evaluation (e.g. groups 1-9 to predict for group 10; groups 2-10 to predict for group 1, etc.). We assessed model performance by comparing R2 and the variables selected, and pooled the 10% of sites held-out each time to produce an overall R2 to assess model robustness. The transferability of models was tested by applying single- and two-district models to the other districts. Results: For the ten models for single districts, R2 ranged from 60% to 73% and model variables were similar and all included nearby road traffic. Model performance and structure was different between areas. R2 for two districts ranged from 50% to 68%. Pooled holdout validation had R2 values of 37%- 59%. Performance in transferring models to other districts within London was overall lower with R2 ranging from 10% to 39%. Conclusion: UFP LUR models may perform well within the confines of a monitoring network but transferring models within a city may have substantially lower performance.

Keywords: exposure models

WE-PL-D3-67
NO2 Air Pollution Exposure Assessment in Urban Mysore, India

Nori-Sarma A. [1], *Thimulappa R.K., Venkataraman G., Fauzie A., Venkareddy L.* [2], *Berman J.* [3], *Lane K.* [4], *Fong K.* [1], *Bell M.* [1]; [1]*Yale University, New Haven, CT, USA,* [2]*University of Illinois at Chicago, Chicago, IL, USA,* [3]*University of Minnesota, Minneapolis, MN, USA,* [4]*Boston University School of Public Health, Boston, MA, USA*

Introduction: In developing countries, rapid urbanization has led to decreased air quality. Few studies have applied spatially heterogeneous sampling to assess ambient air pollution levels in cities in India, to understand the important local predictors of air pollution exposure, which may be unique to Indian urban environments. Methods: Seasonal sampling was conducted in four campaigns (2016-2017) at 150 sites throughout Mysore, a small and rapidly growing city in south India. Nitrogen dioxide (NO2) levels were assessed using passive Palmes tubes and Ogawa badges. Results: Annual average NO2 ranged from 0.8 to 31.2ppb (1.5 to 58.7 ug / m3) for the study area, with higher concentrations in the center of the city. In the final Land Use Regression (LUR) model, proximity to major roads, point sources of pollution, and religious complexes, as well as high human activity land uses and population density were positively associated with air pollution, while proximity to minor roads and low human activity land uses were inversely associated with air pollution (R-squared = 0.535). Cross-validation of the results confirmed the reliability of the model. Kriging improved estimates by accounting for spatial correlation. Conclusions: The combination of passive NO2 sampling and LUR / Kriging modeling techniques allowed for characterization of NO2 patterns in Mysore. While traffic pollution has been indicated in the past as a major contributor to ambient air pollution levels in urbanizing centers of Asia, results indicate that other pollution factors (e.g., point sources), as well as highly localized characteristics of the urban environment (such as proximity to religious buildings), influence air pollution exposure patterns in urban India. Despite a reputation as one of the cleanest cities in India, areas of Mysore consistently experience pollution in excess of WHO health-protective guidelines for NO2.

Keywords: air, exposure models, geospatial analysis/GIS

WE-PL-D3-68
Comparison of two air pollution models applied to assess population exposure in Leipzig, Germany

Altug H. [1], *Zeynalova S.* [2], *Löffler M.* [2], *Wiedensohler A.* [3], *de Hoogh K.* [4], *Schikowski T.* [1]; [1]*IUF – Leibniz Research Institute for Environmental Medicine, Düsseldorf, Germany,* [2]*Leipzig University, Institute for Medical Informatics, Statistics and Epidemiology (IMISE), Leipzig, Germany,* [3]*TROPOS - Leibniz Institute for Tropospheric Research, Leipzig, Germany,* [4]*Swiss TPH - Swiss Tropical and Public Health Institute, University of Basel, Basel, Switzerland*

Objectives: We compared the exposure assignment of two air pollution models to all addresses in the city of Leipzig. Methods: Predicted concentrations of nitrogen dioxide (NO2) and fine particulate matter (PM2.5) for 2010 were assigned to all addresses in Leipzig (n=69,662) from two air pollution models with different spatial resolutions: fine spatial resolution (100x100 m) hybrid land use regression models developed for West Europe within the ELAPSE project and models with 2x2 km resolution from German Environmental Agency (UBA) developed for all Germany using the optimal interpolation technique. Results: Modelled annual mean concentrations of NO2 at all addresses were 26.9 ± 4.9 (11.8-56.2) µg/m3 and 20.7 ± 1.0 (15.8-25.6) µg/m3 for ELAPSE and UBA models, respectively. PM2.5 concentrations were 18.6 ± 0.8 (14.5-21.1) µg/m3 and 17.0 ± 0.5 (15.6-17.7) µg/m3 for ELAPSE and UBA models, respectively. The range of the ELAPSE exposures was much larger than for the UBA for both pollutants. Pearson's correlation coefficients between ELAPSE and UBA exposures were 0.29 and 0.50 for NO2 and PM2.5, respectively. NO2 ELAPSE exposures correlated poorly with NO2 UBA, possibly explained by the fact that NO2 is a locally varying pollutant, strongly associated with traffic, which is better picked up by the ELAPSE model (100x100m) than by the UBA model ((2x2km). On the other side, PM2.5 ELAPSE exposures correlated better with PM2.5 UBA as PM2.5 is a more regional pollutant so less variation at the local scale and hence a better correlation. Conclusion: We assigned two different models developed for Western Europe and Germany to a very defined spatial area in Germany. The model with higher spatial resolution fitted better to explain within city variation of pollutants related to local sources to describe exposures of residents of Leipzig and the LIFE (The Leipzig Research Centre for Civilization Diseases) cohort.

Keywords: air, geospatial analysis/GIS

WE-PL-D3-69

Impact of the deployment of electric vehicles on population exposure to air pollution

Minet L.[1], Gai Y.[1], Wang A.[1], Posen D.[1], Hatzopoulou M.[1]; [1]University of Toronto, Toronto, Canada

The transportation sector contributes 35% of Ontario's total greenhouse gas (GHG) emissions. Similarly, it is also a large contributor to criteria air contaminants. For example, on-road vehicles contribute 34% of total nitrogen oxide (NOx) emissions. Since electric vehicles do not generate any operating emissions, vehicle electrification has been promoted to reduce traffic-related air pollution and GHG emissions. However, in the context of a scenario analysis of the impact of electric vehicles on air quality, it is necessary to consider emissions from surrounding power plants. We propose an integrated methodology to investigate the exposure of the Greater Toronto and Hamilton Area (GTHA) population to air pollution under different scenarios expressing various penetration rates for electric vehicles, which would impact the production of ten power plants present in the GTHA. Expected emissions from these power plants are modeled using current data from the National Pollutant Release Inventory (NPRI) database. Marginal emission factors are generated to take temporal variations into consideration, as only certain types of power generators respond to a marginal increase in demand (e.g. electric vehicle charging). Traffic operating emissions are modeled by combining regional travel demand and traffic assignment results with speed-based emission factors. Finally, a plume-in-grid model, which combines a chemical transport model and a Gaussian puff dispersion model, is used to analyse air quality under various meteorological conditions. Different scenarios are examined in terms of equity with respect to air pollution exposure.

Keywords: aggregate exposure, air, environmental justice, exposure models

WE-PL-E3-CPC: Building Characteristics, Energy Efficiency, Health and Satisfaction with IEQ

WE-PL-E3-70

The effects of the total floor area of a building on building-related symptoms in air-conditioned office buildings: a cross-sectional study

Azuma K., Kagi N.[2], Yanagi U.[3], Kim H.[4], Hasegawa K.[5], Simazaki D.[4], Kaihara N.[4], Kunugita N.[6], Hayashi M.[4], Kobayashi K.[4], Osawa H.[4]; [1]Kindai University Faculty of Medicine, Osakasayama, Japan, [2]Tokyo Institute of Technology, Meguro, Japan, [3]Kogakuin University, Shinjuku, Japan, [4]National Institute of Public Health, Wako, Japan, [5]Akita Prefectural University, Akita, Japan, [6]University of Occupational and Environmental Health, Kitakyushu

Since the early 1970s, building-related symptoms (BRSs) have emerged as an occupational and environmental health issue. We previously conducted three phase studies with two cross-sectional studies using questionnaires and environmental measurements and one longitudinal study to investigate the environmental factors associated with the BRSs. Maintenance of office environment may differ based on the total floor area of a building. Based on our previous studies, we conducted a cross-sectional study concerning the effects of the total floor area of a building on the BRSs, particularly in air-conditioned office buildings. The surveys were conducted from January 2018 to March 2012, involving 216 offices and 1891 employees during the winter season. We administered questionnaires to these employees for obtaining information on BRSs, indoor air quality (IAQ), workplace environment perceptions, and occupational stress. The office buildings were categorized into small buildings (<2000 m2, 93 offices with 654 employees), middle buildings (2000–3000 m2, 22 offices with 293 employees), and large buildings (≥3000 m2, 101 offices with 944 employees) according to the total floor area of the buildings. The complaints of indoor temperature and humidity were significantly lesser in small buildings compared with large buildings. Multivariate analysis revealed that upper respiratory symptoms, although their associations with the buildings were not strong, were significantly lesser in small buildings compared with large buildings (for trend, $p < 0.05$). The BRSs for which the associations with the buildings were strong were not significantly associated with the total floor area of the buildings, although the overall prevalence of BRSs was reduced for small buildings compared with large buildings. In the three categories of buildings, indoor thermal comfort and dryness were significantly associated with strong BRSs. This study suggests that adequate maintenance of office environment and control of IAQ are particularly crucial in large buildings.

Keywords: built/indoor environment, environmental health, epidemiology, health, symptoms, workplace

WE-PL-E3-71 WITHDRAWN

WE-PL-E3-72
Enhancement of thermal comfort in summer conditions by the use of personalized ceiling fans

Rissetto R.[1], Wagner A.[1], Schweiker, M.[1]; [1]Karlsruhe Institute of Technology, Karlsruhe, Germany

Due to increasing summer outdoor temperatures, the cooling demand for active cooling of buildings is expected to increase. Alternatively, the implementation of low-energy cooling strategies has shown great potential, especially by the use of ceiling fans. Additionally, to increase thermal comfort while maintaining energy consumption as low as possible, the attention has been drawn to the use of personalised comfort systems. This study focusses on the impact of different air direction, occupants' activity level and control opportunities of a personalized ceiling fan on occupants' thermal comfort and satisfaction in a working environment. We conducted a 3-week laboratory experiment where 52 participants took part. Two age groups were examined at different ceiling fan's configurations (six different air directions), two indoor temperature settings (28°C and 31°C), two subject's activity levels (sitting and standing) and two air speed control possibilities (fixed and adjustable by the subject). Participants' thermal comfort responses were investigated. Results show that 70% of the subjects perceived the environment at 28°C as comfortable when the fan was in use – in comparison to 22% without fan use. According to both the fan configuration and the control possibility over the air speed, no significant difference in terms of improved thermal comfort was found, while significant differences were found between the sitting and the standing position, under different temperature conditions and between morning and afternoon conditions. The results suggest that the cooling effect of personalized ceiling fans can fulfill participants' thermal comfort at moderately high indoor temperatures. Moreover, this technology could adapt to different building configurations, especially in existing buildings, due to its flexibility and guarantee of thermal comfort, as the air blowing from different directions showed no negative impact on users' thermal satisfaction. On the other hand, occupants' activity level, indoor temperature and subjects' thermal history influence significantly occupants' thermal comfort.

Keywords: air, behavior, built/indoor environment, ventilation, workplace

WE-PL-E3-73

Multifactor Transportation Noise Index for Health-Risk Exposure Metrics and Applications

Leaffer D.[1], Suh H.[1], Durant J.[1], Tracey B.[1], Roof C.[2], Hastings A.[2], Gute D.[1]; [1]Tufts University, MA, USA, [2]Northeastern University, Boston, MA, USA

Background: Noise ranks second after air pollution as the most important environmental exposure (WHO). Chronic exposure to transportation noise is of increasing concern with regards to population health effects. Conventional measures of noise exposure including A-weighted decibels (dBA) and day-night average sound levels (DNL) provide assessment of community impacts from noise annoyance but do not capture other characteristics of noise (impulse, tonal, temporal and spectral variations) that must be evaluated for health effects. Noise metrics encompassing frequency content, acoustic power, time and duration of occurrence, in addition to dBA and DNL, would offer a more robust index of transportation noise exposure. Methods: We implemented a noise monitoring data collection program in two Greater Boston (Massachusetts, USA) communities impacted by transportation-related noise and air pollution. At both sites, a class 1 sound level meter and digital audio recorder were deployed for 150 days. Data collected included A-weighted sound pressure levels (SPL), DNL, raw sound files and 1/3 octave band spectra. Results: Samples of recorded transportation-source noise measurements were analyzed by Discrete Fourier Transform (DFT) to compute power spectral density (PSD). Spectral signatures were identified by transportation class - truck, train, aircraft, which exhibited differential PSD and acoustic power, and linked to duration and time of event and number of occurrences. These factors, Θi were aggregated in a preliminary algorithm to quantify a noise exposure index as a weighted average, tau =∑ωiΘi, where ωi is a weighting coefficient. Implications: We characterized transportation noise exposure by employing a broader range of noise analytics than are traditionally used. Our preliminary statistical algorithm is anticipated to provide a more complete metric for noise exposure assessment and a novel composite index of environmental noise. This index will be beneficial to the development of local ordinances and policies to mitigate impacts from transportation noise for exposed populations.

Keywords: acoustics/noise, aggregate exposure, environmental health, exposure factors, exposure models

WE-PL-E3-74

Energy retrofits, housing satisfaction and health, a 3-year follow-up

Haverinen-Shaughnessy U.[1], Leivo V.[1], Martuzevicius D.[2]; [1]Tampere University, Tampere, Finland, [1]Kaunas University of Technology, Kaunas, Lithuania,

The INSULAtE project (http://www.thl.fi/insulateproject) aimed to develop a comprehensive protocol for assessment of the effects of improving energy efficiency of buildings indoor environmental quality and health; and to demonstrate the effects in two European countries (Finland and Lithuania). As a part of the project, data were collected from the occupants of 39 Finnish multifamily buildings both before (N=199) and about a year after (N=166) energy retrofits (e.g. improving thermal insulation, heating and/or ventilation systems) using questionnaires and diaries. Data were also collected from seven non-retrofitted control buildings. The questionnaire included questions on respondents' satisfaction with indoor temperature, indoor air quality (IAQ), and health. In a sub-sample of ten buildings, questionnaires were repeated at about two years (N=52) and three years (N=10) after the retrofits. This presentation focuses on the long-term follow-up, noticing that such results is rarely published in the scientific literature. Results from occupant diaries will also be presented. Previously published results showed statistically significant improvements in the respondents' satisfaction with IAQ (OR 2.2), absence of upper respiratory symptoms (OR 1.7) and not missing work or school due to respiratory infections (OR 2.4) at the first follow-up one year after the retrofits were completed. Also, a non-significant increase in satisfaction with indoor temperature (OR 1.1) was observed. In the subsequent follow-ups, preliminary results indicates OR 1.8 (2nd follow-up) and OR 3.6 (3rd follow-up) for satisfaction with IAQ; OR 1.3 (both 2nd and 3rd follow-ups) for absence of upper respiratory symptoms; and OR 1.7 (2nd follow-up) for not missing work or school due to respiratory infections. Satisfaction with indoor temperature increased to OR 1.7 (2nd follow-up) and OR 2.2 (3rd follow-up). Whereas these associations are not statistically significant (due to diminishing sample size), they indicate that some positive effects could be seen several years after retrofits were competed.

Keywords: built/indoor environment, environmental health, health, symptoms, risk assessment, sustainability

WE-PL-E3-75

A comparison of the performance of commonly-used sensors for measuring indoor carbon dioxide (CO2) concentration levels

Huang R.[1], Andamon MM.[1], Rajagopalan P.[1]; [1]RMIT University, Melbourne, Australia

The rise of ambient carbon dioxide (CO2) and lower ventilation have implications on indoor environments. CO2 concentration levels is one of the indoor air quality (IAQ) indicators used to evaluate the ventilation requirements in buildings. IAQ is typically addressed through compliance with requirements based on industry consensus standards such as the ANSI/ASHRAE Standard 62.1. However, IAQ investigations are often conducted in an ad-hoc manner. There is still a lack of formal methods and tools to support IAQ investigations, particularly the use of instruments to monitor indoor conditions. This paper presents the preliminary work on verifying the performance of two (2) commonly and readily available instruments which monitor indoor CO2 concentration levels: HOBO MX1102 CO2 data logger and the IAQ probe of Testo 480. These easy-to-use instrumentations allow the measurement, analysis and logging of relevant indoor air quality and ventilation parameters. To understand the performance of the instruments in different settings of CO2 concentration levels, these were tested with the reference grade Vaisala CARBOCAP carbon dioxide probe GMP343. All the sensors were deployed in five case scenarios: (1) an outdoor condition; (2) indoor - with high volume of occupants, and (3) low volume of occupants; (4) un-occupied university classroom; and (5) un-occupied university staff office. The initial results indicate that both the MX1102 and Testo 480 had a similar trend in measured CO2 concentration levels of an unoccupied air-conditioned office in the range of 350-450ppm. However, both sensors underestimated CO2 concentration levels by 50-100ppm as measured by the reference. This paper will report on the preliminary results of the verification study on the performance of commonly used sensors across a range of CO2 concentration levels. The objective is to support an IAQ investigation process that allows the identification of performance variable(s) to aid in the selection of suitable and appropriate instrumentation.

Keywords: air, built/indoor environment, sensor technology, ventilation

WEDNESDAY POSTER SESSIONS

WE-PO-01
A Model To Predict Radon Transport From Building Material To Indoor Air In Building

Lee B.H.[1], Kim S.D.[1], Chun C.Y.[1]; [1]Land & Housing Institute, Sejong City, Korea

Building materials are one of the important radon sources in indoors, second only to soil, so surface radon emission from building materials have been extensively measured. Based on this measured data, several researchers have attempted to predict the indoor radon concentration attributable to radon emitted from building materials such as walls, floors and ceilings. However, an important aspect not considered in this methodology is the correlation between radiation such as r-ray from building materials and radon gas concentration in indoors. This relationship occurs mainly because of the change in the radon diffusion process from the former to the latter configuration. To predict the radon concentration in indoors based on the radiation emission data of building material samples, this paper proposed a semi- empirical model involving radon transport depending on indoor floor plan and the physical characteristics of the building materials as well as surface area, thickness as other input parameters. This model has been established by statistically fitting the ratio of the solution to radon transport equations for the cases of three-dimensional zonal model to a simple mathematical function based on mass balance equation. The model predictions have been validated against the measurements made at a mock-up housing. This model provides an alternative tool to estimate radon emission from building materials without relying on radon emanation factor and bulk density of the materials. Moreover, it may be very useful in the context of developing building codes for radon regulation in new buildings.

Keywords: built/indoor environment

WE-PO-02
Determinants of Household Energy Emissions and Transitions

Bond, T.

Households and their energy demands are complex systems that demand complex description. In this presentation, I introduce the notions of energy service characteristics, interoperability, and robustness to describe household energy choice and their likely transitions. In turn, these choices govern the premature mortality and substantial regional and global climate impacts of household energy emissions. Consumers' choice of energy for cooking is a subset of household services, and is likely affected by other service needs. I outline a framework that describes energy services with five key characteristics: energy quantity, maximum power, transformation specificity, portability, and timeliness. Some of these services can be met with the same energy source and device, termed "interoperability." I argue that these two concepts—interoperability and energy service characteristics—explain many of the currently low adoption rates and the inability to reduce exposures. Robustness is sensitivity to uncontrollable variations in operating conditions. In addition to adoption of improved solutions, robustness of emission rates affects the ability to attain low-emission energy provision for households. From about 150 measurements of in-home cooking and heating events in seven countries, our team has demonstrated that this variability is dominated by user practice rather than by appliances, for devices that burn solid fuel. To reduce overall impacts, designers and manufacturers should focus on increasing robustness by decreasing sensitivity to user behavior.

Keywords: built/indoor environment, environmental health, exposure factors, source control, thermal environment, ventilation

WE-PO-03

Sensor Stories: Outreach materials to support appropriate community use of low-cost particle monitors

Korfmacher K.[1], Hopke P.[1]; [1]University of Rochester Medical Center, Rochester, NY, USA

As the public's understanding of health impacts from air pollution increases, interest in monitoring personal, household, and local air quality grows. Low-cost air quality sensors empower community members to collect air quality data to identify local pollution issues and to investigate environmental health concerns. Particle sensors can serve as the basis for citizen science projects, outreach initiatives, education, and community-engaged research. However, it can be challenging for non-experts to plan, use, and interpret results appropriately. Many community members lack an understanding of air pollution, the health effects of particle pollution, and the limitations of low-cost sensors. A partnership of four Community Engagement Cores from NIEHS-funded research centers developed educational materials to help community groups better understand and use low cost air quality sensors to monitor particle pollution. The materials aimed to help non-expert "facilitators" (health educators, outreach staff, or others) lead workshops teach others about air quality monitoring. The materials include a PowerPoint Presentation, Facilitator Guide, Action Plan, Interactive Activities, Air Quality Index Guide, and Feedback Forms (www.sensorstories.urmc.edu). These materials help community members understand the sources, nature, and health impacts of particle pollution, the diversity and limitations of sensors, and range of uses. Sensor Stories materials prepare community participants to utilize technical resources, such as the EPA's Air Sensor Toolbox and the Air Quality Index. It explains how to choose a sensor, plan monitoring, document findings, and interpret data they collect. The workshop is appropriate for diverse audiences including community group members, health agencies, local government, and the general public. Pilot testing with 8 diverse groups from across the U.S. indicates the workshop can be easily adapted based on the background knowledge, interests, and literacy of the target audience. Users have included smoke free housing advocates, occupational settings (restaurants, auto mechanics), community health workers, transportation planners, and the general public.

Keywords: environmental health, environmental justice, environmental policy

WE-PO-04 WITHDRAWN

WE-PO-05

Title Recent strategy and regional problems of energy saving buildings in Japan

Tsutsumi J.G.[1], Nakamatsu R.[1], Ananda R.U.[1], Matsuda M.[1], Kobayashi F.[1]; [1]University of the Ryukyus, Okinawa, Japan

The Act on the Improvement of Energy Consumption Performance of Buildings was established in 2015 and gradually took effect from 2016. Total land area of Japan extends from 24 to 45.5 degrees North. It is divided into 8 regions to consider the regional characteristics of climate. The 8th region is occupied by only the islands of the Ryukyus, it is Okinawa prefecture, located in subtropical zone and southernmost of Japan. It is a warm and humid region and thermal insulation is not required for the buildings in this region, because thermal insulation is only considered for heating in the act. However, indoor thermal sensation is not so comfortable in winter and heating appliances are used occasionally in this region. Indoor thermal environment in this region should be re-examined precisely for the thermal comfort and energy saving. Practical thermal measurement was done in an existing RC building. The measurement results show that the indoor air temperature is sometimes cooler than outdoor and the indoor surface of walls are cooler than indoor and outdoor air. Even in subtropical zone, indoor thermal environment in winter should be considered carefully. The thermal insulation is effective for heating in winter and it is also working well for cooling in summer. Energy saving of air conditioning system in buildings will be improved by thermal insulation in winter as well as summer even in the 8th region.

Keywords: Thermal insulation, Subtropical zone, Indoor thermal environment, Energy saving

WE-PO-06 MOVED TO WE-PL-A2-31

WE-PO-07
Environmental Fate and Transport Analysis of Low Global Warming Potential (GWP) Refrigerant, R-514A

Le H.[1], Lander D.[1], Claessens M.[1], Glatt C.[1], Russell M.[2]; [1]The Chemours Company, Wilmington, DE, USA, [2]Environmental Insights, LLC, Wilmington, DE, USA

Background The growing international emphasis on climate change has stimulated interest in low global warming potential (GWP) refrigerants. The potential adverse impacts on human health through the life cycles of low-GWP refrigerants at a global scale has not been evaluated. We conducted an environmental fate and transport analysis of new low-GWP refrigerant R-514A, a mixture of two components, (Z)-1,1,1,4,4,4-hexafluoro-2-butene (HFO-1336mzzZ) and trans-1,2-dichloroethylene (t-DCE), 74.7% and 25.3% by weight, respectively. Methods Literature was reviewed and extracted for available physical and environmental properties for each component of R-514A. The environmental fate of each component was assessed separately to determine potential distribution of concentration in air, soil, and water following releases to the environment. In anoxic terrestrial and aquatic environments, t-DCE transforms into vinyl chloride (VC); therefore, VC was also evaluated as one of the components. Potential releases of R-514A components were evaluated by a series of simulations using the environmental model PEARL and the ECETOC TRA environmental model. Results With the worst-case scenario, the 80th percentile of the average annual concentrations of HFO-1336mzzZ and t-DCE in soil pore water were in a range of 0.009-0.066 μg/L (9-66 ppt), and 0.005-0.049 μg/L (5-49 ppt), respectively. For VC, the range was 0.0001-0.005 μg/L (0.1-5 ppt). Discussion Based on the conservative estimates, the potential concentrations of the two constituents of R-514A, and the potential transformation of t-DCE to VC, infiltrating through the soil profile present a negligible impact on the quality of groundwater resources. The worst-case concentrations of t-DCE and VC resulting from soil deposition of localized releases of R-514A are 10-to-100-fold lower than the typical concentrations currently reported in drinking water supplies, and local oral exposures based on soil pore water are well below general population DNELs. The PEARL model is significantly more conservative than the ECETOC TRA model and the differences are discussed.

Keywords: climate change, cumulative exposure, exposure models, lifecycle analysis, risk assessment

WE-PO-08 WITHDRAWN

WE-PO-09
Synoptic features of atmospheric particle in urban environment of India

Pipal A.S.[1], Satsang P.G.[2], Taneja A.[1]; [1]Department of Chemistry, Dr. Bhimrao Ambedkar University, India, [2]Department of Chemistry, Savitribai Phule Pune University, India

Atmospheric aerosols are a manifold mixture of airborne particles from various sources and play a significant role in influencing the weather, air quality and climate on regional and global scales. They contain a range of oxidising aerosol components (Reactive Oxygen Species), which are potentially toxic as they can oxidise biological molecules at the liquid lung surface layer or generate oxidising components once deposited on the lung surface. Here, we examined different aspects of atmospheric aerosols in ambient environment in Indian context in term of carbonaceous aerosols (OC, EC, SOC and POC), morphology (size, shape, mixing states, circularity and aspect ratio etc), trace metals and their carcinogenic outcomes with respect to human health. Surface observation studies indicated that Particulate mass (150-200 μg m-3) levels are higher than Indian and WHO standards limit values. Carbonaceous aerosol studies have revealed that being a high loading of OC (50-70 μg m-3) and EC (10-15 μg m-3) associated with fine particles in Indian cities. The role of individual chemical species of aerosols in light extinction was estimated, the contributions of light scattering coefficient by particles (bsp) were OC (45%), $(NH_4)_2SO_4$ (17%), NH_4NO_3 (8%) and coarse mass (12%) while, the contribution of light absorption coefficient by particle (bap) was 18% (EC). Morphological features of atmospheric particles indicated the synoptic changes in environment. Overall this attempt has been made to study the impact of atmospheric aerosols on regional climate as atmospheric aerosol is one of the most uncertain aspect of aerosol science in estimating climatic and health impact of aerosols.

Keywords: Particle exposure, morphology, sources, urban air quality

WE-PO-10

Changing Characteristics of Particulate Matter in the Eastern United States

McDow S.[1], Chan E.[1], Long R.[2], Richmond-Bryant J.[1], Hemming B.[1]; [1]US Environmental Protection Agency, Research Triangle Park NC, USA [2]University of North Carolina, Chapel Hill NC, USA

Changing particulate matter (PM) composition has the potential to affect exposure measurement errors when estimating PM exposure in air pollution epidemiology studies. Monitoring data for PM2.5, PM2.5 species, and PM10 show remarkable differences in PM2.5 and PM-10 characteristics in the Eastern United States (U.S.) since 2000. During 2000-2015, national average PM2.5 concentrations decreased by 5 ?g/m3. In the Eastern U.S., major changes in regional PM characteristics accompanied this decline. In 2000, PM2.5 accounted for approximately two-thirds of PM10 mass in some cities in the Eastern U.S., and sulfate was the most abundant contributor to PM2.5 mass. Sulfate concentrations decreased more rapidly than other major PM2.5 species concentrations, corresponding to declining SO_2 emissions from 13.9 million metric tons (MMT) in 2006 to 4.8 MMT in 2014. In New York City, the sulfate fraction of PM2.5 fell from 42% in 2003-2005 to 24% in 2013-2015. Organic carbon or nitrate have replaced sulfate as the most abundant PM2.5 component in most Eastern U.S. monitoring locations, and composition changes are reflected in a shift in seasonal PM2.5 concentration patterns because sulfate concentration maxima occur in summer, but not for organic carbon or nitrate concentrations. The changing PM2.5 composition also impacted particle size distribution. Recent data indicate that PM2.5 accounts for 50-60% of PM10 on average, down from two-thirds of PM10 mass before 2000, because PM2.5 concentrations decreased while PM10-2.5 concentrations remained relatively constant. On average, PM is now larger, more organic, and relatively more abundant in winter than in 2000. Because other PM2.5 components and PM10-2.5 are more spatially variable than sulfate, increased PM2.5 and PM10 spatial variability is expected, potentially resulting in greater exposure measurement error. The views expressed in this abstract are those of the authors and do not necessarily reflect the views or policies of the U.S. Environmental Protection Agency.

Keywords: PM2.5, PM10, composition, sulfate, organic carbon

WE-PO-11

Concepts in inhalation dosimetry for trichloroethylene from vapour intrusion

Turczynowicz L.[1], Pisaniello D.[1]; [1]University of Adelaide, Adelaide, South Australia/ Australia

Vapour intrusion is the process of the migration of volatile soil contaminants from subsurface sources into dwellings where occupant inhalation exposure may occur. Exposure estimates are undertaken using vapour intrusion modelling or measurement and are used to estimate human health risks. These estimations are currently based on time-averaged exposures. While the literature recognises the uncertainties and variability of such estimates there is a need to better understand such exposures as they reflect time-dependent processes. These estimates are also dependent on an understanding of the inhalation dose-response relationship and the resultant adverse toxicology.

Keywords: other (specify)

WE-PO-12

Risk assessment of chronic exposure to antibiotics in rural population

Ali A. [1], Malik A. [2], Uddin I. [1], Rashid A. [1]; [1]Department of Environmental Sciences, Arid Agriculture University, Rawalpindi, Pakistan, [2]Civil Aviation Authority, Islamabad International Airport, Islamabad, Pakistan

The rural environment receives antibiotics and their residues from agriculture, livestock and human that may persist for long-time. Hence public exposure to antibiotic residues is multifaceted and need health risk assessment. This study aimed to quantify antibiotic usage, their residues in environment and assess chronic exposure of rural population consequent health risks analysis. Field soil, fecal and irrigation water samples were analyzed for fluoroquinolones and quinolone antibiotics. Target rural population (n=234) were surveyed which includes livestock handlers, agricultural farmers, field workers, female households and children, to assess extent of exposure to antibiotic persist in their immediate environment and epidemiological symptoms. Significant use of ciprofloxacin as self-medication (unethical) was strongly correlated with flu and gastrointestinal infections. Among females gastro (OR=1.51; 95% CI=1.16–1.95) whereas in livestock workers both flu (OR=1.39; 95% CI=1.07–1.80) and gastro (OR=2.11; 95% CI=1.62–2.74) incidences were recorded as consequent impact of chronic exposure to antibiotics. In feces wide range of ciprofloxacin was detected (97.5–188.2 mg kg–1) whereas amount of levofloxacin and ofloxacin were within the WHO regulatory limits. Wastewater channels from rural dwellings and livestock farms were identified as potential hotspots for chronic exposure. Indiscriminate use of antibiotics to enhance dairy produce has forced rural people to indulge in unethical drug administration which has increased their susceptibility to further exposure to antibiotics and their residues in rural environment. We conclude that chronic exposure of rural population to antibiotics provoke health risks and this exposure is on rise due to continuous add-up of sub-lethal levels linked with self-administrated usage which must be strictly checked to minimize their release in environment.

Keywords: community

WE-PO-13

Urinary pesticide metabolite concentrations in pregnant women from Suriname

Alcala C.; Tulane University, New Orleans, LA, USA

Pesticides are widely used in Suriname, a middle-income Caribbean country. The primary agricultural area is district Nickerie. Exposure to pesticides leads to adverse health effects and pregnant women and children are vulnerable. The Caribbean Consortium for Research in Environmental and Occupational Health (CCREOH) assesses the impact of environmental exposures on birth outcomes and neurodevelopment in 1000 mother/child dyads. This study represents a subset of the CCREOH cohort and is aimed at comparing pesticide metabolite concentrations in pregnant women from Nickerie and the capital Paramaribo. Urine samples from 220 participants were collected. A total of 50 samples (42 from Paramaribo and 8 from Nickerie) were recently analyzed at the US Centers for Disease Control and Prevention's environmental health laboratory for three pesticide classes, including the commonly used 2,4-D, organophosphate insecticides, and pyrethroid insecticides. Pregnant women in Nickerie had the highest geometric mean urinary metabolite concentrations of 2,4-D and pyrethroids: 3-phenoxybenzoic acid compared to those living in Paramaribo, (0.56 vs. 0.16 ug/L), (0.52 vs. 0.24 ug/L), respectively ($p<0.05$). Pregnant women living in Paramaribo had significantly higher geometric mean urinary metabolite concentrations of organophosphates: para-Nitrophenol, diethylthiophosphate and diethylphosphate compared to those living in Nickerie, (0.44 vs. 0.22 ug/L), (0.73 vs. 0.22 ug/L), (1.54 vs. 0.54 ug/L), respectively. The above-mentioned levels are higher in Suriname compared to those in the United States based on data from the National Health and Nutrition Examination Survey. Urinary metabolite concentrations of pesticides varied in Paramaribo and Nickerie. This could be due to residential use, intensive spraying during mosquito control programs in Paramaribo and excessive use in Nickerie, as it is a rice production area, respectively. Analysis of an additional 170 urine samples is underway.

Keywords: environmental health, epidemiology, occupational, pesticides, prenatal

WE-PO-14
Measurement of ambient ozone (O3) levels, correlation with the NOx levels and the development of the monitoring network for O3

Turos O.[1], Petrosian A.[1], Kobzarenko I[1]., Kharchenko K.[1], Maremukha T.[1], Ananyeva O.[1]; [1]State Institution – "O.M. Marzeiev Institute for Public Health of the National Academy of Medical Science of Ukraine", Kyiv, Ukraine

Background/Aims. The link between increase O3 ambient air pollution and health issues has been well documented, prompting the introduction of strict rationing standards of O3. This study aims to conduct a quantitative assessment of the level of the ambient air pollution with O3 in order to prove that there is a necessity of organizing of the modern governmental system of monitoring of ambient O3 and substances from which it derives, as well as systems for notification of the population about high level of ambient air pollution in Ukraine. Methods. Concentrations of the O3 and NOx in the air were measured by the gas analyzers (Horiba APOA-370 and APNA-370). The obtained data was processed using the software package Statistica, results were compared with the limit values recommended by WHO. Results. According to the data of the completed research concentrations of tropospheric O3 in Kyiv significantly exceed the established norms. Daily concentration of the O3 for the summer period varies from 127.7818±3.76 μg/m3 to 202.4±20.7 μg/m3, and during spring period – from 200.87±3.61 μg/m3 to 222.83±6.72 μg/m3. Data from semi-annual monitoring indicates that the chemistry of ambient O3 in Kyiv confirm to NOx-sensibility mode, what is verified by the occurrence of photochemical smog in Kyiv with high O3 concentrations. What is more, there was found a stable anticorrelation between O3 and NOx, what is also an evidence of the above-mentioned. Conclusions. The continuation of the monitoring for ambient air quality and prompt analysis of the monitoring data will allow not only to fix the dangerous concentrations of the tropospheric O3, but also justify the organization of the activities for the decreasing of its level, and forecast the ecological situation in the city.

Keywords: industrial hygiene

WE-PO-15
Elevated lead concentration in air of closed shooting ranges – a public health concern to be addressed in future

Matisane L.[1], Martinsone I.[1], Komarovska L.[1], Klavina A.[1]; [1]Riga, Lativa

Background Lead (Pb) is a toxic substance with well-known, multiple, long-term, adverse health effects. Although Pb-free bullets are available, due to lower costs most of bullets used in shooting ranges in Latvia contain high amount of Pb. Shooting at fire ranges can be related to work (e.g. security, police, military) or done as hobby (e.g. training for hunting, shooting as stress remover). Methods The objective of the study was to assess the air Pb concentration in closed shooting ranges in correlation with shooting during sampling and existence of proper ventilation. In total between January 2016 and February 2019 23 samples were taken with individual samplers either from breathing zone or stationary in 3 different ranges. Pb concentration was determined using atomic absorption spectrophotometer. Results Out of 23 samples, in 6 cases Pb concentration in air exceeded 0.05 mg/m3 which is currently the occupational exposure limit value in Latvia. All of the exceeded levels were identified in 2 shooting ranges with highest level 5.4 mg/m3 in one range (ventilation working in normal regime), and 2.55 mg/m3 – in another. 5 of those samples were taken during shooting, 1 - outside the shooting range (in a corridor available to other members of public). In the 3rd shooting range all Pb levels were below the limit - samples were taken in recently renovated facilities when shooting was not performed and the ventilation was working. Conclusions: Extremely high levels of Pb in air can be identified while shooting even in cases when exhaust ventilation has been installed and working; this can raise public health concern as premises located close to shooting ranges available to non-shooters can be contaminated. Further investigations regarding shooter behavioural aspects should be carried out in order to assess the exposure of general public from discharge of Pb in shooting ranges.

Keywords: industrial hygiene, metals, occupational, ventilation

WE-PO-16

Challenges applying activity-based sampling (ABS) in Sibaté, a Colombian asbestos contaminated town

Cely-García M.[1], Lysaniuk B.[2], Pasetto R.[3], Mazzeo A.[4], Magnani C.[5], Ramos-Bonilla J.[1]; [1]Department of Civil and Environmental Engineering, School of Engineering, Universidad de los Andes, Bogotá, Colombia, [2]IRD (MàD par le CNRS), UMR Prodig, Bogotá, Colombia, [3]Environment and Primary Prevention, and WHO Collaborating Centre for Environmental Health in Contaminated Sites, Istituto Superiore di Sanità, Rome, Italy, [4]Department of History and Cultures, University of Bologna, Bologna, Itlay, [5]Unit of Medical Statistics and Cancer Epidemiology, Department of Translational Medicine, University of Eastern Piedmont, and CPO-Piedmont, Novara, Itlay,

Introduction: Based on reports from inhabitants of Sibaté (Colombia), between ~1975 to ~1985 asbestos containing materials were dumped at different urban areas of the town. Starting in ~1986, landfilled zones were built with dumped materials, and on top of these areas different facilities were constructed. In 2017 and 2018, geographic information systems combined with soil sampling campaigns resulted in the identification of an underground friable asbestos layer in the urban area of Sibaté. Aim: Adapt and apply the US EPA Activity Based Sampling (ABS) methodology in potential landfilled areas of Sibaté, to determine the current risk of asbestos exposure and excess lifetime cancer risk (ELCR) at the surface of landfilled zones. Methods: Generic-ABS in a football stadium and specific-ABS in both a playground and a school, were conducted. Personal, area, and blank samples were collected and analyzed by PCM and TEM following NIOSH 7400 and ISO 13794 methods, respectively. Exposure point concentrations (EPCi) were determined and compared against the level of concern (LOCi), and was also used to calculate the ELCRi. Results: Twenty-two air-asbestos samples (12-personal, 7-area, 3-blanks) were analyzed by PCM, and 6 (4-personal, 2-area) were analyzed by TEM. Total asbestos structures concentration of personal samples ranged from ND to 0.109 S/cc (i.e., total asbestos structures counts ranged between 0 and 5). All samples had PCM-equivalent asbestos structures concentrations below the analytical sensitivity, and no ELCRi was found. Seventeen of the 22 samples analyzed were overloaded. Discussion: Although no ELCRi was found, chrysotile and actinolite structures were found at the surface of landfilled zones, suggesting a potential risk of asbestos exposure. In the framework of the methodological, analytical, and statistical challenges encountered, this initial effort can guide future ABS studies, not only in Sibaté, but also in other areas with asbestos contaminated soils.

Keywords: community, ecological exposure, environmental health, risk assessment, sampling methods

WE-PO-17 MOVED TO MONDAY POSTER SESSION (MO-PO-89)

WE-PO-18

Global Temporal Trends in Preschool Children Blood Lead Levels by UN Human Development Index

Hwang Y.H.[1], Hsiao C.H.K.[1], Lin P.W.[1]; [1]National Taiwan University College of Public Health, Taipei, Taiwan, ROC

This study was aimed at examining the global temporal trends in the preschool children blood lead levels by the category of UN Human Development Index (HDI). Blood lead records were obtained from 51 articles searched from PubMed and Google Scholar, with study subjects aged up to 8 years old. In total, 103 blood lead records retrieved from the reviewed articles were employed to plot chronologically by HDI category. The reciprocals of these blood lead records were used in regression analysis against calendar year to establish the global temporal trends in the past 30 years. It is shown that the modes of preschool children blood lead level decreased from 5-7 μg/dL to 0.7-1.6 μg/dL, from 6-13 μg/dL to 3-7 μg/dL and from 13-16 to 5-7 μg/dL, respectively, for the very high HDI countries, the high HDI countries and the medium/low HDI countries. High correlations were found between the reciprocals of preschool children blood lead level and the calendar year, with correlations coefficients of 0.863, 0.696, 0.779, respectively, for the very high HDI countries, the high HDI countries and the medium/low HDI countries. However, there were persistent differences in blood lead level among countries of different HDI category, which implied the effects of disparities and inequalities at the state level on blood lead levels. To further reduce children's exposure to environmental lead should base upon the efforts on (1) identification of individual local factors for lead exposure and (2) averting health disparity and inequalities at the state level.

Keywords: environmental health

WE-PO-19

Role of surface charge in nanoparticle toxicity: implications for indoor air quality

Scheepers P.; Radboud University Medical Center, Nijmegen, Netherlands

There are strong indications that surface charge changes the disposition and toxicity of nanoparticles (NPs). A scoping review was conducted to describe current knowledge in this field. The aim was to extrapolate health implications of real-life exposure to charged NPs from biomedical studies in the field of drug targeting. A structured search in PubMed and Medline resulted in 56 relevant studies. In vitro tests indicated higher toxicity of positively and negatively charged NPs (NP+ and NP-) compared to their neutral counterparts (NP0). Especially NP+ interact with critical cell structures, resulting in cellular toxicity. Interaction of NP+ with negatively charged cell membranes explain the observed higher cytotoxicity due to disruption of the lipophilic bilayer, including embedded proteins and ion channels. Results from animal studies also indicate higher cytotoxicity of NP+ relative to NP0 and NP-. Adverse effects of NP+ and NP- were observed on lungs, internal organs, cardiovascular and central nervous system. Model predictions indicate higher deposition in the airways of NP+ and NP- compared to neutral NPs. Currently, it is difficult to interpret the relevance of toxicity testing for real-life exposures. Laboratory studies with ionizers indicate enhanced surface deposition ('soiling') of charged NPs, but the influence of surface charge on the life-span of airborne particles in an indoor environment is unknown. Also, it is unknown to what extent NPs could lose their charge when translocated from air to the lung lining fluid. If NPs would remain charged, the recent knowledge on the impact of charge on NP toxicity should be considered. There are still important knowledge gaps that hamper estimation of the human health risks. As a precaution the release of charged particles from air cleaners should be prevented and the use of air ionizers near indoor sources of NPs, such as open fire and cooking fumes is not recommendable.

Keywords: nanoparticles

WE-PO-20
Levels of Mercury in Fish from the Northeast US

Melnyk L.[1], Lin J.[1], Lazorchak J.[1], Kusnierz D.[2], Pugh K.[3], Stover M.[4]; [1]U.S. Environmental Protection Agency, Cincinnati, OH, USA, [2]Penobscot Indian Nation - DNR, Indian Island, ME, USA, [3]Agency for Toxic Substances and Disease Registry, Atlanta, GA, USA, [4]U.S. Environmental Protection Agency, Boston, MA, USA

Rivers in the Northeast are home to spawning anadromous and non-migratory fish that are caught and consumed by subsistence fisherman. Fish have been barred from entering some rivers by dams constructed during an industrial boon that occurred over 100 years ago. With dam removal as an effort to restore ecological health of the river system, anadromous fish are again entering rivers and are available for consumption. The safety of human and wildlife consumers is paramount, so the levels of mercury were measured in tissue of six species of migratory fish (Alewife, American Shad, Blueback Herring, Rainbow Smelt, Striped Bass, and Sea Lamprey), as well as, in the roe of one species (American Shad) for two consecutive years (2017 and 2018). The resultant mercury levels were compared to reference doses as established in the US IRIS and wildlife values. Mercury concentrations ranged from 10 μg/kg in roe samples to 650 μg/kg in Sea Lamprey. An average person of body weight 70 kg could safely consume between 0.4 and 6 oz (11 and 170 g) of fish fillet daily, depending on species. The roe could be safely consumed at levels over 20 oz per day. Sea lamprey contained the highest amounts of mercury for both seasons of sampling. Current health advisories are set at sufficient levels to protect fishermen from harmful consumption of the fish species, except for sea lamprey. Based upon a consumption wildlife value of 70 mg/kg for Mink and 100 mg/kg for Otter, consumption of Alewife, Striped Bass and Sea Lamprey pose a risk to Mink. Consumption of Rainbow Smelt, Striped Bass and Sea Lamprey pose a risk to Otters. Sea Lamprey, with consistently higher concentrations of mercury than the other species analyzed, pose a health risk if consumed at levels greater than 0.5 oz/d (14 g/d).

Keywords: SVOCs

WE-PO-21
Simultaneous Determination of Alternative Flame Retardants in Car Seats and Implications for Dermal Exposure

Tokumura M.[1], Terao K.[1], Miwa H.[1], Wang Q.[1], Miyake Y.[1], Amagai T.[1], Makino M.[1]; [1]University of Shizuoka, Shizuoka, Japan

Car indoor environment is one of important indoor environment, in which a variety of flame retardants is used in car interiors (e.g., car seat) to avoid the risk of fire. Flame retardants in car interiors can be exposed to a driver and passengers via various exposure pathways (e.g., vaporization of flame retardants into the car indoor air, migration to dust, and direct contact). In recent years, dermal exposure to flame retardants via direct contact with a product has been paid great attention as an important exposure pathway. However, dermal exposure to flame retardants via direct contact with car seat has been not studied well. As alternatives to hexabromocyclododecane (HBCD), novel brominated flame retardants (BFRs) and phosphorus flame retardants (PFRs) have been used. Due to their varied physico-chemical properties, it is difficult to simultaneously determine their concentrations in a product. In this study, we developed the simultaneous determination method for a variety of alternative flame retardants, BFRs and PFRs, (more specifically, the optimum extraction method and analytical parameters were determined) and estimated the potential dermal exposure rates of the flame retardants via direct contact with car seats. The experimental results revealed that the ultrasonic extraction with tetrahydrofuran was preferred to extract the flame retardants from the car seat. Tris(1,3-dichloro-2-propyl) phosphate (TDCPP) and cresyldiphenyl phosphate (CsDPhP) were detected at high concentrations, while tris(2-chloroisopropyl) phosphate (TCPP) was detected at low concentrations. The potential dermal exposure rates of the flame retardants detected from the car seats were likely to be higher than those via other exposure pathways (e.g., dust ingestion and inhalation). The results in this study revealed that the dermal exposure to the flame retardants via direct contact with the car seat could be significant exposure pathway.

Keywords: SVOCs, flame retardants, risk assessment

WE-PO-22

Public Health Risk attribution due particulate matter from Biomass Burning Emissions

Pinto Herrera LC.[1], Hernandez Florez LJ.[1], Morales Betancourt R.[1]; [1]University of Los Andes, Bogota, Columbia

Background: Biomass burning emissions have a substantial impact on regional air quality and climate, we evaluated the short-term effect of particulate matter on acute respiratory disease morbidity and mortality, determinated by the risk of childs under 5 years old and adults over 60 years old in three n cities at 2016. Methods: Daily Emergency Department (ED) visit records for respiratory illnesses and selected diagnosis, besides daily concentrations of PM2.5 obtained in Bogota as a high altitude city and other cities as Bucaramanga and Medellin, therefore we selected the back-trajectories of air masses reaching the receptor, that were linked with fire spots. Colombia: Bogota, Medellin, Bucaramanga during 2016. Using conditional Poisson time series analysis we assessed the effect of air pollutants on health outcomes using single pollutant, controlling for meteorology and time trends. The percentages of change in the rate of ED visits and their 95% confidence interval were estimated for the joint effect of particulate matter. Results and Conclusions: On one hand in dry season, the amount of firespots increases as single-pollutant model increases particulate matter were associated with 15% increases in ED visits for respiratory diseases. On the other hand the during the wet season the incidence of ED visits decreases. The health charge of concentrations above 50 Micrograms per cubic meter of air represent the 17,6 % of Chronic Respiratory illneses. The cost linked with air pollution is US $1431, this amount increases when the poor communities are included.

Keywords: soil

WE-PO-23

Simultaneous determination and exposure assessment of novel alternative flame retardants in indoor dust

Miyake Y.[1], Furukawa M.[1], Tokumura M.[1], Wang Q.[1], Takahashi Y.[1], Amagai T.[1]; [1]Japan

Flame retardants play an important role in reducing the risk of fire hazard, and therefore, are added in several consumer products. However, the production and use of some flame retardants such as hexabromocyclododecane (HBCD) were banned due to their bioaccumulation, toxicity and persistence. This led to an increase in the use of other flame retardants, including novel brominated- and phosphorus-based flame retardants. We previously determined the concentrations of 18 brominated and 15 phosphorus-based flame retardants in 40 flame-retardant impregnated curtains purchased in Japan in 2014. As the result, we identified the novel flame retardants in curtain samples such as tris(2,3-dibromopropyl)isocyanurate (TDBP-TAZTO), (5-ethyl-2-methyl-2-oxido-1,3,2-dioxaphosphorinan-5-yl)methyl-methyl-methylphosphonate (PMMMP), naphthalen-2-yl diphenyl phosphate (NDPhP), and 6-benzyl benzo[c]-[2,1]benzoxaphosphinine 6-oxide (BzIDOPO). Although flame retardants present in curtains can be released into indoor air and concentrate in indoor dust, the information on concentrations of novel flame retardants in indoor dust is limited. The aims of this study were to develop an analytical method for simultaneous determination of novel flame retardants by use of liquid chromatography interfaced with tandem mass spectrometry (LC-MS/MS), and to determine the concentrations of novel flame retardants in dust samples collected from residential houses in Japan in 2015. In addition, the exposure doses of flame retardants via indoor dust were calculated to assess the human health risks. This study elucidates the occurrence and risk level of novel flame retardants in indoor dust samples.

Keywords: flame retardants

WE-PO-24

Risk Assessment of Polycyclic Aromatic Hydrocarbons and their Chlorinated Derivatives Unintentionally Produced During Cooking via Exhaust Gas

Masuda M.[1], Wang Q.[1], Tokumura M.[1], Miyake Y.[1], Amagai T.[1]; [1]University of Shizuoka, Shizuoka, Japan

Cooking exhaust gas includes polycyclic aromatic hydrocarbons (PAHs) unintentionally generated during cooking foods, which is exposure to a cook and residents via inhalation. PAHs are carcinogens and/or mutagens. As a typical example, benzo[a]pyrene (BaP) is classified as Group1 by International Agency for Research on Cancer (IARC). On the other hand, chlorinated PAHs (ClPAHs), which are compounds whose one to several hydrogen atoms of PAHs are substituted to chlorine atom, some studies reported that some ClPAHs have more toxicities than their parent PAHs. Given that the generation mechanism of ClPAHs is similar with that of PAHs and cooking processes fulfill the conditions to generate ClPAHs (i.e., sufficient organic substances, heat, and chlorine source), ClPAHs can be generated during cooking processes such as grilled foods. Pacific sauries, which are a typical fish consumed in Japan, were targeted in this study and gas-grilled in a typical fish oven in an indoor kitchen. In this study, the PAH/ClPAH concentrations indoors from cooking exhaust were estimated using National Institute of Advanced Industrial Science and Technology – Indoor Consumer Exposure Assessment Tool (AIST-ICET). The risk assessment for the exposure to them was carried out. The exposure amounts of PAHs/ClPAHs from oral exposure and respiratory exposure were comparable. The results revealed that the inhalation of indoor air including cooking exhaust was the major pathway of PAHs/ClPAHs in addition to the consumption of cooked food.

Keywords: food

WE-PO-25

Health impact assessment using taxies to monitor air pollution based on sensors

Yang W.[1], Cho M.[1], Kim Y.S.[2], Sung K.H.[2]; [1]Department of Occupational Health, Daegu Catholic University, Kyongsansi, Korea, [2]Environmental Health Service Center, Daegu Catholic University, Kyongsansi, Korea

Air pollution has drawn a lot of interests over the past few years in Korea. Therefore government have put tremendous efforts on air pollution monitoring system (APMS). However, the APMS is not sufficient the personal exposure to air pollution. Urban outdoor air in big city such as Deagu in Korea is polluted with a variety of air pollutants that can result in increased morbidity and mortality through short and long term exposures. Meanwhile, many studies have demonstrated that air pollutants are associated with a wide range of adverse health outcomes. In this study, we have estimated a large impact on population health with real-time measurements of air pollutants by internet of things (IoT) sensing platform and prevalence rate adjusted by age from National Statistical Office such as tracheitis, pneumonia, and asthma. Air pollutants (NO2, CO, SO2, VOCs, PM2.5, PM10) concentrations were measured using taxies cab with IOT and global positioning system (GPS) from August to October, 2017, Daegu, Korea, and Daegu city was classified into 8 sections according to administrative district. Although the relationship between the measured air pollution concentrations and health effects was definitely not shown, the method of mobile sensors with IoT will indicate the potential. This method by taxi can fill the gap between the conventional monitoring systems and air quality model, and can evaluate the exposure and risk to air pollutants. Therefore, we need the air pollution monitoring systems by combining the low-cost portable ambient sensors and the wireless network. In this study, we may provide the new method on association between air pollution and health effects, and need the further research.

Keywords: risk assessment

WE-PO-26

Internal Dose from radioactive ceasium in Indoor Dust collected in Ibaraki and Chiba Prefectures, 150-200 km South from the Fukushima Daiichi Nuclear Power Plant

Takagi M. [1]*, Tanaka A.*[1]*, Nakayama F.S.*[1]*;* [1]*National Institute for Environmental Studies, Tsukuba, Japan*

The radiation dose effects due to the Fukushima Daiichi Nuclear Power Plant (FDNPP) accident are of public concern. Although indoor dust is an important exposure route for some hazardous chemicals, the radiation exposure dose from the radioactive caesium (Cs) in it has not been well investigated yet. In this study, we measured the radioactive Cs in the indoor dust of private houses in order to estimate the resulting human exposure. We collected vacuum cleaner dust twice (in April and June 2012, n = 250 and 241, respectively) in the prefectures of Chiba and Ibaraki, from houses located 150–200 km south from the FDNPP. The activity concentrations of Cs-134 and Cs-137 in the indoor dust were measured using germanium γ-ray detectors, and the exposure dose was calculated and the household-to-household variations (e.g., in terms of surrounding environment and spontaneous decontamination activities) were analysed. The median activity concentrations of Cs-134 + Cs-137 in the indoor dust collected in April and June were 1,470 and 1,340 Bq/kg, respectively. The median committed effective doses via indoor dust ingestion were estimated as 0.73, 0.66, and 0.43 × 10^−3 mSv/year, and those via inhalation were estimated as 0.003, 0.004, and 0.006 × 10^−3 mSv/year for one- and ten-year-old children and adults, respectively. The univariate analysis revealed that the radioactive Cs concentration was higher in samples collected from detached houses than from multiple dwellings. Within the detached houses, the multiple regression analysis showed that wet wiping was a significant reduction factor of the Cs concentration; however, the small determination coefficient implies the presence of factors that were not considered in this study. In conclusion, in this study area, the estimated committed effective doses via indoor dust related to the FDNPP accident were comparable to those via soil and much smaller than the long-term target dose of 1 mSv/year.

Keywords: radiation

WE-PO-27

A community-based probabilistic risk assessment for the Diné people impacted by the Gold King Mine Spill

Ornelas Van Horne Y. [1]*, Chief K.* [1]*, Canales R.* [1]*, Begay M.* [2]*, Beamer, P.* [1]*;* [1]*The University of Arizona, Tuscon, AZ;* [2]*Nation Department of Health, Window Rock, AZ, USA*

On August 5th, 2015, the Gold King Mine released 3 million gallons of acid mine drainage-containing arsenic and lead into a tributary of the San Juan River, which flows through the Navajo Nation. The Diné (Navajo) rely on this River and use it for a variety of purposes lending to unique exposure pathways. We administered questionnaires in three Navajo communities in collaboration with the Navajo Nation Community Health Representatives to obtain frequency and duration data on 42 activities (e.g., praying with River water, fishing). These activities were utilized to conduct a community-based probabilistic risk assessment from exposure to As and Pb: 1) before the Spill, 2) during the peak of the Spill, and 3) one year after the Spill. Utilizing the Lifeline Community-Based Assessment Software we incorporated distributions for different exposure factors (e.g., activity duration, activity frequency, hand-to-mouth contacts) along with Diné specific activities (e.g., using the sediment as sunscreen) to estimate a distribution of dose. There is a significant difference in exposures (mg/kg/day), with the highest occurring during the peak Spill period (mean= 1.1 x 10-8, sd = 3.9 x 10-9) followed by post (mean= 9.4 x 10-9, sd = 3.9 x 10-9) and pre (mean= 6.9 x 10-9, sd = 2.5 x 10-9) Spill periods. While exposure from non-dietary activities is below the arsenic reference dose (RfD = 3 x 10-4 mg/kg/day), the exposures for the Diné community are higher when compared to those who would only recreationally interact with the San Juan River (mean= 1.6 x 10-13, sd = 9.6 x 10-4). This research provides the first documentation of unique exposure pathways of the Diné people and raises the need to understand and document unique exposure pathways of indigenous communities with subsistence and land-based livelihoods.

Keywords: risk assessment

WE-PO-28 WITHDRAWN

WE-PO-29
Evaluation of Air Quality in Hospitals using Low-cost Sensors

Ferro A.[1], Palmisani J.[2], Di Gilio A.[2], Viana M.[3], Hänninen O.[4], Taimisto P.[4], de Gennaro G.[2]; [1]Clarkson University, Potsdam, NY, USA, [2]Università degli Studi di Bari 'Aldo Moro', Bari, Italy, [3]Institute of Environmental Assessment and Water Research Spanish Research Council (IDAEA-CSIC), Barcelona, Spain, [4]THL Public Health Solutions, Kuopio, Finland

Oncology areas of hospitals are of interest with regard to environmental exposures due to the presence of patients with weakened immune systems. Yet, few studies have investigated the Indoor Air Quality (IAQ) of these microenvironments and how it is impacted by human activities. Recently, many low-cost air quality monitors have become available on the consumer market, and the scientific community has begun to apply these monitors as tools for exposure assessment. While these monitors have a wide range of accuracy and precision estimates, adjustment of the readings using direct field comparison with scientific grade instruments can extend their effectiveness. For this study, we evaluated IAQ in multiple oncology units in Europe and United States using low-cost monitors combined with information on human activity patterns and outdoor air quality. PM monitoring was conducted using low-cost, infra-red LED particle sensors for particle monitoring in the range 0.5-3 µm (Speck DSM 501 Series, Airviz Inc., USA or equivalent). For one location, I.R.C.C.S. Oncology Institute "Giovanni Paolo II" of Bari (Italy), additional monitoring was conducted using a total volatile organic compound (TVOC) sensor with microclimatic parameters (i.e., temperature and relative humidity) (Corvus IAQ monitor, Ion Science, UK) and a CO_2 monitor (ZyAura, Taiwan). Field comparisons with scientific-grade reference instruments were performed. While the low-cost PM sensors showed different degrees of correlation with the reference instruments (R^2 = 0.3 - 0.6), the sensors were highly correlated with one another ($R^2 > 0.8$). In addition, strong correlations were observed for PM and occupancy, although the highest peaks were observed during cleaning activities. Mitigation measures could be used to improve IAQ in these locations, including more efficient scheduling, improved waiting room design that limits crowding and activities that result in particle resuspension, and better ventilation strategies.

Keywords: susceptible/vulnerable

WE-PO-30
Exposure to traffic related pollution in buildings – a risk factor approach

van Dijken F.[1], van Doorn W.[2], Jansen H.[2]; [1]BBA Binnenmileu MVO, Netherland, [2]Wageningen University, Wageningen, Netherlands,

Air pollution is a growing concern among the general public. Air quality is not only an issue outdoors. The local outdoor air quality also affects the air quality indoors, where we spend about 90% of our time. In buildings close to busy roads indoor particle concentrations can reach high levels, which is a risk for the health of the occupants. Poor air quality mainly affects young children, since their lungs are still developing. Moreover, the health of vulnerable elderly is affected. In order to improve their living environment understanding of the exposure risk in kindergartens, schools and nursing homes near busy roads is required. We developed a methodology to estimate the risk of exposure to traffic related pollution in buildings. Based on literature survey and practical experience we defined risk factors for exposure to traffic related pollution indoors. These risk factors can be grouped in the following six themes: a)situation and outdoor sources, e.g. traffic intensity, congestion, urban canyon, urban canopy, other sources. b) exposure paths - outdoor spaces, e.g. distance from the road, type of use. c) exposure paths - building envelope, e.g. air tightness. d) exposure paths - ventilation, e.g. filter quality, location of the outdoor air inlet. e) maintenance, e.g. frequency of filter changes. f) use, e.g. use of the ventilation system, airing, awareness of the outdoor air quality. The risk factor approach has been used to investigate the risk of exposure to traffic related pollution in 61 kindergartens, schools and nursing homes in the region of the Hague that are situated at a location with a high intensity traffic. The results were used to give protective measures, including simple improvements focussed on user behaviour or maintenance, as well as more complex technical interventions.

Keywords: aging/elderly, air, built/indoor environment, children, exposure factors

WE-PO-31
Seasonal changes in bacterial and fungal concentrations in floor dust from Lublin region homes

Staszowska A.[1], Dudzińska M.[1], Siuta-Olcha A.[1], Dumałą S.[1], Połednik B.[1]; [1]Lublin University of Technology, Lublin, Poland

Numerous studies have demonstrated that floor dust can play a role of bacterial and fungal depository in indoor environments. Typically, microbial species originate from sources other than the dust itself like the outdoor environments, occupants and pets. Most notably indoor sources of bacteria and fungi are biological corrosion of building materials, house plants, food and HVAC systems. This study presents seasonal changes in bacterial and fungal concentrations in floor dust from 50 households located in Lublin area, Poland. During one year survey, floor dust samples were collected twice a month from vacuum cleaner bags. No moisture or mold problems were reported in studied houses, and the residents did not use any air cleaners. Moreover, temperature and relative humidity in studied households were recorded during the measurements. The bacterial and fungal concentrations were determined by the dilution plate method. Additionally, the bacterial and yeast strains were taxonomically identified after Gram staining based on their morphology and biochemical analytical profile index (API) tests. Filamentous fungi were identified according to their morphology using identification keys. Average concentrations of bacteria varied from 1.3x104 to 2.4x106 CFU/g dust and for fungi from 1.1x104 to 6x106 CFU/g dust, respectively. Microbial concentrations varied significantly between different seasons and hence the similarity of samples within a home was mainly low. The most commonly isolated cultivable bacterial genera included Bacillus, Staphylococcus, and Pseudomonas. Fungi were dominated by Penicillium, Aspergillus and Penicillium sp.

Keywords: other (specify)

WE-PO-32

Can exposure to cobalt from leather furniture enhance the risk for allergic contact dermatitis?

Kolarik B.[1], *Larsen P.B.*[2], *Jensen A.*[1], *Kampmann K.*[1]; [1]*Danish Environmental Analysis, Vedbaek, Denmark,* [2]*DHI, Horsholm, Denmark*

Metals – including chromium and cobalt - are the most common causes of allergic contact dermatitis (ACD). Recent studies have recognized leather articles as a new source of exposure to cobalt. Furthermore, two Danish case-studies have identified upholstery leather with cobalt content of 800-1250 mg-Co/kg as the cause of ACD. The purpose of the study was to obtain more information on the use of cobalt in leather goods and assess risk of developing skin allergy by use leather furniture available on Danish marked. In total 34 samples of leather in different colors have been obtained from three producers of furniture. Readings above the detection limit revealed the occurrence of cobalt in 5 among the 34 leather samples, with the two highest concentrations of 153 mg-Co/kg and 54 mg-Co/kg. In order to investigate leaching of cobalt from the leather and assess the exposure, those two samples have been subjected to migration analysis in artificial sweat. Derived Minimal Effect Level (DMEL) for cobalt in contact with skin of Co-sensitized persons was determined from the literature data to be in the range of 0.441 – 1.95 μg-Co/cm2. Those values were used to calculate the risk characterization ratio (RCR). The migration of cobalt from upholstery containing highest and second highest cobalt content was 0.09 μg-Co/cm2 and 0.013 μg-Co/cm2 respectively. The calculated RCR was below 1 for both samples (0.05-0.20 and 0.07-0.29, respectively), which indicates that there is no risk of inducing cobalt allergy nor triggering of reactions among cobalt allergics by use of investigated samples. The maximum concentration detected in upholstery leather in this study was far lower than values reported in the literature, where the furniture had been investigated as a potential source of cobalt allergy. Conclusions from this study can therefore not be generalized to any kind of leather furniture available worldwide.

Keywords: building materials, health, chronic diseases, metals, risk assessment

WE-PO-33

Assessment of particulate matter concentrations in indoor air of some educational spaces from urban area of Bucharest, Romania

Vasile V.[1], *Petcu C.*[1]; [1]*National Institute for Research and Development in Constructions, Urbanism and Sustainable Spatial Development "URBAN-INCERC", Bucharest, Romania*

Numerous adverse effects on cardiovascular and respiratory system has been associated with long-term exposure of adults to particulate matter. However, due to their immature immune system and faster breathing rate, children are more susceptible to these adverse health effects. One of the most important sources of airborne particles, in urban areas, are vehicle emissions. Long-term harmful effects on health including asthma and wheezing has been linked with childhood exposure to vehicle emissions. The potential of these particles to cause those effects on human health is determined by their ability to penetrate into the lungs and carry with them a number of toxic compounds. The school environment in which the educational process takes place is a site that can explain a significant part of the exposure of children to the emissions of different air pollutants, given that this category of people spends much of their daily time in school areas. It is expected that schools within a city have different levels of pollutants resulting from local urban characteristics such as placing school buildings near high traffic road arteries. In this context, the aim of our experimental study was to monitor the concentrations of fine particles fractions PM0,3-0,5; PM0,5-1,0; PM1,0-2,5 and of the coarse particles PM2,5-5,0; PM5,0-10,0 and PM≥10 from two kindergarten, three general schools and one high school, and to correlate the obtained values with the indoor comfort parameters (temperature and relative humidity). The monitoring of the particulate matter concentrations has led to obtain useful information on educational spaces, and to underline the acute necessity for action to make healthier the spaces where our children spend an important part of their life.

Keywords: children

WE-PO-34
Secondhand smoke exposure in rural households in northwest China: a pilot investigation

Wang D.[1], Yao Y.[1], Ma H.[1], Li C.[1], Wang J.[1], Liu S.[2]; [1]School of Public Health, Lanzhou University, Lanzhou, Gansu Province, China, [2]School of Health Sciences, Purdue University, West Lafayette, IN

Objectives: The prevalence of exposure to secondhand smoke (SHS) in rural China is high. However, few studies have focused on measuring SHS exposure in rural households in China. This abstract reports SHS levels in rural village households, measured as a potential confounding factor in a study conducted to assess health effects of exposure to household air pollution from using solid fuel in rural northwest China. Methods: Thirty-three village women were enrolled in the study. The field sampling was conducted in September, 2017 and January, 2018. Questionnaires were administered to collect information on demographics, household smoking status and the awareness of SHS exposure. Passive samplers were used to measure nicotine air concentrations over a period of one month in these women's home. Urine samples were collected and analyzed for 8-hydroxy-2-deoxyguanosine (8-OHdG, as a marker of oxidative DNA damage) and 1-hydroxypyrene (1-OHP, as an indicator of exposure to household air pollution from solid fuel use). Results: Seventy-six percent (25/33) of women live with active smokers, with 52% living with one, 18% with two and 6% with three in their households. The median (average; range) households' nicotine concentration (as one-month average) was 0.126 μg/m3 (0.326; 0.019 - 1.88) in summer and 0.731 (1.25; 0.010 – 5.79) in winter. The standard deviation was 0.44 μg/m3 in summer and 1.61 in winter. Paired t test indicated that nicotine levels were significantly higher in winter than in summer ($p = 0.0016$). However, nicotine concentration was not associated with 8-OHdG measured in the village women in summer nor in winter, after controlling for 1-OHP. Conclusion: Household SHS exposure was high in the rural village in northwest China. Variations are also large among households and from season to season. SHS exposure was not the main contributor to the oxidative stress among the village women.

Keywords: secondhand smoke

WE-PO-35
Use of the natural compounds to inhibit growth of the indoor isolated toxigenic Aspergillus westerdijkiae

Mikkola R.[1], Castagnoli E.[1], Alapieti T.[1], Andersson M.A.[1], Salonen H.[1]; [1]Aalto University, Department of Civil Engineering, Espoo, Finalnd

Toxigenic fungi are regularly found in the indoor environment. Such fungi may pose a health risk to occupants. Insulation material could be one of the possible places where the harmful fungi could accumulate. Thus, preventing the growth of harmful fungi in insulation material is one key factor to guarantee healthy indoor air. Earlier we isolated fungal strain Aspergillus westerdijkiae PP2 as the dominant fungus in settled dusts from the building where occupants suffered from indoor air associated illness. Aspergillus westerdijkiae PP2 produces avrainvillamide, stephacidin B and ochratoxin which are cytotoxic at low concentrations. Furthermore, Aspergillus westerdijkiae PP2 is highly tolerant to boric acid, used in insulation materials as a fire retardant and an antimicrobial agent, as well as to numerous biocides like polyhexamethylene biguanide and polyhexamethylene guanidine. This makes the control of such fungus in the indoor environment almost impossible using traditional biocides. The application of natural nontoxic compounds, thus, could be an attractive alternative to the synthetic biocides. The aim of this study was to investigate the fungicidal activity of natural nontoxic agents toward

Aspergillus westerdijkiae PP2. Resveratrol, magnolol, curcumin, betulinic acid, isorhapontigenin, pinosylvin and eugenol were selected based on their antimicrobial properties, low toxicity and their beneficial effects. The antifungal activity was studied using agar diffusion method and 1 µg of each compound. Out of the tested compounds only magnolol resulted in a clear inhibition zone whereas pinosylvin only slightly inhibited the growth of Aspergillus westerdijkiae PP2. Magnolol (5,5'-Diallyl-2,2'-dihydroxybiphenyl) is lignan isolated e.g. from the bark of Magnolia officinalis and stilbenoid pinosylvin (trans-3,5-Dihydroxystilbene) is polyphenolic compound originating from pine. Magnolol and the related allyl dihydroxybiphenyls compounds as honokiol 5,3'-Diallyl-2,4'-dihydroxybiphenyl as well as stilbene polyphenolic compounds might be used as potential nontoxic agents to prevent the growth of harmful fungi in indoor air environment.

Keywords: built/indoor environment, dampness/mold, environmental health, microbial agents, respiratory health

WE-PO-36
Ambient concentration of airborne microbes and endotoxin in rural households of southern India

Muthalagu A. [1], Eerappa R. [1], Qureshi A. [1]; [1]Hyderabad, Telangana, India

Investigation of microbial communities in the inbuilt environment is critically important as 90% of human life is indoor. Airborne microbial flora and their pyrogenic substances in the living spaces can impact human health, and only little is known of their identity and the mechanism of their assembly. Since 70% of Indian population lives in villages, this study provides baseline information about the primary biological aerosol particles (PBAPs) concentration, diversity and abundance, indoor-outdoor relationship and endotoxin concentration from the particulate matter in the indoor and outdoor region of thirty-six rural household settings. While most of the indoor air quality studies focused on the concentration of the bioaerosol in different household settings, this study provides the first-hand information about the bioaerosol concentration and diversity, endotoxin concentration along with the microbial risk assessment for pathogenic bacteria B. anthracis for the study region. Bacterial and fungal aerosols concentration in both the sites of all the thirty-six homes are ranged from IB-822 CFU m-3, IF– 650 CFU m-3, OB-937 CFU m-3, OF– 616 CFU m-3 with a size specific fungal and bacterial composition comprising of Agaricomycetes, Dothideomycetes, Euascomycetes, Eurotiomycetes, Leotiomycetes, Sordariomycetes, Zygomycetes for fungi and Actinobacteria, Alphaproteobacteria, Bacilli, Betaproteobacteria, Flavobacteria and Proteobacteria for bacteria. The I/O ratio for total bacteria and fungi was 1.38 and 0.790. The results suggest in addition to outdoor sources, indoor sources such as home ecology and metrological factors play a role in elevated bioaerosol concentration. Endotoxin concentration ranged from 2-21 EU m-3 and the population living in higher endotoxin sites have asthma and allergy related issues. Quantitative microbial risk assessment (QMRA) model was developed for the pathogen B. anthracis which was found in 8 homes among 36. The probability of infection by B. anthracis was found 11% in one sampling site where already people have been hospitalized due to fever.

Keywords: built/indoor environment

WE-PO-37
Fungal assemblages on indoor moldy surfaces in homes in Korea and Thailand

Yamamoto N.[1], PrechaN.[1], KliengchuayW.[1], WooC.[1], TantrakarnapaK.[1]; [1]Seoul National University, Seoul, Korea, [2]Walailak University, Nakorn Srithammaras, Thailand, [3]Mahidol University, Bangkok, Thailand

Indoor visible molds are known to be linked to allergies and respiratory illnesses. However, knowledge is limited about their identities and compositions. In particular, scarce information is available about how they differ across climate zones (e.g., temperate vs. tropical countries). Here, we investigated and compared fungal assemblages on indoor moldy surfaces in homes in Korea and Thailand. Indoor surface samples were collected by sterile swabs and analyzed by high-throughput DNA sequencing of fungal internal transcribed spacer 1 (ITS1) region. The dothideomycetes were abundant in Korean homes while the sordariomycetes as well as the dothideomycetes were abundant in Thailand homes. Statistical analyses showed significant differences in the fungal assemblage memberships (P-test; $p < 0.001$) and structures (P-test; $p = 0.0005$) between the two countries, suggesting that fungi in homes geographically pattern by the climate zones. Additionally, several allergy-related genera were detected, which included Alternaria and Trichoderma from Thailand homes, and Cladosporium, Aspergillus, Rhodotorula, Cryptococcus and Alternaria from Korean homes. High-throughput sequencing allows for accurately characterizing fungal assemblages and diversities on indoor visible molds. Future studies are expected to investigate the relationships between fungal assemblages and diversities of indoor visible molds and health outcomes of allergies and respiratory illnesses in countries that are situated in different climate zones.

Keywords: built/indoor environment, dampness/mold, environmental health, microbial agents

WE-PO-38
Experimental study on evaporation mechanism of microbial droplets

Liu X. [1]*, Liu L.* [2]*, Wang Y.* [1]*, Zhang X.* [3]*;* [1]*State Key Laboratory of Green Building in Western China, Xi'an, China,* [2]*School of Architecture, Tsinghua University, Beijing, China,* [3]*School of Environmental and Municipal Engineering, Xi'an University of Architecture and Technology, Xi'an, China*

Infectious diseases such as influenza and SARS can be spread by droplets exhaled by humans. Evaporation of exhaled droplets is affected by its composition and environment, droplet temperature, pH, and evaporation also affect its activity of carrying microorganisms. The pathogens carried in the droplets affect its evaporation or not is not clear. In this study, different genus species bacteria were used to replace motile pathogens, and high-precision balances were used to measure mass changes during droplet evaporation. The effect of bacterial species on the evaporation rate of droplets was studied. The experimental results illustrate that after adding bacteria of different genus in 0.9% (w/v) NaCl droplets, bacterium can slow down the evaporation rate and prolong the evaporation time. Bacillus accelerates the evaporation rate and shortens the evaporation time. The evaporation law of Bacillus aeruginosa is similar to that of NaCl solution, and its evaporation rate is between the globular bacteria and Bacillus, but the evaporation rate of the bacterial liquid droplets is faster.

Keywords: infectious disease

WE-PO-39
Using the fluorescence data to speed up discrimination between different mycotoxin producing toxins

Castagnoli E.[1]*, Andersson M.*[1]*, Mikkola R.*[1]*, Kurnitski J.*[2]*, Salonen H.*[1]*;* [1]*Aalto University, Department of Civil Engineering, Espoo, Finland,* [2]*Tallinn University of Technology, Tallinn, Estonia*

The presence of toxigenic fungi in the indoor environment may trigger health symptoms of building occupants. Current knowledge on difference of toxicity between toxigenic fungal strains is limited. The aim of this study was to use fluorescence data to help discriminate between the difference of toxicity of toxigenic Aspergillus and Chaetomium strains isolated from buildings where occupants reported indoor-air related symptoms (in Finland). The selected strains were Aspergillus versicolor Asp2 and SL/3, Asp. westerdijkiae PP2/09, Asp. calidoustus MH34, Chaetomium globosum MTAV 35 and MH 52, and Chaetomium cochliodes OT7 and OT7b. A UV lamp (360 nm) and a fluorescence reader were used to evaluate the fluorescence. Toxicity was evaluated using boar spermatozoa (motility inhibition and membrane disruption assays) and inhibition of cell proliferation assays. Orange (biomass of strains Asp2, and SL/3), blue (PP2/09, MH34, MTAV 35 and MH 52 strains) and no fluorescence (OT7 and OT7b strains) were observed. The fluorescence spectra and toxicity profile of Aspergillus versicolor Asp2 (biomass) was similar to the sterigmatocystin- and averufin-producing Aspergillus versicolor SL/3. Interestingly the exudate of strain Asp2 exhibited blue fluorescence, instead of its orange biomass extract, and was non-toxic (in a 1/20 dilution) towards boar-spermatozoa based assays and ICP assay. This indicates that the composition of its biomass and exudate were different. The Chaetomium strains exhibited similar toxicity to boar spermatozoa; however, the non-fluorescent chaetomin-producing Chaetomium cochliodes strains were 50 times more toxic in the ICP assay than the blue-fluorescent chaetoglobosins-producing Chaetomium globosum strains. Our study revealed that the fluorescence profile helped discriminate between the highly toxic chaetomin-producing Chaetomium cochliodes strains and the less toxic chaetoglobosins-producing Chaetomium globosum strains. The distinctive fluorescence of Aspergillus versicolor Asp2 (biomass vs. exudate) revealed that either different or no toxins were present in its exudate.

Keywords: built/indoor environment, microbial agents

WE-PO-40

The Indoor Air Quality of the Respiratory Care Wards of a Regional Hospital in Northern Taiwan

Lin Y-W.[1], Tang C-S.[2], Chen W-Y.[2]; [1]Fu Jen Catholic University, New Taipei City, Taiwan, [2]New Taipei City, Taiwan

Introduction Respiratory care wards (RCW) are responsible to care the patients who cannot spontaneously breathe without ventilators. These patients are easily to be infected and hypersensitive to the indoor air quality. This research monitored the IAQ of the RCW and illustrated the influential factors of IAQ. Method The IAQ of two five-bed wards and one nurse station of a RCW in a regional hospital were monitored. The monitoring was proceeded for 16 hours or 24 hours during the sampling days with a sampling frequency of 20 minutes per hour. The concentrations of carbon dioxide (CO2), carbon monoxide (CO), particulate matter (PM), nitrogen oxide (NOx), volatile organic compounds (VOCs) , relative humidity (RH) and temperature were measured by direct reading instruments. The bioaerosol samples were taken hourly. The hourly air change rate (ACH) was calculated by measuring the air ventilation volume. Human activities were recorded during the sampling period. The linear mixed model (LMM) was applied to illustrate the impact factors of IAQ. Result The concentrations of CO, CO2, PM, bacterial and fungi exceeded the Taiwan EPA recommendation levels. The major factors affecting the concentrations of CO, PM1 and PM2.5 were location and the number of inpatients. The significant factors to alter the CO2 and VOC concentrations were location and the numbers of in-and-out staff and inpatients. The number of in-and-out staff and the level of activity affected the PM10 value statistically. The level of activity and the numbers of in-and-out staff and inpatients are the significant factors in changing the bacteria and fungi concentrations. Different modules of ventilators did not affect the IAQ significantly. Conclusion The results of LMM can be utilized to predict the pollutant concentrations under various environmental conditions. The results of this study would be a valuable reference for air quality management of RCW.

Keywords: built/indoor environment

WE-PO-41

Noise exposure from personal and recreational activities: implications on occupational hazard prevention

Huang Y.; National Kaohsiung University of Science & Technology, Kaohsiung, Taiwan

Noise may be present in work and non-work environment. While frequency, duration and loudness affect the hearing functions, unpleasant sound may induce other physical and mental effects. On the other hands, the definition of "unpleasant" sound differ greatly among individuals, and it may be difficult to determine how people perceived a given sound as "too loud." Occasionally, people may even seek out purposely to attend activities where noise is prevalent. In this study, sound levels at various outdoor events were evaluated for their potential noise hazards to the audience. Two outdoor events were measured: a punk rock concert, and a college sports event followed by an evening outdoor concert. A level 2 noise dosimeter (Svantek SV 104 noise dosimeter, Svantek, Warsaw, Poland) was used to measure sound levels at various locations during the events. In all, the sound levels varied substantially over the event area, ranging from below 75 to over 110 dBA. Both events occurred at an open area with limited access to attending audience. Higher sound levels was detected at near the loud speakers. The peak noise level was near 110 dBA at both concerts for nearly one hour. Since attending concerts and loud events (such as singing karaoke) may be a result of personal choice, even though such sounds may not be perceived as noise, frequent exposures could nonetheless result in long-term hearing losses. Moderation on attending such loud events may be necessary, especially for those with occupational noise exposures.

Keywords: acoustics/noise, aggregate exposure, industrial hygiene, occupational

WE-PO-42

Estimation and validation of exposure fraction to workplace noise among Korean workers

Kim S.W.[1], Yang, S-H., Ryu, H-W., Kim, E-A.; [1]Keimyung University, Daegu, South Korea

Exposure fractions (sometimes, exposure prevalence) to workplace hazards is important to establish polices and to evaluate programs in occupational health. This study estimated and validated the exposure fraction of employees occupationally exposed to noise by their industry type and business size in South Korea. A Delphi panel consisting of 15 occupational health experts estimated the exposure prevalence of noise in workplace. Data on Industrial Accident Compensation Insurance provided from Occupational Safety and Health Research Institute were combined to produce the exposure population. The results were compared to exposure estimations from other studies. In South Korea, 16.0% of workers, 2,540,000 out of 15,840,000, was estimated to exposed to workplace noise. The rate was 32.7% and 10.3% in manufacturing sector and non-manufacturing sector, respectively. The highest rate, 52.5%, was reported in manufacturing industries of wood and of products of wood and cork (except furniture) and followed by manufacturing industries of other transport equipment as 52.4%. Sorted by their size of business, the rate was higher in workplaces hiring more than 50 than that of workplaces hiring less than 50 in both sector (for example, 35.1% and 30.4% in manufacturing sector). Our estimation was slightly lower for small size businesses (less than 5 employees) than the rate reported from a field survey in 2001 (25.5% and 26.6%, respectively). Compared to the same rate estimated in the US, 17.2%, the result of this study, 17.8%, seems to be in a reasonable range.

Keywords: acoustics/noise, industrial hygiene, occupational, workplace

WE-PO-43
Burden of disease from road and rail traffic noise in Oslo, Norway

Aasvang G., Krog N.[1], Sulo G.[2]; [1]Department of Air Pollution and Noise, Norwegian Institute of Public Health, Oslo, Norway, [2]Center for Disease Burden, Department of Mental and Physical Health, Norwegian Institute of Public Health, Bergen, Norway

Background: Traffic noise is the second largest environmental risk factor in terms of disease burden in Europe. Environmental noise is of increasing public health concern. Methods: The burden of disease from environmental noise was expressed as Disability Adjusted Life Years (DALY). The health outcomes were noise annoyance and noise induced sleep disturbances from road traffic and railroads. In addition, we estimated DALYs due to ischemic heart disease (ICD 10 I20-25) associated with road traffic noise. We used exposure data from the strategic noise mapping for Oslo for 2016. The recent exposure-response functions, based on meta-analyses from the systematic review performed as the basis for the development of the WHO Environmental noise guidelines (WHO, 2018), were used. To estimate DALYs for ischemic heart disease due to noise we estimated Population Attributable Fraction (PAF) from population exposure (Lden) and associated risk estimates. Furthermore, we estimated years lived with disability (YLD) and years of life lost (YLL) due to ischemic heart disease attributable to noise in Oslo based on age and sex specific estimates for Norway from the Global Burden of Disease (GBD) study. Results: Traffic noise contributes with considerable disease burden in Oslo. Sleep disturbances came up with the highest disease burden due to noise (YLD = 2 164) followed by noise annoyance (YLD = 1 862) and ischemic heart disease (DALY = 387). Road traffic noise constitutes the major part (77 %) of the total burden due to transportation noise in Oslo. Methods, assumptions and more results will be presented at the conference.

Keywords: acoustics/noise, environmental health, epidemiology

WE-PO-44
Occupational Noise Burden of Disease

Chen Y.[1], Styliano K.[1], Jolliet O.[1], Neitzel R.L.[1]; [1]Environmental Health Sciences, School of Public Health, University of Michigan, Ann Arbor, MI, USA

According to NIOSH, approximately 22 million U.S. workers are exposed to hazardous occupational noise leading to irreversible hearing loss. Therefore, there is a need to quantify occupational noise impacts and noise-induced hearing loss (NIHL) on a national scale. This study aims to estimate health impacts induced by occupational noise in the US economy by occupation and sector. We developed a novel burden of disease framework for occupational noise, based on specific job occupations. NIHL burden rates (BRs) were estimated in 443 occupations accounting for 6 disease severities and 13 age groups, expressed in disability adjusted life years (μDALY/pers- yr). To provide comparable estimates and identify those with the highest impacts, we multiplied occupation and sector specific BRs by the number of persons employed to yield the annual occupational health impact due to noise, in DAYLs/yr. From the 126 million US workers, 51% were exposed to 80-85 dBA (minimum), 14% to 85-90 dBA (moderately high) and 1.1% to >90 dBA (high). From an occupational perspective, NIHL BRs range from 400 to 7200 μDALY/pers-yr. Accounting for employment, health impacts by occupation ranged from 3.6× 〖10〗 ^(-1) to 1.0 × 〖10〗 ^4 DALYs /yr. The total national burden in the US was estimated 1.7 × 〖10〗 ^4 DALYs/yr. The highest cumulative annual impact was experienced by "Building Cleaning Workers," a group exposed to a moderately high level, which had the 6th largest worker population. From a sector perspective, NIHL BRs range from 160 to 3500 μDALY/pers-yr. The highest impact was experienced by the "Agriculture, Forestry, Fishing and Hunting" sector, representing 13% of total cumulative impacts for only 0.3% of total employment. This study provides a reference for minimizing the NIHL burden by pinpointing areas to reduce the amount of noise in priority.

Keywords: acoustics/noise, health, other/general, lifecycle analysis, occupational, risk assessment

WE-PO-45

Exposure to traffic-related air pollutants during physical exercise outdoors

Monteiro T.[1], Slezakova K.[1], Madureira J.[2], Costa C.[2], Pereira M.[2], Teixeira JP.[2]; [1]LEPABE, Departamento de Engenharia Química, Faculdade de Engenharia, Universidade do Porto, Porto, Portugal, [2]Environmental Health Department, National Institute of Health, Porto, Portugal; EPIUnit - Instituto de Saúde Pública, Universidade do Porto, Porto, Portugal

Various national and international organizations recommend conducting aerobic activities, as exercising on regular basis, since it improves the overall health and well-being of the individuals. Outdoor exercising represents one of the most accessible forms of physical activities. In urban areas, exercising outdoors may lead to individuals' increased exposure (on short-terms) to high levels of air pollutants. This study aims to characterize the exposure to traffic-related air pollutants (TRAP) in different urban areas under several physical exercise scenarios (e.g. cycling, running, and walking). Traffic-related air pollutants exposure will be estimated for PM2.5, PM10 and NO2, both in summer and winter seasons. Pollutants data will be retrieved from network of national monitoring stations of Lisboa and Porto, as these cities are the most polluted urban areas in Portugal and based on the high-profiles sites that are frequently used to conduct physical exercise in the respective places. The individual inhaled doses and associated health risks for PM2.5, PM10 and NO2 will be quantified using the USEPA methodology Quantitative Health Risk Assessment for Particulate Matter. Furthermore, the obtained results will be compared with other Portuguese and international urban areas. Data here obtained will provide relevant information on the potential risks associated with exposure to air pollutants while exercising. Results may be also used to support the development of new risk management strategies to minimize TRAP exposure, as large health and societal benefits can be obtained from healthy urban environments that can promote outdoor physical exercise. This work is supported by FCT-FAPESP (FAPESP/19914/2014) and by UID/EQU/00511/2019 - Laboratory for Process Engineering, Environment, Biotechnology and Energy – LEPABE, funded by national funds through FCT/MCTES (PIDDAC); Carla Costa and Joana Madureira are supported by FCT grants (SFRH/BPD/96196/2013 and SFRH/BPD/115112/2016, respectively).

Keywords: activity patterns, air, particles, risk assessment

WE-PO-46

Evaluating the RAIDAR-ICE model with monitoring data of emerging SVOC compounds in the indoor environment

De Brouwere K.[1], Arnot J.[2], Li L.[3], Geerts L.[1], Lamoree M.[4]; [1]VITO - Flemish Institute for Technological Research, Mol, Belgium, [2]ARC Arnot Research and Consulting Inc, Toronto, Canada, [3]University of Toronto Scarborough, Toronto, Canada, [4]Vrije Universiteit, Amsterdam, The Netherlands

Humans are exposed to a wide variety of chemicals that originate in "near-field" environments (i.e., indoors). Chemicals of concern for human health include dirt and water repellents, flame retardants, and plasticizers which can be emitted from construction materials, electronic equipment, carpets, textiles, flooring and furniture through evaporationor abrasion.. Multi-media measurements for chemicals of concern are being obtained through various monitoring and biomonitoring projects and programs. For example, the Cefic-LRI SHINE project is conducting targeted and non-targeted analyses of dust and air samples in indoor environments in 5 EU countries, coupled with a literature review for biomonitoring data for the target chemicals of concern. . Mechanistic, mass-balance, multi-media models serve as powerful tools to understand important exposure pathways and to predict exposures in the absence of measured data for human health assessment. Such models that link external exposures with internal exposures can also be used to integrate monitoring data (air, dust, surface concentrations) with biomonitoring data (blood, urine). There is a need to test (evaluate) exposure models to foster confidence in their application in regulatory programs in both data rich and data poor contexts. The Risk Assessment IDentification And Ranking – Indoor and Consumer Exposure (RAIDAR-ICE) model combines an indoor mass balance fate model with a physiologically-based toxicokinetic model for an adult male to simulate chemical exposures from direct (e.g., personal care products) and indirect (e.g., vaporization of chemicals from building materials) sources of chemicals. RAIDAR-ICE simulates human exposure from the near-field environment through inhalation of indoor air, dermal absorption, non-dietary object-hand-mouth contact. In this study we evaluate the RAIDAR-ICE model with monitoring and biomonitoring datasets for a range of chemicals, including new data obtained in the SHINE project. The case study chemicals used in the model evaluations comprise a broad range of use categories, partitioning properties, and biotransformation half-lives.

Keywords: SVOCs, built/indoor environment, exposure models, multimedia

WE-PO-47

Evaluation of an Improved Receptor Model for Analyzing Data with Spatial Variation: A Simulation Study

Lee C-L.[1], Liao C-F.[2], Chen H-T.[1], Wu, Y-D.[1]; [1]Institute of Occupational Medicine and Industrial Hygiene, College of Public Health, National Taiwan University, Taipei, Taiwan, [2]Research Center for Environmental Changes, Academia Sinica, Taipei, Tawain

The concentrations of air pollutants show horizontal and vertical variations. In previous receptor model studies, data collected from multiple sites would be analyzed individually or be pooled before analysis. However, it is still controversial which method is better because it would cause smaller sample size if data from different locations was analyzed separately. In contrast, analyzing pooled data may cause unrealistic results because profile variation in horizontal domain could be large. In addition, correlations of source contributions in vertical aspect were not considered. This study was conducted to evaluate the performance of a modified source apportionment model that is suitable for data with spatial variation. The evaluation was achieved by creating synthetic data sets that simulated PM2.5 and VOCs from five sources: petroleum refinery, vehicle exhaust, industrial coating, coal combustion, and natural gas. Daily PM2.5 and VOCs in 3 floors were simulated for a 45-day period. Positive matrix factorization (PMF) is commonly used to quantify the contribution of the sources at a receptor site. The Multilinear Engine (ME) program, a more flexible tool for solving PMF problems, was applied to deal with the vertical issue. It was set that time series of resolved factor contributions in different floors of a building should have similar trends. The average absolute error (AAE) is calculated to examine how profile variation affected the model performance. The correlation coefficients between the input and retrieved source contribution and profiles were above 0.98 when data from different floors of a building was analyzed. Sensitivity analyses results indicated that analyzing data pooled from 3 buildings was more fitted than analyzing them separately with one source having profile variation below 20% or with all source profiles varied below 10%. Therefore, it is recommended that the profile variation should be considered before analyzing data with spatial variations.

Keywords: VOCs, air, particles, statistical methods

WE-PO-48 WITHDRAWN

WE-PO-49

TREXMO plus: an advanced self-learning model for occupational exposure assessment

Savic N.[1], Lee E.G.[2], Gasic B.[3], Vernez D.[1]; [1]Center for Primary Care and Public Health, University of Lausanne (Unisanté), Lausanne, Switzerland, [2]National Institute for Occupational Safety and Health (NIOSH), Health Effects Laboratory Division (HELD), Exposure Assessment Branch (EAB), Morgantown, WV, USA, [3]Chemicals and Occupational Health Unit, Swiss State Secretariat for Economic Affairs (SECO), Bern, Switzerland

Within the context of the EU's REACH regulation, several occupational exposure models are recommended for chemical exposure assessment. These models require exposure parameters to be entered from a few to many number, depending on the level of models' complexity. As of today, the performance of these models is not fully evaluated for a wide range of exposure conditions because of the lack of adequate exposure data. In this study, we developed and tested a new model for airborne exposure, named "TREXMO Plus (+)", that uses three REACH models, i.e. ART, Stoffenmanager (version 4.0), and ECETOC TRAv3, as the independent predictors. This model considers that the performance of the different exposure models may vary over different exposure conditions (such as for lower and higher vapor pressure values). TREXMO+, therefore, is conceptualized by applying weights developed from the training data sets into the independent predictors of three models to derive a refined exposure estimate. The exposure data, which counted 1058 exposure measurements, was split in two sets, where 80% was used to develop the model and the remaining 20% to test its performance. Compared with the three REACH models individually, TREXMO+ was the least biased model and the most accurate. It was found that, on average, the TREXMO+ estimates differ by a factor between 2 and 3 from the measurements. The model was significantly correlated against the measurements and explained best the variance (R-squared, 0.44-0.71). Although more data is required to test further this model, the concept of TREXMO+ is expected to provide better estimates in comparison with the use of the exposure models individually.

Keywords: exposure factors, exposure models, industrial hygiene, occupational, risk assessment, statistical methods

WE-PO-50

Using continuous exposure in-vitro model to estimate percutaneous absorption of different characteristics chemical exposure

Zhu G. [1], Nien Y-T. [1], Wang Y-H. [2], Hsu J-F. [3], Chou T-C. [1]; [1]Department of Public Health, China Medical University, Taichung, Taiwan, [2]Department of Nursing, Yuanpei University of Medical Technology, Hsinchu, Taiwan, [3]National Institute of Environmental Health Sciences, National Health Research Institute, Miaoli, Taiwan

The skin is the main barrier to protect the body and continuous exposure of chemicals to the skin is extremely common in daily life or occupational settings. Continuous exposure refers to the exposure of chemicals in a sequential manner at different times. In the previous in-vitro studies, co-exposure of two or more chemicals were common exist, but few studies have been conducted on the assessment of percutaneous absorption kinetic parameters after continuous exposure in-vitro. The purpose of this study was to evaluate and to compare the percutaneous absorption of target exposed substances with different characteristics after the pre-exposed substance in the continuous exposure in-vitro model. In this study, the pre-exposed substance was linalool (hydrophobic), and the target exposed substances were salicylic acid (hydrophobic) and caffeine (hydrophilic). We used the pig's side abdomen skin and the dynamic diffusion cell was the testing materials. The pre-exposure time was 18 hours, and the sample collection time of the target exposed substance was 24 hours. All samples were analyzed by HPLC-UV. Cumulative mass and kinetic parameters are presented to compare the different characteristics exposed substances. Under 18 hours pre-exposure conditions, we found that the flux of salicylic acid was 25.0±13.6 μg/cm2/h, the permeability coefficient was 0.02 cm/h. Moreover, the flux of caffeine was 180.7 ± 63.8 μg/cm2/h, the permeability coefficient was 0.05 cm/h. It shows that the flux and permeability coefficient of caffeine greater than that of salicylic acid. After exposure to linalool, the absorption of caffeine is higher than salicylic acid. The results show that the pre-exposure to hydrophobic substances can cause the absorption of hydrophilic target substances to be higher than that of hydrophobic target substances on skin. Based on the above research results, the mechanism of percutaneous absorption status of this phenomenon should be further studied.

Keywords: cumulative exposure, exposure factors, exposure models, occupational

WE-PO-51

Can inexpensive sensors be used to improve indoor air quality?

Justo Alonso M.[1], Gram O.K.[1], Mathise H.M.[1]; [1]NTNU, Trondheim, Norway

In most developed countries, buildings account for about 40 % of the total energy use, of which a significant fraction is consumed by HVAC systems. Demand control ventilation (DCV) is often proposed as a measure to reduce HVAC energy use, while maintaining adequate levels of outdoor air ventilation for indoor air quality control. In DCV applications, outdoor airflow rates are reduced during periods with low occupancy, often based on measured CO2 concentrations. However, when reducing outdoor airflow rates, concentrations of other pollutants otherwise controlled by dilution, may rise to unacceptable levels. For example, formaldehyde or other volatile organic compounds (VOC) emitted by building materials, do not necessarily have the same emission schedules as CO2, i.e., their emission rates are not reduced when occupancy is low. When only relying on occupant-generated CO2 for ventilation control, these other pollutants may achieve levels that can negatively impact occupant health, comfort and productivity. Usually, these "other" pollutants are not incorporated into DCV applications, as their measurement is too complex and expensive. In recent years, new technologies have enabled for less expensive sensors, though with reduced accuracy compared to research instruments. This paper evaluates four equal sensors that measure formaldehyde, VOC and PM2.sensors with costs that are less likely to hinder their use in building ventilation controls. Sensor aging and its effect on accuracy will also be evaluated by calibrating their performance after one month of measurements in a school. The study will evaluate the sensors' performance in laboratory tests and by measurements in classrooms. Their potential use for DCV regarding improvements on the indoor air quality, based on their accuracy and ability to react to changes in indoor concentrations will be discussed.

Keywords: sensor technology

WE-PO-52

Study on Optimum Dimension of Negative Pressure Type Public Exhaust System with Equal Section

Zhu A.[1], Zhang Z.[1], Zhou M.[1], Zhang L.[1], Yang P.[1], Gao B.[1]; [1]The China National Engineering Research Center for Human Settlements (CNERCHS), Beijing, China

The public exhaust system is a common system of high-rise residential buildings in China, aiming to reduce the impact of cooking smoke on the facade, atmosphere and neighbors. Due to the unreasonable system design, equipment selection and construction, problems such as refluxing, distinguish odor and blocking occur. And the kitchen pollution has become the most prominent problem in China's indoor environment. Negative pressure type public exhaust system puts power devices at the top with negative pressure in its flue, eliminating these problems. The purpose of this study is to reduce the dimension of public flue under the premise of remaining the smoke exhaust effect in order to improve the space utilizing efficiency through experiments and simulation analysis. The experiments were conducted in Dongguan Super High-rise Full-scale Experimental Tower (122.9m) with air as the experimental medium, and 550*450mm flue size. The performance was tested under four working conditions with 40% and 60% operating rate, respectively. The Fluent model is modified by experimental results. Referring to the national standard atlas 16J916-1, the performance under 450*550mm, 550*400mm, 460*400mm and 430*300mm is simulated and analyzed respectively. The results show that when the main flue size changed from 550*450mm to 450*550mm, the exhaust air volume of each layer decreases by an average of 0.8%, indicating that the layout direction of the main flue has little effect. With the reduction of flue size from 550*450mm to 430*300mm, the exhaust air volume decreased by 42%~51%. And when the size is 460*400mm, the exhaust air volume of four conditions basically maintained within the standard range (300~500m3/h). In conclusion, on the premise of guaranteeing the smoke exhaust effect, the main flue size used in the experimental can be optimized to 460*400mm (saving up to 25% space). This study could guide the design of similar practical projects.

Keywords: built/indoor environment, environmental health, source control, ventilation

WE-PO-53
Addressing Radon in Energy Retrofits

Francisco P. ; University of Illinois at Urbana-Champaign, IL, USA

Because many residential energy retrofit programs include substantial air sealing, there has been concern over the impacts on indoor air quality. Radon, specifically, has become a major concern. Long-term exposure to radon, which primarily comes from soil, is associated with lung cancer. These programs, with a primary focus on energy and with limited budgets, focus on a "do-no-harm" strategy for health & safety as opposed to targeting full mitigation of pre-existing problems. However, radon concentrations in homes vary widely with time and location, so short-term monitoring before and after an energy retrofit may not provide an accurate indication of the impact of the retrofit, including whether or not radon levels increased or decreased. Long-term monitoring, on the other hand, does not fit the work flow of these programs. As a result, a significant amount of research has been undertaken to, first, understand the real impacts of energy retrofits on radon levels; and second, evaluate the potential for precautionary measures to limit the entry of radon into homes that have been made tighter. This presentation provides an overview of the work done in this area in the last eight years. This includes a summary of an initial study on the effects of energy retrofits on radon levels, which showed a statistically-significant increase; a small-sample follow-on study evaluating the impact of exhaust ventilation on indoor radon levels; another study on a larger sample of homes that included the effects of exhaust ventilation on indoor radon levels; and a study that evaluated the efficacy of strengthened policies regarding precautionary measures. These precautionary measures include ventilation requirements, well-installed ground covers in foundation spaces, and sealed sump pump covers. The results of these studies show that these precautionary measures can reduce the effects of energy retrofits on indoor radon levels.

Keywords: built/indoor environment, other (specify), ventilation

WE-PO-54
Rapid identification and antibiotic susceptibility determination for anthrax (Bacillus anthracis) using lethal factor endopeptidase activity coupled with MALDI-MS

Barr J.[1], Rees J.[1], Williamson Y.[1], Boyer A.[1], Gallegos-Candela M.[1], Lins R.[2], Woolfitt A.[1], Pirkle J.[1]; [1]Centers for Disease Control and Prevention, Atlanta, GA, [2]Battelle Atlanta Analytical Services

Anthrax is a zoonotic disease caused by Bacillus anthracis that has rapid onset and a high fatality rate unless diagnosed and treated early. The toxins of B. anthracis are major virulence factors that promote disease progression by disabling host immune defenses. These toxins are composed of three proteins, lethal factor (LF), protective antigen (PA), and edema factor (EF). Additionally, there is concern regarding the emergence of naturally-occurring or human-directed antibiotic resistance. LF is a zinc-dependent endoprotease, and is the biomarker for a rapid confirmatory mass spectrometry-based method at the CDC. In addition to rapid diagnosis, identification of the B. anthracis bacterium and antibiotic susceptibility testing (AST) is important for ensuring prompt and effective treatment. Current methods require prior culture, colony isolation and subsequent culture before antimicrobial sensitivity testing (AST), resulting in days to results and delays in patients receiving the most effective antibiotic treatments. We present a rapid, sensitive MALDI-based detection method for anthrax lethal factor (LF) that both determines the presence of anthrax in a biological matrix and establishes antibiotic resistance profiles to a panel of routinely used antibiotic agents. As LF is secreted only from metabolically active B. anthracis, we found a measurable decrease in the measured amount of LF above the minimum inhibitory concentration (MIC) of a given antibiotic. When B. anthracis is incubated for 6 hours in an AST plate structured with a twofold antibiotic dilution series, the MIC for each respective antibiotic can be accurately estimated by measuring the significant decrease in LF activity as determined by MALDI-MS quantification. Quantification of LF demonstrates the presence of B. anthracis and its antibiotic resistance profile, which can be established at a concentration range from 102 cfu/mL - 107 cfu/mL with total sample preparation and analysis time of 12 hours.

Keywords: Antibiotic susceptibility testing, antibiotic resistance, anthrax

WE-PO-55
The Feasibility of Identifying and Quantifying Worker Exposures to Volatile Organic Chemicals in Beauty Salons and Auto Shops in the Southwestern USA

Lothrop N. [1]*, Sandoval F., Cortez I., Wagoner R., Lopez-Galvez N., Parra K., Bell M., Griffin S., Beamer P.* [1,] *Wolf A., Lee A., Carvajal S., Ingram M.;* [1]*Mel and Enid Zuckerman College of Public Health, University of Arizona, Tucson, AZ, USA*

Workers in beauty salons and auto repair shops are regularly exposed to volatile organic chemicals (VOCs), which have well-documented, adverse health effects. This occupational risk is increased in small businesses due to their limited access to occupational health resources, resulting from a complex interaction of physical, social, and economic factors. Through a community-university partnership, this pilot study examined the feasibility of identifying and quantifying VOCs exposures to predominantly Spanish-speaking workers in beauty salons and auto repair shops in Tucson, AZ. Businesses were recruited door-to-door by trusted community health workers (CHWs). Using a bilingual (Spanish-English), web-based form to minimize shop inconvenience, CHWs documented workplace practices, chemical usage, and controls. CHWs photographed products, which were later referenced by university personnel to build chemical inventories. To quantify personal total VOC exposures, workers wore a VOC monitor in a sling-pack or on a belt custom-fitted for comfort during their shift (total of 4 worker-shifts/business). On a paper form, university personnel recorded worker activities and relevant exposure factors. CHWs visited 16 beauty salons and 23 auto shops, enrolling 10 of each. Participating businesses had 2-3 persons working at a time, often necessitating 2 visits per business. All beauty salons and 9/10 auto shops completed the study. Workers from both shop types changed from sling-pack to belt (or vice versa) more than once per shift, with 3 beauty salon workers hanging the monitor next to them. This pilot study offered valuable lessons for future work. Tracking participation activity through observation was problematic, necessitating asking the participant directly, when appropriate. Entering paper activity logs into an electronic database produced substantial data entry issues, and auto shop recruitment proved difficult. Nonetheless, identifying and quantifying worker VOC exposures is feasible through a community-university partnership, and this pilot study will inform a randomized trial aimed at reducing VOC exposure.

Keywords: VOCs, air, built/indoor environment, community, environmental justice

WE-PO-56 WITHDRAWN

WE-PO-57

Environmental measurements in the Sub-Cohort Study of the Japan Environment and Children's Study (JECS)

Nishihama Y.[1], Isobe T.[1], Tamura K.[1], Iwai-Shimada M.[1], Michikawa T.[1,2], Yamazaki S.[1], Nitta H.[1], Takeuchi A.[1,3], Kobayashi Y.[1], Suda E.[1], Sekiyama M.[1], Ono M.[1], Yonemoto J. [1], Kawamoto T.[1,4], Nakayama S.[1,4]; [1]National Institute for Environmental Studies, Ibaraki, Japan, [2]Department of Environmental and Occupational Health, School of Medicine, Toho University, Tokyo, Japan, [3]Department of Preventive Medicine and Public Health, Keio University, Tokyo, Japan, [4]Japan Industrial Safety and Health Association, Tokyo, Japan

Exposure to particulate matter (PM) is one of the major environmental risks to children's health. World Health Organization (WHO) reported that acute lower respiratory infection from outdoor/indoor PM exposure accounted for almost 10% of the total mortality of children under 5 years of age in 2016. Although some studies reported the significant associations between prenatal PM exposure and childhood asthma or child behavioral problems, only few birth cohort studies have been carried out in Japan to investigate the association between PM exposure and children's health. The Japan Environment and Children's Study (JECS), which is a nation-wide birth cohort study, was launched in 2011 to evaluate the relationships between environmental factors and children's health and development, involving 100,000 mother-child dyads. The Sub-Cohort Study was formed in JECS consisting of approximately 5,000 subjects randomly selected from the main cohort. In the Sub-Cohort Study, home visits were conducted when child participants became 1.5 and 3 years old along with the face-to-face medical and developmental examinations. This presentation introduces the progress status of the JECS Sub-Cohort Study, especially of environmental measurement. Indoor/outdoor air and house dust samples were collected and analyzed for PM, volatile organic compounds (VOCs), aldehydes and acidic gases. PM (PM10-2.5 and PM2.5) sampling was conducted with an active pump operated intermittently for 7 days (repeating 5 min move and 30 min rest). The field staff conducted dwelling observations and interviewed the participants about daily use of commodity and chemical substances. We also collected the information on the frequency of cleaning and the predictors of PM concentration, such as candles and incense burning, smoking and cooking. Out of 5,018 Sub-Cohort participants, indoor/outdoor air samples were collected from 5,014 (99.9%) at age 1.5 and 4,712 (93.9%) at age 3.

Keywords: air, children, environmental health, particles, respiratory health

WE-PO-58

ASSESSMENT OF PERSONAL NOISE EXPOSURE IN DIFFERENT MICRO-ENVIRONMENTS

Miškinytė A.[1], Dėdelė A.[1]; [1]Vytautas Magnus University, Kaunas, Lithuania

Introduction: Noise pollution in urban areas is a major environmental issue. There is evidence that noise exposure is associated with adverse health effects in children and adults. The aim of this study was to assess personal exposure to noise in different microenvironments (MEs) where individuals spend their time. Methods: The study included 54 randomly selected individuals aged 20-74 living in Kaunas city, Lithuania. Data about demographic, socio-economic, health, behavioural and other factors were collected using validated questionnaires. We used global positioning system (GPS) devices and daily dairies to track individual-level time-activity patterns of adults. We collected time and location data recorded every 10 seconds for one week. The analysis of time-activity patterns was performed using geographic information system (GIS). Five different MEs (home, work, other indoor locations, walking and transit) were distinguished for each participant of the study. To assess personal noise exposure in different MEs, a spatial join was performed using GIS and modelled noise map of Kaunas city. Results: Personal noise exposure varied significantly between different MEs. The results of the study showed that the lowest personal exposure to noise was determined in home ME (48.9 dBA). Our findings revealed that the highest personal noise exposure was assessed in transit ME (61.8 dBA). Individual characteristics of participants had significant impact on personal noise exposure. Men, employed individuals, those with higher income and the use of a car were exposed to higher personal noise levels. Conclusion: The present study confirms the importance of considering objective environmental exposure assessment methods using modern sensing technologies. It is important to consider time-activity patterns as well as individual characteristics, which have a significant effect on personal noise exposure.

Keywords: acoustics/noise, activity patterns, behavior, exposure models, sensor technology

WE-PO-59

Polycyclic aromatic hydrocarbons (PAHs) skin permeation rates change with simultaneous exposures to solar ultraviolet radiation (UV-S)

Hopf N.B.[1], Spring P.[2], Hirt-Burri N.[3], Jimenez S.[4], Sutter B.[5], Vernez D.[1], Berthet A.[1]; [1]Center for Primary Care and Public Health (Unisanté), Epalinges-Lausanne, Switzerland, [2]Centre Hospitalier Universitaire Vaudois (CHUV), Department of Dermatology, Lausanne, Switzerland, [3]Centre Hospitalier Universitaire Vaudois (CHUV), Department of Musculoskeletal Medicine (DAL); Lausanne, Switzerland, [4]Institute for Work and Health (IST), Universities of Lausanne and Geneva, Lausanne, Switzerland, [5]Institut National de Recherche et de Securite (INRS), Vandoeuvre Cedex, France

Road construction workers are concomitantly exposed to two carcinogens; solar ultraviolet (UV-S) radiation and polycyclic aromatic hydrocarbons (PAHs) in bitumen emissions. We hypothesized that a combined exposure could lead to enhanced PAH skin permeation rates. Skin permeation rates (J) for selected PAHs in a mixture (PAH-mix) or in bitumen fume condensate (BFC) with and without UV-S co-exposures were measured with a flow-through diffusion cell system. Freshly excised human abdominal skin was dermatomed and mounted on the cells. The PAH-mix or BFC was applied onto the skin and receptor fluid underneath skin was continuously sampled for 24 hours. PAHs were quantified with a HPLC-Fluorescence detection. The steady-state permeation rate (J) and lag time (tlag) across the skin were calculated. Permeation rates (Js) were greater with UV-S for naphthalene, anthracene, and pyrene in BFC (0.08-0.1 ng/cm2/h) compared to without (0.02-0.26 ng/cm2/h). This was again observed for anthracene, pyrene, and chrysene in the PAH-mix. Naphthalene and benzo(a)pyrene (BaP) in the PAH-mix had greater Js without (0.97-13.01 ng/cm2/h) compared to with UV-S (0.40-6.35ng/cm2/h). Time until permeation (Tlags) in the PAH-mix were generally shorter compared to the BFC, and had a large range (1 - 13h). The vehicle matrix could potentially be the reason for this discrepancy as BFC contains additional not identified substances. In conclusion, PAHs permeate human viable skin faster and to a greater extent with simultaneous exposures to UV.

Keywords: exposure models, industrial hygiene, radiation, workplace

WE-PO-60

HCHO, NO2 and SO2 vertical column density across the Caspian Sea: The contribution of five neighbor countries

Yerkenov T. [1], Kuspangaliyeva B. [1], Altazhanova K. [1], Amouei Torkmahalleh M. [1], Darynova, Z. [1]; [1]Nazarbayev University, Kazakhstan

Vertical column density (VCD) of HCHO, NO2, and SO2 over the atmosphere of Caspian Sea during 2005-2017 years period was investigated in order to identify the possible sources of these pollutants and to assess their potential transport to five surrounding countries across the sea. The most polluted coastal area of each country was chosen in this study. The VCD of gases were retrieved from the satellite data derived by Ozone Monitoring Instrument (OMI) which has been built in Aura Satellite board. This study addresses the impact of industrial, social and geopolitical development in the five Caspian littoral states on HCHO, NO2 and SO2 concentrations. Overall NO2 and HCHO VCDs in all coastal countries over the Caspian Sea increased gradually by time due to the elevated oil and gas productions and their consumption, also a boost of population and rise of a number of vehicles. For instance, from 2005 to 2013 NO2 increased up to 4.5×1015 molecules/cm2. However, SO2 VCD showed opposite behavior. The value decreased from 2.5 x 10-6 kg/m2 to 2x10-6 kg/m2 over the study period. HCHO VCD was found to be higher in Sari city located North of Iran compared to other countries due to the present of biogenic VOCs emitted from a forest land located in North of Iran. Among multiple factors affecting the trends, the temperature has an essential effect on the formaldehyde level: the countries closer to the equator have higher temperature thus larger HCHO VCD trend. The results of this study could be a reference for the ongoing negotiations among the five Caspian Sea countries that assigns the Sea share.

Keywords: VOCs

WE-PO-61
Contribution of Aerosol Optical Depth (AOD) Fine Particulate Matter (PM2.5) Fused Surfaces in Assessing Risk of Respiratory-Cardiovascular Hospitalizations and Emergency Room Visits in Baltimore

Hall E.; US Environmental Protection Agency, Durham, NC, USA

Advancements in air pollution monitoring using (satellite) remote sensing of fine particulate matter (i.e., Aerosol Optical Depth PM2.5 [AOD PM2.5]), along with ground-based air quality monitors and air quality models have been used together in Hierarchical Bayesian Models (HBMs) to create a 'fused' air quality model surface. This study examined Baltimore, Maryland during 2004 – 2006, and complements a previous study conducted in the New York City metropolitan area, presented at the 82nd NEHA Annual Educational Conference (AEC), (Weber et al., 2016: DOI: 10.1016/j.envres.2016.07.012; http://www.sciencedirect.com/science/article/pii/S0013935116302961). The objective was to determine if combining these three data sources enhanced understanding of the influence of PM2.5 concentration on the risk of respiratory and cardiovascular diseases. This study used a case-crossover design and conditional logistic regression analysis to determine the contribution of PM2.5 to asthma related emergency department (ED) visits and inpatient hospitalizations for asthma, myocardial infarction (MI), and heart failure (HF) in Baltimore. The analysis used five HBM air pollution concentration surfaces. Four surfaces contained AOD PM2.5 measurements, monitor measurements, and air quality model estimates, and one surface represented the 'baseline' (monitor measurements and air quality model estimates only). The Baltimore study area covered an 11 (north-south) by 9 (east-west) Community Multiscale Air Quality Modeling System (CMAQ) 12 km grid region. Of the 99 CMAQ grids in Baltimore, 17 grids contained PM2.5 air pollution monitors. There were two major differences between the previous New York City study and the Baltimore study. First, there were more asthma, MI and HF cases in New York City than in Baltimore, which was expected, since New York City's population is larger than the population of the state of Maryland. Second, the mean PM2.5 values (in µg/m3), for all five surfaces, were significantly higher in Baltimore than in New York City.

Keywords: air, environmental health, health, chronic diseases, respiratory health, statistical methods

WE-PO-62
Using AOD retrievals and enhanced modeling techniques to estimate ground-level PM2.5 in Southern California

Stowell J. [1], Alhamadan M. [2], Meng X. [1], Bi J. [1], Lee S. [3], Garcia C. [4], Kinney P. [5], Freedman F. [6], Liu Y. [1]; [1]Rollins School of Public Health, Emory University, Atlanta, GA, USA, [2]National Space Science and Technology Center, Huntsville, AL, USA; [3]Air Quality Management District, [4]Diamond Bar, CA, USA, California Air Resources Board, Sacramento, CA, USA, [5]Boston University School of Public Health, Boston, MA, USA, [6]San Jose State University, San Jose, CA, USA

Background: Increasing economic growth and climate change add to concerns regarding the effects of anthropogenic pollution. Studies of PM2.5 are limited by accuracy of exposure levels. Hence, the use of satellite data in exposure science has increased, with studies highlighting the benefits of using AOD to estimate PM2.5. However, few studies have addressed this issue in southern California. Methods: We estimated the PM2.5/AOD relationship using a two-stage model. The first stage estimated temporal PM2.5/AOD relationships using a linear mixed effects model. The second stage accounted for spatial variation using geographically weighted regression. Goodness of fit for the final model was evaluated by comparing the fitted and observed results using R2 and RMSE statistics. Results: Mean predicted PM2.5 concentration for the study domain was 12.04μg/m3. The first stage performed well with R2=0.74 and RMSE=3.29μg/m3. The second stage yielded a R2 increase to 0.77 and decreased RMSE to 3.11μg/m3. Overall, the second stage improved performance (R2 increase=0.03, RMSE decrease=0.18μg/m3). During model fitting, the presence of larger sizes of PM proved an important variable in the AOD/PM2.5 relationship (β=14.88, p<=0.00). Including a ratio of these parameters improved performance, resulting in 0.06 increase in R2 and a 0.34 decrease in RMSE. Discussion: Utilizing the high-resolution MAIAC AOD, fine–resolution exposure data can be estimated where measurements are sparse. The inclusion of a second stage increases accuracy, accounting for spatial variance in the dataset. This study adds to the current research using remote sensing data to achieve better exposure data in the understudied region of southern California. Overall, we demonstrate the usefulness of MAIAC AOD and the importance of considering larger particulate matter in airborne dust prone areas.

Keywords: air, environmental health, exposure models, geospatial analysis/GIS, particles

WE-PO-63
Comparison of AOD from CALIPSO, MODIS, and AERONET

Falah S., Haifa. Israel

Numerous studies used satellite-based Aerosol Optical Depth (AOD) as an explanatory variable in different statistical models that predict ground level particulate matter (PM) concentrations. This work examines relationships between satellite-sensed and ground-observed AOD, and will use them in the future for predicting PM at the surface. At first, we evaluated the satellite AOD retrievals against ground-based AOD measurements by the AERONET sun photometers. Specifically, AOD retrievals from MODIS onboard Aqua and CALIOP onboard CALIPSO were evaluated against AERONET data. The AOD from the latest MODIS (MAIAC) AOD at 470nm and 550nm, retrieved for 3x3, 5x5, 9x9, 20x20, and 30x30 pixels (each of 1x1 km in size) and Calipso (CALIOP) AOD at 532nm retrieved for 5x5, 10x10, 20x20, and 30x30 pixels (each of 0.1x5 km in size) were centered over each of the AERONET sites in the study areas (Northern Africa, California, Germany and IGP, India) for the years 2007-2017. The AOD in the MAIAC-AERONET and CALIOP-AERONET collocated observations were compared, with the intercept of the linear regression indicating the uncertainty in the surface reflectance estimation and the slope associated with errors in the aerosol model assumptions. We analyzed the root mean square error, mean absolute error, relative mean bias, and the fraction of MAIAC and CALIOP AOD that falls within the reported expected error (EE). Good correlations were observed between MODIS AOD and collocated hourly AOD measurements from two AERONET stations in Tunisia. The different between the two AERONET stations can be attributed to variations in surface reflectance, differences in meteorological conditions (e.g. planetary boundary layer height and relative humidity profile along the atmospheric column), presence of aerosols aloft (e.g. long-range transport of mineral dust), and geographical differences in PM composition. All these will be tested in the following stages of the research.

Keywords: air

WE-PO-64

Building up PM2.5 personal exposure model based on time-weighted universal kriging

Lyu Y.[1], Xu D.[1]; [1]National Institute of Environmental Health, Chinese Center for Disease Control and Prevention, Beijing, China

Personal exposure assessment to PM2.5 is necessary for accurate evaluation its health effects. We chose Beijing as the research site to develop a time-weighted model incorporating universal kriging to evaluate personal PM2.5 exposure level. First, we analyzed the main factors affecting PM2.5 concentration and collected related information to help fit a multivariable liner regression model, which includes road network data, traffic performance index (TPI) and land cover types as main predictors. We identified the optimal form for each predictor through scatter diagrams and calculated within buffer lengths from 50m to 2km. Daily PM2.5 concentrations (2015-2017) were obtained from routine monitoring sites. Other potential predictors include population density and meteorological data. Stepwise regression was used to build the model and the variables selected criteria were made. At the second stage, we used kriging to interpolate the residuals obtained from the former stage, then integrated the two stages building the final outdoor PM2.5 concentration predicting model. The indoor PM2.5 concentration was obtained by multiplying outdoor PM2.5 concentration by I/O ratio of indoor and outdoor concentrations from our former studies. Meanwhile we recruited 60 subjects for GPS tracking. When they were located outside, we used the outdoor model to predict the PM2.5 concentration of their location. When they were indoors, we used the aforementioned formula to calculate their indoor PM2.5 concentrations. The personal PM2.5 exposure was predicted by time-weighted concentrations of different microenvironments. The outdoor PM2.5 concentration model is tested by leave-one-out cross-validation (LOCV), and the PM2.5 exposure predicted by the model can be verified by directly obtained the PM2.5 concentration through wearing personal sampler. We believe the personal PM2.5 exposure by our prediction model is more accurate than using the PM2.5 concentration from nearby monitoring sites directly, which can be useful for health impact and risk assessment in epidemiology researches.

Keywords: aggregate exposure, exposure factors, exposure models, geospatial analysis/GIS, particles

WE-PO-65

Evaluation of a plant uptake model for BTEX by three plant species under accidental release scenarios

Choi, S.-H.[1], Kown, J.-H.[1]; [1]Division of Environmental Science and Ecological Engineering, Korea University, Seoul, Korea

Benzene, toluene, ethylbenzene, and xylenes (BTEX) are widely used in the chemical industry, and they are often episodically released by chemical accidents or spills. When an accidental release occur, plants may uptake released chemicals via various routes. We evaluated a plant uptake model to predict time-course changes of the internal concentration of BTEX in plant leaf from air under accidental release scenarios. Three plant species, i.e., rice (Oryza sativa), Areca palm (Dypsis lutescens) and Hoya (Hoya carnosa), were chosen as model species to cover plants both in outdoor and indoor. The accidental release scenario assumes a large amount of gaseous BTEX release in 10 min. Partition coefficients for BTEX between test plant leaves and water (KPW) were measured in batch tests and subsequent plant-air partition coefficients (KPA) were obtained by multiplying KPW with the Henry's law constant. The performance of the uptake model was evaluated using small chamber tests. Gaseous concentration of BTEX agreed very well with predicted values under the accidental release scenarios. It was found that the overall uptake is dominated by the mass transfer resistance in plant leaf. The logarithm of mass transfer coefficient was proportional to log KPA for a given plant species. However, differences in mass transfer coefficients among plant species suggest that parameters describing plant physiology should be included to enhance the predictability of the generic uptake model.

Keywords: VOCs, exposure models

WE-PO-66

Association between overweight and human gut microbiota composition in shift-work security guards

Chen Y-J.[1], Wu C-J.[2], Lai C-H.[1]; [1]School of Public Health, National Defense Medical Center, Taipei, Taiwan, China, [2]Department of Family and Community Medicine, Division of Family Medicine, Tri-Service General Hospital and School of Medicine, National Defense Medical Center, Taipei, Taiwan, China

Introduction: Shift work is prevalent throughout the world. When so many employees engaged in shift work, it is important to determine its possible health-related effects. Previous studies indicated that the interruption of circadian rhythm influence the composition of the microbial community in the human gut and have been proposed to promote metabolic disturbances that also occur after irregular sleep time that may lead to obesity. Objectives: We examined association between overweight and gut microbiota composition in shift workers. Methods: Two hundred twenty-three security guards, excluded 207 participants because: no fecal samples (n=106), female (n=36), smoking or quit smoking <1 year (n=39), eating betel nuts (n=3), drinking (n=6), shift-work security guard working year <1 (n=6), had medical history (n=10), administration staff (control group) had dyslipidemia (n=1). Total sixteen participants (12 shift-work security guards and 4 administration staffs) enrolled in this study. Fecal microbiome composition was profiled via 16S rRNA gene sequencing. PICRUSt analysis, alpha diversity and Firmicutes/Bacteroidetes (F/B) ratio were employed to assess the variabilities in richness, diversity, organism-level microbiome phenotypes and molecular functions to find out association between overweight and human gut microbiota. Results: Microbiota composition analysis (16S rRNA gene sequencing) revealed that shift-work security guards exhibited an increased Firmicutes/Bacteroidetes ratio than administration staffs. And the content of Firmicutes was gradually increased while the content of Bacteroidetes was decreased with increasing Waist-Hip Ratio in the shift-work group ($P<0.05$). However, not significant effect between shift-work security guard and administration staffs on beta diversity was found due to the limited sample size. Conclusion: Analysis revealed that shift workers tend to have more anaerobic and less facultatively anaerobic bacteria, indicating a reduced resistance to oxidative stress that lead to overweight.

Keywords: health, other/general

WE-PO-67

Relationship between AGEs and Framingham Cardiovascular Risk Score in Predicting Cardiovascular Disease

Tsai YH., Lai C-H.

Introduction: In recent years, Taiwan has paid more and more attention to labor health, especially the long-term sedentary and standing of security personnel has become a major worry for health. Among them, cardiovascular disease is the primary disease of health. Framingham Cardiovascular Risk Score (FCRS) is currently being used to predict the incidence of cardiovascular disease in the clinic. With the emergence of new long-term chronic indicators of Advanced Glycation End Products (AGEs) in recent years, it can also be used to predict the possibility of cardiovascular disease in the future. The measurement of AGEs is through the Skin Autofluorescence reaction, which is quick and easy to measure and can improve the acceptance of the general public. Objectives: To explore relationship between Skin Autofluorescence AGEs and FCRS. Methods: The study design is cross-sectional study, subjects were collected 163 security personnel. To predict the risk of cardiovascular disease not only can use AGEs-reader but also can calculate the FCRS. Measure skin autofluorescence with AGEs-reader, AGEs can show the long-term condition of health. FCRS using five risk factors including gender, age, HDL, SBP and TC. Multivariate linear regression was used to examine the relationship between FCRS and AGEs. Results: The result showed as AGEs increasing 1 unit, FCRS increased 2.42 points ($\beta=2.42$, 95%CI 0.88~3.97, $p=0.002$) after adjusting gender and group. Gender was also associated with FCRS ($\beta=-7.96$, 95%CI -10.78~-5.14, $p=0.001$). Conclusion: There is significant association between AGEs and FCRS. It is recommended to implement exercise and dietary intervention for security workers that can greatly enhance the health awareness. Skin autofluorescence AGEs reader is not only a promising indicator to predict cardiovascular risk but also for its non-invasive measurement efficiency.

Keywords: health, chronic diseases

WE-PO-68
Analysis of Thermal Comfort in Residential Structures for the Elderly in Atlantic Climate

Torres P.[1], Aguiar L.[2], Pereira C.[1,2], Forcada N.[3], Tejedor B.[3], Teixeira J.P.[1,2], Mendes A.[1,2]; [1]Instituto de Saúde Pública, Universidade do Porto, Porto, Portugal, [2]National Health Institute - Environmental Health Dept., Porto, Portugal, [3]Group of Construction Research and Innovation, Department of Project and Construction Engineering, Universitat Politècnica de Catalunya, Terrassa, Barcelona, Spain

Europe has the highest percentage of people aged 60 or over (25%), expecting that the number of people 80 and older will triple by 2050, explaining the growing demand for long-term care services, such as elderly care centers (ECC). Determining influential parameters in thermal comfort (TC) is necessary both to design pleasant spaces and to ensure the well-being and health of buildings occupants. Efficient design and management of the buildings not only offer a comfortable thermal sensation to the occupants but also the amount of energy consumed by the building's cooling and heating systems. This ongoing project focus on the analysis of TC in ECC in the Atlantic climate zone and on the prediction of acceptable and preferred thermal conditions for older people. For this, environmental parameters such as air temperature, average radiant temperature and air velocity, external conditions such as relative humidity and atmospheric temperature, metabolic rates, thermal insulation of the clothing and thermal sensation of the users, will be analyzed in 8 ECC, making a longitudinal analysis across all seasons over a year. The ECC have been analyzed for TC 3 entire days by each season during normal activities, starting from February to November 2019. At the same time, residents will be surveyed weekly to determine the thermal sensation within the ASHRAE range. In order to determine TC, the predicted mean vote (PMV) and the adaptive model will be taken into consideration. These data will later be analyzed and used in the application of environmental comfort models for the elderly. From the resulting analytical model, it will be possible to create more efficient TC control practices, both at the level of building structures and at the level of implementation of new practices, improving people's quality of life and the energy efficiency of buildings.

Keywords: aging/elderly, analytical methods, built/indoor environment, environmental health, thermal environment

WE-PO-69
The Association Between Particulate Matter Exposure and Problem Behaviors in a Highly Lead-Exposed Group of Korean Children

Lee J.E.[1], Lee C.W.[1], Yu S.D.[1], Ha M.N.[2], Kim S.J.[1]; [1]Environmental Health Research Department, National Institute of Environmental Research, Incheon, South Korea, [2]Environmental Health Bureau, Ministry of Environment, Sejong City, South Korea

For the last decade, literature on the detrimental impacts of air pollution on brain, cognition and behavior has exponentially increased. It is acknowledged to interfere with the development of the nervous system along with lead, and is therefore particularly toxic to children, causing potentially permanent neural and cognitive impairments. Our aim of this study was to investigate the relationship between air pollution and the problem behaviors of children in a highly lead-exposed group. The study assessed all 2,308 children, aged 3–18 years, were included in the study after excluding missing values from the Korean Environmental Health Survey in Children and Adolescents (2012–2014). Korea-Child Behavior Checklist(K-CBCL) was used to evaluate each child's behavior. Daily averages of particulate matter(PM10) were predicted from a spatiotemporal model for South Korea and linked to surveyed school's administrative divisions. The life-time average PM10 exposure amount was calculated for each child. A logistic model was used to estimate adjusted odd ratio(OR) with 95% confidence intervals(CI) for problem behaviors in children exposed to particulate matter air pollution with high blood lead levels. A significant association was found between 10μg/m3 increment in the PM10 concentration and total behavior problem behaviors with adjusted OR of 1.07(95% CI 1.01-1.14) among all-aged children. More specifically, internalizing problems, externalizing problems, obsessive-compulsive problem and attention deficit hyperactivity disorder(ADHD) for PM10 exposure were 1.05(95% CI 1.01-1.20), 1.04(95% CI 1.02-1.06), 1.07(95% CI 1.03-1.12), 1.09(95% CI 1.02-1.17), respectively. The association between PM10 and total behavior problem was stronger for the high blood lead level group (1.11, 95% CI 1.03-1.18) than the low blood lead level group. There were also stronger associations between PM10 and all specific behavior problems. The findings suggest higher PM10 exposure affects problem behaviors in children especially in the high blood lead level group.

Keywords: air, biomarkers, children, environmental health, epidemiology

WE-PO-70
The Impact of Cooking Aerosol on Human Brain Wave Pattern

Gabdrashova R.[1], Naseri M.[2], Jouzizadeh M.[3], Tabesh M.[3], Malekipirbazari M.[4], Nurzhan S.[1], Farrokhi H.[3], Khanbabaie R., Mehri-Dehnavi, H.[3], Bekezhankyzy Z.[1], Gimnkhan A.[1], Dareini M.[2], Kurmangaliyeva A.[1], Islam N.[1], Crape, B.[5], Buonanno G.[6], Cassee F.[7], Torkmahalleh M.[2,8]; [1]Nazarbayev University, Nur-Sultan, Kazakhstan, [2]Chemical and Aerosol Research Team, School of Engineering, Nazarbayev University, Nur-Sultan, Kazakhstan, [3]Department of Physics, Babol Noshirvani University of Technology, Babol, Iran, [4]Department of Industrial Engineering, Bilkent University, Ankara, Turkey, [5]School of Medicine, Nazarbayev University, Nur-Sultan, Kazakhstan, [6]Department of Civil and Mechanical Engineering, University of Cassino and Southern Lazio, Cassino, Italy, [7]National Institute for Public Health and the Environment, Bilthoven, The Netherlands, [8]The Environment and Resource Efficiency Cluster, Nur-Sultan, Kazakhstan

The cooking emission in residential or commercial kitchens is one of the main indoor PM sources that results in human exposure to high concentrations of ultrafine particles (UFPs) (Amouei Torkmahalleh et al., 2017). He et al. (2004) found cooking as the largest source of indoor air pollution with the highest level of PM2.5. So far there exist only two exposure studies in literature that investigated the impact of combustion fumes from cigarette smoking (Wilbanks et al., 2016) and diesel exhaust (Cruts et al., 2008) on the human brain. The main objective of this study was to investigate the brain's wave pattern changes during and after acute exposure to cooking aerosol. Twelve non-atopic, non-smoking, and healthy adults (18-46 years old) were monitored for brain wave pattern before, end of, and 30 minutes after cooking, using electroencephalography (EEG). Frying ground beef meat in sunflower oil using electric stove was conducted. UFPs, particulate matter, CO2, indoor temperature, indoor RH, oil and meat temperatures were monitored continuously throughout the experiments. The UFP peak concentration was recorded to be approximately 2.0×105 part.cm−3. Brain activity had statistically significant changes during the post-exposure compared to the before exposure suggesting the translocation of UFPs to the brain. This change mainly occurred in the frontal and temporal lobes of the brain. The brain slow-wave band (Delta) decreased while the fast-wave band (Beta2 and Beta3) increased similar to the pattern found in the literature for the exposure to smoking fumes and diesel exhaust. The brain pattern observed in this study showed no similarity to neurodegenerative diseases such as Alzheimer, Parkinson, dementia and schizophrenia which were found to decrease the fast–wave band Beta. The second phase of this study was conducted with 30 volunteers and measurements up to 120 minutes after cooking. The results will be presented during ISES/ISIAQ 2019.

Keywords: air, biomonitoring, built/indoor environment, health, other/general, particles

WE-PO-71

Effect of ambient air pollution on gestational diabetes mellitus and potential biological linkage: A review

Zhang H.; Department of Health Policy and Management, School of Public Health, Sun Yat-sen University, Guangzhou, China

Gestational diabetes mellitus (GDM) is a serious complication of pregnancy, which could have adverse health effects on both mothers and fetuses. Air pollution is a growing global issue. Plausive links has been established between air pollution and GDM. To better understand the association between maternal exposure to air pollution and GDM, we summarized the recent data on how ambient air pollution impact GDM and its potential mechanisms. Pollutants include particulate matter ≤2.5 μm (PM2.5), particulate matter ≤10 μm (PM10), nitrogen dioxide (NO2), nitrous oxides (NOX), black carbon, sulfur dioxide (SO2), carbon monoxide (CO) and ozone (O3). Studies varied refer to assessment of exposure to pollution, definition of GDM, and the variables selected in the model. The results showed that air pollution, especially for PM2.5, has adverse effect on the incidence of GDM in first trimester, second trimester and the entire pregnancy. It indicates β-Cell dysfunction, chronic insulin resistance, neurohormonal dysfunction, inflammation, oxidative stress and gut microbiome may all play important roles in air pollution induced GDM.

Keywords: air, climate change, environmental health, epidemiology, prenatal

WE-PO-72 WITHDRAWN

WE-PO-73

Maternal exposure to air pollutants and childhood autism spectrum disorder in Taiwan

Guo H-R., Wang S-Y., Cheng Y-Y., Tseng Y-CT.

Autism spectrum disorders (ASD) is a group of neurodevelopmental disorders. Its etiology is not well known. To evaluate associations between maternal exposure to air pollutants during pregnancy and childhood ASD, we conducted a study of infants born between 1996 and 2000 identified from a cohort of one million individuals randomly sampled from the National Health Insurance Research Database of Taiwan. After excluding those who had missing data on insurance date and who lived outside the Taiwan Island, we included 62 919 children and followed them till 2013. We obtained data on air pollutants (ozone [O3], carbon monoxide [CO], nitrogen dioxide [NO2], sulfur dioxide, and particles with a diameter less than 10 nm [PM10]) measured by air quality monitoring stations established by the Environmental Protection Administration of Taiwan near the maternal residence during pregnancy. After adjusting for other risk factors, we observed positive associations between ASD and CO during the whole gestation period (hazard ratio [HR]=1.88, 95% confidence interval [CI]: 1.49-2.36, per ppm) as well as in the first (HR=1.93, 95%CI: 1.55-2.39), second (HR=1.77, 95%CI: 1.41-2.22), and third (HR=1.75, 95%CI: 1.39-2.21) trimesters individually. Similarly, positive associations were observed between ASD and NO2 during the whole gestation period (HR=1.42, 95%CI: 1.22-1.66, per 10-ppb) as well as in the first (HR=1.39, 95%CI: 1.22-1.58), second (HR=1.25, 95%CI: 1.10-1.42) and third (HR=1.18, 95%CI: 1.03-1.34) trimesters. Negative associations were observed between ASD and O3 during the whole gestation period (increase HR=0.74, 95%CI: 0.56-0.97, per 10-ppb) and between ASD and PM10 during the whole gestation period (HR=0.90, 95%CI: 0.85-0.96, per 10-μg/m3) as well as in the second (HR=0.94, 95%CI: 0.90-0.99) and third (HR=0.91, 95%CI: 0.87-0.96) trimesters. The results suggested that air pollution during pregnancy may affect the development of the nervous system of the fetus and that ASD may be related to exposures to environmental pollutants.

Keywords: air, children, environmental health, epidemiology, prenatal

WE-PO-74

Research on Human Thermal Adaptive Behavior in the Vernacular Housings: A case study in Sandaoguai Traditional Settlements, Chongqing, China

Zhao Y. [1], *Liu K.* [1], *Lin B.* [1]; [1] *Tsinghua University, Beijing, China*

Vernacular architecture, unlike many modern buildings, mainly relies on various bioclimatic design strategies and passive methods to create and regulate the indoor physical environment rather than mechanical equipment. Therefore, the types and patterns of human thermal adaptive behavior are different and varied, which is worthy of further study. However, most studies that related to thermal adaptive behavior are focused on modern buildings, such as offices, schools and apartments, only few aimed at vernacular architecture. Based on the theoretical framework of thermal adaptation comfort, this paper, taking Sandaoguai traditional settlements （ Chongqing, China） as a case study, conducting a long-term field investigation on the behavioral adjustments in three aspects: personal adjustments, technical/environmental adjustments and cultural adjustments. Firstly, in the aspect of personal adjustments, the average thermal resistance of clothes in this area is counted. Combined with the calculations in other studies in the same area, this paper reveals the characteristics and causes of clothing adjustments. Secondly, in terms of technical/environmental adjustments, more than 10 behavior patterns that related to cooling and heating, as well as the regulation rules of typical modes such as door opening, window opening and air conditioning use are analyzed. In addition, this paper finds there are different patterns of reconstruction behavior, which are unique in vernacular architecture. It shows that the technical/environmental adjustments in Sandaoguai vernacular architecture are not only acted on the use of the building, but also on the partial renovation of the building. Thirdly, in cultural adjustments, the patterns and regularities of work and rest time, spatial use and social activities are discussed. Finally, based on the characteristics and patterns of human thermal adaptive behavior, this study also puts forward suggestions for the sustainable design and renovation of vernacular housings.

Keywords: behavior, built/indoor environment, statistical methods, sustainability, thermal environment

WE-PO-75

A review on light in the built environment and its relationship to mental health outcomes

Kohl N. [1], *Hoisington A.*[1]; [1] *Air Force Institute of Technology, Dayton, OH, USA,* [2] *2Military and Veteran Microbiome: Consortium for Research and Education (MVM-CoRE), Dayton, OH, USA*

With an increased awareness of public mental health issues, more investigation is warranted in the mechanisms of how the built environment may influence an individual's mental health. Specifically, light research has shown progress in reducing negative mental health outcomes and lessons learned in that research may be informative in the built environment. For example, bright light therapy, bright light in the morning, low light in the evening, and exposure to natural light over artificial light positively influences mental health outcomes. Bright light in the morning and low levels of light in the evening reduces depression symptoms within a week. In contrast, bright light in the evening has an adverse effect on depression through modulation of individual's melatonin levels that alters the natural circadian rhythm affecting mood, sleep, and mental agility. The excretion of melatonin in individuals can be suppressed by light when its intensity exceeds approximately 2500 lux. In addition to the intensity, individuals are affected psychologically by the different spectrums of light. Artificial lighting does not usually contain the full spectrum lighting required for positive mental health outcomes, although full spectrum fluorescent lights contain the necessary color spectrum to cause positive mental health outcomes. Research suggest that natural light and full spectrum fluorescent lights have roughly the same color spectrum and both produce positive mental health outcomes. Due to the rise in urbanization and time in the urban built environment, architects and engineers may need to consider optimizing the built environment to include proper light exposure influencing positive mental health outcomes. Prior to design and modifications, it is recommended that can controlled studies to target specific light parameters and their relation to mental health outcomes be conducted.

Keywords: behavior, built/indoor environment, health, other/general, workplace

WE-PO-76

Characterization of environmental multi-exposure to pesticides in pregnant women

Regnaut L.[1], Mercier F.[1], Le Bot B.[1], Chevrier C.[1], Bonvallot N.[1]; [1]Univ Rennes 1, EHESP, INSERM, Irset, Rennes, France

Pesticides are largely used to control pests and diseases in crops. Biological measurement is a mean of assessing environmental exposures by an integrated approach. In this context, characterization of pesticide exposure still represents a challenge since amounts of biological specimens available are low, and pesticides are generally detected as metabolites whose chemical structures may be unknown. The objective was to develop a multi-residue targeted approach to measure different insecticides, fungicides and herbicides and their metabolites in urine from pregnant women, and to assess the determinants of exposure to these pesticides. Methods: 338 women from the PELAGIE study (mother-child cohort, Brittany) were selected according to the availability of a urinary sample collected at the early pregnancy. 296 urine samples were analyzed by UHPLC/QTOFMS after a sample preparation step: enzymatic hydrolyze and solid phase extraction. 42 molecules corresponding to 23 pesticides and 6 groups of pesticides belonging to different chemical families and nicotine and cotinine were quantified. Bayesian statistical analyses were used to study the determinants of exposure to these pesticides. Results: 4 groups of pesticides (pyrethroids, ethyl- and methyl-organophosphorous and phenoxypropionic acid derivatives) and 2 pesticides (chlorpyrifos and fluazifop) were detected in > 60 % of samples; 4 pesticides (prochloraz, bromoxynil, diazinon and procymidone) were detected between 10 and 50%. Urinary concentration levels range between < LOD until 600 µg/L for methyl-organophosphorous. The determinants of exposure were: location in rural zone (bromoxynil), proximity with vegetable crops and presence of cereal crops in the town of residence (chlorpyrifos), proximity with wheat crops (fluazifop); certain types of diet (organophosphorous); smoking (diazinon); and surprisingly, the absence of peas and potatoes crops in the town of residence (pyrethroids). This study allowed to characterize many pesticide metabolites that are not routinely measured in environmental health.

Keywords: SVOCs, biomarkers, biomonitoring, pesticides

WE-PO-77

Association between Risk of Metabolic Syndrome and the Level of the AGE in Security guards- a cross-sectional study

Chen B-R., Lai C-H.

Introduction: The working characteristics of security guards include shift work and work long hours, stay in monotonous work and sedentary for a long time, which can easily lead to lower physical activity, risk factors for cardiovascular diseases, little knowledge exists about the association between risk of Metabolic Syndrome (Mets) and the Level of the advanced glycation end products (AGEs) in security guards. Many previous studies show the contribution of AGEs in the origin and progression of certain chronic diseases, as well as Mets. Mets is a high-prevalent metabolic disorder characterized by hypertension, central (abdominal) obesity, hypertriglyceridemia, low HDL-cholesterol, and/or increased fasting glucose. Objectives: The objective the present study was to examine the association between skin autofluorescence (SAF) AGEs and the prevalent of Mets in security guards.

Study population was composed of 144 security guards and 66 administration workers. The components of metabolic syndrome were measured by physical examination, and the level of AGE were evaluated by SAF. Data analyses were performed in 2018 using SPSS software version 22. Continuous variables were measured by t-test and categorical variables were measured by chi-square test. The odds ratios(OR) of risk component of metabolic syndrome were measured by logistic regression. Results: There was a substantial difference in the prevalence of Mets between security guards and administration staff. Notably, 45.1% of the security guards have metabolic syndrome while 16.7% of the administrative staff have metabolic syndrome. Additionally, in the mean of level of AGE, security guards had higher mean skin autofluorescence than the administrative staff (2.7±1.0 AU versus 2.2±0.7 AU). Conclusion: The security guards have higher risk of metabolic syndrome and the higher level of AGEs than the administration staff. There are a positive association between skin autofluorescence (SAF) AGEs and the prevalent of Mets.

Keywords: environmental health

WE-PO-78

Review of environmental and occupational cohort studies related to the up- and mid-stream oil and gas industry

Galea K., Steinle S.[1], Basinas I.[1], Sleeuwenhoek A.[1], Loh M.[1], Pronk A.[2], Stierum R.[2], Fransman W.[2], Kuijpers E.[2], Makris K.[3]; [1]IOM, Edinburgh, United Kingdom, [2]TNO, Zeist, The Netherlands, [3]Cyprus University of Technology, Limassol, Cyprus

The documented global burden of disease and injury for the oil & gas sector sector is likely to be appreciable, due to a number of important occupational risk factors including carcinogens, asthmagens, gas and particulate matter, etc.). The EXPOSOGAS project (www.exposogas.eu) will use the integrated approach of the human exposome concept to assess, manage and communicate information about the overall health risk to workers and the general population from exposures encountered throughout the oil & gas industry lifecycle. A systematic literature review of worldwide studies related to the health impacts to workers and the general population from up-and mid-stream oil & gas industries is currently ongoing. Search terms and strings were developed, with relevent titles being identified in ProQuest Dialog, PubMed and Web of Science databases. Two reviewers will independently assess the titles and abstracts for their compliance with the devised inclusion criteria, with the full manuscript then being reviewed. Data from the identified articles will be extracted into a template developed with due consideration of the STROBE checklist, with the quality of each study being assessed in accordance to the Newcastle Ottawa Scale criteria. The systematic review is in progress - of the 768 articles identified in ProQuest, 60 titles were judged suitable for inclusion, which was reduced to 24 articles upon review of the abstract. Of the 637 articles identified in PubMed, 157 titles were considered suitable for inclusion, which was reduced to 115 upon review of the abstracts. The presentation will describe the findings of the review, with these results helping to inform the prioritization of future EXPOSGAS project activities.

Keywords: epidemiology

WE-PO-79

The plants affected by air pollution have impact on human health

Budreviciute A.[1], Plakys G.[2], Katileviciute A.[2], Kodzius R.[2]; [1]Lithuanian University of Health Sciences, Kaunas, Lithuania, [2]Kaunas University of Technology, Kaunas, Lithuania

The outdoors air pollution affects both humans and other life forms such as plants. The grass, especially it's kind, the food crops are of enormous importance to our human lives, as we daily consume the crops (such as wheat) as food. The burning fossil fuel generates sulfur dioxide and sulfuric acid, which in its turn affects leaves and plant tissues. The same effect is observed for the ozone influence. The car combustion engines produce nitrogen oxides, slowing growth of the plants. However, generated nitric acid is a source of nitrogen, which is essential for plant growth. The carbon monoxide from combustion engines halts the growth of the plants as well. The secondary pollution effect on plants is the diminished immune system, causing easier infestation by various viral, bacterial or fungi infections, including the insect infestation. The small particles generated by anthropogenic or non-anthropogenic sources are especially damaging to the plants. The plant growth is challenged, and the small particles are absorbed by leaves. In this way, the pollutants are removed from the air. Some of the contaminants, such as sulfur dioxide, can be metabolized by the plants. Other pollutants, however, remain in the plants and can enter the food chain (i.e., wheat). Processing the food crops, theoretically, it is possible to mitigate or completely remove the pollutants by modifying the industrial processes. Human health is of enormous importance, and the pollutions ways are affecting us through the plants should be investigated and followed.

Keywords: biomonitoring, ecological exposure, food, health, other/general, particles, sustainability

WE-PO-80

Exposure to household air pollution associated with adverse immune responses in rural women in northwest China

Yao, Y.[1], Wang D.[1], Ma H.[1], Li C.[1], Wang, J.[1], Chang X.[1], Low P.[2], Liu S.[3]; [1]School of Public Health, Lanzhou University, Lanzhou, Gansu Province, China, [2]School of Public Health, University of California, Berkeley, Berkeley, CA, USA, [3]School of Health Sciences, Purdue University, West Lafayette, IN, USA

Background: The associations between impairments of immune functions and exposure to polycyclic aromatic hydrocarbons (PAHs) in air pollution remain inconclusive due to the limitations of previous studies. In this study, we aimed to examine the direction and magnitude of immune changes related to PAH exposure from household air pollution among rural women living in Gansu, China. Methods: Healthy village women (n = 34) were recruited and enrolled in the study. Blood and urine samples were collected and analyzed during non-heating and heating seasons. Urinary 1-hydroxypyrene (1-OHP) was quantified as the biomarker of PAH exposure. Immunoglobulin E (IgE), percent of T-regulatory (Treg) cells, and gene expression of forkhead box transcription factor 3 (Foxp3), transforming growth factor-β (TGF-β), interleukin 10 (IL-10), and interleukin 35 (IL-35) were analyzed to evaluate immune responses. Urinary 8-hydroxy-2-deoxyguanosine (8-OHdG) was measured as a marker of oxidative DNA damage. Results: the concentration of 1-OHP increased almost 20 times from 0.90 to 17.4 μmol mol-Cr -1 from summer to winter ($p < 0.001$). Meanwhile, average percent of Treg cells decreased from 5.01% to 1.15% ($p < 0.001$); IgE and mRNA expressions of Foxp3, TGF-β, IL-10, IL-12α and EBi3 all significantly decreased ($p < 0.001$); Urinary 8-OHdG increased from 12.7 to 30.3 ng mg-Cr -1 ($p < 0.001$). The changes in 8-OHdG, Foxp3 and TGF-β were significantly associated with the increase of 1-OHP. Conclusions: We observed a substantial increase of PAH exposure in winter, which was significantly associated with the repression on Treg cell function and oxidative DNA damage. Exposure to PAHs in household air pollution possibly induced immune impairments among rural women in northwest China.

Keywords: epidemiology

WE-PO-81

The connection between indoor air quality and mental health outcomes

Taylor W.[1], Hoisington A.[1]; [1]Air Force Institute of Technology, Wright Patterson AFB, Dayron, OH, USA

In the United States, nearly one in five adults live with a mental illness, and 44,000 individuals committed suicide in 2016. Neuropsychiatric disorders trail only cancer and cardiovascular diseases as the highest cause of disability adjusted life years in Europe. Such profound health outcomes drive a need for greater understanding of mechanisms that might influence mental health, including possibly the built environment. To better understand the connection between mental health and the built environment, we administered a self-reported survey to United States Veterans. The survey included aspects of the built environment such as: sources of indoor particulate matter, residence proximity to a busy roadway, and recent occurrence of water damage. Particulate matter is generated by common household activities like smoking, cooking, and burning candles or incense. Studies of outdoor particulates have shown that individuals have an increased risk of suicide when exposed to increased levels of particulate matter. This opens the possibility that indoor generated particulate matter could have an affect on mental health outcomes as well. Additionally, the location of a residence can strongly influence particulate matter concentrations within the home, with homes within 100 meters of a busy roadway having increased concentrations. Finally, water damage in a residence can also have harmful mental health repercussions. Housing quality has been shown in literature to be inversely related to psychological distress, suggesting that individuals unable to afford the repair of water damage within their homes will suffer from increased psychological distress. The literature surrounding the physical health effects of indoor air quality is thorough and well studied, yet the connection to mental health is lacking. Future studies of indoor air quality should seek to identify factors that have a direct impact on mental health.

Keywords: built/indoor environment, health, chronic diseases, health, other/general, ventilation

WE-PO-82
Perfluoroalkyl Substances and Metabolic Syndrome in Elderly

Lin C-C.[1], Tsai M.S.[1], Chen M.H.[1], Chen P.C.[1,2,3]; [1]Institute Of Occupational Medicine And Industrial Hygiene, National Taiwan University College Of Public Health, Taipei, [2]Department of Occupational Safety and Health Office, National Taiwan University Hospital [3]Taiwan Public Health Association

BACKGROUND/AimS Recent epidemiology studies found the perfluoroalkyl substances exposure were associated with chronic diseases. The Keya River which located in the Hsinchu, Taiwan and its aquatic products have been discovered in high concentration of perfluoroalkyl substances (PFASs). Our study investigated the association between contaminants of emerging concern (CECs) exposure and metabolic syndrome in the elderly. METHODS We recruited 397 elderly residents over 60 years old in Hsinchu, Taiwan and measured the serum concentrations of PFASs. The components of the metabolic syndrome included waistline, hypertension, fasting blood glucose, hypercholesterolemia, hypertriglyceridemia, mixed hyperlipidemia, and dyslipidemia. RESULTS We found the serum levels of PFASs including perfluorooctane sulfonate (PFOS, C8) and perfluorooctanoic acid (PFOA, C8) were positively associated with dyslipidemia or low density lipoprotein (LDL). PFOS had significantly increased LDL levels (mg/dL) at the quartile 2 [β (95% CI) = 16.08 (7.17, 25)], the quartile 3 [β (95% CI) = 13.24 (4.23, 22.25)], and the quartile 4 [β (95% CI) = 14.06 (4.97, 23.15)] with p for trend 0.0067. PFOA also had similar findings at the quartile 3 [β (95% CI) = 9.14 (0.12, 18.16)] and quartile 4 [β (95% CI) = 9.34 (0.29, 18.39)] with p for trend 0.0221. The male elders had stronger association with the PFAs exposure than female elders. CONCLUSIONS Exposure to the PFASs had the impact of metabolic syndrome, especially in the male elder. The health risks of those exposures need further exploration.

Keywords: perfluoroalkyl substance, Metabolic Syndrome, Elderly

WE-PO-83
Polycyclic aromatic hydrocarbons (PAHs) exposure and birth outcomes

Agarwal P.[1], Taneja A.[1]; [1]Department of Chemistry, Dr. B. R. Ambedkar University, Khandari Campus, India

Polycyclic aromatic hydrocarbons, by disturbing redox status or by altering the endocrine activities are reported to cause poor pregnancy outcomes. Hence, present study was investigated to estimate possible association between placental PAHs levels and birth size among North Indian pregnant women. A total of 84 healthy non-smoking mothers participated in the study, whose placental tissues were analyzed for the presence of 16 US-EPA listed PAHs with the help of GC-MS. Detailed summary of socio-demographic characteristics of studied population were also administered during the study. Multivariate regression models were performed with PAHs levels as independent variables and birth size (birth weight, length, head circumference, ponderal index and cephalization index) as dependent variables. Results demonstrated the abundance of benzo(a)anthracene (13.71%) and naphthalene (12.76%) in studied population. For birth weight, there was a significant decrease of 0.34 and 0.57 units with unit increase of fluoranthene and benzo(a)pyrene respectively. A marginal significant decrease of 0.29 unit in birth length and enhancement of 0.54 units in cephalization index was found for unit increase in benzo(a)pyrene. These results reveal that in-utero exposure to some PAHs could impair the anthropometric development of the fetus, reducing birth weight and length. As neonatal anthropometric measures are considered as an important tool to identify newborns at higher risks of morbidity and impaired fetal growth. Thus, it is an urgent need of hour to lessen exposure of harmful contaminants to child bearing population and to evaluate maternal PAHs exposure in relation to long term impact on infant health.

Keywords: Polycyclic aromatic hydrocarbons, birth weight, placenta

WE-PO-84
Database of Residential Building Characteristics, Occupant Behaviors, and Reported Symptomology

Grimes C.; Hayward Healthy Home Institute, Monterey, CA, USA

A database of over 37,000 sets of residential building characteristics, occupant behaviors, and reported symptomology has been initially analyzed. Although specific symptoms are not being associated with specific characteristics, of the 80 factors included in the model, the housing and behavioral characteristics are providing a reasonable level of confidence for predicting the number of reported symptoms. Ten major and a variety of minor effects are significant when predicting number of symptoms. The R-Squared ~50% is sufficient for further investigation. Additional analysis included a lack of support for several housing conditions expected from field experience. However, the results varied according to the method of analysis. This potentially indicates that the differences between the traditional study design for building characteristics, for occupant behaviors, and for reporting of symptomology may not be sufficiently consistent between them. Requiring an adjustment to analysis and interpretation.

Keywords: Residential Buildings, Characteristics, Behaviors, Symptoms

WE-PO-85 WITHDRAWN

WE-PO-86
Assessing the impact of (non-) built environment on mental health in Brussels

Pelgrims I.[1], Keune H.[2], Thoms I.[3], Bastiaenes H.[4], Remmen R.[4], Nawrot T.[5], De Clerco EM.[1]; [1]Sciensano, Brussels, Belgium, [2]Nature and society, Own-Capital Research Institute for Nature and Forest, Brussels, Belgium, [3]Center for Operations Research and Econometrics université catholique de Louvain (UCL), Belgium, [4]Primary and Interdisciplinary Care, University of Antwerp (UA), Antwerp, Belgium, [5]Center for Environmental Sciences University of Hasselt (UH), Hasselt, Belgium

Title : Assessing the impact of (non-) built environment on mental health in Brussels Introduction Mental illnesses are a growing problem in modern societies. While the impact of demographic or socioeconomic factors on these pathologies is acknowledged, the interaction with urbanized environment is little understood. This quantitative research is part of the interdisciplinary study "NAMED" which aim to investigate the impact of the (non-)built environment on mental health in Belgium, one of the most urbanized countries in Europe. Methods Research focuses on the inhabitants of Brussels and older than 15 years. The epidemiological study was carried out based on the coupling between data from the national health surveys (2009 and 2013) and specifically developed indicators describing each participant's surroundings in terms of (non-)built environment, air quality and noise. These data are based on the X,Y coordinates of the participant's residence and processed using Geographical Information Systems (GIS). Participant's mental health was approached through different indicators from validated questionnaires (GHQ, SF36 vitality scale, etc.) and isolated variables (reported depression, suicidal ideation). Multiple logistic regression models were run for each environmental indicator and mental health outcome, taking into account socio-economic factors. Results Adjusted measures of association between mental health and environmental indicators will be presented at three levels : the residential level (view of green, garden coverage, building characteristics), the street level (vegetation coverage, tree density, canyon/corridor effect) and the neighborhood level (green coverage with a buffer of 600 and 1km, typology of urban intensity). Same results will be provided for air quality/noise and subjective measures of the environment. Discussion The results permit to better understand the impact of urban environment on mental health and to formulate concrete, evidence-based policy regarding public health, urban planning and green space management.

Keywords: VOCs

WE-PO-87

Duty of care of Employers - Air quality control inside a work space

Green J.; JN Bank Limited, Kingston, Jamaica

International health and safety standards have established that employers have a duty of care to their employees; a duty not to expose employees to occupational hazards at the workplace. In adequate indoor air quality in Call Centres is one health hazard that can negatively impact the comfort and productivity of the staff in the workplace. As such, health and safety legislations in developed countries have made provisions for air quality control inside the work space. This legislation protects Call Centre Workers who are particularly at risk to oxygen deficiency, thermal discomfort, unnatural ventilation and artificial lighting. These risks exists because Call Centres are generally small and overcrowded; with staff being subjected to odours from various different perfumes and food scents to name a few sources.

Keywords: workplace

WE-PO-88

BSim simulation studies on indoor air quality of the sports hall – a case study

Siuta-Olcha A. [1], Dudzinska M.[1], Polednik B.[1], Staszowska A.[1], Cholewa T. [1]; [1]Lublin University of Technology Faculty of Environmental Engineering, Lublin, Poland

In recent years simultaneous assurance of the required indoor air quality, high building's thermal performance and reduction of energy consumption for heating and cooling has been an important problem for architects and engineers. It is important when it comes to premises occupied by children, when the number of occupants vary. Therefore, there is a need for thermal modeling and dynamic energy simulation in both designed and modernized buildings in order to optimize indoor thermal conditions. The simulation research of indoor air quality with the use of the BSim program were carried out for the school sports hall. The analyzed room has a volume of 4 413.1 m3. The sports hall is heated by means of heaters with a total power of 32 kW and mechanically ventilated with a heat recovery efficiency at the level of 80%. Air exchange rate was assumed at the level of 1 h-1. It was found that in many cases the indoor temperature exceeded 30C. This was due to solar heat gains through windows and internal heat gains from people. The indoor air relative humidity in the winter months was lower than the minimum permissible level of 30%. The CO2 concentrations were repeatedly exceeded (maximum values in the range 630 - 1495 ppm). In order to improve the indoor air quality, a cooling device with a total power of 50 kW was introduced to the thermal model of this room. The influence of solar shading devices, building orientation and air exchange rates was determined on the quality of the indoor environment, as well as the energy balance of the sports hall. It was established that external movable shading provisions are most effective among the devices analyzed and allowed to reduce heat gains by 48% and to decrease the total energy demand by 27% yearly.

Keywords: thermal environment

WE-PO-89

Evaluation of PM2.5 Exposures for an Environmental Justice .1Community Using a Low-Cost PM Sensor

Cho S-H.[1], Cicutto L.[2], Siripanichgon N.[1], McCombs M.[1], McCullough M.[2], Crews K.[2], Chartier R.[1], Thornburg J.[1], Chang C.[3], Harshfield G.4, Rink B.4, Thomas G.5, Ogletree M; [1]RTI International, Research Triangle Park, NC, USA, [2]National Jewish Health, Denver, CO, USA, [3]Groundwork Denver, Denver, CO, USA, 4Colorado Department of Public Health and Environment, Denver, CO, USA, 5City of Denver, CO, USA

As low-cost air sensors become readily available, communities have opportunities to collect air quality and exposure data to improve their knowledge and contribute to environmental health research and policy decisions. Citizen scientists in Globeville, Elyria Swansea (GES), an environmental justice community in Denver, with help of researchers, conducted air quality and exposure measurements for PM2.5 in the Summer and Fall of 2018. RTI MicroPEM sensors were placed at 12 sites within the community as a neighborhood-scale sensor network, including near-road sites (NRS) collocated with reference monitors. Nineteen citizens measured their exposures wearing MicroPEM and completed time-activity diaries for two 72-hour monitoring periods. MicroPEMs data were highly accurate (>90%) after adjustment using filter-based gravimetric data and correcting temperature-associated bias. Ambient PM2.5 data showed variability with up to 3 times difference between site pairs. We observed up to 2.5 times higher PM2.5 levels in Summer. The temporal measurements among the most sites were highly correlated with NRS data ($r > 0.7$), suggesting a large influence of traffic on the community air quality. However, few sites such as parks showed a significant difference from NRS during weekends. Individual's personal exposure levels varied by day, influenced by activity types. Real-time peak concentrations matched with participant-reported exposure events such as proximity to combustion sources and cleaning. On average, personal PM2.5 exposures were higher than the nearby ambient sensor measurement and lower than the NRS. However, for the individuals with high exposure events, the personal levels were 6-24 times higher than nearby ambient sensor levels. This study shows the value of low-cost sensors in assessing the spatial and temporal variabilities of air quality and the degrees of exposure measurement error, which is critical for accurate epidemiologic models. Additionally, the sensors are useful tools for behavioral intervention to reduce air pollution exposure. EPA Grant Number: RD83618701

Keywords: thermal environment

THURSDAY SESSIONS

Thursday, August 22nd

TH-PL-A1-EHR: Prioritization and Assessment Frameworks/Models for Chemicals

TH-PL-A1-01

Framework on exposure-informed adaptation of REACH hazard information requirements

Hahn S.[1], Dudzina.T.[2], Escher S.[1], Bitsch A.[1]; [1]Fraunhofer ITEM, Hannover, Germany, [2]Exxon Mobil, Machelen, Belgium

The European regulation on Registration, Evaluation and Authorization of Chemicals (REACH) requires generation of information on the intrinsic properties (including hazard) of registered substances depending on their tonnage. In situations where human or environmental exposure is absent or not significant, the so called exposure-based adaptation (EBA) may be considered. EBA comprises all types of modifications of the standard information requirements, of which the exposure based waiving (EBW) is the only option that allows to waive higher tier mammalian toxicity studies. All other adaptations require some form of higher tier experimental data on the substance itself or an analogue. With that, the EBW is the least successfully utilized option (based on analysis of compliance check decisions). In order to justify for a certain endpoint the omission of the standard information requirement, a high level of confidence is needed to demonstrate no or no significant exposure. In practice, REACH registrants often lack explicit technical guidance precluding them from consistently and transparently documenting the evidence of low exposures. As part of the PetCo initiative/discussions, Concawe and Fraunhofer ITEM developed an advanced version of the workflow given in ECHA R.5 EBA guidance. The project started with review of available material related to EBA, followed by developing a set of toxicological and non-toxicological cut off criteria to provide the basis for robust science-based EBA justifications and identify the most relevant exposure route(s) for new testing proposals. The framework discusses various approaches that could be applied to justify a weight of evidence on low exposure and risk in the absence of sufficient hazard characterization data.

Keywords: consumer and personal care products, other (specify), workplace

TH-PL-A1-02

US EPA's Science-Based Approach to Understanding and Managing Environmental Risk from PFAS

Gillespie A.[1]; [1]US Environmental Protection Agency, Research Triangle Park, NC, USA

Per- and polyfluoroalkyl substances (PFASs) are a growing concern among many communities in the US due to increasing reports of potential human exposures. The science required to protect public health and the environment from PFAS exposure cuts across many applications and disciplines. The risk assessment/risk management paradigm provides a useful means to assess the state of the science available for informing decisions, and to identify gaps in knowledge needed to support risk management. US EPA's initial scoping of information available for assessing and managing PFAS risks revealed deficiencies in all key areas of the risk paradigm including hazard and toxicity (there are likely hundreds of PFAS in the US environment, and most lack sufficient suitable toxicity information for informing understanding of the potential for human or ecological effects); exposure (information about different PFAS sources, fate and transport, and human and ecological exposure is sparse, both spatially and temporally); and suitable treatment and remediation approaches (there is little information on effective methods and costs for treating or removing different PFAS from drinking water, groundwater, wastewater, air, soils, and sediments). US EPA is conducting an integrated set of research activities aimed at filling gaps in our current ability to conduct scientifically rigorous risk assessment and risk management activities. This research program is designed to address these data gaps and enable stakeholders to begin making effective decisions for identifying and mitigating risk from PFAS in the environment. Exposure science plays a critical role in the US EPA's research program because the complexity of PFAS chemistry and use leads to a multitude of possible exposure pathways and both human and ecological receptors. Understanding exposure is key to identifying the key risk management actions needed to achieve US EPA's mission of protecting human health and the environment.

Keywords: chemical prioritization, ecological exposure, risk assessment, water

TH-PL-A1-03
Hazard assessment of chemical substances in indoor environments: what are the high-priority compounds for French monitoring campaigns?

Achille, J.[1], Ramalho, O.[1], Boulanger, G.[1], Perouel G.[1], Garnier R.[1], Mandin C.[1]; [1]Scientific and Technical Center for Building (CSTB), Paris, France

Identifying potentially hazardous substances in indoor air is essential to ensure relevancy of the measured parameters in monitoring campaigns. To prepare the future nationwide survey of the French Indoor Air Quality Observatory in dwellings, an inventory of substances possibly found in indoor environments was performed and a ranking method was created, based on long-term health criteria. A literature review was conducted, focused mainly on European and northern American countries, to identify substances possibly found in indoor environments based on i) the compositions of commercial mixtures, articles, products or materials used indoors, ii) their emissions from materials and products or iii) their concentration measured in indoor air or settled dust. A scoring method was then developed and applied to rank the substances according to their potential health hazard. Human hazard was primarily based on the EU classification and labelling regulation: carcinogenicity, mutagenicity, reproductive toxicity, specific target organ toxicity – after repeat exposure, skin and respiratory sensitizing properties. When possible, other information sources were consulted such as carcinogenic classifications established by IARC or US EPA. Furthermore, endocrine-disrupting properties were investigated for each substance using six distinct classification lists. Finally, substances were ranked depending on their assigned scores, on a range from 0 to 28. A threshold of 10 was set, above which the substances are considered as high-priority substances. A list of 2,741 substances likely to be present in indoor environments was established. The ranking method lead to the identification of 206 high-priority substances. This study allowed the identification and prioritization of substances with potential human health concerns regarding exposure in indoor environments. This method is currently used to select substances to be measured in indoor air and/or settled dust in the next housing survey.

Keywords: SVOCs, VOCs, chemical prioritization, health, chronic diseases

TH-PL-A1-04
A personalised real-time air quality informatics system for exposure and health risk communication

Che W.[1], Lau A.[1]; [1]The Hong Kong University of Science and Technology, Hong Kong, [1]The Hong Kong University of Science and Technology

Air pollution is the single largest environmental health risk, contributing to 7 million pre-mature deaths every year. Prevention of diseases caused by air pollution requires enhanced knowledge on exposure to air pollution and its sources, which are subject to high spatial and temporal variability. Advances in sensor and mobile communications technologies are generating large amount of air quality monitoring and human activity data in wide scale with relative low cost. Meanwhile, the development of air quality modelling enables the capability of tracking air pollution from sources to receptor with high temporal and spatial resolution down to street level. The big data generated from low-cost sensor and high-resolution models represents ubiquitous opportunities to address individual and integrated exposure to air pollutants and related health risks. We present an example of integrating such technologies to provide Personalised Real-Time Air Quality Informatics System for Exposure – Hong Kong (PRAISE-HK, http://praise.ust.hk/index.php/team/). The PRAISE-HK was built based on information of real-time emission sources and urban morphology in fine resolution, the start-of-art knowledge of physical and chemical process in meteorology and pollution transportation, as well as big data integration of sensor monitoring, location tracking and crowd sourcing to effectively analyze and forecast the air quality, personal exposure and health response in wide-scale with high resolution. The first phase of the system has been accomplished and released as a mobile application with the capability to provide real-time air quality simulation down to street level and forecast up to 48 hours. The second and third phase of the system will enable the prediction of personal exposure and health response, respectively, which will be released in 2019 and 2021. The PRAISE-HK demonstrated an example of collective intelligence to enhance evidence for policymakers and to build capacity to address air pollution, exposure and related health impacts.

Keywords: air, community, cumulative exposure, environmental health, risk assessment

TH-PL-A1-05
High Throughput Comparative Exposure and Risks of Chemicals in Multiple Consumer Products

Jolliet O.[1], Meyer C.[1], Huang L.[1], Fantke P.[1]; [1]Environmental Health Science, School of Public Health, University of Michigan, Ann Arbor, MI, USA

Recently, increasing attention has been given to human exposure to chemicals in consumer products. Yet, Alternatives Assessment (AA), Risk assessment (RA) and Life Cycle Assessment (LCA) methods lack flexible approaches to quantify exposure and hazard for the thousand product-chemical combinations currently in use. To address this gap, we used the operational matrix-based high-throughput framework coupling multi-pathway consumer with far-field general population exposures and with hazard data. We applied it to multiple product categories as follows: a) Chemical composition was obtained from the Pharos database for building materials, from published suspect screening analysis and composition analyses for toys, and from the Stochastic Human Exposure and Dose Simulation Model (SHEDS-HT) for a series of personal care, houshold maintenance and cleaning products. b) Using seven models (skin surface layer, object surface layer, article interior, indoor air, food contact material, pesticide residues, direct emission), we calculate Product Intake Fractions (PiF) for various receptors, and c) we compare total user intake dose with observed or in silico NOAEL data to quantitatively characterize risks via vizualization of margins of exposures. Chemicals in toys identified via the suspect screening analysis had exposures from 10-8 to 0.1 mg/kgBW/d and margin of exposure greater than 100, deeming them relatively safe. In contrast, chemicals in toys with known weight fractions had much lower margins of exposure. For building products, several of the chemical-product combination yielded margins of exposure lower than 100, with doses between 0.001 and 1000 mg/kgBW/d for the first 50 days, with approximately 74% having margins of exposure under 100. For household products, intakes range from 10-8 to 400 mg/kgBW/d, with lowest margins of exposure associated with personal care products, cleaning products, home maintenance and other home products. The developed approach proves a powerful tool able to analyze thousands of product-chemical combinations and prioritize those requiring special scrutiny.

Keywords: building materials, chemical prioritization, consumer and personal care products, exposure models, risk assessment

TH-PL-A1-06

The challenge of assessing exposure to fast-elimination endocrine disruptors - statistical approaches to optimize biomonitoring strategies

Faÿs F.[1], Palazzi P.[1], Zeimet E.[1], Hardy E.[1], Vaillant M.[1], Appenzeller B.[1]; [1]Luxembourg Institute of Health (LIH), Strassen, Luxembourg

Introduction: In epidemiological studies linking exposure and health outcomes, exposure is usually assessed by the concentration of biomarkers in a single biological sample, used as a surrogate of the internal dose. Many endocrine disruptors such as phthalates, are however rapidly eliminated in urine, resulting in a highly variable concentration questioning the reliability of exposure assessment. Although increasing the samples number may provide more reliable information, no rational is available yet to determine the optimal strategy. Methods: In this study, 1-3 urine samples per week were collected from 16 volunteers over 6 months and analyzed for 29 endocrine disruptors including phthalates, pesticides and bisphenols. Two different algorithms using bootstrap-like simulations were used to determine 1) the optimal number of urinary samples needed to properly classify individuals, and 2) the number of samples required to reach a stable mean concentration. Results: Most biomarkers presented high detection rates, with 21 metabolites detected in at least 80% of samples. The first algorithm demonstrated that the maximum of available samples allowed up to 46% of correct ranking. Log-transformation of the data improved the rate up to 65%. The second algorithm demonstrated that between 15 and 20 samples were needed to reach a stable mean. Conclusion: The high detection rate of the urinary biomarkers highlights the frequency of human exposure to endocrine disruptors and their ubiquity in human surroundings. The statistical analyses applied here demonstrate the complexity to classify individuals according to their level of exposure to fast elimination endocrine disruptors. Our findings also highlight the limitation of the classical approaches where assessment is based on a single sample. The results strongly incite to take into account the variability of urinary biomarkers concentration in the design of epidemiological studies or to consider other biological matrices (hair) more adapted to assess exposure to fast elimination pollutants .

Keywords: biomarkers, biomonitoring, environmental health, pesticides, phthalates

TH-PL-B1-SSS: Strategies for Minimizing Exposure to Indoor Pollutants

TH-PL-B1-07

New indoor air quality standards; ISO16000-34: Strategies for the measurement of airborne particles and ISO16000-37: Measurement of PM2,5 mass concentration

Moisio M.[1], Lambaerts P.[1]; [1]Tempere University of Technology, Tempere, Finland

Particulate matter air quality is estimated to be one of the most significant sources of adverse health effects and even very significant direct cause of death. The Global Burden of Disease (GBD) study estimated that 2.9 million deaths in 2015 were associated with household air pollution, whereas WHO estimated 4.3 million related deaths in 2012 (Lancet Commission of Pollution and Health). Indoor air particle concentrations are a major factor for determining particle exposure and possible health risks. The concentration, composition, and size distribution of airborne particulate matter in indoor environments can be challenging to measure reliably. Measurement strongly depend on parameters such as the room size, relative humidity, air exchange rate, air flow conditions, and sink effects on different surfaces (e.g., walls, ceilings, floor coverings, furnishings). Depending on conditions this can result in highly variable levels of indoor PM pollution that are not easily determined or measured in terms of their impact on health. Increased need for reliable and comparable measurements has brought up a requirement for the standardisation of measurement of fine and ultrafine (PM10, PM2.5 and UFP) particles and their real-time behaviour in indoor air. This work has been carried in ISO Technical Committee ISO/TC 146, Air quality,

Subcommittee SC 6, Indoor air. The newly published ISO 16000-34 and 37 standards include for the first time a comprehensive guidance on indoor air quality airborne particle measurement instrumentation and methods. ISO16000-37 presents a EN12341 based method for measuring indoor PM 2.5 together with supplementary real-time measurement. This presentation will review the new ISO 16000-34 and 37 standards, evaluate their meaning for the indoor air exposure studies and give insights and instrument solutions providing both PM2.5 and real-time data.

Keywords: air, built/indoor environment, environmental policy, environmental regulation, nanoparticles, particles, respiratory health, sampling methods

TH-PL-B1-08
Publication of the U.S. EPA's Guidelines for Human Exposure Assessment

Tulve N.[1], Olsen M.[1], Broder M.[1]; [1]United States Environmental Protection Agency, Office of Research and Development, National Exposure Research Laboratory, Research Traingle Park, NC, USA

The mission of the U.S. EPA is to protect human health and the environment by understanding, characterizing, and reducing health risks associated with exposure to environmental contaminants and other agents. Scientifically sound exposure assessments are critical to the risk assessment process and inform the Agency's regulatory decision-making. Exposure science characterizes, estimates, and predicts exposures and provides information for preparing exposure assessments and for developing effective strategies to reduce exposure and risk. When conducting a risk assessment, the assessor needs to understand whether an agent may cause an adverse health effect and how exposure to that agent may be reduced. Advances in the field of exposure science require updated resources for conducting exposure assessments as part of the risk assessment paradigm. The Guidelines for Human Exposure Assessment (https://www.epa.gov/risk/guidelines-human-exposure-assessment) provides an updated resource for exposure assessors, risk managers and decision makers both within and outside the Agency. This document builds on and supersedes the 1992 Guidelines (https://cfpub.epa.gov/ncea/risk/recordisplay.cfm?deid=15263), incorporating advances that have occurred since the Guidelines were originally published. This updated document describes the principles of exposure assessment, provides guidance on approaches to conduct an exposure assessment, presents references for more detailed information, and supplies hyperlinks to exposure assessment tools and technical documents. The Guidelines are arranged into chapters. Each chapter explores a component of the exposure assessment process, including: basic concepts and principles in exposure science; planning and scoping; incorporating lifestages, vulnerable groups, and populations of concern into an assessment; collecting and using data; using models; planning and conducting an observational human exposure measurement study; evaluating uncertainty and variability; and presenting and communicating results. This presentation will showcase many of the updates in the document.

Keywords: other (specify)

TH-PL-B1-09
Air pollution: Can subjective perception be related to objective measures?

Bartonova A.[1], Grossberndt S.[1], Castell N.[1]; [1]NILU Norwegian Institute for Air Research, Kjeller, Norway

Air pollution in Europe is often not visible and does not smell any more, and as such, is hard to assess by own senses outside special high pollution episodes or accidents. But air pollution levels are still posing a health risk. Additional reduction measures are likely to require popular support. Knowledge about air pollution and individual perceptions may play an important role. We investigated relationship between individual air quality perceptions and objective indicators. Data on air pollution perception collected using a smartphone app and an online survey were coupled with (1) air quality measured at reference monitoring stations, (2) individual estimates using an atmospheric dispersion model with high temporal and spatial resolution (matching location and time-stamp), and (3) long-term averaged data for the location where the perception was collected. The survey provided also information about individual's knowledge on air pollution. The studies were done in Oslo. General interest in AQ showed strongest association of all investigated factors in relation to perception. People perceiving AQ as less good showed tendency to be generally more interested in AQ. Comparison with actual AQ measurements shows no relationship between perception and local pollution level the actual hour, however, linking perception to local annual average pollution load shows clear association. There is a relationship between annual average AQ and reported perception "here and now". Long-term average AQ could be seen as indicator of persistent environmental quality (e.g., traffic intensity locally), and other local qualities that can deteriorate e.g., with higher traffic. Absence of associations with short-term pollution is not surprising considering the multitude of factors influencing perception "here and now", and pollution patterns. We argue that for development of measures and their public acceptance, participatory methods allowing to investigate citizens' perceptions can provide valuable input, but need significant development to build replicable environmental perception models.

Keywords: air, community, environmental policy

TH-PL-B1-10
Systematic analysis of dermal exposure to hazardous chemical agents at the workplace (SysDEA): Results and Implications

Roitzsch M.[1], Schlüter U.[1], Meyer J.[1], Poppek U.[1], Rother D.[1], Spaan S.[1], Franken R.[1], Gerritsen-Ebben R.[1], Kasiotis K.[1], Tsakirakis A.[1], Chartzala I.[1], Machera K.[1]; [1]Federal Institute for Occupational Safety and Health (BAuA), Dortmund, Germany

In the SysDEA research project, different sampling strategies for measurement of dermal ex-posure at workplaces were compared: sampling with patches vs. whole body dosimetry (coverall), sampling by handwashing vs. sampling with gloves, sampling by wiping the forehead vs. sampling with a headband. In addition, the project aimed to pinpoint important influencing parameters, such as the viscosity of the applied liquids. In total, 4.000 individual samples were taken and analysed for six common application scenarios of chemical substances, which were investigated in this project: Pouring liquids (high and low viscosity liquid), Dumping of powders (dusty powder), Spreading of liquids on a surface by rolling (high and low viscosity liquid), Spraying liquids on a surface (high and low viscosity liquid), Manually handling of objects immersed in liquids (high and low viscosity liquid), Manually handling of objects contaminated with powder (dusty powder). In order to work out the differences arising from the different sampling methods, all applications have been performed according to detailed operation protocols in a laboratory environment. This aimed to reduce random factors, e.g. deviations in the course of the applications or contact with contaminations already existing at the workplace, to a minimum. This presentation will focus on the most important results from the SysDEA project. Characteristics and patterns of the dermal exposure resulting from the investigated scenarios will be presented, and the resulting differences seen for the investigated sampling methods will be pointed out. In light of these results, we will discuss scientifically sound recommendations for standardisation of the measurement of dermal exposure to chemicals at the workplace, which can be substantiated with the project results.

Keywords: exposure factors, occupational, sampling methods, workplace

TH-PL-B1-11
A unique ecosystem for healthy indoor environment

Du L.[1], Salmela H.[1], Luuppala A.[1], Taipale A.[1], Erkkilä J.[1]; [1]Tamlink Innovation-Research-Development, Tampere, Finland; Lappeenranta-Lahti University of Technology (LUT), Lappeenranta, Finland

As a response to the climate change, national policies and programs exist in almost all European countries to fulfill the Energy Performance of Buildings Directive (EPBD) in both new and existing buildings. The improved energy efficiency can affect indoor environmental quality (IEQ) and occupants' health and wellbeing. Energy efficient buildings with healthy indoor environment requires systematic co-operation with different partners within the value chain, including motivated and committed users. In Finland, we have launched an advanced co-operation network with joint efforts, namely Indoor Air Quality Ecosystem (IAQe*). This platform connects all parties of indoor air quality and its development such as real estate owners (e.g. municipalities), operators, contractors, design engineers, authorities, research and technology organizations and technology providers. The starting point is to find new comprehensive solutions targeting elevated indoor air quality and energy efficiency through joint projects, knowledge and partnership. An example of a joint concept co-creation is Indoor air Quality as a Service (IAQaaS). The key idea is that in the future high-quality indoor air is purchased and produced as an overall service, including criteria specification, tendering and contract processes, building services, advanced technological solutions, and information and communication with the stakeholders through the contract period. The aim is to demonstrate the concept as a mutual innovation partnership between the service purchasers and the providers in the near future. * IAQe project (https://www.iaqe.fi) is partially sponsored by Business Finland, and coordinated by Tamlink Ltd.

Keywords: ventilation

TH-PL-B1-12
Low Exposures: European collaborative research efforts and future focus areas

Dudzina T., Meijster T., Tozer S., Henschel O., Rodriguez C., Bachler G., Zaleski R., Money C., ter Burg W., Fantke P.

In the last two decades, sustainable and safe use of chemicals became a focal point of various international policy efforts and has been addressed through launching of chemical management systems worldwide. In the EU, the regulation on Registration, Evaluation and Authorization of Chemicals (REACH) primarily concentrates on generating substance intrinsic properties data (including hazard) depending on the tonnage being registered. However, using the chemical production/import volume alone as a proxy/indicator of true exposures undermines risk management decision making and deprives exposure information of its real power (e.g. smart studies design, more efficient prioritization). The European Centre for Ecotoxicology and Toxicology of Chemicals (ECETOC) together with the European Chemical Industry Council Long Range Initiative program streamline their research efforts towards gaining wider regulatory acceptance of exposure-informed testing and risk management practices. This presentation provides an overview of recent industry-funded research building a knowledgebase on tools and methods for assessment of low exposure. The problem definition of low exposure in various contexts will be discussed, focusing on aspects/factors that make low exposure very important in certain situations. A detailed outline will be provided for multiple projects that consider low exposure from different perspectives. The examples will cover comparative exposure analysis of alternative exposure sources to the same substance, assessment of intentional and non-intentional exposure, exposure based waiving and workplace risk assessment. The presented material should inform the workshop participants about the ECETOC's perspective on the importance/weight of (low) exposure in regulatory risk assessment and set the stage for a follow-up panel discussion to help identify remaining knowledge gaps and future focus areas.

Keywords: chemical prioritization, ecological exposure, environmental health, exposure factors, risk assessment

TY-SY-C1: Towards a European Exposure Science Strategy – Developing a roadmap 2020-2025-2030

TH-SY-C1-13
Goals and key thematic areas of ISES Europe

von Goetz N.[1], Bruinen de Bruin Y.[2], Bessems J.[3], Connolly A.[4], Fantke P.[5], Schlueter U.[6]; [1]Swiss Federal Office of Public Health, Bern, Switzerland, [2]European Commission, Joint Research Centre, Ispra, Italy, [3]Vito, Moi, Belgium, [4]National University of Ireland Galway, Galway, Ireland, [5]Technical University of Denmark (DTU), Kgs. Lyngby, Denmark, [6]BAUA, Dortmund, Germany

European legislations create unique demands for the European exposure science community. Human and environmental exposure assessments for chemicals are required as part of the risk assessments undertaken in the context of various legislations. In addition, regulations on general product safety, classification, labelling and packaging, control of air quality and major-accident hazards require input on exposure. Moreover, security-driven exposure assessments are being increasingly requested related to the misuse of chemical, biological, radiological and nuclear materials. Knowledge gaps, method and tool limitations, new needs together with longer-term EU strategies and several new trends all add additional challenges to the field of exposure science calling for the development of a European Strategy for Exposure Science. As a first step of strategy building, ISES Europe organized its first workshop in Germany aiming at (1) to design the backbone of The European Exposure Science Strategy with a roadmap 2020-2030; (2) to create working groups with their own goals and agenda in alignment with the overall strategy; (3) to identify actions for further research and policy needs in Europe; and (4) to attract new ISES Europe members. The workshop was structured around six thematic areas identified via two stakeholder surveys on needs for exposure science in Europe being 1. Data repositories and analytics, 2. Regulatory exposure assessment, 3. Exposure data production and monitoring, 4. Building partnerships and collaboration, 5. Exposure assessment methods and tools, and 6. Exposure science education and communication. In July 2019 ISES Europe organizes its second workshop to develop thematic action plans with roadmaps until 2030. The themes overarch the classical silos of occupational, consumer and environmental exposure and stimulates discussion and cooperation among ISES Europe members. This presentation will give an overview of the key thematic areas of ISES Europe and how ISES Europe intends to shape exposure science in Europe.

Keywords: exposure factors, exposure models

TH-SY-C1-14
Toward a European Exposure Science Strategy

Fantke P.[1], von Goetz N.[2], Bessems J.[3], Connolly A.[4], Schlüter U.[5], Dudzina T.[6], Bruinen de Bruin Y.[7]; [1]Technical University of Denmark, Kgs. Lyngby, Denmark, [2]Federal Office of Public Health, Switzerland, Liebefeld, [3]Flemish Institute for Technological Research, Moi, Belgium, [4]National University of Ireland, Galway, Galway, Ireland, [5]Federal Institute for Occupational Safety and Health, Dortmund, Germany, [6]ExxonMobil, Brussels, Belgium, [7]European Commission, Joint Research Centre, Ispra, Italy

In Europe, exposure science is closely related to various regulations. However, recent scientific advances face difficulties in finding their way into regulatory common practices. This is a major driver for the European Chapter of the International Society of Exposure Science (ISES Europe) to initiate developing an urgently needed «European Exposure Science Strategy» by soliciting exposure scientists from all relevant stakeholder communities. In building this European strategy for exposure science, we will focus on the following set of six overarching objectives: (i) To build a European Programme and related infrastructure to generate, assess, exchange, and communicate experimental and model-based exposure information; (ii) To advance exposure science with focus on closing existing knowledge gaps, building science-based and operational data and methods, fostering integration and interaction of disciplines involved in exposure science; (iii) To provide guidance and recommendations for exposure assessment in science and policy; (iv) To foster the inclusion of realistic exposure information in various regulatory and non-regulatory assessment and management frameworks and related decision-making processes; (v) To strengthen the impact of exposure science on human health and environmental policies; and (vi) To support trans-agency and trans-institutional coordination, education, and engaging a broader European stakeholder community including triggering. These objectives will be addressed through establishing dedicated working groups from the wider European exposure science community to foster a focused and consolidated effort that will ultimately provide guidance for improving and increasing the use of exposure science in in European regulation and practice. Initial focus will be on objectives with highest priority and capacity within the related stakeholder community.

Keywords: aggregate exposure, chemical prioritization, cumulative exposure, environmental regulation, exposure models

TH-SY-C1-15
Establishing exposure science in education

Connolly A.[1], Fantke P.[2], Von Goetz N.[3], Bessems J.[4], Schlüter U.[5], Dudzina T.[6], Bruinen de Bruin Y.[7]; [1]National University of Ireland Galway, Galway, Ireland, [2]Technical University of Denmark, Kgs. Lyngby, Denmark, [3]Federal Office of Public Health, Switzerland, Liebefeld, [4]VITO, Belgium, [5]BAUA, Dortmund, Germany, [6]ExxonMobil, Brussels, Belgium, [7]European Commission, Joint Research Centre, Ispra, Italy

Exposure science provides fundamental knowledge at the interface between chemicals and other stressors in the environment including workplaces and consumer goods, and their negative effects on humans and ecosystems. This knowledge is crucial for our understanding of the complex human-environment-stressor interactions in support of a safe and sustainable future. Education and training of the various exposure science topics yields important expertise required for regulatory risk assessment, for product evaluations and for scientific advances essential for health and environmental science, as well as market and societal demands. The European Chapter of the International Society of Exposure Science (ISES Europe) has initiated and reported on exposure science requirements as input for building a European Exposure Science Strategy, collated from all relevant stakeholders and the scientific community. In support of building this strategy, various topics to strengthen exposure science in Europe were discussed at the first ISES Europe workshop in 2018 along a set of thematic themes addressing: 1. data repositories and analytics, 2. regulatory exposure assessment, 3. exposure data production and monitoring, 4. building partnerships and collaboration, 5. exposure assessment methods and tools and 6. exposure science education and communication. To contribute to ISES Europe's objectives, a working group focusing on exposure science in education, training and communication was established, with an overarching aim to anchor exposure science in academic research and education, and to establish a defined career pathway for graduates. An important building block in the European Exposure Science Strategy is to develop a tiered education/training scheme with ECTS equivalent points/certificates, and to identify ways to support and promote existing courses with focus on exposure science. The preliminary analysis, objectives and expected outcomes will be presented in this talk, alongside the action plan to achieve these outcomes.

Keywords: other (specify)

TH-SY-C1-16
Exposure models for Europe: Towards harmonization and standardization of exposure modelling across regulatory sectors in Europe

Schlueter U.[1], von Goetz N.[2]; [1]BAuA, Dortmund, Germany, [2]Federal Office of Public Health, FOPH - BAG, Bern, Switzerland

A number of exposure models and tools are available in Europe and elsewhere representing a valuable exposure science component in various assessment contexts. However, the diversity and complexity, as well as the lack of transparency and documentation for many of the existing exposure modelling approaches limit a consistent and harmonised use in regulation and elsewhere. This leaves important synergies from the different available approaches unexplored. At the ISES Europe Workshop in 2018, it was emphasised during two different break-out sessions that improving our knowledge and understanding of exposure modelling is crucial for advancing exposure science. In this presentation, we report the key needs and strategies associated with models and tools that have been identified by the newly established ISES Europe Working Group "Exposure Models". The working group has the overarching aim to facilitate and promote the use of existing exposure models, identify possibilities for harmonisation, standardisation and exchange between different science and policy fields, and advance the development of new tools and approaches. To reach this aim a number of activities are proposed, including -Contributing to the development, maintenance and external review and evaluation of exposure models and tools. -Cooperating with bodies working on similar topics, e.g. Exchange Network on Exposure Scenarios or REACH Exposure Expert Group. -Developing an inventory of exposure models that identifies, describes and presents selected models to advance exchange of modelling approaches between different regulatory and non-regulatory processes. -Establishing knowledge transfer e.g. as guidance, training, cooperation, development of best practice. -Identifying options for improvement, innovation and invention, e.g. addressing exposure to mixtures, use of big data and artificial intelligence, and exposure related to release from articles. The preliminary analysis and expected outcomes will be presented, along with an action plan to achieve these outcomes.

Keywords: chemical prioritization, environmental regulation, exposure models, risk assessment

TH-SY-C1-17
Have your say in the strategy for exposure science in Europe

Bessems J.; VITO Health, Mol, Belgium

The format of this presentation is a moderated plenary discussion. A teaser presentation of about 5 minutes will summarize ISES-Europe's strategies for advancing Exposure science in Europe in the next 10 years and stimulate the following discussion around its key thematic areas and building blocks.

Keywords: exposure models

TH-PL-D1-ECS: Air Cleaning for Exposure Control

TH-PL-D1-18
Acceptable Indoor Air Quality in Childcare Centres is Achievable Even During Haze Period

Sankaran G.[1], Shen J.[1], Nazeem M.[1], Lim M.[1], Tan R.[1], Yap J.[1], Gutiérrez R.[1], Wang B.[2], Jia, S.[2], Chang W.C.[2], Fang M.L.[2], Ng L.C.[1]; [1]Environmental Health Institute, National Environment Agency, Singapore, [2]Division of Environmental and Water Resources Engineering, School of Civil & Environmental Engineering, Nanyang Technological University, Singapore

Singapore has experienced instances of poor outdoor air quality due to recurring transient periods of transboundary haze from forest fires in the region. Elevated particulate matter (PM) levels during haze can compromise indoor air quality (IAQ) and have respiratory-related health effects in adults and children. To understand the impact of haze on Singapore's childcare centres, four case studies were conducted during the 2015 haze episode at three centres served by centralized air-conditioning systems and one naturally ventilated centre. Among the three air-conditioned centres, one centre was found to have time-weighted- average PM 2.5 and CO2 levels that exceeded the respective acceptable limits (specified by the Singapore Standard Code of Practice for IAQ in Air-Conditioned Buildings, SS554:2016). Subsequent measurements during a non-haze period showed PM 2.5 levels reduced by 33% and within acceptable limits, although CO2 levels still exceeded the acceptable limits. Data suggested poor IAQ due to inefficient filtration of air supply with inadequate ventilation. In contrast, both CO2 levels and PM 2.5 levels were within the acceptable limits at another centre, even during the haze period. The results from this centre shows how best practices of maintaining adequate ventilation and use of particulate filters of at least MERV 14 efficiency can provide good IAQ for occupants even during haze period in Singapore. For naturally ventilated centres, it is common to keep windows closed during haze and manage thermal comfort by the use of portable air-conditioners. In the centre we studied, PM 2.5 levels exceeded the acceptable limits during haze, despite closure of doors and windows, as gaps in openings are usually unavoidable. These case studies showed that air quality in childcare centres could be compromised during haze. However, a combined strategy of adequate ventilation and effective air cleaning could help achieve good IAQ, even during the haze period.

Keywords: air cleaning, built/indoor environment, children, particles, ventilation

TH-PL-D1-19

A Crossover Study of In-Vehicle Air Filtration and Acute Changes in Heart Rate Variability and Cognition Among Healthy Adults

Mallach G.[1], Shutt R.[1], Thomson E.M.[1], van Rijswijk D.[1], Valcin F.[2], Kulka R.[1], Weichenthal S.[2]; [1]Health Canada, Ottawa, Ontario, Canada, [2]Health Canada, Montreal, Quebec, Canada

Background: Traffic-related air pollution is a well-recognised and much studied contributor to smog and is linked to a number of adverse health outcomes. Although traffic pollutants can travel long distances, exposure to the highest levels of the raw emissions can occur closest to the source; e.g. in a car in dense traffic conditions. Time spent in-vehicle may contribute up to half of commuters' daily exposure to certain air pollutants. Most new cars now have or allow for a cabin air filter, but it is not known how well cabin air filtration can reduce exposure to traffic-related air pollution. Methods: We measured commuters' exposure to air pollutants in rush hour traffic, and evaluated whether this exposure could be reduced by cabin air filtration. Also, we looked at effects of commuting during rush hour on heart health (i.e. heart rate variability), and cognition (using a computer-based cognitive assessment) in an environment where any deficit could be important to safety. We examined whether cabin air filtration using an electrostatic filter reduced exposure to traffic-related air pollutants and improved short term heart and cognitive function. Results: Preliminary results show that participants' heart rate variability and cognitive function were impacted by in-vehicle air pollution exposures. Cabin air filtration reduced in-vehicle particulate exposures by approximately one third. In-vehicle pollutant concentrations were notably elevated in tunnels. Conclusions: While commuting, Canadians are exposed to air pollutants that may have a measurable impact on their health. It may be possible to reduce in-vehicle exposures by installing an electrostatic air filter. Relevance: This study provides information about a potentially valuable and economical means of reducing exposure to traffic-related air pollution during daily commutes. This study also examined the impacts of air pollution exposure on cognition, an area of growing concern.

Keywords: air cleaning, epidemiology, health, chronic diseases, health, other/general

TH-PL-D1-20

Effectiveness of portable indoor air cleaners to reduce indoor pollutants: Impact of open windows

Kipen H.[1], Black K.[1], Laumbach R.[1], Alimokhtari S.[1], Black T.[1], Cohen J.[2], Gallegos E.[2], Weschler[1], Ryan P.[2]; [1]Rutgers University, Piscataway, NJ, USA, [2]Emory University, Atlanta, GA, USA

Rationale: Indoor air pollution has been implicated in a number of adverse health outcomes. One approach to reduce indoor pollutant exposure is to use portable indoor air cleaners. However, the impact of window opening on exposure reductions is not established. Methods: Portable air cleaners (Honeywell HPA300 with HEPA and charcoal filters) were deployed in 36 occupied apartments in New Jersey and Atlanta, Georgia in May through

October, 2016 and 2017. Within each home, air cleaners were operated with/without filters for two days. Indoor concentrations of PM2.5, black carbon and total particle counts were measured continuously in the living room. Participants recorded window use daily. Results: Living room windows were reported to be always closed in 23 homes (closed homes). In 9 homes, windows were opened (more than 1 hour in each period; open homes). In the remaining 4 homes, window use was mixed; these homes were excluded from this analysis. In the open homes, duration of window opening ranged from 5.4 h to 46.9 h per 48 h treatment period. Closed home reductions of pollutants during filtration were significant for all 3 pollutants (PM2.5: 5 v. 8 µg/m3; p<0.01; black carbon: 78 v. 443 ng/m3, p<0.01; particle number: 831 v.3532 #/cm3, p<0.001). In open homes black carbon concentrations were significantly lower (151 v. 472 ng/m3, p<0.001); indoor particle counts were borderline significantly lower (1921 v. 3715 #/cm3, p=0.06) and PM2.5 levels were lower on days with filtration (6 µg/m3 v. 11 µg/m3, p=0.09). Conclusions: Portable air cleaners reduced indoor pollutant concentrations during real-world window usage, however the number of open homes was small. Further, the analysis relied on self-report of window use. Future analyses will include AER, room volume, duration of window opening, and the effect of window opening in other rooms in the home as additional factors.

Keywords: air cleaning

TH-PL-D1-21

Gravimetric and spectrometric methods of assessing VOC interactions for improved air quality

Hunter Sellars E.[1], Parkin I.P.[2], Williams D.R.[2]; [1]Department of Chemical Engineering, Imperial College London, United Kingdom, [2]Department of Chemical Engineering, Imperial College London, United Kingdom

Volatile organic compounds, or VOCs, are a prominent class of chemicals present in indoor air environments and have been found to be a major contributor towards 'sick building syndrome' and the associated decrease in indoor air quality. The use of adsorbent materials provides a method of reducing the concentrations of volatiles that humans are exposed to, potentially leading to improved health and quality of life for those individuals. Unlike many industrial applications, the levels of humidity in indoor and outdoor air can exceed volatile concentrations by many orders of magnitude, and lead to reduced adsorbent performance. The impact of process humidity on the adsorption of volatiles with solid materials is the focus of this study. Two-component gravimetric methods were used to qualitatively assess the impact of humidity on the removal of ethanol and toluene by adsorbent materials. Hydrophilic sorbents such as molecular sieves and amorphous silica had negligible performance following exposure to water activities as low as aw = 0.1 compared to materials with more hydrophobic surface chemistry. These initial measurements provided important information for the performance of adsorbents in realistic conditions typical of those in indoor air, and guided the development of more in-depth techniques. The construction of a novel breakthrough curve analysis approach, utilising a selected-ion flow-tube mass spectrometer (SIFT-MS), has allowed for the generation and study of volatile interactions in the part per billion concentration range. SIFT-MS has been designed and commonly used in medical and air quality fields to handle complex mixtures of volatiles in the presence of high levels of humidity, two process conditions that industrial adsorbent characterisation methods lack. The flexibility of this method has allowed the study of interactions between complex mixtures of humidity and VOCs with a range of both industrial and household materials, including cotton and human hair.

Keywords: VOCs, air cleaning, analytical methods, built/indoor environment

TH-PL-D1-22
Applications for Ultraviolet Germicidal Irradiation (UVGI) in Hospital Isolation Anterooms

Grosskopf K.[1], Mousavi-Rizi E.[2], Lau J.[1]; [1]University of Nebraska-Lincoln, Lincoln, NE, USA, [2]Clemson University, Clemson, SC, USA

Certain pathogens are transmitted through air by respiratory droplets that desiccate shortly after emission and form droplet nuclei. Droplet nuclei are small (<5.0μm) and can remain suspended in the air, creating a pathway between an infected and susceptible person. This process is called 'airborne transmission'. Patients with an airborne infectious disease (e.g. tuberculosis, varicella, etc.) are separated from the healthcare environment by an Airborne Infection Isolation Room (AIIR). AIIRs have special features to effectively contain and remove airborne pathogens, including a -2.5Pa air pressure relationship and directional airflow away from adjoining healthcare spaces. Directional airflow, however, may be disrupted and even reversed by door opening motion and the movement of people and equipment into and out of the AIIR. Ultraviolet germicidal irradiation (UVGI) is a disinfection method that uses short-wavelength ultraviolet (UV-C) light to kill or inactivate airborne microorganisms. UVGI has proven to be among the least energy intensive and most cost-effective methods of disinfection. However, germicidal UV wavelengths can produce ozone and prolonged exposure can cause damage to the skin and eyes. As a result, continuous exposure to germicidal UV may be limited to 0.2μW/cm2 for a 8-hour period, or in some cases, prohibited entirely. This paper explores the use of UVGI in unoccupied anterooms as an additional airborne infection control measure. Specifically, this research highlights the results of both experimental and computation studies that explore air exchange and particle movement between isolation rooms and adjacent hospital corridors, and, the potential use of occupancy controlled UVGI in isolation anterooms and vestibules.

Keywords: ventilation

TH-PL-D1-23
Time-dependent efficiency of air cleaners for gas-pollutants: an attempt at validating a predictive model

Berne P.[1], Brouard C.[1], Golanski L.[1], Roussey A.[1], Wayser B.[2], Guiot A.[1], Clavaguera S.[1], Delléa O.[1]; [1]Univ. Grenoble Alpes, Grenoble, France, [2]Univ. Grenoble Alpes, Montreuil, France

Many portable air cleaners (ACs) use adsorbents for the treatment of gas pollutants. A particular feature of adsorbents is their gradual saturation. This makes the assessment of ACs performance difficult: the popular "Clean Air Delivery Rate" index (CADR), for instance, only gives information on the initial efficiency of an AC. In a previous work, we have proposed a method based on the measurement of the breakthrough curve (BTC) of pollutants in the adsorbent. Coupling the BTC with a balance equation for pollutant mass in a room allows building a model capable of predicting the effect of an AC as a function of time. The performance of ACs can then be compared in terms of both initial efficiency and long-term behaviour. In this work, we shall present an attempt at validating this concept against experimental data acquired on commercial equipment. The BTCs can in theory be measured on a small-scale sample or on the full-size adsorbent assembly of an AC. Both approaches will be compared. Characterisation of a test chamber by tracer gas will be reported. Pollutant concentration histories measured in this chamber will be compared to model predictions. Aknowledgement: The authors are grateful to Guard Industrie for financial support.

Keywords: VOCs, air cleaning

TH-PL-E1-24
Measurement of Local Clothing Properties for Human Thermal Comfort Evaluation

Nomoto A.[1], Takahashi Y.[1], Yoda S.[1], Ogata M.[1], Tanabe S.[1], Ito S.[2], Aono Y.[2], Yamamoto Y.[2], Mizutani K.[2]; [1]Waseda University, Tokyo, Japan, [2]Tokyo Polytechnic University, Kanagawa, Japan

To accurately simulate skin, core temperatures, and thermal comfort, certain human physiology and comfort models categorize the human body into multiple body parts. Most of these parts are normally clothed, which must be quantified in the simulation. Unfortunately, existing databases of clothing evaporative resistance only characterize the evaporative resistance for the whole body and not for individual body parts. In other words, each body part has the same level of clothing evaporative resistance. However, using a constant value as input to each body part of a multi-segmented human physiology and comfort model may be one of the causes of simulation error. Therefore, while evaluating the human thermal physiology and comfort in detail, it is desirable to evaluate it by measuring the local clothing evaporative resistance. In this research, we measured local clothing insulation and clothing evaporative resistance for 8 typical summer clothing ensembles in a climate chamber using a 20-segment sweating thermal manikin and presented the clothing insulation and evaporative resistance value for each body part as well as those for the whole body. Further, the moisture permeability index for each body part was calculated using the obtained two values. As a result, it was observed that the local clothing insulation, clothing evaporative resistance, and clothing moisture permeability index had comparative part differences for any clothing ensemble measured in this study because of the material of clothing and the state of overlaying. Hence, it was quantitatively shown that when we evaluate the human thermal physiology and comfort in detail for each part, it is necessary to consider the clothing thermal characteristics of each body part. The data of the local thermal characteristics of clothing obtained via this study is useful for the multi-segmented model of human physiology and comfort model.

Keywords: thermal environment

TH-PL-E1-25
Physiological and psychological responses to ambient humidity from young and elder subject experiment

Kawakami R.[1], Yamaguchi M.[2], Otsuka T.[1]; [1]Institute of Technology, Shimizu Corporation, Koto, Tokyo, Japan, [2]Department of Architecture, Faculty of Engineering, Daido University, Nagoya, Aichi/Japan

The purpose of this research is to verify the necessity to control indoor humidity for human wellness and well-being. Since too high or low humidity condition are considered to cause discomfort feeling, heatstroke, dry integument and expansion of infectious diseases. Accordingly, human physiological and psychological responses to ambient humidity were studied using human subjects experiment in an artificial climate chamber where indoor air condition (temperature, humidity, velocity and etc.,) can be controlled at discretion. For easily finding the characteristics of generation, two subject groups, young subjects whose average age is 25 and elder subjects whose age are over 65, participated in the test. As a result, this study revealed an existence of the variation in the physiological and the psychological reactions to a change in ambient humidity, regarding only young subjects. In particular, a surrounding with high absolute humidity tends to increase stratum corneum moisture content on inner forearm. Furthermore, young subject group is able to perceive the change in humidity. These circumstances could not be found in the elder subject group. On the other hand, contrary to expectation, it is also specified that young people have more dry skin than elder people. Therefore, the conclusion drawn from the human experiment is that it is necessary to auto and premediated control indoor humidity regardless the age of occupants. In detail, the value of indoor humidity should maintain to fall in the range which does not impede healthful activities, without resort to occupants' awareness. However, this study acquired the data regarding some body parts from human experiments throughout only one hour acclimatization time. That is to say, it remains a challenge for future researches to investigate physiological and psychological responses of the whole body and the influence of long-term acclimatization for ambient temperature and humidity upon human wellness and well-being.

Keywords: aging/elderly, built/indoor environment, health, other/general, thermal environment

TH-PL-E1-26
Impact of a Uniform on a Human Being's Level of Comfort

Borodinecs A.[1], Geikins A.[1], Zemitis J.[1], Daksa G[1]; [1]Riga Technical University, Riga, Latvia

This paper focuses primarily on the impact of clothing insulation in relation to the wellbeing of building occupants and building energy performance. A thorough understanding of how these factors interact with each other is crucial in order to improve one's feeling of comfort and as a result, increase their productivity at work. This leads to believe that in buildings where employees are obliged to wear uniforms with at increased thermal insulation level it might not be necessary to maintain standard temperature levels. Employees comfort level may stay optimal in a room with air temperature 2-3 degrees lower, thus saving resources by using less thermal energy. Theoretically, this yields two types of positive effects at once: employees increased productivity and savings on heating. The opposite effect may occur if buildings are equipped with air conditioning units and would require additional cooling during the summer. A survey regarding indoor air quality thermal comfort of people in uniforms was conducted in three different office where people wearing military dress.

Keywords: built/indoor environment

TH-PL-E1-27
Research to Practice: Translating the Science of the Microbiology of the Built Environment Program

Shaughnessy R.[1], Haverinen-Shaughnessy U.[1], Peccia J.[2]; [1]1University of Tulsa, Tulsa, OK, USA, [2]Yale University, New Haven, CT, USA

An A.P. Sloan funded effort was initiated in 2017 with specific goals to extract and distill practical implications of Sloan funded research on the microbiology of the built environment (MoBE), provide a forum to share and exchange information with practitioners, coordinate efforts with the US Environmental Protection Agency Indoor Environments Division and other government agencies (e.g. NIOSH and HUD) to promote dissemination of information to practitioners and the public, and for Sloan research outcomes to be integrated into their respective research agendas. This project puts into effect a logical progression to distill and extract practice-based implications and begin the process of transfer of the science to practice. It serves an endpoint that directly parallels the Sloan objective of encouraging US government funding agencies to fund research on the built environment. By participating in a timely forum into the immediate and potential future implications of the Sloan research, government agencies can be better informed of commonalities which may be infused into their research agendas and outreach efforts. This presentation expounds upon progress to date on formulation of a document focused on insights into practical implications of Sloan MOBE Research.

Keywords: practitioners, translating research to practice, Sloan MOBE research

TH-PL-E1-28
The SATAKUNTA intervention trial - results of the first twenty towns and municipalities

Putus T.

In Finland, a nation-wide program was launched in 2018 aiming towards better indoor air quality in public premises, such as day-care centers, schools and health care facilities. According to previous studies, at least every third school building has problems with indoor air quality. As much as 50% of the teachers are not satisfied with the ventilation system and at least 20% has observed microbial odor. Similar results were found in the day care centers.In the project, intensive action has been taken to remediate school buildings. In all, 100 communities will be recruited into the study. To each study population a similar reference town will be chosen. The intervention comprises of training, quality control and the s.c. 'Dry Chain' principle during the construction. Additionally, specific attention is paid to select construction materials so that allergenic and irritative substances are avoided. The health status of the occupants will be followed-up before the remediation and after. Costs and benefits of the investment will be calculated. The follow-up will last for ten years and thus different methods and materials will be compared. In the reference populations no specific action is taken (other than normal maintenance). Health data will be collected with surveys (children and employees), lung function measurements and immunological tests in serum samples (employees). The intervention has been piloted in several towns during the past two years and results from these will be presented in the congress. The expected results will be better indoor air quality, less respiratory and general symptoms, less absence from the work / school and less chronic respiratory diseases among children and adults. Especially voice problems among the teachers have been on a high level among Finnish teachers during recent years and the absenteeism from work is expected to be reduced.

Keywords: workplace

TH-PL-E1-29
THE EFFECT OF PLANTS ON THE INDOOR AIR QUALITY OF TWO SMALL OFFICES

Berger J.; University of Reading, Reading, United Kingdom

Previous studies have proposed that indoor plants can benefit the health and well-being of office workers by improving the indoor air quality. Bench scale studies have shown that plants can reduce carbon dioxide (CO2) levels, remove airborne pollutants and increase the relative humidity (RH) of the atmosphere. The majority of research has been carried out under controlled laboratory conditions on a small scale, and few studies have been published on the effectiveness of the plants in improving air quality in real life office buildings. This case study investigates the impact of plants on the indoor environmental quality (IEQ) within two small offices. The study monitored two comparable single office rooms; IEQ conditions were measured before and after the introduction of plants. Bench scale tests, using sealed chambers were carried out to determine the impact of plants on the rate of CO2 decay. Six plant species differing in structure and function, and a range of plant numbers within the species, were tested. Plants from the two species with the highest CO2 removal rates (Ficus benjamina and Asplenium nider) were introduced into the offices. The room CO2 concentration was increased to approx. 2500 ppm, to equal that of peak occupancy levels and the impact of the plants on the rate of CO2 decay, temperature and RH were measured. The chamber tests show that there are considerable differences between plants species and quantities and their effect on IEQ. The preliminary results of the office tests show the plant activity and CO2 decay rate are affected by conditions within the office. The plants had a positive impact on the rate of CO2 decay within the offices at the 90% statistical confidence level. This research is ongoing.

Keywords: productivity

TH-PL-A2-EHR: Conceptual Approaches to Exposure and Risk Assessment

TH-PL-A2-30
The Role of Environmental and Public Health Factors in Children's Social Mobility

Baxter L.[1], Rappazzo K.[1], Weaver C.[1]; [1]United States Environmental Protection Agency, Research Triangle Park, NC, USA

Previous studies have established that the neighborhood where a child grows up can have substantial effects on their upward income mobility. The Opportunity Atlas, constructed by researchers at Harvard University, Brown University, and the U.S. Census Bureau, is a publicly available atlas with census-tract-level estimates of children's earnings distributions, incarceration rates, and other outcomes, such as teenage birth rates by race, gender and parental income for the entire United States. These estimates allowed the researchers to trace the roots of mobility outcomes back to the childhood neighborhoods. However, only about half of the tract-level variance in children's outcomes was explained by neighborhood disadvantage characteristics. In this study, we use data on environmental exposures (e.g., location of industrial sources), public health factors, (e.g., access to medical care), and ecosystems good and services (e.g., green space) from the U.S. Environmental Protection Agency's EJScreen to explore whether such environmental determinants of health and welfare have the potential to explain the remaining variance. Various factors related to census tract proximity to point pollution sources (e.g., national priority list sites, Risk Management Plan facilities, and Treatment Storage and Disposal facilities) were positively correlated with variables from The Opportunity Atlas. Pearson correlations ranged from 0.10 to 0.16, depending on gender, for the fraction of children incarcerated and a variable related parent's income level. Income rank was also correlated (0.11 to 0.15) with house units built before 1960 (an indicator of lead). These results suggest that environmental indicators may potentially contribute to adult outcomes, in conjunction with social determinants of mobility. Therefore, a more comprehensive research program is needed to understand the relationship between a child's total environment (environmental, social, and building) and long-lasting health and social impacts.

Keywords: children, cumulative exposure, environmental health, environmental justice, susceptible/vulnerable

TH-PL-A2-31
Alternatives to Bisphenol A – toward safer substitutes?

Reale E.[1], Hopf N.[1], Vernez D.[1]; [1]Center for Primary Care and Public Health (Unisanté), Lausanne, Switzerland

Exposure to Bisphenol A (BPA) is a major concern in environmental and occupational health because of the growing evidence of its endocrine-disrupting effects. BPA is present in numerous consumer products, including polycarbonate plastics, epoxy resins, and thermal papers such as cashiers receipts, labels and tickets. Skin contact with these papers can contribute to overall BPA body burden. The increasing BPA restrictions led the industry to alternative substances such as Bisphenol S (BPS), D-8, and Pergafast201® (PG201). Toxicokinetic (TK) models have been developed for BPA and BPS focusing on the oral route of exposure. Our objectives were to include skin absorption kinetics into existing TK models and use these models to predict body burden for the alternative substances. We modeled different exposure scenarios of cashiers handling thermal paper with any of the four substances. The influence of exposure parameters and biological variability for the systemic dose with and without skin absorption was also investigated. D-8 and BPA skin absorption contributed to the overall body burden. This can be explained by their similar skin permeability coefficients. Although the BPS structure is similar to BPA's, its contribution was much lower probably due to its greater polarity. PG201 was only dependent on oral route of exposure. BPS and PG201 in thermal papers are interesting substitutes in reducing overall exposure. TK models can be a useful tool in predicting overall body burden for less known substances.

Keywords: biomonitoring, exposure models

TH-PL-A2-32 WITHDRAWN

TH-PL-A2-33

Review on technologies and their applicability in an exposome-based assessment of hazards and risk in the petrochemical industry

Kuijpers E.[1], Loh M.[2], Galea K.[2], Makris K.[3], Stierum R.[1], Fransman W.[1], Pronk A.[1]; [1]TNO, Zeist, The Netherlands, [2]IOM, Edinburgh, United Kingdom, [3]Cyprus University of Technology, Limassol , Cyprus

Oil & gas workers have been shown to be at increased risk of various types of diseases including several types of cancers, asthma, chronic obstructive pulmonary disease and hearing loss. Exposome technologies for assessing the external (e.g. sensors, use of smart phone apps and online platforms, big data) and internal exposome (e.g. omics, pharmacokinetic modelling, biomarker technologies) offer new possibilities for effective prevention of these diseases. The objective of this study was to review the literature on these technologies and evaluate their applicability in an exposome-based assessment of hazards and risk related to the petrochemical industry. Both peer reviewed literature and grey literature focused on the petrochemical industry and the most frequently-occurring stressors (including particulate matter, volatile organic compounds, and noise) were identified by searches in Medline, Web of Science, and Google Scholar, for 1) identifying current occupational challenges, 2) identifying current exposome-based methods used, and 3) recommend on exposome-based methods to be used in future exposome studies on the petrochemical industry, which can help to overcome current challenges. Based on preliminary results, current challenges in the petrochemical industry include limited information about exposure at the workplaces, working history and life styles to establish causal relationships between exposure and disease. In addition, knowledge on hazards improved, which has led to a (further) reduction of exposure limits for many stressors (e.g. benzene, noise), requiring new exposure assessment methods capable of meeting the new detection limited. Finally, there is a need for more personal risk assessment as individual susceptibility to multiple stressors is becoming increasingly important. Currently, exposome-based methods are still largely missing in the petrochemical industry. This presentation will provide an overview on promising technical applications (including chemical sensors, context sensors, smart devices, biomarkers, omics, adverse outcome pathways and non-invasive monitoring) used in other occupational settings, recommended to overcome the identified current challenges.

Keywords: occupational

TH-PL-BW-SSS: Children's Exposure: Chemical and Non-Chemical Stressors

TH-PL-B2-34
Children's exposures to chemical and non-chemical stressors: what have we learned?

Tulve N.; United States Environmental Protection Agency, Office of Research and Development, National Exposure Research Laboratory, Research Traingle Park, NC, USA

Children are exposed to a wide variety of chemical and non-chemical stressors at each lifestage throughout their lifecourse, potentially impacting their lifelong health and well-being. To understand potential chemical exposures, we collect and analyze multimedia (e.g., air, water, soil, dust, food, urine) samples for many chemicals including, but not limited to, pesticides (organophosphates, pyrethroids), polybrominated diphenyl ethers, bisphenol A, nanosilver, and other consumer product ingredients. Additionally, we collect supporting information such as time activity data and household inventories. Over time, our research has evolved to consider a holistic approach to children's environmental health. Using a systems approach, we collect and analyze data and information on exposures to both chemical and non-chemical stressors, recognizing that these interrelationships may impact the biological response to a chemical agent. Examples of non-chemical stressors include access to parks and recreational amenities, food types and access, health care options, family dynamics, educational attainment, and/or neighborhood characteristics. Our conceptual framework was designed to show how children's health and well-being should consider exposures to chemical and non-chemical stressors from the built, natural, and social environments, activities (e.g., type and duration of physical activity level and locations) and behaviors (e.g., children's mouthing frequency), and inherent characteristics. Our research approach includes distinct research activities with a recent emphasis on a combined analysis approach that considers how non-chemical stressor information may be used to interpret chemical exposure data. Our research has shown that both childhood obesity and general cognitive ability are influenced by a myriad of chemical and non-chemical stressors. Furthermore, tribal children may be disproportionately affected by the combined influences of chemical and non-chemical stressors as compared to children from the U.S. general population. This presentation will explore our research strategy, showing how data and information can be considered within the conceptual framework and highlight public health examples.

Keywords: activity patterns, aggregate exposure, behavior, children, environmental health

TH-PL-B2-35
Towards a better management of early-life exposures through the enhancement of the knowledge on air pollution in households

Gabriel M.[1], Felgueiras F.[1], Ramos E.[2], Mourão Z.[1], Oliveira Fernandes E.[3]; [1]INEGI, Institute of Science and Innovation in Mechanical and Industrial Engineering, Porto, Portugal, [2]EPIUnit, Institute of Public Health and Department of Clinical Epidemiology, Predictive Medicine and Public Health of the University of Porto, Porto, Portugal, [3]University of Porto, Porto, Portugal

Exposure to air pollutants in early childhood can exacerbate the risk of development of noncommunicable diseases (NCDs), including respiratory disorders, cardiovascular diseases and cancer, throughout childhood and the entire life course. Since children younger than 3 years spend most of their time in indoor environments, particularly at home, exposures occurring indoors are expected to have an important contribution in the definition of the individual's "exposome". In this study, we aimed to assess the indoor air quality (IAQ) of households of newborn children in order to identity opportunities for promotion of childhood health. More than 300 parents of newborn children, from four public maternities located in the Porto Metropolitan Area, were invited to fill out a questionnaire to disclose data on potential air pollutant sources in their households. In addition, IAQ audits were conducted in 30 homes of participant families with twins aged 3 to 12 months. These included an extensive building survey and monitoring and sampling work for assessment of a comprehensive set of environmental parameters, including carbon monoxide (CO) and dioxide (CO2), nitrogen dioxide (NO2), ozone (O3), aldehydes, ultrafine particles (UFP), PM2.5, PM10, volatile organic compounds (VOC) and airborne bacteria and fungi. The results of the analysis of each household were used to propose a list of recommendations and best practices to empower families to actively participate in the control of their own exposures at home. The initial outcomes show that the main concerns linked to early life exposures at home are related to emissions from use of household solid fuels, indoor tobacco, household cleaning products, fragranced consumer products (e.g. air fresheners, incense and candles), moisture-related pathologies, new furniture, recent remodeling/painting works and nearby traffic-related pollution sources. Overall, this work provides important insights to establish evidence-based educational/awareness campaigns to promote public health by creating healthy households for children.

Keywords: built/indoor environment, children, exposure factors, neonatal, source control

TH-PL-B2-36

Impact of Parental Migration on Environmental Exposure Related Time-Activity Patterns of Left-behind Children in China: Results from the First National Population-based Survey

Xiangyu X.[1], Qian G.[1], Beibei W.[2], Suzhen C.[1], Ning Q.[1], Xiuge Z.[2], Ying L.[2], Chunrong J.[3], Xiaoli D.[1]; [1]School of Energy and Environmental Engineering, University of Science and Technology of Beijing, Beijing, China, [2]Chinese Research Academy of Environmental Sciences, Beijing, China, [3]School of Public Health, University of Memphis, Memphis, TN

Internal migration in China has separated large numbers of rural children from their parents, with the advancement of China's urbanization process. Understanding the patterns of time activity of left-behind children in China helps to make reliable exposure assessments that have not been noticed or confused in the past. Our study based on the first national wide survey, the Children's Environmental Exposure Related Activity Patterns (0-5 years), investigates the distribution of left-behind children in China, outdoor and indoor activity time, traffic use and transportation of left-behind children We found that nearly one third rural children(29.7%) under 6 years old were left-behind children separated with one or both parents. After controlling the confounding factors we found that separation from parents significantly affects children's time activity patterns in rural areas. For different age groups, the increase effect and 95% confidence interval of separation from parents for the outdoor activity time of children in rural are: <1 year: 16.1min/day (95%CI:10.6, 21.6) , 1 year: 16.1min/day (95%CI:10.6, 21.6) , 2 year: 16.6min/day (95%CI:8.9, 24.4) and 3~5year: 16.6min/day (95%CI:8.9, 24.4), respectively. The reduce effect on the indoor activity time are: <1 year: -16.1 min/day(95%CI:-21.6, -10.6), 1 year: -10.3min/day (95%CI:-17.7, -2.9) , 2 year: -16.0min/day (95%CI:-23.7, -8.3) and 3~5year -19.8min/day (95%CI:-24.2, -15.5), respectively. Separation from parents increases the incidence of traffic behavior in rural children which means left-behind children emerge 1.15 (1.02, 1.30)(OR) more times than non-left-behind children on traffic and 1.17 (1.09, 1.33)(OR) more times in vehicle on transportation. The proportion of stunted children left behind is statistically significantly higher than that of non-left-behind children. We quantified the impact of the separation of parents and children on their time activity patterns, which can help us to improve the assessment of children's air pollution exposure without neglecting left-behind children.

Keywords: activity patterns

TH-PL-B2-37
Results of a Meta-Analysis Exploring Chemical and Non-Chemical Stressors Affecting Childhood Obesity

Lichtveld K.[1], Viguria U.[1], Clinger J.[1], Thomas K.[2], Tulve N.[2]; [1]The University of Findlay, Findlay, OH, USA, [2]United States Environmental Protection Agency, Office of Research and Development, National Exposure Research Laboratory, Research Triangle Park, NC, USA

According to the World Health Organization, childhood obesity affects an estimated 41 million children aged 0-5 years globally (2016 reporting period). Childhood obesity is associated with a wide range of lifelong health consequences, and without intervention, childhood obesity results in obesity and poor health in adulthood. While much research has focused on individual behaviors impacting obesity, little research has emphasized the complex interactions of numerous chemical and non-chemical stressors found in a child's total (built, natural and social) environment and how these interactions affect a child's overall health and well-being. The objectives of this meta-analysis were to identify the chemical and non-chemical stressors associated with childhood obesity that are detrimental or protective in regards to child-specific environments within a community setting. This meta-analysis identified twenty-three factors (chemical and non-chemical stressors and activities/behaviors) related to childhood obesity. Aspects of twelve factors were associated with significant increases in childhood obesity, including 5 non-chemical stressors (access to recreational facility, rural/urban environments, stress, access to supermarkets, family income), 4 chemical stressors (exposure to smoke during pregnancy, exposure to bisphenol A, antibiotic use early in life, and smoking in the home), and 3 activities/behaviors (diet, sedentary behavior, hours of television viewing). Breastfeeding was identified as the lone protective factor. Ten factors included in our analysis were not statistically significant, including 5 non-chemical stressors (proximity to convenience stores, parental education, daycare attendance, social attitudes, distance to fast food establishments), 2 chemical stressors (DDT/DDE, perfluorinated compounds), and 3 activities/behaviors (computer use, active transport, general activity level). This analysis suggests that the interactions between chemical and non-chemical stressors and activities/behaviors are important for understanding a child's overall health and well-being. By considering these complex interrelationships, the research community can better design and implement strategies to reduce childhood obesity in a community setting.

Keywords: children

TH-SY-C2: National and International IEQ Guidelines

TH-SY-C2-38
National and international IEQ guidelines

Haverinen-Shaughnessy U.[1], Hägerhed L.[2], Lee K.[3], Caulfield S.[4], Cull I.[5]; [1]Tampere University, Tampere, Finland, [2]University of Borås, Borås, Sweden, [3]Seoul National University, Seoul, South Korea, [4]Turner Building Science & Design, Harrison, ME, USA, [5]Indoor Sciences, Inc., Chicago, IL, USA

Since early 2017, a number of ISIAQ members have participated in collecting and sharing information about international and national guideline values concerning IEQ parameters, such as indoor temperature, relative humidity, and concentrations of various pollutants such as carbon dioxide, radon, particulate matter, and various biological and chemical substances. An email account (ieqguidelines@gmail.com) was set up and the database constructed as a Google Sheet. Sharing is done by sending out a link to the spreadsheet, and anyone with the link can get access to view the database, but only those logged in and approved by the administrator can make changes to the database. The database structure has been reformatted by dividing the guidelines into four categories: 1) International Standards & Guidelines (including WHO and EU), 2) Government Regulations (information available from nine countries), 3) Government Guidelines (available from 10 countries), and 4) Professional Societies (available from two counties). Two countries have information in all three categories, and three countries have information in two categories (both government regulations and guidelines). In total, information is available from fourteen countries: Brazil, Canada, Denmark, Finland, France, Germany, Lithuania, Norway, Portugal, Singapore, South Korea, Spain, Sweden, and USA. Based on the information gathered so far, there are differences between countries in terms of existing guidelines, but direct comparison is difficult due to methodological issues. For example, guidelines for formaldehyde exist in many countries, but some of them are based on short term (0.5-2 hours) measurements, whereas some are based on 8-hour and some annual averages. Some countries have numerical values for certain parameters whereas some other countries have qualitative descriptions for required conditions. Subsequent to data collection, other relevant societies could join ISIAQ in assessing the guidelines, possibly aiming to provide recommendations towards more harmonized guidelines worldwide.

Keywords: VOCs, air, built/indoor environment, particles

TH-SY-C2-39
Guidelines for IAQ in the United States

Caulfield S.[1], Cull I.[2]; [1]Turner Building Science & Design, LCC, Harrison, ME, USA, [2]Indoor Sciences, Inc., Chicago, IL, USA

The presentation will focus on IEQ guidelines and standards in non-industrial settings in the United States. The presentation will outline the limited codified standards and the best practice documents that are used by many practitioners in the absence of legal requirements. In the United States, there are very few national laws that govern exposure to indoor contaminants in buildings like offices or schools. Laws or regulations with any enforcement power are mainly limited to industrial exposures. Many standards exist from various organizations (ASHRAE, ANSI, etc.) and some of these have been adopted into building codes. These standards then apply to new construction or renovation of buildings, but do not affect the vast stock of existing structures. However, much of the indoor air community operates on the basis of "best practice" standards to avoid litigation. Therefore, when indoor air is assessed in a non-industrial setting, good practitioners often cite the lowest applicable standard for comparison. This is of particular interest in the area of vapor intrusion from ground sources of industrial contaminants (solvents from dry cleaners, leaks from petroleum facilities, etc.). Some states have begun adopting standards for these chemicals based on the long term exposure of children and house-bound occupants. These regulated exposure levels in affected residential properties can be hundreds or even thousands of times lower than the industrial standards for the same chemical contaminants. Standards and guidelines to be discussed include those for dampness/mold, ventilation, particles, and some chemical exposures.

Keywords: VOCs, built/indoor environment, dampness/mold, industrial hygiene, particles, ventilation

TH-SY-C2-40
Scandinavian IEQ Guidelines

Hägerhed L.[1], Kolarik B.[2], Holos S.[3]; [1]University of Borås, Borås, Sweden, [2]Danish Environmental Analysis, Vedbæk, Denmark, [3]SINTEF Building and Infrastructure, Oslo, Norway

Scandinavian, i.e., Sweden, Norway and Denmark, building guidelines and requirements are not harmonized, but there are many similarities. Most IEQ requirements are functional requirements with qualitative letters. The chapter including IEQ in Swedish regulations, BBR, is formulated "Buildings and their installations must be designed so that air and water quality as well as light, moisture, temperature and hygiene conditions will be satisfactory during the life of the building and thus disadvantages for human health can be avoided." Those are complemented with so-called General Advices, where limit values are expressed as a guideline to achieve the qualitative regulations, but these are generally used as the requirements, even though there is a possibility to achieve the demands in other ways. The reason for this arrangement is that the governments do not want to impede technological progress through overly detailed rules and any requirements refer to national and international standards. Details quantified requirements as limit values is expressed for some criteria as for ventilation and radon. Regarding ventilation, fresh air inlet, in dwellings, the Swedish BBR: 0,35 l/s per m2 floor area, the Norwegian 1,2 m3 / hour per m2 floor are, and the Danish regulations are 0,30 l/s per floor area. Those requirements are connected to national international standards (Norway: NS-EN15241, Denmark: DS 447). Regarding radon, the Swedish and Norway regulations are maximum 200 bq/m3 and Denmark 100 bq/m3. Harmonization is ongoing.

Keywords: built/indoor environment

TH-SY-C2-41
Indoor Environmental Quality Regulation in Korea

Lee K.; Seoul National University, Seoul, Korea

[Background/Aim] Indoor environmental quality is important for protection of environmental health. The purpose of this presentation was to introduce indoor air quality regulation in Korea. In addition, the temporal variation of CO2 and PM in daycare centers was compared with the official test method managed by one-time measurement.

Keywords: analytical methods, built/indoor environment, environmental regulation

TH-PL-D2-ECS: VOCs, SVOCs and Particles: Toxicity and Effects on IAQ

TH-PL-D2-42
Airborne particles inside occupied residences more toxic than outdoor particles. Results from 16 residences in Sweden

Wierzbicka A.[1], Omelekhina Y.[1], Thoustrup Saber A.[2], Bloom E.[3], Strandberg B.[4], Pagels J.[1], Jacobsen N.R.[2]; [1]Ergonomics and Aerosol Technology, Lund University, Lund, Sweden, [2]The National Research Centre for the Working Environment, Copenhagen, Denmark, [3]RISE Research Institutes of Sweden, Stockholm, Sweden, [4]Occupational and Environmental Medicine, Lund University, Lund, Sweden

Toxicity and physicochemical characteristics of airborne particles found indoors remain not well understood, even though we spend the majority of our time indoors. There is a clear need to better understand effects of exposures in our homes. The aim of this study was to assess the differences in physicochemical properties and toxicity of particles found indoor and outdoor occupied residences in Sweden. Weeklong, simultaneous indoor and outdoor measurements were conducted in sixteen occupied residences in southern Sweden during winter 2016/17.

Residences comprised of detached single-family houses and apartments with natural or mechanical ventilation. Two identical sets of instruments for indoor and outdoor measurements were used to assess physical characteristics of particles and collect particles for off-line chemical analysis and toxicological studies in mice. Measured time-resolved physical characteristics were number concentration and mean size, PM2.5 mass and black carbon concentrations. Off-line analysis comprised determination of polycyclic aromatic hydrocarbons, metals and soluble ions. Particles collected for toxicological studies were extracted in methanol and vacuum evaporated. Mice received a single intratracheal instillation of 18, 54 and 162 µg of each of the pooled particle samples, namely indoor, outdoor and blanks. Lung inflammation and DNA damage in broncheoalveolar lavage cells, lung and liver tissue were evaluated 1, 3 and 28 days after intratracheal instillation. Average particle number concentration (10-300 nm) was higher indoors than outdoors. Gravimetrically assessed PM2.5 mass concentration was higher outdoors than indoors. Higher concentrations of metals Cr, Mn, Fe, Co, Ni, Cu, Cd, and Ba were measured indoors than outdoors. On average soluble ion levels Cl-, NO3-, SO4- were higher outdoors than indoors. Measured endotoxin levels were generally low, but with higher concentrations indoors. Results of toxicological studies in mice showed higher toxicity of particles collected indoors compared to outdoors. Swedish Research Council FORMAS financed this project.

Keywords: particles

TH-PL-D2-43
Identifying new substances in indoor air from a retrospective analysis of chromatograms

Achille J.[1], Nicolas M.[1], Ramalho O.[1], Thiry P.[1], Boulanger G.[2], Keirsbulck M.[2], Pernelet-Joly V.[2], Mandin C.[1]; [1]Scientific and Technical Center for Building (CSTB), Paris, France, [2]French Agency for Food, Environmental and Occupational Health and Safety (ANSES), Maisons-Alfort, France,

When conducting air monitoring campaigns, the list of target substances to be analyzed is usually established beforehand using literature reviews and previous field experiments. The risk of this approach is to miss specific compounds that have never or rarely been documented but may have health effects. Even though non-target analysis is being developed and allows identification of a much larger range of compounds, it remains expansive for the moment. An alternative approach was tested to identify untargeted compounds within gas chromatograms from previous sampling and analysis. An indoor air monitoring campaign was performed in 2016-2017 in recently built dwellings in Paris metropolitan area. Volatile organic compounds were sampled during seven days through passive sampling using Carbograph 4 sorbents and analyzed by gas chromatography-mass spectrometry. The chromatograms were reanalyzed in two steps. First, a subset of 40 chromatograms was manually screened and peaks were identified. The identification threshold was set at 1 µg/m3 in toluene equivalent. Secondly, the resulting list of newly identified substances was used as targets for an automatic screening of all available chromatograms (n = 81). The occurrence in dwellings was calculated for each identified substance, along with semi-quantitative data for concentration value. The analysis of 40 chromatograms out of the 81 available allowed the identification of 168 substances that were not targeted previously, and acquisition of semi-quantitative data for the entire studied sample. Most frequently found substances will be presented along with their respective concentration range. These results provide a new insight on volatile substances in indoor air. The next step will be the hazard assessment of the 168 substances to help to identify substances of concern. Another prospect is to compare results with outputs from an automated identification software that could allow big data treatment without prior knowledge of analytical conditions.

Keywords: VOCs, air, analytical methods, health, other/general

TH-PL-D2-44

On-line coupling of thermal extraction with gas chromatography / tandem mass spectrometry for the analysis of semi-volatile organic compounds (SVOCs) in a very small amount of indoor settled dust

Mercier F.[1], Raffy G.[1], Gilles E.[1], Soulard P.[1], Mandin C.[2], Dassonville C.[2], Le Bot B.[1]; [1]EHESP/Irset-Inserm UMR, Rennes, France, [2]CSTB, Marne-la-Vallée, Paris, France

A large number of studies dealing with the contamination of indoor settled dust by SVOCs have been published in recent years and many analytical methods have already been proposed. A new multi-residue method based on thermal extraction (TE) and gas chromatography/tandem mass spectrometry (GC/MS/MS) was developed to simultaneously quantify, from a very small amount of sample (a few milligrams), a wide range of semi-volatile organic compounds (SVOCs) of health concern including polycyclic musks, organochlorines (OCs), organophosphates (OPs), oxadiazolones, polycyclic aromatic hydrocarbons (PAHs), polybromodiphenylethers (PBDEs), polychlorobiphenyls (PCBs), phthalates and pyrethroids, in indoor settled dust. Method LOQs ranged from 5 ng/g for PCBs, oxadiazon, 4,4'-DDE and 4,4'-DDT to 2000 ng/g for DEHP for a 2 mg sample of sieved dust. The proposed method was successfully validated in terms of accuracy and precision via replicate analysis of four different standard reference materials (SRMs 1649b (Urban Dust), 2585 (Organic Contaminants in House Dust), 2786 and 2787 (Fine Atmospheric Particulate Matter)) supplied by the National Institute of Standards and Technology (NIST) and then applied to five real indoor settled dust samples collected in French schools. In addition, its performances were compared to those of a previously published method based on pressurized liquid extraction (PLE) and GC/MS/MS. The different results obtained demonstrate the advantages of the proposed method over conventional methods and illustrate its two main features: i) its ease of use and rapid implementation (only three steps: sieving, weighing and analysis), making it therefore particularly appropriate for environmental monitoring programs or large-scale studies and usable by the largest number of laboratories and ii) its ability to precisely and accurately quantify a wide range of SVOCs from trace (a few ng/g) to highly concentrated (several mg/g) compounds from a very small amount of sieved dust sample (2 mg).

Keywords: SVOCs, analytical methods, built/indoor environment, particles

TH-PL-D2-45

Indoor environmental quality in multi storey office buildings and its implication on the health and safety of workers. Evaluation of Lagos State Government Administrative buildings in Nigeria

Kadiri S.[1], Dakwak L.S.[1]; [1]University of Nigeria, Nsukka, Nigeria

The problems associated with indoor environment quality (IEQ) and the possible subsequent effects thereof on workers, as experienced by the employees of Lagos state government in multi storey office buildings in Lagos, Nigeria, were investigated and are discussed. To find a possible association between these poor physical environmental indoor conditions and the potential adverse affects on the health and motivation of workers, a cross sectional survey was conducted to assess the physical, biological and chemical quality of indoor environmental quality (Refer to as IEQ – Indoor Environmental Quality). The data were collected between Feb and April 2012. A total of 1000 Questionnaires were randomly distributed among employees working in the twelve multi storey buildings within the Secretariat out of a total of sixteen. Each building consists of six floors each and with an average of sixty offices in each block. Occupants in each office vary between one and maximum of six. Eight hundred and ninety seven (897) responded to the questionnaires. The questions addressed how poor air quality, insufficient awareness on effects of poor IEQ i.e. training on health and safety as regards to IEQ, lack of access to daylight, unpleasant acoustic conditions, and control over lighting and thermal comfort leads to dissatisfaction with the IEQ of the buildings. Observations from the data led to the view that the satisfactory level of IEQ awareness is low among the occupants. Organisational structure needs to be formed that will enlighten occupants about factors that contribute to poor indoor air quality (IAQ).

Keywords: VOCs

TH-PL-E2-CPC: Ventilation and Indoor Air Quality Studies

TH-PL-E2-46

Humanity moved from "Ecosystem Nature" into "Sealed Building Biotopes"

Hugentobler W., Former lecturer at Inst. of Primary Care, University Zurich

In the very short timeframe of the last two hundred years, industrial revolutions shifted our living and working spaces more and more into buildings, which became increasingly airtight. Spending more than ninety percent of our lifetime in sealed rooms, often mechanically ventilated with filtered air, limited our interactions with the diversified microbiomes of nature. Our building habitats have become sealed biotopes. For hygienic reasons, natural and locally available mainly porous materials for construction, interior finish and accessories have increasingly been replaced by fabricated, easy to clean, nonporous materials. These materials lack any water storage capacity and are either wet (if undesired condensation occurs) or dry, independent of the relative humidity in the surrounding air. From the perspective of surface attached bacteria and mould, the consequence is a permanent, year round stress by deprivation of the essential resources water and nutrients. This stress scenario, intended to efficiently prevent any proliferation of bacteria and mould for hygienic reasons, turned out to have dangerous side effects on human beings. The provoked stress adaptations (morphologic, metabolic and genetic) of bacteria result in multi-resistance, including antibiotic resistance. This issue is becoming particularly evident in hospital environments where we had to recognize that the multi-resistant gram positive and negative bacteria causing health care associated infections share the common feature of resistance to water and nutrient deprivation. We need to reconsider disinfection, sterilisation, dryness and cleanliness as hygienic concepts in the light of updated knowledge of microbial stress reactions. Instead of bulk reduction of bacterial load as a proxy, we should as "those actions and practices that reduce the spread or transmission of pathogenic microorganisms, and thus reduce the incidence of disease."

Keywords: indoor living, nonporous building materials, water shortage, stress, microbial resistance

TH-PL-E2-47

The impact of cooling and ventilation strategies on exposure in lecture rooms using CO2 as surrogate

Yang J.[1], Tham K.W.[1], Santamouris M.[2], Sekhar C.[1], Mohan Kumar D.I.[1], Horák O.[3]; [1]National University of Singapore, Singapore, [2]University of New South Wales , Sydney, Australia, [3]Czech Technical University in Prague, Prague, Czech

Air conditioning and ventilation strategies in lecture rooms impact energy and indoor exposure with consequential effects on occupant satisfaction with the indoor environmental quality. Optimizing thermal comfort, ventilation effectiveness and energy efficient is challenging. Few field studies on the penalty of over ventilation or under ventilation in lecture rooms with real occupant data and the associated occupant exposure have been reported. In this study, intervention was performed on the temperature setpoints in the lecture room operating with a Variable Air Volume Air-Conditioning system. There was uneven exposure to thermal and air quality of the occupants due to the stepped-incline seating design of the lecture room. We present a systematic analysis of the compromises among air conditioning energy, ventilation, CO_2 exposure and occupants' satisfaction in indoor space. In particular, there was diversity in the associated perception, satisfaction and health (SBS) symptoms.

Keywords: ventilation

TH-PL-E2-48
Thermal comfort in a deep plan naturally ventilation atrium: a two year case study.

Hathway A.[1], Ramsay T.[1], Hacker J.[2], Salgado A.[2], Rana D.[2]; [1]University of Sheffield, Sheffield, United Kingdom, [2]Arup, London, United Kingdom

Natural ventilation can provide an energy efficient approach to achieving thermal comfort in many parts of the world. The provision of tall atriums, utilising the buoyancy effect, is a common approach to providing large natural ventilated spaces for circulation or informal meeting spaces. Alternatively they are used to assist the movement of air through the surrounding rooms by utilising the atrium to increase the pressure differences driving the flow. A lot is understood about the performance of atriums from a substantial body of theoretical work and scaled water bath experiments. More recently this has been complimented with CFD studies. Within practice CFD is a common tool to aid the design of such spaces, ensuring the risk of draughts in cold weather is minimised and that fresh air is well mixed, avoiding areas of short circuiting and poor air quality. However, detailed analysis of built atriums is limited, largely due to the complexities of observing the internal conditions over such a large volume, particularly when occupied. In this study the availability of 58 temperature sensors and 18 CO_2 sensors within a 4 storey (20m), 70m long, atrium allows the conditions in a real atrium to be investigated in a wide range of external conditions and window opening scenarios. Clear differences between the summer displacement scenario (opening façade and roof windows) and the winter mixing condition (only roof windows) are shown demonstrating the importance of having different design approaches for each season. Further the distribution of temperature and CO_2 readings across the floor plate allow the ability of air to penetrate into the centre of the space to be understood.

Keywords: built/indoor environment, green buildings, thermal environment, ventilation

TH-PL-E2-49
The Microbiome Of Occupants And The Built Environment At The United States Air Force Cadets

Hoisington A.[1], Maestre J.[2], Kinney K.[2], Brenner L.[3], Lowry C.[4]; [1]Air Force Institute of Technology, Wright-Patterson AFB, OH, USA, [2]University of Texas Austin, Austin, TX, USA, [3]Department of Veteran Affairs, Denver, CO, USA, [4]University of Colorado Boulder, Boulder, CO, USA

The microbiome of the built environment has important implications for human health and wellbeing; however, bidirectional exchange of microbes between occupants and surfaces can be confounded by lifestyle, architecture and external environmental exposures. Here, we present a longitudinal study of United States Air Force Academy cadets which have substantial homogeneity in lifestyle, diet, and age, all factors that influence the human microbiome. We characterized bacterial communities associated with: (1) skin and gut samples from roommate pairs; (2) four built environment sample locations inside the pairs' dormitory rooms; (3) four built environment sample locations within shared spaces in the dormitory; and, (4) roommatched outdoor samples from the window ledge of their rooms. Linear convergence of microbial composition and structure was observed between an occupants' skin and the dormitory surfaces that were only used by that occupant (i.e., desk). Conversely, bacterial community beta diversity (weighted Unifrac) convergence between the skin of both roommates and the shared dormitory floor between the two cadet's beds was not seen across the entire study population. Cohabitation was significantly associated with increased skin microbiota similarity, but did not significantly influence the gut microbiota. Following a departure from the occupied space of several weeks, the skin microbiota, but not the gut microbiota, showed a significant reduction in similarity relative to the building. In addition to the cohabitation, results showed interesting confounding factors that have not been well defined by other research to include influence on skin and desk microbiomes from predictive mean vote analysis, correlation between salivary cortisol levels and skin microbiomes, and influence in young adult guy microbiomes depending upon childhood locations of metro or non-metro areas. Overall, longitudinal observation of these dynamics enables us to dissect the influence of occupation, diet and lifestyle factors on occupant and built environment microbial ecology.

Keywords: microbial agents

TH-PL-A3-EHR: Multi-Chemicals' Assessment Indoors

TH-PL-A3-50
Can we really promote air quality by essential oil diffusion?

Verriele M.[1], Angulo Milhem S.[1], Nicolas M.[2], Thevenet F.[1]; [1]IMT Lille Douai, Douai, France, [2]CSTB, Grenoble, France

Essential oil blend diffusers create the buzz on "green lifestyle" or "diy" blogs! Inconsistent posts promote air cleansing and toxin-neutralizing by the diffusion of "purifying essential oil blends". Although their antibacterial and sedative properties can be undoubtedly acknowledged, the terpenoid compounds they contain are characterized by significant allergenic and irritant effects. These species are highly reactive with air oxidants and may even lead to fine particle formation which is worrying regarding indoor air quality and occupant exposure. Our study addresses the question of terpene emissions from the use of essential oil diffusers indoors. Experiments implementing several types of diffusers have been designed to assess pollutant concentration levels resulting from the vaporization of tee-tree oil in a 1-m3 experimental chamber. The emission dynamics and the duration of the impact on air quality have been recorded using a custom compact GC-FID analyzer. Contrasted behaviors have been evidenced. (i) Diffusing oil by an ultrasonic mist diffuser or by a soft oil warmer induces similar terpene concentration trends: compounds are efficiently and rapidly emitted leading to a concentration peak, before being purged by the chamber air renewal. The maximum concentration is reached between 2 and 5 hours depending on the compound volatility; however the concentration peak is dependent on the compound volatility and on its mass fraction in the pure oil. (ii) In contrast, a capillary diffuser leads to long-term emissions: after several hours, steady state concentrations are achieved and maintained for several days, depending on the volume of the flask. Moreover, analyzing the relative abundances between emitted terpenes, it was noticed that the footprint of the pure diffused oil is here more consistent using capillary diffusers. The final purpose of this kind of study is to propose effective recommendations on the use of such substances indoors.

Keywords: VOCs, activity patterns, consumer and personal care products

TH-PL-A3-51
Indoor VOCs, aldehydes and particle number concentration measurements in the German Environmental Survey for Children and Adolescents (GerES 2014-2017)

Birmili W.[1], Daniels A.[1], Bethke R.[1], Neumann K.[1], Schechner N.[1], Miethig S.[1], Brenske K.-R.[1], Kura J.[1], Pietsch A.[1], Niemeyer T.[1], Rucic E.[1], Kolossa-Gehring M.[1], Conrad A.[1]; [1]German Environment Agency, Berlin, Germany

Various indoor air pollutants, including volatile organic compounds (VOC), aldehydes, and particle number concentrations were measured during the German Environmental Survey for Children and Adolescents (GerES V, 2014-2017). Residential indoor air was characterised to obtain representative pollutant concentrations for the 3-17 years old participants and to identify factors associated with elevated pollutant levels. The survey was performed jointly with Wave 2 of the German Health Interview and Examination Survey for Children and Adolescents (KiGGS W2) of the Robert Koch Institute (RKI). More than 600 rooms were examined for VOC and aldehydes, and more than 2,100 for total particle number concentration. VOC were collected over 7 days using Perkin Elmer/Markes tubing-type passive samplers using Tenax TA® as a sorbent. Thermodesorption and coupled gas chromatography/mass spectrometry were used for quantification. Aldehydes were collected over 7 days using UMEx 100 passive samplers, with DNPH derivates being quantified by coupled liquid chromatography/mass spectrometry. Total particle number concentrations were measured during one hour with condensation particle counters (CPC, type TSI 3007). We observed, amongst others, the following approximate geometric mean concentrations: 275 µg Total VOC/m^3, 22 µg formaldehyde/m^3, and 5900 particles/cm^3. We present detailed results including a comparison to previous surveys. We further document how pollutant concentrations can vary by age, sex, type of residential area, socio-economic status, migration background, the type of furniture and flooring present in the room of measurement, as well as the proximity to traffic sources. We thank all participants and the RKI for making GerES V possible. Funding by the German Ministries for the Environment, Nature Conservation and Nuclear Safety (BMU) and of Education and Research (BMBF) is gratefully acknowledged.

Keywords: VOCs, built/indoor environment, children, particles, sampling methods, statistical methods

TH-PL-A3-52

Bioaerosols in indoor households post-2018 Kerala (India) floods

Muthalagu A.[1], Qureshi A.[1], Bhargava A.[1]; [1]Hyderabad, Telangana, India

Floods account for about 40% of all natural disasters. Their frequency and intensity are expected to increase under climate change scenarios. Flood water brings along various microbes from the outdoor environment into the indoor environment, and increases the moisture within the indoor environment, increasing the risks of residents to infection. Here, we report the occurrence of bacteria and endotoxins, fungi and viruses in homes flooded during the 2018 Kerala floods, designated as a highest level disaster by the Indian government. Samples were collected from twenty-one flooded houses from the (declared) severely affected Kadapra village, one month after the onset of floods. Leptospirosis and H1N1 outbreaks have occurred in the area after the floods. The indoor water level in affected houses reached as high as 1.8 m. Four houses in the same village that did not experience any flooding were used as 'flood' control. Average bacterial and fungal concentrations in indoor air of flooded homes were 3053 CFU (colony forming units)/m3 and 2258 CFU/m3, respectively, approximately three times higher than in control houses (1035 CFU/m3, 690 CFU/m3). Higher concentrations in affected households were likely a result of higher relative humidity and mold growth compared to the 'control' houses. Opportunistic pathogens of genus Staphylococcus, Roseomonas, Candida and Penicillium were detected in all the affected houses and accounted for more than 20% of total CFUs. Actinobacteria, Alpha, Beta and Gammaproteobacteria, Bacilli, Flavobacteria and Cocci were the major detected class of bacteria (Sanger sequencing), and Agaricomycetes, Dothideomycetes, Eurotiomycetes, Saccharomycetes, Sordariomycetes, Zygomycetes were the major detected class of fungi. Results on the occurrences of endotoxins and viruses (using qRT-PCR) in indoor air, and in soil and water just outside all the houses will also be presented. Our study will present new data on the presence of bioaerosols in India and viruses in flooded environments.

Keywords: built/indoor environment, environmental health, infectious disease, microbial agents, respiratory health

TH-PL-A3-53

The IMAGE project: A human biomonitoring assessment to evaluate glyphosate exposures among families in the Irish population

Connolly A. [1], Koch H.M.[2], Kolossa-Gehring M.[3], Conrad A.[3], Coggins M[1]; [1]Centre for Climate and Air Pollution Studies, School of Physics and the Ryan Institute, National University of Ireland, University Road, Galway, Ireland, [2]Institute for Prevention and Occupational Medicine of the German Social Accident Insurance, Institute of the Ruhr-Universität Bochum (IPA), Bochum, Germany, [3] German Environment Agency (Umweltbundesamt), Berlin/Dessau-Roßlau, Germany

Glyphosate is the highest volume used herbicide globally, widely and extensively used in the agricultural industry including for genetically modified crops as well as in horticultural industry and for amateur use, that has resulted in this chemical being ubiquitous in the environment. In 2015, the International Agency for Research on Cancer's classified glyphosate as a '2A probably carcinogenic to humans', intensifying the international controversy over this chemical. Glyphosate exposure can occur through occupational and amateur use, as well as diet and environmental exposure. The Human Biomonitoring for Europe (HBM4EU) initiative, which has an objective of advancing human biomonitoring in Europe, has recently classed glyphosate as a priority substance. Previous human biomonitoring studies have identified a potential for glyphosate exposures among the general population. A pilot study among the Irish population found detectable glyphosate levels of 0.80 – 1.35 µg L-1 in 20% of the urine samples collected, warranting further investigation. The frequency and levels of detectable glyphosate concentrations were comparable to results from German studies, e. g. time trend data from the German Environmental Specimen Bank (ESB). To fully evaluate glyphosate exposure among Irish families, the IMAGE project has been initiated – Ireland's bioMonitoring Assessment of Glyphosate Exposures: An environmental assessment of glyphosate exposures among the Irish population using human biomonitoring sampling strategy. This follow-on human biological study will collect and analyse urine samples from farm and non-farm families', from parents and children, in Ireland for glyphosate, and its main metabolite, Aminomethylphosphonic acid. Questionnaires will be used to collect detailed contextual information, which will align with the HBM4EU protocols, to support the human biomonitoring data including information on diet, occupation, lifestyle and health. Preliminary results of the IMAGE project will be presented at the conference including the methodology, sampling protocols and a novel chemical analysis method developed for the project.

Keywords:

TH-PL-A3-54

The benefit of multi-residue analysis in biomonitoring – an example of polycylcic aromatic hydrocarbons exposure assessment with hair analysis

Appenzeller B.[1], Palazzi P.[1], Mezzache S.[2], Bourokba N.[2], Hardy E.[1], Schritz A.[1], Bastien P.[2], Emond C.[1], Li J.[2], Soeur J.[2]; [1]Luxembourg Institute of Health, Strassen, Luxembourg, [2]L'Oréal Research and Innovation, Aulnay sous Bois, France

Background Due to technical limitations, the assessment of exposure to PAH has often been conducted with a limited number of biomarkers or even with a single one (e.g. urinary 1-hydroxypyrene). The relevance of this "one for all" approach is however questioned by several authors, and recent studies made significant efforts to increase the number of biomarkers in order to provide a more comprehensive assessment of the global exposure to PAH. By means of a multi-residue method, the present study aimed at documenting the cumulative exposure to PAH and testing the capacity of each biomarker to be representative of the global exposure. Methods In this work, hair samples were collected from 204 women aged 25-45 y living in two Chinese cities (Baoding and Dalian) and analyzed with a multi-residue method allowing for the detection of 71 biomarkers simultaneously: 15 parent PAH and 56 PAH metabolites. Inter-biomarkers associations were tested with regression analysis and Kendall's τ correlation coefficient for censored data. Results The 15 parent PAH and 31 of the metabolites were detected in at least one subject. Eight parent PAH and 7 metabolites were detected in all the samples. Although highly positive correlations were observed between some pollutants and groups of pollutants, several biomarkers only presented low levels of association with the others. On the whole, no biomarkers presented a high level of association with all the other ones. Conclusions The multi-residue method used here demonstrated the presence of at least 15 biomarkers in all the subjects and clearly highlighted the cumulative exposure of the population under study. The statistical analysis revealed that no biomarker out of the 71 tested here (being parent or metabolite) could be considered representative of all the other ones. The results highlight the benefit of multi-residue methods over single "surrogate biomarker" to better document individual's exposome.

Keywords: aggregate exposure, biomarkers, biomonitoring, cumulative exposure

TH-PL-A3-55

Exposure to recovery workers and residents from chemicals released during floods - risk mitigation and applications for waterfront communities.

Chaisson C.F.[1], Chari R.[2], Madrigano J.[2], Osorio J.C.[3], Diskin K.[1]; [1]The LifeLine Group, Annandale, VA, USA, [2]RAND Corp, Arlington, VA, USA, [3]NYC-EJA , NEW York NY, USA

After Superstorm Sandy and dozens of floods throughout the US, it is evident that floodwaters can free previously contained chemical contaminants for indiscriminate distribution throughout affected neighborhoods. We studied this scenario and the risks these fugitive chemicals pose for residents of Brooklyn, New York where industries, transportation routes, small businesses, schools and residences are closely amassed—much like the dynamics of most waterfront community areas. We applied new data collection techniques, including social media sources, and adapted exposure assessment models to consider people's activities unique to disasters and those common to everyday life post storm events. Potential exposures were calculated with and without mitigation equipment and/or strategies to reveal most risky scenarios and most efficacious mitigations. This evaluation can be applied to any community. It entails identification of chemicals from the 6 chemical source categories and applying existing geospatial and exposure assessment models. Key to these projects was community-based research methods, which informed the assessment of source categories, generation of activity profiles, and dissemination of results. Use of data from industry, business records and social media greatly inform the assessments. We'll show models and calculations used to calculate potential chemical concentrations across city areas under different storm conditions and modeling of exposure as a function of concentration in the media and activities by which people are exposed. Depending on activities that individuals were engaged in, and personal protective equipment used, our results demonstrate that exposures can be mitigated significantly. Our results from two chemical sources in the Brooklyn waterfront community (industrial and small chemically intense businesses throughout residential areas) will be presented for exposures to non-professional recovery workers and post-recovery general populations returning to their homes, businesses, and public areas. We'll present key findings and the template for application to any community considering risks from flooding events.

Keywords: susceptible/vulnerable

TH-PL-B3-56
Neonicotinoid exposure in the U.S. general population

Calalfat A.[1], Wang L.-Y., Baker S., Bishop Serafim A., Morales-Agudelo P., Ospina M.; [1]Centers for Disease Control and Prevention, Atlanta, GA, USA

Background: Neonicotinoids are agricultural insecticides, also used for flea control in household pets. Neonicotinoids have become popular replacements for other insecticides (e.g., organophosphates, carbamates), and use may be on the rise. Methods: We measured urinary concentrations of six neonicotinoid biomarkers: acetamiprid, clothianidin, imidacloprid, thiacloprid, N-desmethyl-acetamiprid, and 5-hydroxy-imidacloprid in 3,038 samples from participants 3 years and older from the 2015–2016 National Health and Nutrition Examination Survey. We calculated distribution percentiles and used regression models to evaluate associations of various demographic parameters with urinary concentrations above the 95th percentile (a value that represents higher than average concentrations) of neonicotinoid biomarkers. Results: The weighted detection frequencies were 35% (N-desmethyl-acetamiprid), 19.7% (5-hydroxy imidacloprid), 7.7% (clothianidin), 4.3% (imidacloprid), and <0.5% (acetamiprid and thiacloprid). Median and 95th percentile concentrations were <0.40–1.37 μg/L (5-hydroxy imidacloprid), <0.20–1.29 μg/L (N-desmethyl-acetamiprid), and <0.20–0.396 μg/L (clothianidin). Children 3–5 year old who fasted <8 hours were more likely to have N-desmethyl-acetamiprid concentrations above the 95th percentile than adolescents (adjusted odds ratio (OR) = 3.12; 95% confidence interval [CI], (0.98-9.98) and adults (OR = 4.29; 95% CI, (2.04-9.0)). Asians were two times more likely than non-Asians to have concentrations above the 95th percentile of N-desmethyl-acetamiprid and 5-hydroxy-imidacloprid (OR = 1.94; 95% CI, (1.08-3.49) and 2.25; 95% CI, (1.44-3.51)), respectively. The probabilities of having N-desmethyl-acetamiprid and 5-hydroxy-imidacloprid concentrations above the 95th percentile in the summer were 1.55 and 2.43 times higher than in winter, respectively. Conclusions: These data suggest human exposure to select neonicotinoids in the U.S. general population 3 years of age and older. Furthermore, metabolites appear to be better biomarkers of background exposure than the parent compounds themselves. Compared to others, young children and Asians may experience higher exposures although reasons for such differences remain unknown.

Keywords: pesticides

TH-PL-B3-57
Cigarette Smoking, Smokeless Tobacco Use Prevalence, and Secondhand Smoke Exposure Evaluated Using Serum Cotinine Levels in the Time-frame of the Cheyenne River Sioux Tribal Clean Air Ordinance

Erdei E.[1], O'Donald E.[1], Miller C.[1], O'Leary R.[2], Foos K.[1], Pacheco B.[1], Nez-Henderson P.[3], O'Leary M.[2], Lewis J.[1], Henderson J.[3]; [1]University of New Mexico Health Sciences Center, Albuquerque, New Mexico, [2]Missouri Breaks Industries Research Inc. , Eagle Butte, NM, USA, [3]Black Hills Center for American Indian Health, Rapid City, SD, USA

Cheyenne River Sioux Tribal partners and communities sought academic partnerships as early as in 2000 in order to address concerns of immunotoxicity associated with mining waste metals to public water sources and PM exposures. The Cheyenne River Sioux Tribe is among the first Tribal Nations in the U. S. actively controlling tobacco exposures in public since 2015. Cigarette smoking is also a contributor of metal exposures. We collected cigarette smoking, smokeless tobacco and second hand smoke (SHS) information from 225 CRST adults (53% males, 47% females; mean age: 42.9 years (range: 18-77 yrs) enrolled to our Tribal environmental health study in addition to fishing and land-use surveys. Blood and urine samples were obtained for personal biomonitoring of metals and metalloids. We explored the effects of the new tobacco policy among participants enrolled into our study by describing tobacco use, ETS prevalence, and identified predictors of serum cotinine (SC) concentrations. Quantile regression modeling (QRM) of SC natural log values was utilized to provide estimates robust against SC outliers. Interaction term between active smoking and tobacco chewing was also included in the QRM. From all participants, 51% (N=114) reported current smoking and 58% (N=129) had current SHS exposure, of those 35 participants were non-tobacco users. Among tobacco chewers, 44% also smoked cigarettes. Significant differences in SC median concentrations were found among participants with and without current SHS exposure. Extremely high SC concentrations (~100 times larger than the median) were also detected in some non-tobacco users. After implementing the smoke-free air policy, SC decreased in CRST participants with intermediate (3-15 ng/ml) and high (>15 ng/ml) SC levels showing association with an abatement of opportunities for SHS exposure. Significant predictors of SC levels were sampling year, current smoking, and smokeless tobacco use. No gender differences were observed in SC.

Keywords: biomarkers, biomonitoring, community, environmental policy

TH-PL-B3-58
Moisture damage related microbiota of classrooms and associations with respiratory health.

Adams R.I.[1], Karvonen A.M.[2], Leppänen H.K.[2], Jacobs J.[3], Vepsäläinen A.[2], Borrás-Santos A.[4], Valkonen M.[2], Krop E.[3], Zock J.-P.[4], Heederik D.[3], Hyvärinen A.[2], Pekkanen J.[5], Täubel M.[2]; [1]California Department of Public Health, Richmond, CA, USA, [2]Environmental Health Unit, National Institute for Health and Welfare, Kuopio, Finland, [3]Institute for Risk Assessment Sciences, Utrecht University (IRAS), Utrecht, The Netherlands, [4]IS Global, Barcelona, Spain, [5]Department of Public Health, Helsinki University, Helsinki, Kuopio, Finland

While associations between moisture damage and dampness in indoor environments and ill health are well established, the contribution of individual exposure agents to the adverse health effects remains largely elusive. The HITEA (Health Effects of Indoor Pollutants) school study was designed to investigate the role of microbes in the link between moisture damage and health. Here, we explore associations between moisture damage in school buildings, the classroom microbiota, and respiratory symptom prevalence in pupils. Settled dust samples were collected with electrostatic dustfall collectors from more than 150 classrooms in 10 Dutch and 6 Finnish primary schools, categorized as 'index' and 'reference' based on observations of moisture damage, dampness, and visible mold. Microbial measurements included qPCR targeting Gram positive and negative bacteria, total fungal DNA and Penicillium/Aspergillus spp. group, and bacterial 16S rRNA and fungal ITS amplicon sequencing. Differences in microbial levels were tested using Wilcoxon two sample test; moisture damage indicator taxa were identified using ANCOM (Analysis of composition of microbiomes); and associations with respiratory symptom (wheeze, congestion/phlegm, dry cough, nasal symptoms) prevalence are being tested in multivariable logistic models. Moisture damage was significantly ($p<0.05$) associated with elevated levels of most microbial qPCR markers in Finnish and Dutch classrooms and with bacterial and fungal species richness in Finland. Preliminary results indicate positive associations between qPCR levels as well as bacterial species richness and respiratory symptoms (mostly wheeze and nasal symptoms) in Finnish but not Dutch pupils. The moisture indicator taxa analyses performed on sequence data identified two fungal and one bacterial taxa in Finnish, and 21 fungal and 11 bacterial taxa in Dutch dust samples, including several sequence types of Sphingomonas (bacteria) and Vishniacozyma (yeast) species. Association of these indicators with respiratory symptoms are currently being tested; results will be presented in Kaunas.

Keywords: built/indoor environment, children, dampness/mold, microbial agents, respiratory health

TH-PL-B3-59

Quantifying the Short-Range Airborne Transmission of Expiratory droplet Nuclei by Two Breathing Thermal Manikins

Liu L.[1,2,3], Fu L.[2,3], Wang Y.[2,3], Zhang Y.[2,3], Meng X.[2,3], Wang L.[2,3], Duan M.[1], Nielsen P.V.[4], Jensen R.L.[4], Li Y.[5]; [1]Department of Building Science, Tsinghua University, Beijing, China, [2]State Key Laboratory of Green Building in Western China, Xi'an, China, [3]School of Building Services Science and Engineering, Xi'an University of Architecture and Technology, Xi'an, PR China, [4]Department of Civil Engineering, Aalborg University, Aalborg, Denmark, [5]Department of Mechanical Engineering, the University of Hong Kong, Hong Kong, China

Considerable uncertainty exists regarding the relative importance of short-range/long-range airborne, large droplet and contact transmission of respiratory viruses. Consequently, leading authorities have made contradictory recommendations about intervention methods for influenza, SARS and other respiratory diseases. In this study, we conduct both full-scaled measurements and numerical simulation to quantify the short-range airborne transmission. We locate two breathing thermal manikins in an ISO5/Class 100 cleanroom, release monodispersed particles from 1 to 6 μm from one manikin, and count the number of inhaled particles by the other manikin. The cleanroom is ventilated by displacement with an air change of 2.3. The two standing manikins simulate average females of 1.7 m height that have an activity level of 1.2. Each release 90W heat, and the preheating period is over 12 hours. Impact of particle size, person-to-person distance, thermal stratification level were examined by triplicate or quadruplicate measurements, while the sampling time in each test is 1 hour. We find that the susceptible manikin inhales less than 2% of the particles exhaled by the index manikin, when the person-to-person distance is 0.5 m. The average inhaled number of 1.5 μm particles is 1.64 times of 5.5 μm particles. Adding to the volumetric potential of carrying pathogens, the relative importance of the 5.5 μm particles is about 30 times of 1.5 μm particles. The inhalation exposure are at least 5 times more than the average environmental exposure level, and the difference enlarges with the particle size. Ambient thermal stratification can alter the trajectory of exhalation airflows. It can boost the short-range transmission within 0.5 m by a maximum of 10 times, but it has almost no effect when the person-to-person distance is more than 1.5 m. Our findings could be used to improve understanding of the mechanisms of disease transmission and ventilation control method.

Keywords: Interpersonal exposure, large droplet transmission, building ventilation, respiratory infection, CFD

TH-PL-B3-60

Home environment and health: Domestic risk factors for rhinitis, throat symptoms and non-respiratory symptoms among adults across China

Zhang X., Norbäck D., Zhang Y., Li B., Zhao Z., Huang C., Deng Q., Qian H., Yang X., Sun Y., Sundell J.

Few studies exist on associations between the home environment in China and adult health. We studied associations between home environment factors (other than dampness and mould) and rhinitis, throat symptoms and non-respiratory symptoms among young parents in six cities across China (N=36,541). They were recruited as parents from randomly selected day care centers in the cities and answered a questionnaire on medical symptoms and the home environment. Multilevel logistic regression was used to calculate odds ratios (OR), adjusting for gender, atopy, smoking, home size and dampness/mould. Overall, 3.1% had weekly rhinitis, 2.8% eye, 4.1% throat and 4.8% skin symptoms, 3.0% headache and 13.9% fatigue. Redecoration was associated with rhinitis, eye and skin symptoms, headache and fatigue. New furniture was associated with eye, throat and skin symptoms and fatigue. Gas cooking was associated with eye and throat symptoms, headache and fatigue. Biomass cooking was associated with eye and throat symptoms and headache. Burning incense was associated with eye, throat and skin symptoms, headache and fatigue. Presence of cockroaches and mosquitos or flies was associated with all six symptoms. Rats or mice were associated with eye and dermal symptoms. Cat keeping was associated with eye symptoms while dog keepers had less fatigue. Living near major roads was associated with rhinitis, eye, throat and skin symptoms and fatigue. Daily cleaning, a mechanical ventilation system in the kitchen or in the bathroom, living in older buildings and living in less urbanized areas were protective factors. In conclusion, urbanization, traffic exhaust, indoor emissions from redecoration and new furniture, gas cooking and air pollution from burning incense and biomass may cause mucosal and dermal symptoms, headache and fatigue among adults in China. Indoor animals (cats, mice/rats, cockroaches) were other risk factors. Daily cleaning, mechanical ventilation and living in older buildings can be protective.

Keywords: built/indoor environment, dampness/mold, environmental health, epidemiology, health, symptoms

TH-PL-B3-61

An Improved Approach for Highly-Efficient, Concentrated Sampling of Aerosol Particles from 10 nm to 10 µm: Bioaerosol Applications

Savage N.[1], Keady P.[1], Hering S.[2]; [1]Aerosol Devices Inc., Fort Collins, CO, USA, [2]Aerosol Dynamics Inc., Berkeley, CA, USA

Bioaerosols have the potential to negatively impact human health. However, regular monitoring of bioaerosols are a difficult challenge. Current bioaerosol sampling methods are limited resulting in a lack of quality information about the airborne nature and effects of bioaerosols. Active sampling methods are common tools for the collection of bioaerosols in indoor environments. However, most active air sampling methods have strong particle size dependencies for their collection efficiency and impart mechanical stresses on the collected microorganisms. The condensation growth tube collector (CGTC) uses a three-stage, moderated laminar-flow condensation approach to enlarge airborne particles as small as 5 nm at moderate temperatures. The high humidity of the sample flow and the warm water blanket encapsulating the particles is advantageous for viable microorganism collection. Live laboratory-generated virus, bacteria and yeast particles tested a prototype high-flow (8 L/min) condensation growth sampler and compared the results to commonly used liquid impingers. The collection of infectious H1N1 influenza virus particles was 13 times higher than the commonly used SKC BioSampler (Lednickly et al., 2016). Pan et al. (2016) observed the fraction of MS2 particles that remained viable after capture was 100 times greater for the CGTC then the SKC BioSampler. The CGTC had a detection limit lower than one order of magnitude for the observed yeast aerosol compared to the SKC BioSampler, regardless of the media used (Pan et al. 2018). We also present a new collection technique using the CGTC that instantly preserves microbial genetic material upon capture. This innovative method is directly compatible with molecular biology analyses including quantitative PCR, targeted genomic amplifications, metagenomic DNA sequencing, RNA sequencing as well as targeted biopolymer analyses and proteomics (Nieto-Caballero 2019).

Keywords: air, analytical methods, biomonitoring, infectious disease, respiratory health

TH-PL-C3-PCP: Community Involvement in Air Quality Studies

TH-PL-C3-62

Community-engaged exposure assessment: a case study using toenails as biomarkers of metal exposures

Wilson J. [1], Levy J. [1], Korrick S. [2,3], Vieira V., [4], Hebert K. [5], Fabian M.P. [1]; [1]Department of Environmental Health, Boston University School of Public Health, Boston, MA, USA, [2]Channing Division of Network Medicine, Department of Medicine, Brigham and Women's Hospital and Harvard Medical School, Boston, MA, USA, [3]Department of Environmental Health, Harvard T.H. Chan School of Public Health, Boston, MA, USA, [4]Program in Public Health, Susan and Henry Samueli College of Health Sciences, University of California, Irvine, CA, USA, [5]NorthStar Learning Centers, New Bedford, MA, USA

Background Population-based exposure monitoring studies can be costly and logistically challenging, especially when working within vulnerable communities. Strong partnerships with community organizations, including local businesses, can improve recruitment and provide value to the community. In this study, we focus on community-engaged strategies o recruit participants for an exposure assessment study in New Bedford, Massachusetts, a racially diverse low-income city with extensive contamination from a Superfund site, industrial sources, and hazardous waste sites. Methods Our goal was to recruit a sample of sociodemographically diverse reproductive-age women living in New Bedford, for both toenail collection to characterize metal exposures and administration of a short demographics and metal exposure risk factors questionnaire. Recruitment strategies included flyer advertisements, an established partnership with a family-oriented community organization, and developing partnerships with local nail salon owners. Our anticipated recruitment end date is May 31, 2019. Results We tested multiple recruitment strategies and met our recruitment target through the combined outreach and advertising efforts of our community partners, the nail salon owners, and the research team. Recruitment numbers increased substantially when systems were established to facilitate optimal engagement of nail salons and community partners by better aligning research and community partners' goals. For participants to date, 54.8% identified as white, 12.9% as black, and 22.6% as "other" (Cape Verdean, Greek, Hispanic, and Portuguese). The primary languages spoken were English (93.5%), Spanish, (22.6%) and Portuguese (19.4%), and the primary neighborhoods represented were North (48.4%), South (25.8%), and West (19.4%). Metal biomarkers have been heterogeneous across the population. Conclusions To optimize participation in exposure assessment studies, community members and relevant small businesses should be included in research study planning. In particular, establishing mutually beneficial partnerships with nail salons allowed for robust recruitment. The study reinforced the importance of community-engaged research strategies in which research goals are aligned with community partners' goals and researchers gain cultural competency.

Keywords: Exposure Assessment; Community Partnerships; Cultural Competency

TH-PL-C3-63

The effects of global efforts of promoting cleaning cooking practices on solid fuel use and early child development

Nazif-Munoz J.I.[1], Oulhote Y.[2], Spengler J.[1]; [1]Harvard University, Boston, MA, USA, [2]University of Massachusetts, Amherest, MA, USA

Background: Household air pollution from solid fuel burning is a contributor to delay early child development (ECD). A global effort to decrease solid fuel use (SFU) by promoting cleaner burning stoves, cleaner fuels, and/or providing chimneys or other ventilation changes, has been introduced since 2010. This program, called the 'Clean Cooking Alliance' (CCA), has targeted 100 countries, eight of which have received a more in-depth treatment. Thus, countries are not randomly exposed to global efforts to tackle SFU, and thus SFU variation may be dependent on their relative position to this program, and in turn affecting differently ECD. Objective: To investigate the effect of SFU on ECD and explore the possible influence of CCA's global efforts on SFU and ECD. Methods: A cross-sectional analysis of 30 countries with data on their relative position to CCA efforts--used as a proxy for CCA's differentiated efforts--linked to socio-demographics and early development of 76576 3-4 years children from the Multiple Indicator Cluster Surveys. A standardized global ECD index (ECDI) was calculated. Two-stage least squares regression models were conducted, using countries' relative position to CCA's efforts as an instrumental variable, to study the effect of SFU on ECDI. Results: After adjustment for covariates, absence of SFU was associated with an increase of 0.16 in ECDI (95% confidence interval, CI: 0.10-0.20). Countries relative position to CCA's efforts was an independent predictor of SFU (beta: 0.54, CI:0.25-0.84), F value= 14.52. When the relative position to CCA's efforts was used as an instrumental variable, the association between SFU and ECDI weakened but remained significant (beta: 0.15, CI:0.07-0.24). Conclusion: The use of solid fuels in the household is an independent predictor of a child's early development. Global efforts of the Clean Cooking Alliance appear to be improving early childhood development by facilitating better cleaning cooking practices.

Keywords: environmental policy

TH-PL-C3-64
Indoor Air Quality and Thermal Comfort in Elderly Care Center in Lithuania

Dobravalskis M.[1], Didžiariekytė U.[1], Leonavičiūtė G.[1], Gražulevičiūtė-Vileniškė I.[1], Prasauskas T.[1], Martuzevičius D.[1], Seduikyte L.[1]; [1]Civil Engineering and Architecture, Kaunas University of Technology, Kaunas, Lithuania

The ratio of people aged 65 or above to the ones aged 15-64 is going to increase in the EU in the coming decades. Elderly people living in elderly care centers (ECC) usually spend most of their time indoors, especially if the person has trouble mowing around easily. As a result of that, indoor air quality (IAQ) in their rooms can have a noticeable impact on their well-being and respiratory health [1]. This study presents thermal comfort (TC) and IAQ results carried out in February – March of 2019 in one of ECC located in Kaunas, Lithuania. TC parameters were measured in 15 naturally ventilated rooms (6 belong to an intensive care section) and in two rooms of the day care center. Employees of ECC were asked to evaluate IAQ and TC by filling in questionnaires. Results show that an average temperature in a room was 20.36°C and the average relative humidity was 37.38%. Both of these parameters were within the reference levels [2]. However, in some rooms values were outside national reference norms, as temperature was 17.52°C and relative humidity was lower than 24%. Questionnaire survey showed the different thermal perception of employees. Acknowledgment This work was supported by COST Action CA16226 "Indoor living space improvement: Smart Habitat for the Elderly (SHELD-ON)". References: 1. SIMONI, M., JAAKKOLA, M.s., CARROZZI, L., BALDACCI, S., PEDE, F. Di and VIEGI, G. Indoor air pollution and respiratory health in the elderly. European Respiratory Journal. 2003. Vol. 21, no. Supplement 40p. 15–20. DOI 10.1183/09031936.03.00403603. 2. HN 125:2011 "Suaugusių asmenų stacionarios socialinės globos įstaigos: Bendrieji sveikatos saugos reikalavimai". 2011.

Keywords: aging/elderly, air, respiratory health, thermal environment

TH-PL-C3-65
Report back of low-cost air particulate sensor data to community participants

Korfmacher K.[1], Hopke P.[1]; [1]University of Rochester Medical Center, Rochester, NY, USA

Community members, environmental health professionals and researchers have expanded the use of low-cost air particle monitors in recent years. However, sharing the results of these projects is challenging. Many people do not understand the limitation of these sensors, the sources and impacts of air pollution, or the challenges of linking particle pollution and health. Communicating patterns of air pollution when data is nearly continuous (reported each minute), but of limited accuracy, is particularly complex. Community engagement experts worked with air quality researchers to develop a readable report of indoor and outdoor particulate data from low-cost sensors located in and around homes. This presentation will share the resulting user-friendly report template and discuss the challenges of meaningfully synthesizing the health implications of findings for community members. Subsequently, the team developed "sensor stories" – interactive educational materials to build communities' understanding of monitoring data (www.sensorstories.urmc.edu), including the ability to appreciate uncertainty and interpret results appropriately. This presentation will highlight how monitoring projects with different goals may require different kinds of report-back and dissemination of data. Report-back may be useful for promoting retention, engagement, and education with diverse audiences.

Keywords: particles

TH-PL-C3-66

Fractured effects of solid fuel use in early child development: analyses of urban and rural territories in Ghana

Nazif-Munoz J.I.[1], Oulhote Y.[2], Spengler J.[1]; [1]Harvard University, Boston, MA, USA, [2]University of Massachusetts, Amherest, MA, USA

Background: About 41% of households globally, mainly in developing countries, rely on solid fuels for cooking with consequences for health outcomes in populations. In Ghana, more than 77% of the population depends on biomass fuels for cooking and heating. Previous studies posit that solid fuel use (SFU) in Ghana has had adverse effects on child health outcomes, stillbirth, and pregnancy. Yet these studies have not considered the effect on early child development (ECD), nor how SFU' potential effects vary across rural and urban areas. Objective: To investigate the potential effect of SFU on ECD, and whether it varies across rural and urban territories in Ghana. Methods: Using the 2011-2012 Ghana's Multiple Indicator Cluster Surveys – UNICEF (N = 2972 3-4 years children). A standardized global ECD index (ECDI) was calculated. A two-multilevel logistic regression model conditioning on household characteristics (second level), and individual-level covariates, corresponding to child and mother's characteristics (first level) were used to estimate the association between SFU and ECDI. Results: After adjustment for covariates, SFU was associated higher odds of not being developmentally on track (Odds ratio [OR]= 2.6; 95% confidence interval, CI:1.2, 5.6) compared to non SFU in all children of Ghana. This association was significantly different between rural and urban children, with an OR of 3.8 (95% CI:1.2, 12.3) in rural areas, compared with an OR of 1.5 (95% CI: 0.5, 4.1) in urban areas. Discussion: The use of solid fuels is an independent predictor of a children's early development in rural Ghana, however this association in urban territories was not observed. This may point to specific characteristics of rural kids that play a synergistic role with solid fuel use. Policy efforts which tackle solid fuel use should be mindful of these differences. More research is needed to assess what rural characteristics are drive this association.

Keywords: built/indoor environment

TH-PL-C3-67

What can you learn about Indoor Environmental Quality in Homes using Low Cost Sensors?

Miller S.L.[1], Auguste D[1], Wang L.[1]; [1]University of Colorado Boulder, CO, USA

Our research group has been deploying low-cost sensors, which measure VOCs, fine particulate matter, CO2, T and RH, in Denver and Boulder homes to assess indoor environmental quality. Additional data have been collected on respiratory health, asthma, and wellbeing. The goal of these studies have been (1) to engage families in data science in an effort to increase their knowledge and interest in science, technology, engineering and math, (2) to provide information about occupant activities that may degrade their homes' indoor environmental quality, and (3) assess any impacts in the home from a move off of fossil fuel dependence for energy to renewable energy. Results show that the household identified many key actions by occupants that could potentially degrade air quality and trigger asthma episodes, that home natural gas furnaces can generate constant low levels of carbon monoxide, and that interest in science and technology is increased but that identity as a scientist or engineer still needs to be developed.

Keywords: respiratory health, Low-Cost Sensors, PM2.5, VOCs, CO, CO2, asthma, energy

TH-PL-D3-68

Test method for measuring pollutant removal of induction cooktop with integrated downdraft

Jacobs P.[1], Borsboom W.[1]; [1]TNO, Delft, The Netherlands

Cooking fumes can be a major source of indoor exposure to moisture, odours, particulate matter and other contaminants. Local exhaust is the most effective way to capture these cooking fumes before they dilute in the kitchen and living room space. For wall mounted range hoods several test methods have been set up. However for induction cooktops with integrated downdraft, which are becoming more and more popular, no test method exist. In this study we developed a test setup to investigate which parameters have effect on the capture efficiency of these devices. This test setup was used to measure a number of downdraft units with different geometries to explore the most dominant parameters. Several pan positions, geometries and dry and boiling conditions have been tested. Results show that in case the exhaust is at cooktop level flow rates in excess of 450 m3/hour are required in order to capture boiling water fumes for more than 90%. Based on the physical measurements a CFD model will be set up. The CFD model will be use to test the effect of different conditions. Both the measurements and CFD model simulations will be used to set up a future test method.

Keywords: source control

TH-PL-D3-69

An intervention field study to improve ventilation, indoor air quality, and the respiratory health in northern and remote homes

Aubin D.[1], Ouazia B.[1], Poulin P.[2], Levesque B.[2], Tremblay F.[2], Louis L.-P.[3], Duchaine C.[3], Degois J.[3], Maltais F.[3], Brisson M.[4], Savignac M.[5], Boucher P.[5]; [1]National Research Council Ottawa, Canada, [2]Institut national de sante publique du Quebec, Quebec City, Canada, [3]Universite Laval, Quebec City, Canada, [4]Nunavik Regional Board of Health and Social Services, Quebec City, Canada, [5]Kativik Municipal Housing Bureau, Kuujjuaq, Canada

Homes with inadequate ventilation and indoor air quality (IAQ) are particularly common in northern and remote communities. Previous studies have observed that the indoor air in these homes can have elevated concentrations of CO_2, environmental tobacco smoke, and elevated relative humidity (RH) leading to mold issues. These conditions may cause various health problems, such as compromised respiratory health, for the occupants and in particular in children with developing respiratory systems.The objectives of this current study were to measure the effectiveness of targeted preventative maintenance of existing heating and ventilation systems at improving ventilation, IAQ, and the respiratory health of children. This study enrolled homes with children under the age 10 in both an intervention group (55 homes) and control group (24 homes) over the winter and spring of 2017-18 in Kuujjuaq, Québec, Canada. Various IAQ, ventilation, and behavioral characteristics were measured through field sampling and through questionnaires in the intervention group both before and after the intervention. No field sampling was conducted in the control group.Following the intervention we observed large reductions in the median values a number of health relevant IAQ parameters. We observed statistically significant reductions (pre-intervention, post-intervention, Δ%, p-value) in RH (31.4 RH, 28.5 RH, -9.3%, $p<0.05$), CO (2.9 ppm, 1.5 ppm, -46.7%, $p<0.01$), CO_2 (907 ppm, 668 ppm, -26.4%, $p<0.01$), PM2.5 (6.0 μg m^3, 4.4 μg m^3, -26.6%, $p<0.01$), formaldehyde (13.5 μg m^3, 10.3 μg m^3, -23.7%, $p<0.01$) while the measured air exchange rate increased by 68.7% (0.16 1/h, 0.27 1/h, +68.7, $p<0.01$). This study demonstrated that targeted preventative maintenance and optimization of ventilation systems can significantly improve ventilation rates and IAQ as a result. This presentation will discuss the ventilation and IAQ results obtained during this study while the health and microbiological aspects of this study will be discussed in later publications.

Keywords: ventilation

TH-PL-D3-70

Building air quality guide for air polluted environment vis-à-vis the need for real time data

Ogbonda U.; University of Salford, Manchester, United Kingdom

Air quality has been a reoccurring decimal in the recent event of research both in meeting sustainable goals and millennium development goals respectively. Priority placed on the most vulnerable vis-à-vis children and the current rate of out of school incidence due to health issues ranging from nausea, flu, lungs infections and the recent trend in cancer development in children is some of the carcinogenic effects. Research and several experiments have been carried out over a period allowing the specification of an air quality guide initiated by World Health Organisation (WHO), Federal Environmental Protection Agency (FEPA), United Nations (UN), Environmental Protection Agency (EPA) etc. These results based on a one-off experiment and used as a yardstick for different countries notwithstanding differences in economic activities, geographical location and the inherent micro activities undertaken in such places. Based on works of literature and existing air quality experiments undertaken in different locations in the Niger Delta Area of Nigeria where there is a high level of air pollution due to Gas flaring from oil exploration and exploitation activities. The individualistic nature of the experiment carried out and the degree of variance between different results shows the dilemma. Furthermore, the use of coal to cooking and insistent intensification of illegal bunkering makes it more micro sensitive placing more risk on the health of people. These activities make the guide provided less functional as it is not able to capture the immediate data in real time, which allows accurate measures of microclimatic air quality for efficient design of buildings in a polluted environment.

Keywords: environmental health

TH-PL-D3-71

A database to document and retrieve effectiveness test results of risk management measures for workers' exposures to chemicals

Fransman W.[1], Goede H.[1], Urbanus J.[2], Meijster T.[2], Wind T.[3]; [1]TNO, Zeist, The Netherlands, [2]Shell, Den Haag, The Netherlands, [3]Henkel, Düsseldorf, Germany

REACH and other European legislation require that companies demonstrate the safe use and control of hazardous substances. In order to predict the occupational exposure or environmental concentrations it is necessary to evaluate the Conditions of Use that are part of relevant exposure scenarios. As part of this it is essential to evaluate the quantitative effectiveness of Risk Management Measures (RMM). Under the CEFIC LRI B15.2 project an integrated risk management measure library (ECEL v2.0) was recently developed by TNO. The library applies a statistical method to derive and visualize effectiveness values for both occupational and environmental risk management measures (RMM). To this end, it combines data from earlier versions of ECEL and the CEFIC B15 library and provides access to the data through a new ECEL v2.0 online interface. The online tool provides a suitable and stable software platform with various functionalities. It facilitates the search and selection of exposure scenarios related to the RMM of interest (using frequency plots, prompts & warnings). Given the selection, the database applies a statistical approach of calculating a confidence interval. ECEL v2.0 allows the assessment of RMMs in order to derive a quantitative effectiveness value for a specific RMM. This can for example be applied for REACH dossiers and regulatory purposes. The data in the database are based on literature reports and have been assessed for quality, as well as linked to REACH use descriptors to facilitate their use in REACH implementation aspects such as registration dossiers, and on-site verification of Exposure Scenarios. The database is intended to cover occupational and environmental RMM's; the presentation will focus on the occupational part which is currently most developed: the occupational data amounts to 423 records.

Keywords: workplace

TH-PL-D3-72

Streetscape noise attenuation by tree canopy and ground vegetation cover in Toronto, Canada

Oiamo T.H.[1], Stefanova D.[1], Connor C.[1]; [1]Ryerson University, Toronto, Ontario, Canada

Environmental noise is a growing problem in urban centres. With a growing population and increasing proximity of residential areas to other land uses, large cities may fail to offer quiet spaces and peacefulness to its residents. Various studies demonstrate the effect of vegetation on noise attenuation. However, there is a lack of understanding of the magnitude of this effect in urban environments. This study investigated the relationship between tree canopy, ground cover vegetation and modelled traffic noise effects on observed daytime and nighttime noise levels. The analyses compared the effects of tree canopy cover, grass and shrubs in residential and mixed use zoning areas throughout the City of Toronto, Canada. Overall, the results showed a significant and strong effect of tree canopy cover in mixed use areas (e.g., residential-commercial) and a significant effect of vegetation ground cover in both land use categories. Specifically, in mixed use areas we observed a reduction of nighttime noise levels by 2.45, 3.02 and 4.34 dBA for receptors with <25%, 25-50% and 50% or more tree canopy cover within 50 m, respectively. We also observed a 3.00 dBA reduction for receptors with 50% or more tree canopy within 50 m in residential areas. Stronger attenuation by ground cover vegetation in residential areas may have confounded the effect of tree canopy cover, but the results nonetheless demonstrate the benefit of increasing tree canopy cover in mixed use areas where higher noise levels are generally also observed.

Keywords: acoustics/noise, geospatial analysis/GIS

TH-PL-D3-73

An exposome-based approach to compare the health impacts of multi-stressor exposures

Jolliet O.[1], Stylianou K.[1]; [1]Environmental Health Science, School of Public Health, University of Michigan, Ann Arbor, MI, USA

High-throughput exposure strategies comparing exposure to various nutritional, chemical and physical stressors are needed for exposome wide associations. This presentation shows how recent databases and models enable us to predict exposure and impacts for multiple stressors. Occcupational exposures were assessed across the entire US industry, based on a) measured noise levels and b) observed organic chemical concentrations at workpace. c) Dietary exposures and nutritional impacts were assessed using NHANES survey and the decomposition of individual diet into 16 dietary riks factors. d) Physical activity is charaterized using MET. e) Consumer exposures to chemical in multiple products are assessed using stochastic usage and chemical contents databases. f) PM2.5 environmental exposures are assessed based on residence location and a multiscale model predicting local concentrations. All exposures are then coupled with corresponding dose-responses to obtain individual burden in µDALY/pers-day. Potential impacts are highly variables depending on individual behaviour and can typically range a) from 0 to 14 µDALY/pers-day depending on job occupation for hearing impairements due to noise; b) from 0 to 1600 µDALY/pers-day depending on industry sector for impacts of organic pollutants on workers; c) from a typical impact of 250 µDALY/pers-day lost from detrimental food, up to a benefit of 300 avoided µDALY/pers-day depending on selected diet, with main impacts associated with cardiovascular disease, diabetes and cancers; d) from a benefit of 0 to 400 avoided µDALY/pers-day as a function of physical activity; e) from 0 to 300 µDALY/pers-day for exposure to chemicals in consumer products and f) from an impact of 0 to 40 µDALY/pers-day lost associated with fine particulate 2.5µm, depending on residence location. We enter an exciting era in which you can "Tell me when and where you lived and worked, what you consumed, exercised and ate... and I will tell you who you are (your exposome !).

Keywords: aggregate exposure, chemical prioritization, cumulative exposure, environmental health, exposure models, health, other/general, workplace

TH-PL-D3-74

Investigating the in vitro metabolism of the dental resin monomers BisGMA, BisPMA, TCD-DI-HEA and UDMA using human liver microsomes and quadrupole time of flight mass spectrometry

Vervliet P.[1], Van Den Plas J.[1], De Nys S.[2], Duca R.[3], Boonen I.[4], Elskens M.[4], Van Landuyt K.[2], Covaci A.[1]; [1]Toxicological Centre, University of Antwerp, Wilrijk, Belgium, [2]KU Leuven (University of Leuven), Department of Oral Health Sciences, BIOMAT & University Hospitals, Leuven (UZ Leuven), Belgium, [3]Environment and Health, Department of Public Health and Primary Care, Leuven, Belgium, [4]Department of Analytical, Environmental and Geo-Chemistry, Vrije Universiteit Brussel, Brussel, Belgium

Dental resin systems have been used for several decades and have gained a higher market share after the Minamata convention and the encouragement of the World Health Organization (WHO) for a global phase-down of dental amalgam. Many (meth)acrylic monomers are an important part of the dental resin system, and are either based on BPA or lack the BPA core. The degree of conversion/polymerization during restoration is in general between 50-70 %. This may allow leaching from unreacted monomers to the oral cavity where they can be taken up through the pulp or gastrointestinal tract after ingestion with subsequent hepatic metabolism. This study identified the in vitro Phase I and Phase II metabolism of the dental resin monomers BisGMA, UDMA, BisPMA and TCD-DI-HEA, using human liver microsomes (HLM) and human liver cytosols (HLCYT). During Phase I incubation with HLM, the (meth)acrylic acid moiety in the monomers was rapidly cleaved with subsequent oxidative and hydroxylation pathways. For BisPMA, an O-dealkylation pathway occurred resulting in the formation of BPA. The carbamates moieties present in TCD-DI-HEA and UDMA were resistant to biotransformation reactions. Phase II biotransformation products were observed only for BisPMA and included conjugation reactions with sulphate and glucuronic acid. In total 4, 3, 12 and 3 biotransformation products were identified in this study for BisGMA, UDMA, BisPMA, and TCD-DI-HEA, respectively. Several identified biotransformation products are potential biomarkers for biomonitoring studies to assess the exposure to monomers after dental restorations.

Keywords: biomarkers

TH-PL-E3-CPC: Indoor Environment, Occupant Health and Behavior

TH-PL-E3-75

Human lung constructs as a model to study exposure to indoor air particulate matter

Huttunen K.[1], Nordberg M.-E.[1], Täubel M.[2], Hyvärinen A.[2]; [1]University of Eastern Finland (UEF), Kuopio, Finland, [2]National Institute for Health and Welfare, Kuopio, Finland

The cellular processes activated by exposure to indoor air particulate matter (PM) can be studied in model systems utilizing respiratory cells. We tested the suitability of lung tissue constructs differentiated from primary human bronchial epithelial cells in studying the effects of indoor air PM. The lung tissue constructs cultured in air-liquid interface were exposed to increasing concentrations of PM collected from a residential apartment within one week of active, intermittent sampling. After 24-hour exposure, we measured changes in transepithelial electrical resistance (TEER) of the tissue, total protein concentration in apical wash, interleukin (IL) –8 concentration in culture medium and expression of toxicological panel of 386 genes from sequenced RNA of the cells. The exposure to indoor air PM induced a slight increase in the concentration of IL-8 in the culture medium, but did not disrupt the integrity of the tissue nor increase the total concentration of proteins in the apical wash. The analysis of expression of 386 genes associated with toxicological processes revealed that exposure to PM upregulated the transcription of over 50 genes. Interestingly, the four most upregulated genes were related to immunotoxicity. Also several genes associated with heat shock response and DNA repair were targeted by the PM exposure. On the other hand, transcription of over 20 genes was downregulated, including genes associated with responses to oxidative stress. The results show that human lung constructs can be applied in studies of toxicological responses to environmental exposure agents. The studied exposure levels were not acutely toxic to lung tissue, but sufficient for inducing up- and downregulation of several genes associated with toxicological processes. The genes with most upregulated transcription were related to different functions of immune defense. In future studies, this exposure model enables the comparison of toxicological transcriptome associated with exposure to PM samples from different indoor environments.

Keywords: respiratory health

TH-PL-E3-76
Perfluoroalkyl substances interfere with female reproductive and immune health

Feng Y.[1], Wang W., Liang F., Zhang J., Wang Y.; [1]College of Public Health, Shanghai Jiao Tong University School of Medicine, Shanghai, China

Perfluoroalkyl substances (PFASs) are a family of synthetic, fluorinated organic compounds that have been widespread in the environment, wildlife and human. Experimental and epidemiologic evidence suggested that PFASs are capable of interfering with endocrine processes and have potential adverse effects on female reproduction. In the present study, a case-control study was performed with 220 fallopian tube obstruction cases and 397 healthy women in reproductive age. The results showed that serum levels of three PFASs analytes–PFBS, PFHxS and PFOS, were significantly higher in tube obstruction cases than in healthy controls. In addition, logistic regression analysis revealed a positive association between these chemicals and the risk of tube obstruction. Perfluorobutanesulfonic acid (PFBS) is being widely used as an alternative to PFOS in recent years. However, a lot of science evidence about its toxicities rises the concerning about risk assessment of PFBS. We further performed an animal study to investigate the potential toxicities of PFBS on the female reproductive and immune system. C57BL/6 female mice were exposed to PFBS via drinking water for five weeks. Compared to control mice, increased sex hormones in peripheral blood were observed in PFBS-exposed mice, accompanied by decreased percentage of CD8+ T cells in spleen. Expression of CD200 and CD200 receptor (CD200R1), which are presumed to transmit inhibitory signaling in a wide range of inflammatory diseases, were significantly elevated in ovary and uterus of exposed mice, suggested inhibitory immune microenvironment in the upper reproductive tract. Ex vivo cells studies also observed increasing CD200R1 expression and decreasing IL-6 secretion in mouse bone marrow derived macrophages, as well as increasing CD200 expression in mouse ovarian granulosa cells, after the cells were exposed to PFBS. In summary, PFBS was proposed to interfere with CD200-CD200R signaling in female reproduction system, which might regulate site-specific immune response and compromise reproductive health.

Keywords: environmental health

TH-PL-E3-77
Biophilic adaptation at the workplace: Multidimensional impact on human's behavior

Sun S.[1], Ordieres Meré J.[1], Yin J.[2], Zheng X.[1]; [1]Universidad Politécnica de Madrid, Madrid, Spain, [2]Harvard T.H. Chan School of Public Health, Boston, MA, USA

Background: Productivity is vital in the workplace. Numerous studies endorsed by scientific community affirm that biophilic design in workplace improves productivity. In previous conducted research studies, biophilic design was only evaluated with well control Indoor Environmental Quality (IEQ). The World Health Organization (WHO) reported that nine out of 10 people breathe polluted air. However, how biophilic design behaves in an inadequate IEQ workplace is still not well addressed. Objective: To gain additional knowledge about the impact of biophilic design on productivity (creativity) and cognitive performance in a working environment with limited quality. Methods: The dissatisfactory IEQ site was selected through on site data analytics, of which data were collected by a low-cost IoT based indoor air quality monitoring system. Recruited participants will be engaged in this study, with a standard body, non-smokers and non-sensitive groups. Selected plants are applied as biophilic design. The environmental parameters and human behaviours are measured by the developed indoor air quality monitoring system and various wearable devices. The productivity ability is measured by exercises designed by Cambridge University Brain Science lab. The simulation task could be typing (information processing work). The creativity is measured by Taxonomy of Creative Thinking Method. The simulation task could be mind map (knowledge creation work). Cognitive performance is assessed by visual reaction time task, stroop task and visual backward digit span task etc. Results: Multiple data sources collected from various wearable devices are analyzed to measure human reaction to biophilic design. The adopted methodology to assess biophilic design on site will be also defined. An IoT based indoor air monitoring system was developed for environmental assessment. Conclusions: Effects of the biophilic design for the people's effectiveness at work is discussed, by considering multidimensional data collection involving IoT technologies. Limitations for the ongoing research are presented and future research is foreseen.

Keywords: productivity

TH-PL-E3-78

The combined effect of poor perceived indoor air quality and supervisor's social support on long-term sickness absences in workplaces – a follow-up study

Finell E.[1], Nätti J.[1]; [1]1Tampere University, Tampere

This paper examined whether combined effect of employees' perceived indoor air quality (IAQ) and reported social support provided by the supervisors predicted employees' long-term absence from sickness (more than 10 days) between one and three years later. By using negative binomial modelling, we analysed a representative sample of the working-age population in Finland (N = 16,084) from the Finnish Quality of Work Life Surveys (FQWLS) in 1997, 2003, 2008 and 2013 combined with the register-based follow-up data of participants' long-term absences covering a period of between one to three years after the FQWLS was collected. After all necessary background, work, and health-related factors were included in the model, employees who reported both low perceived IAQ and low social support had significantly higher rates of long-term absence than those who reported low perceived IAQ and good social support ($p < 0.01$). In addition, those who reported low perceived IAQ and good social support had significantly higher rates of long- term absence than those who reported only mild or no problems with IAQ and good social support ($p < 0.01$). This finding underlines the importance to take into account the combined effect of environmental and social stressors when analysing the health effects of indoor environments. It also emphasises the essential role of supervisors when solving workplaces' indoor air problems.

Keywords: indoor air quality; supervisor's social support; discrimination; occupational health; sickness absence

TH-PL-E3-79

The effect of indoor air quality in Dutch higher education classrooms on students health and performance

Brink H.W.[1,2], Loomans M.G.L.C.[2], Mobach M.P.[1,3], Kort H.S.M.[2,4]; [1]Hanze University of Applied Sciences, Groningen, the Netherlands, [2]University of Technology, Eindhoven, the Netherlands, [3]The Hague University of Applied Sciences, The Hague, the Netherlands, [4]University of Applied Sciences, Utrecht, the Netherlands

Teachers and students need good learning environments to perform well. In this study it is pre-supposed that the spatial properties of classrooms can contribute to the quality of the educational process. Thermal, acoustic and visual conditions and indoor air quality (IAQ) may be extremely powerful in order to support the in-class tasks of teachers and students. But what are the optimal conditions? And do schools provide optimal indoor environmental conditions? Research shows that adequate ventilation and thermal comfort in classrooms could improve academic performance of students. However, different studies also suggest that poor indoor environmental quality in classrooms are common and, in some cases, even unhealthy. This study investigates the relationship between indoor air quality (IAQ), perceived indoor air quality (PIAQ) and building-related symptoms of students in university classrooms via subjective assessment and objective measurement. This study was carried out in 59 classrooms of a university of applied sciences in the northern part of the Netherlands during heating season. Responses from 366 students were obtained through a questionnaire. Results shows that carbon dioxide concentrations (CO2) exceed minimum Dutch guidelines in 36% of the observed classrooms. Moreover, after a 40 minute class this raised to 45% of the observed classes. Poor IAQ can affect teachers and students level of attention, cause arousal and increase the prevalence of building-related symptoms. A significant correlation was found between CO2 concentrations and PIAQ and between PIAQ and the ability to concentrate, tiredness and dry skin. The research findings imply that increased CO2 concentrations will affect the PIAQ of students and may cause inability to concentrate, increased tiredness and dry skin. These building-related symptoms can cause distraction and affect the academic performance of students negatively. It is highly recommended to improve IAQ in classrooms by offering better indoor environmental conditions through reducing CO2 concentrations.

Keywords: building-related symptoms, carbon dioxide, indoor environmental conditions, quality of learning, university classroom

International Society of Exposure Science (ISES)
c/o Infinity Conference Group
1035 Sterling Road, Suite 202
Herndon VA 20170
USA

CURRAN ASSOCIATES INC.
proceedings.com

ISBN 978-1-5108-9756-4